6.19
12-1-89
DATE DUE
AF796052
526449

Uncertainty in Risk Assessment, Risk Management, and Decision Making

ADVANCES IN RISK ANALYSIS

This series is edited by the Society for Risk Analysis.

Volume 1 **THE ANALYSIS OF ACTUAL VERSUS PERCEIVED RISKS**
Edited by Vincent T. Covello, W. Gary Flamm, Joseph V. Rodricks, and Robert G. Tardiff

Volume 2 **LOW-PROBABILITY/HIGH-CONSEQUENCE RISK ANALYSIS**
Issues, Methods, and Case Studies
Edited by Ray A. Waller and Vincent T. Covello

Volume 3 **RISK ANALYSIS IN THE PRIVATE SECTOR**
Edited by Chris Whipple and Vincent T. Covello

Volume 4 **UNCERTAINTY IN RISK ASSESSMENT, RISK MANAGEMENT, AND DECISION MAKING**
Edited by Vincent T. Covello, Lester B. Lave, Alan Moghissi, and V. R. R. Uppuluri

A Continuation Order Plan is available for this series. A continuation order will bring delivery of each new volume immediately upon publication. Volumes are billed only upon actual shipment. For further information please contact the publisher.

Uncertainty in Risk Assessment, Risk Management, and Decision Making

Edited by

Vincent T. Covello
National Science Foundation
Washington, D.C.

Lester B. Lave
Carnegie-Mellon University
Pittsburgh, Pennsylvania

Alan Moghissi
Institute for Regulatory Science
Alexandria, Virginia

and

V. R. R. Uppuluri
Oak Ridge National Laboratory
Oak Ridge, Tennessee

PLENUM PRESS • NEW YORK AND LONDON

CENTRAL MISSOURI
STATE UNIVERSITY
Warrensburg,
Missouri

Library of Congress Cataloging in Publication Data

Society for Risk Analysis International Workshop on Uncertainty in Risk Assessment, Risk Management, and Decision Making (1984: Knoxville, Tenn.)
Uncertainty in risk assessment, risk management, and decision making.

"Proceedings of the Society for Risk Analysis International Workshop on Uncertainty in Risk Assessment, Risk Management, and Decision Making, held September 30–October 3, 1984, in Knoxville, Tennessee"—T.p. verso.
Includes bibliographical references and index.
1. Risk management—Congresses. 2. Risk—Congresses. 3. Uncertainty—Congreses. I. Covello, Vincent T. II. Society for Risk Analysis. III. Title.
HD61.S625 1984 658 87-2403
ISBN 0-306-42557-2

Proceedings of the Society for Risk Analysis International Workshop on Uncertainty in Risk Assessment, Risk Management, and Decision Making, held September 30–October 3, 1984, in Knoxville, Tennessee

© 1987 Plenum Press, New York
A Division of Plenum Publishing Corporation
233 Spring Street, New York, N.Y. 10013

All rights reserved

No part of this book may be reproduced, stored in a retrieval system, or transmitted in any form or by any means, electronic, mechanical, photocopying, microfilming, recording, or otherwise, without written permission from the Publisher

Printed in the United States of America

HD
61
S625
1984

PREFACE

The subject of this volume--uncertainties in risk assessment and management--reflects an important theme in health, safety, and environmental decision making. Most technological hazards are characterized by substantial uncertainty. Recent examples include nuclear waste disposal, acid rain, asbestos in schools, carcinogens in food, and hazardous waste. Dealing with such uncertainty is arguably the most difficult and challenging task facing risk assessors and managers today.

Four primary sources of uncertainty in risk assessment and management can be identified: (1) uncertainties about definitions; (2) uncertainties about scientific facts; (3) uncertainties about risk perceptions and attitudes; and (4) uncertainties about values. Uncertainties about definitions derive primarily from disagreements about the meaning and interpretation of key concepts, such as probability. Uncertainties about scientific facts derive primarily from disagreements about failure modes, the probability and magnitude of adverse health or environmental consequences, cause and effect relationships, dose-response relationships, and exposure patterns. Uncertainties about risk perceptions and attitudes derive primarily from disagreements about what constitutes a significant or acceptable level of risk. Uncertainties about values derive primarily from disagreements about the desirability or worth of alternative risk management actions or consequences.

The papers in this volume address each of these sources of uncertainty from a variety of perspectives. Reflecting the broad scope of risk assessment and risk management research, the papers include contributions from safety engineers, epidemiologists, toxicologists, chemists, biostatisticians, biologists, decision analysts, economists, psychologists, political scientists, sociologists, ethicists, and lawyers.

All papers in this volume were presented at the fourth annual meeting of the Society for Risk Analysis held in Knoxville, Tennessee. Support for the meeting was provided by the Environmental Protection Agency, the Nuclear Regulatory Commission, Oak Ridge National Laboratory, Union Carbide Corporation, and the U.S. Department of Energy. We would also like to thank Raymond Boykin, Anthony Buhl, George Flanagan, and others at Oak Ridge for their extraordinary work in organizing and running a highly successful meeting.

Vincent T. Covello
Alan Moghissi
V.R.R. Uppuluri

526449

CONTENTS

UNCERTAINTY AND RISK ASSESSMENT:
PERSPECTIVES FROM THE SOCIAL, BEHAVIORAL AND POLICY SCIENCES

UNCERTAINTY AND RISK ASSESSMENT: PERSPECTIVE FROM THE ENVIRONMENTAL AND HEALTH SCIENCES

UNCERTAINTY AND RISK ASSESSMENT: PERSPECTIVES FROM ENGINEERING AND STATISTICS

POLICY ISSUES RELATED TO WORST CASE RISK ANALYSES AND THE ESTABLISHMENT OF ACCEPTABLE STANDARDS OF DE MINIMIS RISK*

Miller B. Spangler
U.S. Nuclear Regulatory Commission
Washington, D.C. 20555

ABSTRACT

Worst case analyses of risks become important whenever there are wide ranges of uncertainty surrounding estimated risks of technological applications, especially if environmental and societal impacts can reach catastrophic proportions. Risk may be defined as the possible consequences of a decision option times the probability of consequences materializing and special circumstances that may be involved. Increasingly severe magnitudes of consequences resulting from a technological option are sometimes accompanied by increasingly small probabilities that can reach de minimis, or insignificant, levels. Thus, a dilemma exists for policy development as to what the de minimis level of probability is for those rare combinations of circumstances that can produce worst case consequences. One possible guide is the probability level of fatal accidents individuals routinely accept in their personal activities. Another is the establishment of safety-cost trade-off criteria that take into account equity considerations of the diverse opportunities for saving lives by the allocation of finite financial resources. On November 29, 1978, the Council on Environmental Quality issued implementing regulations for environmental impact statements (40 CFR 1502.22) requiring a worst case analysis, under certain conditions, and an indication of the probability of its occurrence. Difficulties and controversies in implementing the concept of worst case analysis have led the CEQ to reconsider this requirement (48 FR 36486, August 11, 1983). This paper explores a number of policy issues related to worst case analyses within a spectrum of high-consequence/low-probability events and the establishment of acceptable standards of de minimis risk. Case material is used comparing the catastrophic risk of nuclear power plants with the chronic and accidental risks of using coal to generate electricity.

KEY WORDS: Worst Case Analysis; De Minimis Risk; Probabilistic Risk Assessment; Risk Perception; Environmental Impact Analysis; Nuclear Power Plants; Coal-Fired Plants; Catastrophic Events; Regulation.

1. POLICY ISSUES OF WORST CASE ANALYSIS AND DE MINIMIS RISK

This paper is divided into two parts. The first part encompasses a discussion of policy issues arising from the 1978 requirements of the National Environmental Policy Act (NEPA) for worst case analysis (or the 1986 replacement requirements) and the need for, and policy issues related to, establishing acceptable standards of *de minimis* risk that would provide a useful procedure for limiting the treatment of worst case analyses to reasonable classes and numbers of worst case events.* In the second part, a variety of approaches for establishing acceptable standards of *de minimis* risk will be appraised in terms of technical implementability and ethical or equity considerations in the social management of technological risk with special focus on potentially catastrophic events in the use of uranium and coal fuels to generate electricity.

1.1 NEPA Requirements for Worst Case Analysis of Environmental Impacts

On November 29, 1978, the Council on Environmental Quality (CEQ) issued implementing regulations regarding analyses required under the National Environmental Quality Act of 1969.[1] Section 1502.22 of these regulations provides Federal agencies with the following procedural requirements for worst case analyses in Environmental Impact Statements (EISs).

> "When an agency is evaluating significant adverse effects on the human environment in an environmental impact statement and there are gaps in relevant information or scientific uncertainty, the agency shall always make clear that such information is lacking or that uncertainty exists.
>
> "(a) If the information relevant to adverse impacts is essential to a reasoned choice among alternatives and is not known and the overall costs of obtaining it are not exorbitant, the agency shall include the information in the environmental impact statement.
>
> "(b) If (1) the information relevant to adverse impacts is essential to a reasoned choice among alternatives and is not known and the overall costs of obtaining it are exorbitant or (2) the information relevant to adverse impacts is important to the decision and the means to obtain it are not known (e.g., the means for obtaining it are beyond the state of the art) the agency shall weigh the need for the action against the risk and severity of possible adverse impacts were the action to proceed in the face of uncertainty. If the agency proceeds, it shall include a worst case analysis and an indication of the probability or improbability of its occurrence." (40 CFR 1502.22).

While on its face the above procedural requirements seem reasonable enough, there are numerous issues of practical and legal interpretation. How does one draw the line between "significant" and "insignificant" adverse effects? In principle, there are both fuzzy and robust

---

*Since this paper was presented on October 2, 1984, three important events relevant to the concepts herein presented: (1) the Council on Environmental Quality on April 25, 1986 issue a Final Rule (51 *Fed. Reg.* 15618) rescinding the requirement for a worst case analysis but still requiring (under certain conditions) the evaluation of impacts with catastrophic consequences having low probabilities; the NRC on June 19, 1986 reached a final decision on a "Policy Statement on Safety Goals for the Operation of Nuclear Power Plants" (to be announced in the *Federal Register*) with policy and guidance for the regulatory treatment of severe nuclear accidents of high consequences and low probabilities; and the Chernobyl nuclear accident took place in the Soviet Union on April 26, 1986 with catastrophic consequences whose dimensions are yet to be ascertained. The significance of these events to the concepts of this paper are briefly noted in an Addendum at the end of this paper.

interrelationships between a legion of technological and environmental parameters of cause and effect and also between decisional parameters affecting an unusually broad scope of societal values that are interrelated at least at the level of resource allocation. How then does one define the analytical system's boundaries between "relevant" information and "insignificantly relevant" information? By what standards are the overall costs of obtaining the information to be judged "exorbitant" or "acceptable"? How can the "importance" of the information to the decision be established while certain practices are yet evolving: the scope of direct and indirect considerations to include as decision criteria; the determination of who are legitimate parties of interest; and the limits of spatial and temporal dimensions of these considerations (i.e., the ranges of distance and time horizons of impacts)?

Nor does it seem likely that the lid can be slammed shut on such a Pandora's Box of tormenting issues if actions by the courts and the reluctance of legislatures to modify current environmental and safety regulations are a guide. To judge by the incremental changes in dealing with these issues by the stream of study reports and policy pronouncements of regulatory agencies and the decisions and analyses of the Courts, our society seems bent on a course of muddling through. In part, there is a yielding to societal pressures for a more sensitive consideration of adverse and beneficial effects of technological and policy options and, in part, there is a discrediting of extremist positions where improved information and analysis put to the test of equity considerations exposes the unreasonableness of certain claims while confirming others.

Guidance is provided in Section 1502.22 of the CEQ Regulations that the treatment of worst case analysis "should be understood to contain an initial threshold of reasonably foreseeable impacts or effects which must be established before this provision can be applied." This guidance plus the legal precepts of the Administrative Procedure Act would appear to yield to Federal agencies sufficient administrative flexibility to pursue the principle of establishing an initial threshold of reasonably foreseeable effects so long as this conforms to a "rule of reason." Indeed, a landmark decision in 1972 by the U.S. Court of Appeals (NRDC v. Morton, 458 2d 873) can be drawn upon to support this view:

> Detailed discussion is not required of alternatives that "are deemed only remote and speculative possibilities, in view of basic changes required in statutes and policies of other agencies." And the agencies need not indulge in "crystal ball inquiry" in assessing the effects of alternatives. The agency will have taken the "hard look" demanded by NEPA if it has discussed the reasonably foreseeable impacts with a thoroughness commensurate with their severity and the significance of the action.

Although the Court provided no legal definitions for such terms as "remote" and "speculative," reasonable interpretations would be:

(1) Worst case events of "remote possibility" would be those whose chances of occurrences are so miniscule as to produce an insignificant level of risk even with large consequences (i.e., the significance-of-the-action test in the above quotation).

(2) All assessments of impacts of alternatives wherein scientific methods and available data yield a significant degree of imprecision or unreliability thus require a dimension of subjective judgment or speculation (i.e., risk) in their acceptance as worthwhile considerations for decisionmaking. If the scientific methodology and supporting evidence is regarded as overly flimsy or unrespectable <u>and</u> if the expected penalties to certain parties of interest by ignoring the highly uncertain

estimates of impact are not very significant by virtue of their remote chances of happening, then the speculation is unwarranted. That is to say, it seems logical in the Court's coupling of these criteria that the permissible degree of speculation, or confidence bounds, is intended to be judged by the precedent assessment of the remote chances of the event.

However, the situation is considerably more confused and controversial when the scientific basis for assessing the remoteness of the chances of the event (as well as its consequences) to take place have a highly speculative aspect yielding broad uncertainty bounds. In at least one case, a court (the Fifth Circuit) held in Sierra Club v. Sigler (695 F. 2d 957, 1983) that worst case analyses must be included in EISs even for "low-probability events." Yet the precedential value of any one court decision, especially at the circuit level, is itself suspect. Pertinent to the worst case analysis (WCA) in which uncertainty over the probability of the event is the dominant consideration is the court case of Southern Oregon Citizens Against Toxic Sprays Inc. v. Clark (SOCATS).[2] The proposed action was the U.S. Bureau of Land Management's (BLM's) use of herbicides for vegetation management in the Pacific Northwest.[3] The uncertainty involved the possibility that the spraying program would harm human health. Here, the uncertainty involved the probability rather than the magnitude of the impact. The Ninth Circuit ordered a worst case analysis.

On August 11, 1983, the CEQ published proposed additional guidance on CEQ's regulations dealing with worst case analysis.[4] This was in response to the Council's concern that the worst case analysis requirements are being read to require Federal agencies to conduct such analyses for potential effects that may well be highly remote or unlikely. In particular, the Council was concerned that the guidance provided in their Publication or Memorandum to Agencies Containing Answers to 40 Most Asked Questions of NEPA Regulations, 46 FR 18026 (March 23, 1981), is incomplete with regard to this regulation. The Council concluded that the obligation to prepare a worst case analysis, at bottom is nothing more than the duty to provide information essential to a reasoned choice among alternatives. As described in response to Question 20 (b) in their earlier guidance on this subject: "The purpose of the analysis is to carry out NEPA's mandate for full disclosure to the public of the potential consequences of agency decisions, and to cause the agencies to consider these potential consequences when acting on the basis of scientific uncertainties or gaps in available information." (Emphasis added.) Thus, a worst case analysis is required in those circumstances where significant adverse impacts are possible, and scientific uncertainties or gaps in available information exist regarding those impacts. According to the 1978 guidance of the CEQ, this means that there is an initial threshold of probability which must be crossed before the worst case analysis of Section 1502.22 comes into play.

1.2 The Need for De Minimis Standards in the Treatment of Uncertainty

The CEQ's guidance of 1978 that an initial threshold of probability is the crucial criterion in determining whether worst case analyses are required is also suggestive of the need to develop an acceptable standard of de minimis risk that would limit the number (and also classes) of worst case scenarios or events that it is reasonable to consider in regulatory decisionmaking. A growing number of risk analysts and regulatory officials have recently recognized the importance of basing decisions on a de minimis level of risk. However, centuries ago jurisprudence has recognized the principle of "de minimis non curat lex" (i.e., the law does not concern itself with trifles). Friedell, a member of the National Council of Radiation Protection and Measurements, called attention to the desirability of establishing a de minimis standard of acceptable risk of radiation exposure from the application of technologies, but proposed no standard.[5]

It is to be noted that there is an important distinction between "acceptable risk" and an "acceptable standard of de minimis risk" upon which this paper focuses. All risk levels regarded as trifling (or de minimis) should, by this very same token, be regarded as acceptable risks. However, society finds certain other non-trifling risks to be acceptable whenever the expected net beneficial effects in a risk-cost-benefit (RCB) analysis are perceived to outweigh the expected adverse effects associated with individual or public hazards.

This principle is implicit in the concepts presented in Figure 1 as reproduced from an article by Richard Wilson.[6] The numbers in the left hand margin on risk of individual death were derived from the work of Lord Brian Flowers, Chairman of the Royal Commission on the Environment of the U.K. In Figure 1 the upper zone with risk of individual fatality per year

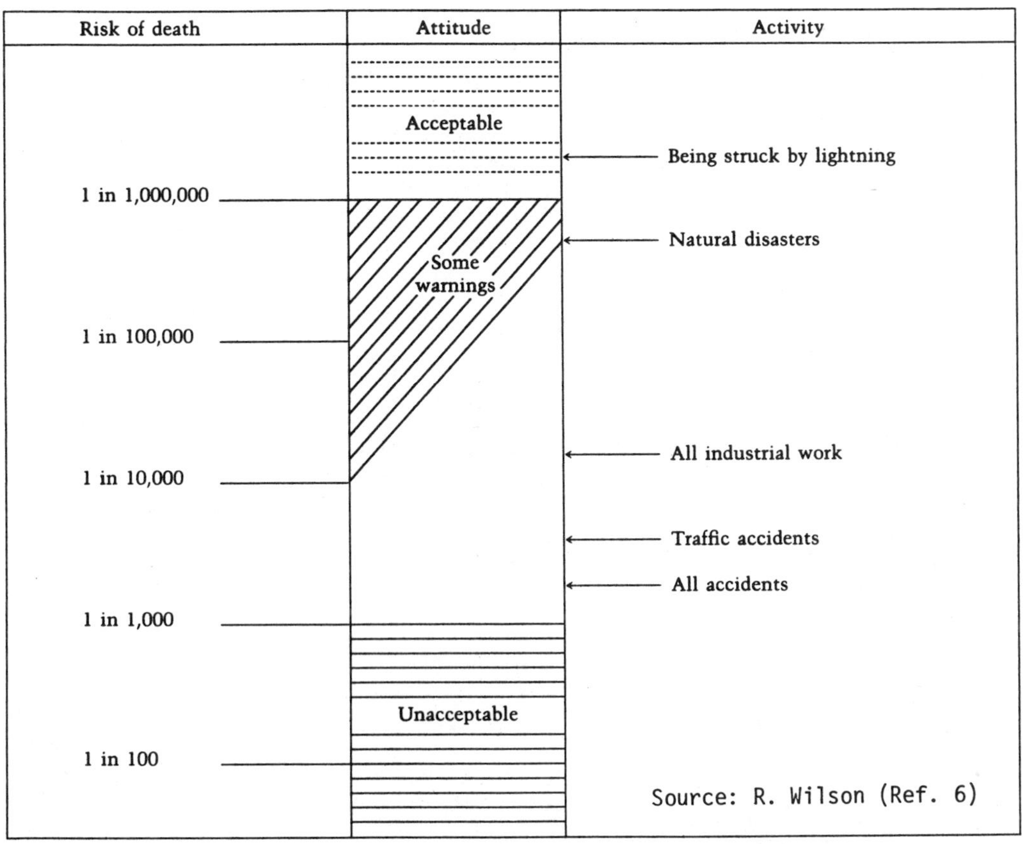

Fig. 1. Probability of death for an individual per year of exposure (orders of magnitude) in terms of acceptable/unacceptable risk.

of exposure that is less than one chance in a million (i.e., 10^{-6} per yr.) might appropriately be regarded as an acceptable standard of de minimis risk. On the other hand, the lower zone with individual risk greater than one chance in a thousand per year of exposure might be regarded as clearly unacceptable. The middle, in-between zone would then be the valid arena of debate for worst case and other RCB analyses to be performed in social decisionmaking as to whether risks are acceptable or not.

In this regard it is important that a distinction be made between "acceptable risks" and "acceptable deaths." As stated by the Nuclear Regulatory Commission (NRC) in publishing its proposed safety goals for nuclear power plant operation:

> We want to make clear at the beginning of this section that no death attributable to nuclear power plant operation will ever be "acceptable" in the sense that the Commission would regard it as a routine or permissible event. We are discussing acceptable risks, not acceptable deaths. In any fatal accident, a course of conduct posing an acceptable risk at one moment results in an unacceptable death moments later. This is true whether one speaks of driving, swimming, flying or generating electricity from coal. Each of these activities poses a calculable risk to society and to individuals. Some of those who accept the risk (or are part of a society that accepts risk) do not survive it. We intend that no such accident(s) will occur, but the possibility cannot be entirely eliminated.[7]

In February 1984 Joyce Davis[8] provided testimony to NRC's Advisory Committee on Reactor Safeguards (ACRS) on the subject, "The De Minimis Regulatory Cut-Off Concept." Her presentation included a wide scope of issues: Why is it needed? How could it be used? Who has recognized the need? What economic benefits would it provide? How could it be established and implemented. As to how it might be used, Davis suggests the following applications of special significance for the NRC and other Federal agencies concerned with regulating radiation hazards:

1. For setting regulatory priorities.
2. As a "floor" for As Low As Reasonably Achievable (ALARA) considerations.
3. As a cut-off level for collective dose assessments.
4. For setting outer boundaries of geographical zones.
5. As a floor for definition of low level waste.
6. As a presumption of triviality in legal proceedings.
7. To foster administrative and regulatory efficiency.
8. To provide perspective for public understanding, including policy judgments.

1.3 What Is An Appropriate Measure of Triflingness?

Beyond establishing a need for establishing an acceptable standard of de minimis risk to limit and facilitate the treatment of worst case analyses and other regulatory purposes, a next useful step is to decide on an appropriate measure or measures of triflingness. One such measure might arise from the situation wherein some accident scenarios are so low in probability that the estimated level of risk falls within the "noise level" of the ranges of uncertainty surrounding the estimates of the more dominant accident scenarios (such as severe nuclear accidents) that contribute most of the overall level of risk. A second measure of trifling impacts is that derived from a comparative risk analysis of viable alternative technologies in meeting common needs of society (e.g., whenever the estimated risk of the base case technology is less by a factor of ten or a hundred of the acceptable level of risk of alternative technologies). A third measure might arise whenever the expected cost of countermeasures to compensate for or reduce the incremental risk to individuals is of relatively trifling proportions. A fourth measure might be that the risk of a worst case scenario falls at a level comparable to that which individuals routinely accept without much, if any concern, in day-to-day activities. These and other optional approaches for establishing and implementing acceptable standards of de minimis risk are discussed in Part 2.

1.4 Trifling To Whom? The Quest For Peace Of Mind

Numerous studies indicate that the perception of risk varies widely between individuals and also between how experts assess risks and how risks are generally perceived by the lay public.[9-18] The problem of ascertaining triflingness is compounded whenever there are wide divergencies in the assessment of risks by the experts themselves. The exacerbation this produces in public reactions to risk policy formulations is not difficult to imagine in the face of the "Let's-Pick-An-Expert" syndrome.[19]

Douglas and Wildavsky[18] note that the current consideration of risk has three peculiarities. The first is that disagreement about the problem is deep and widespread in the Western world. The second is that different people worry about different risks--war, pollution, employment, inflation. The third is that knowledge and action are out of sync: whatever programs are enacted to reduce risks, they conspicuously fail to follow the principle of doing the most to prevent the worst damage. In sum, substantial disagreement remains over what is risky, how risky it is, and what to do about it. They go on to state that most people cannot be aware of most dangers at most times. Hence, no one can calculate precisely the total risk to be faced. How, then, do people decide which risks to take and which to ignore? On what basis are certain dangers guarded against and others relegated to secondary status?

Some gaps in the treatment of technological risks by technical experts and the lay public are shown in Table 1.[11] Regarding the subject of worst case analyses and the establishment of acceptable standards of de minimis risk, there is a tendency among some of the lay public to use different risk assessment methods than the experts. The latter generally assess risk as the possible consequences of a decision option times the probability of consequences materializing, including estimates of different scenarios of sequential events or causal factors producing the events and their consequences. For example, complex modeling of this kind is pursued by experts in assessing risk to the public from using either coal or uranium fuels to generate electricity.[20] In contrast, the perception of risks by the lay public are often formed by intuitive and impressionistic mental processes that may anchor on the searing impressions of worst case scenario consequences without consideration of a numerical estimate of the consequences and, in particular, ignoring the probability coefficient of the consequences. In these mental processes, personal experience, memory, value systems, media accounts, cultural exchanges, and what I have referred to as syndromes, or attitudinal patterns of gathering and using information, are important.[19]

However, both experts and lay persons are subject to judgmental biases in risk assessment/perception since even the most advanced and objective methods for risk assessment of complex interrelationships between causes and effects will include major inputs of subjective judgment and modeling assumptions that may depart substantially from reality in a biased direction.[21] According to Slovic et al.,[9] one

Table 1

Gaps in the Treatment of Technological Risks by Technical Experts and the Lay Public (Some Generalizations Having Notable Exceptions)[a]

Approach	Treatment common to experts	Treatment common to the public
1. Decision criteria for risk acceptance/rejection		
a. Absolute vs. relative risk	Risk judged in both absolute and relative terms	Greater tendency to judge risk in absolute terms
b. Risk-cost trade-offs	Essential to sound decisionmaking because of finite societal resources for risk reduction and impractability of achieving zero risk; tends to ignore nondollar costs in such trade-offs	Since human life is priceless, criteria involving risk-cost trade-offs are immoral; ignores risks of no-action alternatives to rejected technology; gives greater weight to non-dollar costs
c. Risk-benefit comparisons of technological options	Emphasizes total (net) benefits to society, neglecting benefits that are difficult to quantify; also neglects indirect and certain long-term benefits	Emphasizes personal rather than societal benefits; includes both qualitative and quantitative benefits but tends to neglect indirect and long-term benefits
d. Equity considerations	Tends to treat shallowly without explicit decision criteria and structured analyses	Tends to distort equity considerations in favor of personal interests to the neglect of the interests of opposing parties or the common good of society
2. Risk assessment methods		
a. Expression mode	Quantitative	Qualitative
b. Logic mode	Computational • Risk = consequence × probability • Fault trees/event trees • Statistical correlation	Intuitive • Incomplete rationale • Emotional input to value judgments
c. Learning mode	Experimental • Laboratory animals • Clinical data for humans • Engineering test equipment and simulators	Impressionistic • Personal experience/memory • Media accounts • Cultural exchange
3. Basis for trusting information		
a. Source preference	Established institutions	Non-"establishment" sources
b. Source reliability	Qualifications of experts	Limited ability to judge qualifications
c. Accuracy of information	Robustness/uncertainty of scientific knowledge	Minimal understanding of strengths and limitations of scientific knowledge
4. Risk attribute evaluation		
a. Low frequency risk	Objective assessment using conservatisms	Tends to exaggerate or ignore risk
b. Newness of risk	Broad range of high and low estimates	Tends to exaggerate or ignore risk
c. Catastrophic vs. dispersed deaths	Gives equal weight	Gives greater weight to catastrophic deaths
d. Immediate vs. delayed deaths	Diverse views over treatment of incommensurables and discount rate	Gives greater weight to immediate deaths except for known exposure to cancer-producing agents
e. Statistical vs. known deaths	Gives equal weight	Gives greater weight to known deaths
f. Dreadness of risk	Generally ignores	Gives greater weight to dreaded risk
g. Voluntary vs. involuntary risk	Gives equal weight	Gives greater weight to involuntary risk
5. Technological considerations		
a. Murphy's Law (if anything can go wrong, it will)	Stimulus for redundancy and defense-in-depth in systems design and operating procedures; margins of conservatism in design; quality assurance programs	Stimulus for "what-if" syndromes and distrust of technologies and technocrats; source of exaggerated views on risk levels using worst case assumptions
b. Reports of technological failures/accidents	Valued source of data for technological fixes and prioritizing research; increased attention to consequence mitigation	Confirms validity of Murphy's Law; increased distrust of technocrats

[a] Some of the descriptors in this table unintentionally reflect the image that the "experts are always right." Experts, of course, are not without emotions and sources of bias. Indeed, experts could benefit from improved information and scientific advances in reducing uncertainties in assessing technologies and their societal impacts as well as wider appreciation of public attitudes and changing social values.

inferential strategy that has special relevance for risk perception is the availability heuristic.[22] People using this heuristic judge an event as likely if instances of it are easy to imagine or recall. Because frequently occurring events are easier to imagine or recall than rare ones, availability heuristics are important. However, availability is also affected by other factors, namely, a recent disaster or a vivid film that could seriously bias risk judgments. Another problem is that we often see the future as a mirror of past experience, trends, or events.[23]

Another reason why some persons perceive certain risks as de minimis, or trifling, and others regard them with a measure of concern, or even alarm, is the disparities in the bases for trusting different sources of information bearing on risk assessment/perception. Faced with the horrendous problems of information overload, all of us must rely on information inputs provided by others. In the struggle for getting at the truth, persons with an anti-establishment bias, for example, will usually not trust information available from government or industry sources. They prefer instead the popular press for information or the "put-it-in-a-nutshell" limitations of television that fail to provide a balanced treatment of the complexities of risk-cost-benefit assessments. A not wholly unrelated basis for distrusting certain sources of information is the perception of bias due to an assumption of correlation between certain risk assessments and the personal welfare of the risk assessor. One should not trust, following this philosophy, the assessment of the risks of an operation if it comes from the "hand" who will wield the surgeon's knife. Yet, the risks of not trusting such sources of information must also be reckoned with. For in a sense, no one is without self interests or sources of bias. These include the biases of incomplete but relevant information that is known to others and the biases inherent in scientific methods (or funding limitations) for getting at the truth of things.

Moreover, the pursuit of truth has a certain dynamics about it. One can expect changes in values, world and local conditions affecting risk levels, the level of information available and utilized in assessing risks, and personal attitudes toward trust and distrust that can make important differences in perceptions of de minimis risk. Indeed, if an expert is one who possesses a high degree of skill and knowledge of facts on a given subject, there is a serious question whether anyone can be an expert on the future course of these dynamics. Not only is this because there are too many subjects of relevance to the future for any one person to be very knowledgeable on them all, but also because there can be no present facts about the future--only more-or-less-informed speculations. Thus, the pursuit of truth is the pursuit of unconventional wisdom that can only come from the transactional analyses of persons of many disciplines as well as those whose values are affected by the dynamics of truth progression and policy decisions that derive therefrom.

1.5 A Matter Of Life And Money: The Issue Of Equity

Substantial research expenditures and regulatory and industry actions are in progress in the United States and abroad that will reduce the ranges of uncertainty in estimates of the comparative risks of the coal and nuclear options as well as to reduce the levels of the imposed risks. Such investments, however, incur costs to society. The absence of safety-cost trade-off criteria to guide these decisions was commented on by the President's Commission on the Accident at Three Mile Island. The Commission recommended that included in the NRC's general substantive charge "should be the requirement to establish safety cost trade-offs and explain them to the public."[24] In response, the Nuclear Regulatory Commission proposed in 1983 the following benefit-cost guideline (i.e., safety-cost trade-off criterion) for trial use as one consideration in arriving at decisions on safety improvements:[7]

> "The benefits of an incremental reduction of societal mortality risks should be compared with the associated costs on the basis of $1,000 per person-rem averted."

This benefit-cost guideline is roughly equivalent to an investment cost of $10 million per life saved and makes no distinction between prompt or delayed fatalities as is sometimes done in other studies.[25] It should be noted that the value of $10 million per delayed fatality averted that is implicit in the above benefit-cost guideline is 20 to 40 times as large as the range of values referenced by the U.S. Environmental Protection Agency in the Environmental Impact Statement for establishing 40 CFR 190.[26] This disparity of agency references or benchmarks suggests (if NRC does not elect to change the values in this guideline following the trial use period) that the NRC cost-benefit standard might be regarded as a surrogate not only for reduced fatality rates, but also perhaps for other health and genetic effects; the psychic costs of societal risk aversion associated with catastrophic and involuntary risk and societal resiliency penalties of catastrophes; and possibly other socio-economic considerations not included within the scope of the NRC's current practices in regulatory impact analysis.[27]

Including cost as a consideration in risk reduction decisions is not a matter of questionable morality, as some would have it, in putting a price on human life. Life is indeed priceless to those whose lives are cut short and to those who mourn. Rather, the consideration of cost in conjunction with risk is imperative because of the common sense recognition that society has finite resources to devote to life-saving or risk reduction investments. Hence, an imposing ethical issue is whether an excessive expenditure for risk reduction in one area of human activity would inequitably deprive the use of these same resources for risk reduction in other areas with expectations for saving possibly an even greater number of lives (see Table 2). Moreover, society uses risk-benefit trade-offs in achieving a balance between expenditures on behalf of risk reduction and the use of financial resources to enhance the quality and enjoyment of life in numerous ways.[10] Thus, in my view, a superior perspective for looking upon a (dollar) figure of merit for fatality reduction is not one of putting a price on human life (a calloused notion); rather, a more befitting perspective is that such a figure of merit is to be humanely and ethically perceived as an "equity value of saving lives".

The establishment of an acceptable standard of de minimis risk can be performed from considerations external to the technology in question. If such a standard focuses on the risk of individual fatality per year of exposure, as in Figure 1, and there are other adverse risk factors to consider than fatality, these should be accommodated in decisionmaking apart from this de minimis standard. The remaining difficulty, it would appear, lies in the verifiability of whether the actual risk estimates of individual fatality are at, or below, the de minimis level. The cost of the information to validate this proposition to a level that would satisfy a large sector of affected parties might, in some instances, indeed be exorbitant (see the criterion of worst case analysis referenced above from the CEQ regulations Section 1502.22). To conduct accident progression experiments regarding various severe accident scenarios from core-melt experiments of full-scale nuclear plants that would replicate the large variety of generic and custom designs would cost untold billions of dollars.

Table 2

Illustrative Examples of the Cost of Life-Saving Opportunities in Alternative Uses of Financial Resources[a]
(In Dollars Per Life Saved)

Item of Opportunity		Estimated Cost (dollars)
1.	Improved medical X-ray equipment	3,600
2.	Improved highway maintenance practices	20,000
3.	Screening for cervical cancer	30,000
4.	Proctoscopy for colon/rectal cancer	30,000
5.	Mobile cardiac emergency unit	30,000
6.	Road guardrail improvements	30,000
7.	Tuberculosis control	40,000
8.	Road skid resistance	40,000
9.	Road rescue helicopters	70,000
10.	Screening for lung cancer	70,000
11.	Screening for breast cancer	80,000
12.	Automobile driver education	90,000
13.	Impact absorbing roadside device	110,000
14.	Breakaway signs and lighting posts	120,000
15.	Smoke alarms in homes	240,000
16.	Road median barrier improvements	230,000
17.	Tire inspection	400,000
18.	Highway rescue cars	420,000
19.	Home kidney dialysis	530,000

[a]Source: Selected from a list of similar items as catalogued by E. Siddall (Ref. 27).

Such cost expenditures, if spared, have a cost-equivalency of lives saved if the funds were used instead for a variety of opportunities for saving lives as identified in Table 2. For example, a savings of $1 billion in costs of information that is spared could be used to save 2000 lives for a set of opportunities averaging, say, $500,000 per life saved. This may substantially exceed the number of expected fatalities at stake by quite a sizeable error in the risk estimate used to determine whether the de minimis standard of risk has been met. Likewise, vast research expenditures could be required if improved estimation of health risks is insisted upon by using the large number of laboratory animals required to determine cancer incidence rates from increasingly low doses of radiation or chemical carcinogens. Even then, there would be substantial uncertainty regarding the implications of the research results for human cancer incidence rates from low doses.

2. ALTERNATIVE APPROACHES TO ESTABLISHING ACCEPTABLE STANDARDS OF DE MINIMIS RISK

2.1 Consistency With Routine Behavior Of Individuals In Accepting Day-To-Day Risks Of Fatality

In 1979, the late Cyril Comar[29] in a brief editorial in Science pointed out that society is becoming increasingly well informed and anxiety-prone about technology-associated risks, which leads to desire for their elimination. The logical and traditional approach is first to estimate the risk, a scientific task. Then comes the issue of risk acceptance, a most difficult step--moving from the world of facts to the world of values. Ideally, judgments involving risk acceptance should be made on society's behalf by a constitutionally appropriate body. But no such public decisionmaking process exists. We make do with disparate efforts of individuals, special-interest groups, self-appointed public interest groups, and legislative, judicial, and regulatory systems. However, if at least very large and very small risks were dealt with on the factual basis of effects, the individual and social value systems could be accommodated to some degree and much confusion avoided. To an important degree this philosophy is similar to the one illustrated in Figure 1.

For the purpose of discussion, Comar went on to suggest some guidelines, stated in terms of numerical fatality risk to an individual per year of exposure:

(1) Eliminate any risk that carries no benefit or is easily avoided.
(2) Eliminate any large risk (about 1 in 10,000 per year or greater) that does not carry clearly overriding benefits.
(3) Ignore for the time being any small risk (about 1 in 100,000 per year or less) that does not fall into category 1.
(4) Actively study risks falling between these limits, with the view that the risk of taking any proposed action should be weighed against the risk of not taking that action.

It should be noted that the value of de minimis individual risk proposed by Comar at one chance in 100,000 per year of exposure falls an order of magnitude higher than that suggested as clearly acceptable in Figure 1. However, an examination of routinely accepted risks shown in Table 3 provides little evidence that a risk ten times as large as the one chance in a million per year reflected in these numbers would not still be regarded as de minimis by those who accept them. For example, the risks of smoking 30 U.S. cigarettes instead of 3, drinking 5 liters of wine in a year or 300 cans of diet soda, visiting New York or Boston for 20 days, spending 60 minutes in a canoe, etc. appear to be rather negligible levels of risk. Such risk comparisons serve a useful purpose principally as "anxiety benchmarks" involving quite low levels of anxiety--or, indeed, insignificant concern--for many individuals, to judge from their lifestyles.

At least one high level judicial view seems to be in reasonable agreement with Comar's suggested standard for de minimis risk of one chance in 100,000 of an individual fatality per year of exposure. In the case of the standard for benezene exposure in the workplace, decided by the U.S. Supreme Court in 1980, the majority decided that OSHA had not found that the risk is significant (443 U.S. 607, 664. Industrial Union Department v. American Petroleum Institute). Yet the only risk analysis before the court (Richard Wilson's estimate) suggested that the risk is 2×10^{-5} per year.[30] The Chief Justice, Warren Burger, referred to the idea that this risk is de minimis according to the legal maxim de minimis non

Table 3

Actions Which Can Increase the Average Risk of Death by 1 Part in 1 Million or Reduce Life Expectancy by 9 Minutes for Cancer or 15 Minutes for Accident[a]

Quantity	Action	Cause of death
2 (U.K.) 3 (U.S.)	Cigarettes	Cancer, heart disease
2 months	Of living with a cigarette smoker	Cancer, heart disease
½ liter	Wine	Cirrhosis of the liver
40 T.	Peanut butter	Liver and other cancers caused by aflatoxin
1 year drinking	Miami drinking water	Cancer caused by chloroform
30 cans	Diet soda	Cancer caused by saccharin
100	Charcoal broiled steaks	Cancer caused by benzo(α)pyrene (risks of red meat, fattening, etc., additional)
2 months	Visit to Denver	Cancer caused by cosmic rays
6,000 miles	Jet flying at 35,000 ft	Cancer caused by cosmic rays
1,000 miles	Jet flying	Accident
1	X-ray in a good hospital	Radiation cancer
20 years	Living within 5 miles of a polyvinyl chloride plant	Cancer caused by vinyl chloride
2 days	In New York or Boston	Air pollution
3 hours	In coal mine	Accident
1 hour	In coal mine	Black lung disease
150,000 times	Dyeing hair with lead acetate dye	Cancer caused by lead
1,000 times	Drinking from banned plastic bottle	Cancer caused by acrylonitrite
6 minutes	In a canoe	Accident
1 year	At site boundary of nuclear power plant	Radioactive accident
3 weeks	Living below a dam	Accident (dam failure)

[a]Source: R. Wilson (Ref. 6).

curat lex, even though it is 20 times bigger than 1 in a million.

Nevertheless, in my view a proposed standard of one chance in a million of individual fatality per year of exposure would be a more acceptable one for the following situations:

(1) Whenever the risk is involuntary;
(2) Whenever there is a substantial band of uncertainty in estimating risk at such low levels;
(3) Whenever the risk has a high degree of expert and public controversy (the political costs of defending a standard of higher risk are probably not worth the difference); and,
(4) Whenever there seems a reasonable outlook that new safety information is more likely to yield higher-than-current "best" estimates of the risk level rather than lower estimates.

2.2 Signal-To-Noise Relation As A Measure Of De Minimis Risk

In engineering practice, when a number of factors of unequal size are being added together, if the largest of these can only be measured, say, to an accuracy of ±50%, it is prudent to exclude altogether the smaller factors that would add only a negligible percentage to the total. That is to say, the smaller factors can be regarded as within the "noise" level of uncertainty bounds of the dominant factors. They have thus lost any practical significance to decisionmaking relative to the importance of the information being signaled by the dominant factors.

This principle is often used by risk assessment experts on the NRC regulatory staff and elsewhere in preparing quantitative risk assessments of potentially severe accidents in the operation of nuclear power plants.[31] The methodology of these Probabilistic Risk Assessments (PRAs) is based on fault tree/event tree logic and involves highly complex and interactive computer models. These include: the sequences of untoward events that could produce a core melt under certain rare combinations of these events; the modeling of accident management techniques and accident progression phenomena; containment failure modes and their probabilities and timing; the quantity and chemical form of radioactivity released if the containment is breached; modeling of weather and dispersion of radionuclides in the atmosphere; the modeling of emergency response effectiveness involving sheltering, evacuation, and medical treatment; and dose-response modeling in estimating health effects of population exposure.

Many accident scenarios are possible with different combinations of common cause and sequential events yielding a wide range of adverse consequences and their associated probabilities. The "best estimate" probabilities of the scenarios that collectively contribute the bulk of the total accident risk often fall within an order of magnitude of one chance in 100,000 per reactor-year of operation (or 10^{-5}). Some of the lesser accident scenarios may have values in the range of 10^{-7} to 10^{-9}, or 100 to 10,000 less frequent than the dominant risk scenarios. Although there is no policy or regulation that formally has established such low-probability accident scenarios as de minimis, PRA practitioners generally automatically discard scenarios of probability 10^{-7} or lower because they are in the noise level of the broad ranges of uncertainty surrounding estimates of the dominant scenarios that contribute much more importantly to the overall risk level.[32] In effect, this practice constitutes a de facto standard of de minimis risk since the trifling significance of accident scenarios with estimated probability of 10^{-7} or lower is automatically assumed without recourse to a cost-benefit (or value-impact) analysis.

Another concept for viewing radiation hazards as negligible or trifling is to compare radiation doses delivered to humans by man-made activities to the background radiation exposure from natural sources (e.g., cosmic or solar radiation and radiation received from radioactive materials in rock, soil, air and water, or buildings constructed with materials containing minute quantities of radioactivity). In the United States variations in background radiation of these natural sources range from 100 millirem to 200 millirem, depending principally on altitudes and radioactive minerals in the ground. A rather thorough review of natural and man-made radiation doses and their effects on health and safety has been made by Edward Pochin.[33] A case can be made that variations in background radiation are themselves to be regarded as negligible or trifling in view of:

(1) The statistics on cancer fatalities in regions of the highest

background radiation display no statistically significant variations from those with the lowest background radiation, indicating that the effect of this radiation on cancer fatalities is well within the level of noise of the variance of the more dominant contributors to cancer risk.

(2) Information available to the public on variations in natural background radiation, including increased cosmic radiation exposure in high altitude airplane flights, does not seem to have awakened any sign of (non-trifling) concern among the public.

(3) The range of dosages received from natural background radiation is characterized as "low-level" radiation from which no directly observable health effects have been recorded. Health effects are inferred from linear (or sub-linear) extrapolation of effects from observed effects on humans and animals of high radiation doses.

On the latter subject, the 1980 (BEIR III) Report of the Committee on the Biological Effects of Ionizing Radiations[34] does not rule out the possibility of zero cancer fatalities from low radiation dosages. In part this is because of the efficiency of biological repair processes in counteracting carcinogenic effects of low-level radiation exposure. But neither is there proof of zero risk. The Delaney Clause amendment to the Food and Drug Act in 1958 often has been interpreted as requiring absolutely zero health risk from any additive to food that is found to induce cancer when ingested by man or animal; yet in recent years a de minimis (or non-zero) level of such risk is regarded as permissible.[35] A case in point is the Coca Cola bottle made with plastic (acrylonitrile polymer). Traces of acrylonitrile, which as a monomer is carcinogenic in animals, can leach into the acid soft drink and, in 1977, the Food and Drug Administration (FDA) banned the bottle. In reviewing the case, the Court of Appeals[36] stated that the administrator of FDA can ignore de minimis risks, notwithstanding the superficial rigidity of the Delaney Clause.

Since the health risks of natural background radiation are apparently quite small, if not proven to be zero, it would appear that variations in background radiation would be a useful benchmark in establishing a standard of de minimis risk for man-made radiation hazards. Such a standard has been established by the Federal Republic of Germany for nuclear power plants:[37]

> Regarding the health risk attributable to the routine operation releases of radioactive material, the natural radiation exposure has been considered as a standard. By the German Radiation Protection Ordinance the annual whole body dose of a reference person near a plant caused by operational radioactive releases is limited to 30 mrem. This is within the variation of natural background radiation.

A discussion of the possible establishment by NRC of a de minimis standard was presented in a recent paper by Guy Cunningham, Executive Legal Director of the NRC.[38] The NRC has promulgated various regulations under the Atomic Energy Act for protection of the public from radiation hazards. Of particular interest in connection with the de minimis concept are (1) the "Standards for Protection against Radiation in Part 20 of NRC's regulations, (2) the principle that radiation exposures must be kept "as low as is reasonably achievable: (ALARA), and (3) certain specific exemptions from NRC's licensing requirements. Exemptions that have been made that permit use or disposal of exempted products or quantities without regard to their specific radioactivity are not properly considered de minimis levels. This is because they are not expressly based on a finding that the risk is negligible. Rather, exemptions in NRC's regulations are based primarily on a balancing of benefit and cost, of which risk may be a non-trivial element.

Unlike a specific exemption, the de minimis concept represents a level of radiation exposure so low that it may be considered of no regulatory concern. It is not based on the ALARA principle in 10 CFR Part 20 that maintains releases and exposures "as low as reasonably achievable" taking into account the state of technology, the costs and benefits of risk reduction, and other socioeconomic considerations including the utilization of atomic energy in the public interest. Rather, the de minimis concept requires a determination that the health risk to a member of the general public as a result of the presence of radioactive materials or radiation is so trivial that the exposure may be legally and practically disregarded. A de minimis risk can be expressed in terms of either a particular dose rate or a specific quantity of radioactive materials, likely to result in a particular level of exposure. The important point is that the level chosen must represent a negligible risk to human health.

Cunningham notes there is a possibility that the NRC will adopt a generic de minimis level in its regulations[39]. On March 20, 1980, the Commission published an advance notice of proposed rulemaking in which it announced that it was considering a major revision to its radiation protection standards in 10 CFR Part 20 (45 Fed. Reg. 18023). The revision would be consistent with recommendations of the International Commission on Radiation Protection in its Publication 26. In addition, it would be based on explicit quantifications of risk and judgments about their acceptability. The NRC has not yet published a proposed rule for public comment. But a draft prepared by the staff on January 13, 1983 recognizes the need for a de minimis concept in radiation protection and defines de minimis as follows:

> "As applied to radiological protection, de minimis can be a level of risk (or dose rate, as a surrogate measure) so low that it would be a trifle in comparison to the risks which the individual is subjected to daily as part of normal living habits and activities. It would constitute a level of risk so low that no resources could be justified to control it, or to be further concerned with it".

The above definition relates to the first approach discussed above of relating a de minimis risk standard to routine behavior of individuals in accepting day-to-day risks of fatality. However, an NRC Licensing Board finding drew attention to natural background radiation as a comparative benchmark for assessing the negligibility of risk in granting permission to construct a uranium mining and milling facility. The Licensing Board found (8, NRC 87-1978) that the calculated health effect that might result--half a death per year in a population of 300 million--was a minimal impact. Properly stabilized mill tailings piles and reclaimed uranium mines would make the impact 100 times lower. The Board concluded that the best means of characterizing the significance of radon releases attributable to operation of the facility was to compare them with those associated with natural background radiation and its fluctuations. The incremental releases attributable to the facility were so small as to be completely undetectable. Thus, the Board concluded that their impact could not be significant.

2.3 Comparative Risks and Uncertainties of Viable Alternatives: Coal versus Nuclear

Another useful benchmark in ascertaining the significance versus triviality of a variety of risks associated with a technology is to compare these with the risk attributes of the viable alternative technologies. Indeed, the NRC invites a comparison of coal and nuclear risks in one of its safety goals:[7]

> "Society risks to life and health from nuclear power plant operation should be comparable to or less than the risks of generating electricity by viable competing technologies and should not be a significant addition to other societal risks".

The comparative part of this goal is to be interpreted as requiring that the risks from nuclear power plant operation are comparable to or less than the risks of the operation of competing electricity generating plants, particularly coal-fired plants. A comparison of coal and nuclear risks for all elements of the fuel cycle (mining, fuel processing, transportation, power generation, and waste management) was made by Gotchy[40] with estimates of coal risk based principally on sources of information outside NRC. These comparative generic risk estimates have appeared in NRC's Environmental Impact Statements for review of applications for Construction Permits since 1979. These estimates suggested that the overall risks of excess mortality (both occupational and to the public) for the nuclear fuel cycle is only a few percent of the total risks of the coal fuel cycle (i.e., 42 deaths per plant-year for coal versus 1.0 for nuclear). If society does not regard at least the public risks from coal as deserving much concern (see below), then the public risks of nuclear power generation might justifiably be regarded as negligible by this standard.

Thus, a more meaningful comparison than total fuel cycle risks is a comparison of similar elements of the fuel cycle. These may affect similar classes or territorial zones of society; moreover, regulatory issues are not limited to the overall public acceptability of the technologies. They also involve issues over the expenditure of resources for risk reduction. For excess fatalities from mining, the risk from coal mining, according to these estimates, is over ten times as much as for uranium mining, principally because it takes about one ton of uranium fuel to generate as much electricity as 15,000 tons of coal. For the same reason, the risk of excess fatalities for transporting coal to the power station, at 1.2 deaths per plant year, is many times greater than the negligible risk from the transport of nuclear fuel. Indeed, the estimated risk to the public from coal transportation alone is roughly equal to the estimated 1.0 deaths per year for the <u>entire</u> nuclear fuel cycle. The largest component of the coal fuel cycle risk is not from mining or transportation, but from air pollution in the generation of electricity. At a geometric mean estimate of 17 deaths to the public per plant year for power generation using coal, this is about four times the estimated occupational deaths (both accidents and disease) from mining coal.

However, the data on which these comparisons are based are from 9 to 12 years old. Since then, technological and regulatory changes can be assumed to have reduced substantially the risk for both coal and nuclear, especially in the excess fatalities to the public resulting from power generation. In the case of coal-fired plants these are practically all latent or delayed fatalities from cancers or chronic diseases arising from air pollution. Likewise, for nuclear power generation, the large majority of excess fatalities would be delayed cancer deaths (20-40 years) arising from radiation exposure due to routine or accidental releases--although with severe nuclear accidents, some scenarios would yield a significant number of early fatalities.

Another dimension of comparison of coal and nuclear risks is that of the ranges of uncertainties surrounding the risk estimates. Studies of psychological factors in risk perception suggest that if two competing technologies were to have equal estimates of risks, costs, and benefits, but grossly unequal ranges of uncertainty surrounding the risk estimates, then a substantial segment of society would prefer the option with the narrowest bands of uncertainty for the risk estimates. The available frequency data are by far the more robust for inferring excess deaths due to coal transportation (principally accidents at railroad crossings) and the accidental deaths from coal mining. However, in the more important areas of risk impacts, namely, those due to latent deaths from power generation, the ranges of uncertainty of the risk estimates are about equally broad for coal as for nuclear risk.[20,41]

Yet, since many of the estimated fatalities to the public from the risks of either kind of power generation are from low levels of radiation from nuclear plants and low dosages of air pollutants from coal-fired plants, it would appear that the development and application of de minimis standards for these kinds of risks would considerably restrict the territorial range where individual risks would exceed these standards. This is germane to one of the NRC's proposed safety goals which states:[7]

> The risk to the population in the area near a nuclear power plant of cancer fatalities that might result from nuclear power plant operation should not exceed one-tenth of one percent (0.1%) of the sum of cancer fatality risks resulting from all other causes.

A range of 50 miles from the plant site is assumed for the implementation of risk estimates to determine variations from this proposed safety goal. However, it is not difficult to foresee that if all severe nuclear accidents with a probability less than one-in-ten-million reactor years of operation were eliminated from further consideration by such a de minimis standard and if a de minimis standard were established removing from consideration all ranges where the dose received from non-trivial accident scenarios was, say, no greater than the natural background dose received in a year by that population, then the truly non-trifling component of the public risk estimates would likely shrink substantially, especially in territorial terms. Moreover, this approach would place into clearer public perspective the non-trivial elements of nuclear (and of coal) risk that merit priority consideration either for risk reduction or uncertainty reduction in determining estimated variances from safety goals or other standards of acceptable risk. There are other attributes of coal and nuclear risk including long-term storage of high level nuclear wastes and the potentially catastrophic dimensions of risk associated with greenhouse effects and acid rain from using coal and other fossil fuels that are beyond the limitations of this paper. Catastrophic consequences of both nuclear and coal risk extend beyond health and fatality risks. They also include a diversity of tragic economic and social disruptions and losses of societal importance to present and future generations.[20]

2.4 Compensatory Measures Of Personal Risk Reduction

Let us suppose that you live less than ten miles from a nuclear plant and that a severe nuclear accident actually happened with substantial offsite release of radioactivity. Further, that you failed to shelter or evacuate fully effectively to avoid a radiation dose that, although insufficient to cause concern for your early demise, nevertheless was estimated by authorities to be approximately five times the dose you receive annually from natural background radiation. You probably would be highly motivated to learn a great deal about the health effect prospects for persons receiving a radiation dosage of this amount. You would learn that the increased risk of your dying from cancer is not estimated to be large and that it is in the statistical range of an expected life shortening not more than several days.

If you felt this to be a non-trifling effect, what changes in life style, then, could you possibly make or expenditures for health care, increased automobile safety, etc. that could provide a restoration of this expected amount of life shortening, singly or in combination? The information found in Tables 2 and 3 and other sources would obviously provide some helpful clues. For example, recent research findings indicate that there are possible counter-measures such as to reduce intake of dietary foods with known carcinogens and to eat more foods having anticarcinogens.[42] Also, you possibly would learn there is research in progress holding prospects for early detection of imbalances in personal enzyme chemistry that comes with advancing age and thus to restore this balance that is needed to prevent the onset of cancer by strengthening biological repair mechanisms.[43] These and other countermeasures to gain back reduced longevity expectations might not involve substantial expense or life style adjustment. Coupled with a very low probability of such an accident scenario in the first place, the expected costs of these countermeasures would be very much reduced a priori than the a posteriori premises on which this discussion was based.

3. ADDENDUM

The purpose of this addendum is to briefly explore the regulatory implications of three recent events for the concepts and conclusionary thrusts of this paper presented in October 1984 regarding the policy issues of worst case analysis and the related problem of establishing acceptable standards of de minimis risk. The first two of these involve final regulatory actions taken by the Council on Environmental Quality (April 25, 1986) and the U.S. Nuclear Regulatory Commission (june 19, 1986). The third event is the catastrophic (worst case) nuclear accident beginning April 26, 1986 at Chernobyl in the U.S.S.R.

3.1 Final Rule of the CEQ on Section 1502.22, "Incomplete or Unavailable Information"

On August 9, 1985 the CEQ published a proposed amendment to Section 1502.22 of its implementing regulations (40 C.F.R. Part 1502), which addresses incomplete or unavailable information in an environmental impact statement (EIS), (50 Fed. Reg. 32234). After reviewing the comments received in response to that proposal, the CEQ issued on April 25, 1986 the final amendment to that regulation (51 Fed. Reg. 15618). The key substantive provisions of this Final Rule on Section 1502.22 are as follows:

> "When an agency is evaluating reasonably foreseeable significant adverse effects on the human environment in an environmental impact statement and there is incomplete or unavailable information, the agency shall always make clear that such information is lacking.
>
> "(a) If the incomplete information relevant to reasonably foreseeable significant adverse impacts is essential to a reasoned choice among alternatives and the overall costs of obtaining it are not exorbitant, the agency shall include the information in the environmental impact statement.
>
> "(b) If the information relevant to reasonably foreseeable significant adverse impacts cannot be obtained because the overall costs of obtaining it are exorbitant or the means to obtain it are not known, the agency shall include within the

environmental impact statement: (1) a statement that such information is incomplete or unavailable; (2) a statement of the relevance of the incomplete or unavailable information to evaluating reasonably foreseeable significant adverse impact on the human environment; (3) a summary of existing credible scientific evidence which is relevant to evaluating the reasonably foreseeable significant adverse impact on the human environment, and (4) the agency's evaluation of such impacts based upon theoretical approaches or research methods generally accepted in the scientific community. For the purposes of this section, 'reasonably foreseeable" includes impacts which have catastrophic consequences, even if their probability of occurrence is low, provided that the analysis of the impacts is supported by credible scientific evidence, is not based on pure conjecture, and is within the rule of reason."

In the same action, the CEQ withdrew the existing guidance regarding 40 C.F.R. 1502.22, found in Question 20 of Forty Most Asked Questions Concerning CEQ's National Environmental Policy Act Regulations, 46 FR 18032 (1981). Guidance relevant to the amended regulation will be published sometime after the final rule becomes effective on May 27, 1986. In comparing the Final Rule with the 1978 requirements of Section 1502.22 (see section 1.1, *supra*) the most obvious change is that the requirement to prepare a "worst case analysis" has been rescinded. Other changes of more or less subtle significance are that certain conceptual words or phrases appearing in the 1978 rule have been deleted: "gaps in relevant information", "scientific uncertainty", "risk and severity of possible adverse impacts", and "beyond the state of the art". New words or phrases used in the Final Rule are: "reasonably foreseeable" (as a descriptor of significant adverse impacts), "theoretical approaches or research methods generally accepted in the scientific community", "catastrophic consequences" (even if their probability of occurrence is low), "credible scientific evidence", "pure conjecture", and "within the rule of reason". In issuing the Final Rule, the CEQ emphasized that it concurs in the underlying goals of the original 1978 regulation -- that is, disclosure of the fact of incomplete or unavailable information; acquisition of that information if reasonably possible, and evaluation of reasonably foreseeable significant adverse impacts even in the absence of all relevant information. According to the CEQ, these goals are based on sound public policy and early NEPA case law. Rather, the need for amendment is based upon the Council's perception that the "worse case analysis" requirement is an unproductive and ineffective method of achieving those goals, "one which can breed endless hypothesis and speculation" (51 *Fed. Reg.* 15618).

It is fair to ask if this Final Rule on "incomplete or unavailable information" had been available in 1984 what difference would the changes have made in the above presentation and discussion of policy issues in worst case analysis and the establishment of acceptable standards of *de minimis* risk. The answer is really "not much". One obviously desirable change is that I might have substituted for the term "worst case risk analysis" that of "low-probability/high-consequence risk analysis". This is a term used as a title of a workshop sponsored by the Society for Risk Analysis at a meeting in Arlington, Virginia on June 15-17, 1982 and one that I recommended to the CEQ on January 31, 1985 as a replacement for worst-case analysis in my invited comments on the CEQ Advance Notice of Proposed Rulemaking on 40 CFR 1502.22.[44] A low-probability/high-consequence event is an appropriate synonym for the new phrase in the Final Rule requiring analysis of impacts from events "which have catastrophic consequences, even if their probability of occurrence is low". Moreover, my discussion of worst case analysis was cast in the judicial background of the "rule of reason" and the requirement of limiting analysis to "reasonably foreseeable impacts with a thoroughness commensurate with their severity and the significance of the action" (NRDC v. Morton, 458 2d 873). Moreover, the four alternative approaches to establishing acceptable standards of de minimis risk were each in consonance with the "rule of reason". However, in discussing the "signal-to-noise relation" as a measure of de minimis risk, I might have lent more discussion to the requirement of the Final Rule dealing with "credible scientific evidence" in the face of many "trans-scientific" issues, to use a term of Alvin Weinberg's.[45,46] Indeed, it is easy to predict that the credibility of scientific evidence for catastrophic consequences of low probability will remain the primary focus of controversy in implementing the new Final Rule as it was under the former rule requiring worst case analysis.[47]

3.2 NRC's New Policy Statement on Safety Goals for the Operation of Nuclear Power Plants

On June 19, 1986 the NRC voted in favor of adopting a final Policy Statement on Safety Goals for the Operation of Nuclear Power Plants (to be published in the Federal Register). Most of its key provisions are basically the same as those presented in the proposed safety goals issued for public comment in May 1983.[7] Insofar as the discussion of the proposed safety goals in the present paper are concerned, the most significant changes in the recently adopted goals and supporting discussion are as follows:

(1) Although the quantitative safety goal dealing with delayed cancer fatalities remains the same, the range of its applicability has been reduced from 50 miles to 10 miles, with the explanation that the risk to the people beyond 10 miles will be less than the risk to the people within 10 miles. (A substantial fraction of the persons residing in the 10 to 50 mile range would, in the event of a nuclear accident with breach of containment, receive radiation doses at a sufficiently low level as to be characterized as de minimis risk according to the precepts of this paper and that of numerous others.)

(2) The qualitative safety goal stating the societal risks to life and health from nuclear power plant operation should be comparable to or less than the risks of generating electricity by viable competing technologies remains unchanged. However, on the matter of credible scientific evidence the supporting discussion in the new statement notes that the broad spectrum of expert opinion on the risks posed by electrical generation by coal, and the absence of authoritative data, make it impractical to calibrate nuclear safety goals by comparison with coal risks based on what we know today. Yet later, in the discussion of quantitative risk objectives, the new statement notes that individual and societal risks from nuclear power plants are "generally estimated to be considerably less than the risk that society is now exposed to from each of the other activities mentioned above" (includes generating electricity from coal).

(3) The Commission proposed as a performance guideline for further staff examination: "Consistent with the traditional defense-in-depth approach and the accident mitigation philosophy requiring reliable performance of containment systems, the overall mean frequency of a large release of radioactive materials to the environment from a reactor accident should be less than 1 in 1,000,000 per year of reactor operation". This guideline is equivalent to a de minimis risk standard as noted above and also

in keeping with the Commission's own statement in its qualitative safety goals that the individual and societal risks to life and health from nuclear power plant operation "should not be a significant addition to other individual and societal risks". (De minimis risk is insignificant risk.)

(4) The safety-cost trade-off criterion of $1000 per person-rem noted in section 1.5 above has been deleted. However, the new policy states that it is currently envisioned that specific (additional) guidelines will be required by the staff that would address such matters as plant performance guidelines, indicators for operational performance, and guidelines for conduct of benefit-cost analyses.

3.3 Some Implications of the Chernobyl Nuclear Accident

There are at least three identifiable issues germane to the subject of this paper on which the Chernobyl nuclear accident of April 26, 1986 in the U.S.S.R. may have a bearing. One of these is whether the probability of a catastrophic nuclear accident (i.e., a worst case scenario) is as low as scientific estimates state? There is a question whether the differences in design between U.S. lightwater reactors and Chernobyl graphite reactor -- and especially possible differences in containment performance for certain kinds of severe accident scenarios*--plus possible differences in safety practices or regulations could make a substantial difference in U.S. reactor safety vs the Soviet reactor safety. Some experts feel there are significant differences, while others take the position that we don't know enough as yet about the Soviet reactor design as well as their safety practices to draw any firm conclusions.[48] Even if there are important differences, as seems probable, this does not entirely erase the question arising over scientific credibility in estimating the probability of such accidents given the universal weakness of frequency data of rare events contributing to such accidents and other known limitations of risk assessment methodology. While U.S. and other estimates of severe accident risk using light methodology. While U.S. and other estimates of severe accident risk using light water reactor technology are decidedly not based on "pure conjecture", there are, and probably always will be, important elements of subjectivity in these estimates despite the impressive gains in safety research and the mounting volume of operating reactor experience data.

---

*From what I have been able to learn as of this writing, it appears that the Soviets modified the earlier design of the Chernobyl reactor with a compromising "do-the-best-you-can" philosophy while the plant was being fabricated. This expediency did not permit a more desirable safety philosophy addressing the broader range of accident scenarios involved in a more complete systems approach in protecting against design basis accidents. This led to a modified installation of multiple barriers including a distribution of liners, seals, and check valves plus a heat suppression pool. These measures provided a degree of protection against pipe breaks and other failures due to excessive water pressure of certain accident precursor events. However, the containment design did not appear to provide an adequate high-pressure boundary surrounding the reactor core itself. Thus, it did not have the capability to protect against over-pressure resulting from a gaseous explosion of the kind that took place in the Chernobyl accident. Although information on the design features of the Chernobyl/Unit 4 reactor is incomplete, it is clear that this design did not have a containment structure surrounding the primary reactor system such as is used in the U.S. reactors.

A second possible implication of the Chernobyl accident is to what extent will the economic and health effects data that could -- but not necessarily will -- be made available in the aftermath of this accident improve the level of credible scientific evidence that is relevant to evaluating such impacts distributed over space and time as related to the same, or similar, radiation doses or depositions from a nuclear accident should one occur elsewhere? International cooperation in making available this data and in analyzing its impact significance in a policy context needs to be encouraged.

A third implication is the need to explore what improvement in public risk perception and policy development related thereto could result from a determination of what fraction of the surrounding population exposed to measurable amounts of radiation from the Chernobyl accident will likely experience a de minimis level of delayed cancer fatality risk or other "anxiety benchmarks" such as that proposed in Alvin Weinberg's concept of "beyond demonstrable effect"?[47] For example, Dr. Robert Gale, an expert on bone marrow transplant who went to Russia following the Chernobyl accident, is reported to have said, "There are about 50,000 to 100,000 individuals who have received at least some dose that may be of long-term concern."[48] Without accompanying data as to what fractions of the population thus exposed received, for example, whole body doses under 100 millirads, 100 millirads to 10 rads, 10 rads to 50 rads, etc. persons thus exposed (and especially policy makers) have no very meaningful basis for a determination of what level of concern or anxiety is warranted.

However, even with such data the quantitative relationship between dose and response is scientifically controversial (i.e., credibility of evidence is in question) except in the range above 50 rads or rems.* A report of a scientific advisory panel of the American Medical Association provides the following statement of a radiation dose level that is sufficiently high to be regarded as yet in the scientific realm of demonstrable "proof".[49]

> "Studies of various populations, some of which have been ongoing for more than 50 years, provide strong evidence that single radiation doses greater than 50 rems are carcinogenic. Among the populations manifesting this effect have been persons exposed to atomic bombs; persons treated for medical conditions such as ankylosing spondylitis, mastitis, and certain conditions of the head, neck and chest; those having occupational exposures, like underground uranium, tungsten and lead miners, radium dial painters, and radiologists; and people exposed to nuclear weapons fallout in the Marshall Islands........
>
> "There are some limitations in past studies that tend to reduce certainty about low-and intermediate-level effects of radiation. These include inadequacy of the dosage date, lack of long-term follow-up, and failure to take into consideration personal characteristics such as co-existing chronic conditions and the use of tobacco. Even the long-term studies of Japanese atomic bomb survivors cannot provide a definitive estimate of risk in the low dose range, in spite of carefully estimated dosimetry and a population of 55,000 persons exposed to doses less than 10 rads."

Nevertheless, it could be considerable comfort to those individuals

---

*Rems and rads are sometimes used interchangeably as roughly equivalent units in assessing health effects of nuclear plant accidents.

exposed to less than 50 rads (i.e., a crude boundary for demonstrable scientific evidence of the carcinogenic effects of radiation) to know that they had received only a fraction of this amount as well as to learn of the results of the Life Span Study of the A-bomb survivors at Hiroshima and Nagasaki.[50] For these survivors, the total of excess cancers deaths was only 0.78 of 1% out of a total 67,660 recorded deaths from all causes for this group that experienced a mean estimate of 16 rads of radiation dose.

REFERENCES

1. Council on Environmental Quality, National Environmental Policy Act Regulations: Implementation of procedural provisions, 43 Federal Register 55978-56007, November 29, 1978; also, 40 CFR Parts 1500-1508.
2. 720 F. 2d 1475, Environmental Law Reporter, 14, 20061 (1983).
3. K.L. Rosenbaum, Update: The NEPA worst case analysis regulation, Environmental Law Reporter, 14, 10267 (1984).
4. Council on Environmental Quality, Proposed guidance memorandum for Federal agency NEPA liasons, Federal Register, 48, 36486 (August 11, 1983).
5. H.L. Friedell, Radiation Protection: Concepts and Tradeoffs, Lecture No. 3, Lauriston S. Taylor Lectures in Radiation Protection and Measurements, National Council on Radiation Protection and Measurements, Washington, D.C. 20014 (September 1979).
6. R. Wilson, Commentary: Risks and their acceptability, Science, Technology, and Human Values, 9, 2, 11-22 (Spring 1984).
7. U.S. Nuclear Regulatory Commission, Safety Goals for Nuclear Power Plant Operation, NUREG-0880, Revision 1, For Comment (May 1983), p. 12.
8. J. Davis, The De Minimis Regulatory Cut-Off Concept, Testimony before the Advisory Committee on Reactor Safeguards, U.S. Nuclear Regulatory Commission (February 9, 1984), 23 pp.
9. P. Slovic, B. Fischhoff, and S. Lichtenstein, Perceived risk: Psychological factors and social implications, Proc. R. Soc. London, A 376, 17-34 (1981).
10. R. Wilson and E. Crouch, Risk/Benefit Analysis (Cambridge, Mass.: Ballinger Publishing Co., 1982).
11. M. Spangler, The role of interdisciplinary analysis in bridging the gap between the technical and human sides of risk assessment, Risk Analysis, 2, 2, 101-114 (1982).
12. C. Starr, R. Rudman, and C. Whipple, Philosophical basis for risk analysis, Annual Review of Energy, vol. 1 (Palo Alto, CA.: Annual Reviews, Inc., 1976), pp. 629-662.
13. C. Hohenemser, R. Kates, and P. Slovic, The nature of technological hazard, Science, 220, 4595, 378-384 (April 1983).
14. V. Covello, W. Flamm, J. Rodricks, and R. Tardiff, Eds., The Analysis of Actual Versus Perceived Risks (New York: Plenum Press, 1983).
15. H. Otway and M. Fishbein, Public Attitudes and Decision Making, RM-77-54 (Laxenburg, Austria: International Institute for Applied Systems Analysis, 1977).
16. H. Kunruether and J. Linnerooth, Risk Analysis and Decision Processes (New York: Springer-Verlag, 1983).
17. R. Kasperson and J. Kasperson, Determining the Acceptability of Risk: Ethical and Policy Issues, CENTED Reprint No. 41 (Worcester, MA.: Center for Technology, Environment, and Development, Clark University, 1984).
18. M. Douglas and A. Wildavsky, Risk and Culture (Berkeley, CA: University of California Press, 1982).
19. M. Spangler, Syndromes of risk and environmental protection: The

conflict of individual and societal values, The Environmental Professional, 2, 3/4, 274-291 (1980).
20. M. Spangler, An international perspective on equity issues involving involuntary and catastrophic risks of the coal and nuclear fuel options, Journal of Public and International Affairs, 5,1 (Winter 1984, pp. 707-721).
21. M. Spangler, "Heuristic Opinion and Preference Evaluation Research for Assessing Technological Options--A User's View," in Technology Assessment, Environmental Impact Assessment, and Risk Analysis, Eds., V. covello, J. Mumpower, P. Stallen, and V. Uppuluri, NATO Advanced Study Institute (Heidelberg, W. Germany: Springer-Verlag, 1985).
22. A. Tversky and D. Kahneman, Judgment under uncertainty: Heuristics and biases, Science, 185: 1124-31 (1974).
23. M. Spangler, A critique of methods in the quantification of risks, costs and benefits in the societal choice of energy options, Annals of Nuclear Energy, 10, 3/4, 119-151 (1983).
24. J. Kemeny, Ed., The Need for Change: The Legacy of TMI. Report of the President's Commission on the Accident at Three Mile Island, Washington, D.C. (October 1979).
25. Advisory Committee on Reactor Safeguards, An Approach to Quantitative Safety Goals for Nuclear Power Plants, NUREG-0739, Report of the Advisory Committee on Reactor Safeguards, U.S. Nuclear Regulatory Commission (October 1980).
26. Environmental Protection Agency, Environmental Radiation Protection Requirements for Normal Operations of Activities in the Uranium Fuel Cycle, EPA 520/4-76-016, Vol. 1, U.S. Environmental Protection Agency; Washington, D.C. (November 1976), p. 51.
27. U.S. Nuclear Regulatory Commission, Regulatory Analysis Guidelines of the U.S. Nuclear Regulatory Commission, NUREG/BR-0058 (January 1983).
28. E. Siddall, Risk, Fear and Public Safety, Atomic Energy of Canada Limited (April 1981), pp. 39-43.
29. C. Comar, Risk: A pragmatic de minimis approach, Science, 203, 4378 (January 26, 1979).
30. R. Wilson, Letter to the Secretary, U.S. Nuclear Regulatory Commission regarding his comments on the proposed policy on safety goals for nuclear power plants (45 FR 71023), May 17, 1982.
31. U.S. Nuclear Regulatory Commission, NRC Policy on Future Reactor Designs: Decisions on Severe Accident Issues in Nuclear Power Plant Regulations, Draft NUREG-1070, U.S. Nuclear Regulatory Commission (April 18, 1984).
32. U.S. Nuclear Regulatory Commission, Probabilistic Risk Assessment (PRA): Status Report and Guidance for Regulatory Application, Draft NUREG-1050, For Comment (February 1984).
33. E. Pochin, Nuclear Radiation: Risks and Benefits (Oxford, U.K.: Clarendon Press, 1983).
34. National Academy of Sciences, The Effects on Populations of Exposure to Low Levels of Ionizing Radiation, Report of the Committee on the Biological Effects of Ionizing Radiations, BEIR III, National Academy of Sciences (July 1980).
35. C. Marraro, Regulating food additives and contaminants, Quantitative Risk Assessment in Regulation, L. Lave, Ed. (Washington, D.C.: The Brookings Institution, 1982), pp. 213-231.
36. Monsanto, et al., v. Kennedy (FDA), 613 F 2nd 947 (DC Circuit), 1979.
37. P. Gottschalk, E. Hofer, A. Jahns, W. Werner, Regarding PRA and safety goals in nuclear power plant safety issues in the Federal Republic of Germany, A paper published by the Gesselschaft f*r Reaktorsicherheit (GRS) mbH, K*ln 1, Federal Republic of Germany (July 1984).
38. G. Cunningham, The de minimis concept in radiation protection, An invited paper delivered at the 1983 Annual Meeting of the Nuclear Safety Research Association of Japan, Tokyo (June 16, 1983).

39. Cunningham, op. cit., pp. 11-12.
40. R. Gotchy, Health Effects Attributable to Coal and Nuclear Fuel Cycle Alternatives, NUREG-0332, U.S. Nuclear Regulatory Commission; Washington, D.C. (September, 1977).
41. J. Bolten, P. Morrison, and K. Solomon;, Risk-Cost Assessment Methodology for Toxic Pollutants from Fossil Fuel Power Plants, Rand Corporation Report R-2993-EPRI (June 1983).
42. B. Ames, Dietary carcinogens and anticarcinogens: oxygen radicals and degenerative diseases, Science, 221, 4617, 1256-1264 (September 23, 1983).
43. G. Blonston, Prevention: The time to beat cancer is before you get it, Science 84, 5, 7, 36-39 (September 1984).
44. M. Spangler, Invited comments on CEQ Advance Notice of Propose Rulemaking on 40 CFR 1502.22-Incomplete or Unavailable Information, Letter to A, Alan Hill, January 31, 1985.
45. A. Weinberg, Science and trans-science, Minerva, x, 2(April 1972), pp. 209-222.
46. A Weinberg, Science and its limits: the regulators' dilemma. A paper presented at the National Academy of Engineering Symposium on Hazards: Technology and Fairness, Washington, D.C., June 3-4, 1985.
47. M. Spangler, Trans-scientific issues in risk-cost-benefit analysis of energy options, a paper presented at a workshop on Risk Tradeoffs with New Energy Technologies, American Association of Engineering Societies, Washington, D.C. October 10, 1985.
48. S. Diamond, Chernobyl design found to include new safety plans, New York Times, May 19, 1986.
49. J. Mathews, Chernobyl may affect 100,000, Washington Post (May 19, 1986), p. A-1.
50. American Medical Association, Risks of Nuclear Energy and Low-Level Ionizing Radiation, A report of the AMA Advisory Panel to the Council on Scientific Affairs of the American Medical Association, Washington, D.C. (1981).
51. H. Kato and W. Schull, Cancer Mortality Among Atomic Bomb Survivors, 1950-78, Part I, Life Span Study Report No. 9, Radiation Effects Research Foundation, A cooperative Japan-United States Research Organization (July 1982).

STUDIES ON THE APPLICABILITY OF NSF/PRA SPONSORED RESEARCH TO THE MANAGEMENT OF SPECIFIC RISK SITUATIONS

D. Okrent, G. Apostolakis, and W.E. Kastenberg

University of California
Los Angeles, CA 90024

ABSTRACT

The applicability of the results of several NSF/PRA sponsored grants are examined in terms of specific risk situations which, among them, introduce most of the attributes frequently used in characterizing risks or stakeholder concerns and values. The risk situations used for case studies are the following:

- asbestos in schools
- hazardous chemical waste sites
- indoor air pollution
- transportation of common hazardous materials by rail or truck
- contaminants in drinking water
- the proposed construction of a plant which stores or employs large quantities of a chemical like dioxin

A decision analyst, a lawyer, a philosopher, and a psychologist, all former and/or current principal investigators under NSF/PRA grants, have participated in the study first by providing selected papers and specific comments on the manner in which their work relates to the management of the specific risk situations. After two rounds of interchange of ideas by mail, a workshop was held involving the four NSF PIs, two individuals involved in actual management of some of the risks under consideration, and the UCLA group, including a consultant who provided a different psychological perspective. The results of the overall study, including recommendations for future research, are summarized in this paper.

KEY WORDS: NSF/PRA Sponsored Research; Risk Management; Research Needs.

1. INTRODUCTION

Recognizing that one of the most difficult problems faced by health, safety and environmental policy decision-makers is how to assess and balance technological risks and benefits, the House Committee on Science and Technology in 1979 encouraged the National Science Foundation (NSF) to develop a program or research on risk analysis. NSF concurred and assigned program responsibility to the NSF Decision of Policy Research and Analysis (PRA). The Technology Assessment Group in PRA accepted the assignment and was renamed Technology Assessment and Risk Analysis (TARA).

The NSF/PRA research program on risk was initiated in August, 1979, and as of the end of FY 1983 more than twenty proposals had been funded. These include studies on the following: the methodologies used in analyzing and comparing risks; methods of managing risk; alternatives to government regulation; ethical considerations, the perception of risk; economic, legal and institutional considerations; and decision-making.

These NSF-funded studies vary in their depth and breadth. In general, they are designed to provide insight into some aspects of the overall problem and not intended to provide detailed answers or methods for use by a specific regulator on a specific issue. Nevertheless, it is of interest to examine now the results of the NSF/PRA risk studies, both individually and as a group, may be applicable to the resolution of risk issues faced by decision makers, and to evaluate possible bases for improved utilization of the study results by those involved in risk management.

The objectives of this study are threefold:

- To apply the results of research emanating from the NSF/PRA Risk Analysis Program to a selected set of risk management situations in order to identify how the results can collectively be of the most usefulness.
- To develop recommendations for the utilization of completed research emanating from the Risk Analysis Program by those charged with risk management policy responsibilities.
- To develop recommendations for new lines of inquiry for risk analysis research.

The study participants include, in addition to the authors, four individuals coming from different background (M. Baram, a lawyer; D. MacLean, a philosopher; M.W. Merkhofer, a decision-analyst and P. Slovic, a psychologist), each of whom has been the principal investigator or a major participant in a completed NSF/PRA study on risk. W. Edwards, a psychologist, acted in an advisory capacity to the UCLA group.

The risk situations are chosen so that, as an entity, they include all the following characteristics which the literature on the subject has identified as relevant to risk management decision-making:

intergenerational effects
high consequence/low probability effects
chronic, low-level, ubiquitous effects
distribution of risks and benefits
dread, common
federal control
state or local control
voluntary/involuntary
retrofitting/forward-fitting

The risk situations chosen are as follows:

1. Asbestos. What safety policy should school boards adopt concerning possible exposure to asbestos in existing schools? Should expenditure to reduce this risk be compared in efficacy to other risk reduction measures under school board control?
2. Disposal of hazardous chemical wastes. What constitutes acceptable risk for future disposal sites? For existing sites?
3. Indoor air pollution. Should state or local government regulate indoor air pollution or otherwise consider it in regard to various ongoing activities such as conservation of energy?
4. Transportation of a common, hazardous material by rail or truck. What safety should guide the regulation of this hazard?

5. Carcinogens and mutagens in drinking water. What constitutes acceptable risk in this regard and how much is it worth spending to reduce this risk?
6. A manufacturing plant that employs and/or stores large quantities of a material like dioxin. What safety policy should guide the regulation of such a plant, if any?

The UCLA group prepared a preliminary report which provides background information about each risk situation, and which includes issues and questions likely to arise and identifies a few alternate risk management approaches, in outline form.

Each PI was requested to do the following in a letter report:

1. Identify specific aspects of reports completed under his NSF grant that apply to each applicable risk situation.
2. Provide a trial proposal (s) on how to approach risk management/safety policy for at least one of the risk situations.
3. Provide other suggestions, comments and recommendations on how to pursue and achieve the objectives of this study.

The UCLA group carried out analyses of the risk situations, using the recommendations of the PIs, reports from the Risk Analysis Program, and other sources. Trial risk management/safety policy approaches were developed for each risk situation.

Each PI was, then, asked to comment on whether his own prior NSF-supported research had been properly utilized in the UCLA analysis, and to recommend changes in the trial risk management/safety policies based on his own research results.

Additional input was generated at a workshop which was attended by the study participants as well as by a California State official responsible for choosing and implementing actions regarding potential clean-up of existing hazardous chemical waste sites, and a representative of the Safety Division of the Los Angeles School District.

In Section 2 of this paper we discuss several issues that have been raised in attempting to apply the research results to the six case studies. In Section 3 several recommendations for achieving increased use of NSF/PRA sponsored studies by risk manager are offered, as well as suggestions for future research.

2. ISSUES RAISED BY THE RISK STUDIES

2.1 Audience Of The Research

The examination of the applicability of the NSF PI research to the six specific risk situations immediately introduces a dilemma, namely, what is the audience to whom the research is being addressed. Several possible audiences emerge, including the following:

- the academic world working in similar kinds of research
- the very broad policy makers, such as members of Congress, heads of regulatory agencies, governors.
- staff members of regulatory bodies or governmental agencies, such as school boards, who are responsible for formulating and executing risk management approaches for specific, possibly local, problems.

Baram, MacLean and Slovic seem to be addressing the first two audiences above, for the most part. On the other hand, the risk situations and their corollary questions were mostly specific, quantitative or both. The UCLA group chose to take the regulatory staff member, rather than the broad policymakers, as its target audience in examining the applicability of the research, at least in its first go-round. Hence, to some extent, there is a mismatch. And while the UCLA group found the work of Baram, MacLean and Slovic of general interest, they did not find a basis for helping make the detailed judgments implied in the description of each risk situation.

Of course, who the decision maker is defines the whole problem in terms of alternatives for action, values, etc. In this context, the problem of indoor radon (General Accounting Office, 1980; Hurwitz, 1983; Spengler & Sexton, 1983) points out several issues warranting further consideration. What appears to be a potentially large health effect (2,000-20,000 additional lung cancers per year in the U.S.) is not being regulated, and regulators seem reluctant to take it on. Groups advocating conservation, which can involve reducing air exchange and hence an increase in health effects, are not the ones who seem to have identified the potential for such risk and they appear reluctant in many cases to have attention called to the matter.

There appears to be no "public interest group" which has taken this issue as one to pursue. Although indoor radon appears to have the potential for causing far more premature deaths than high level radioactive wastes or nuclear power reactors (Hurwitz, 1983), those who criticize technocratic risk have studied and given much media attention to the latter, but rather little emphasis to indoor radon and indoor air pollution.

This issue and the issue of siting a new factory employing dioxin, for example, seem to point to the usefulness of a State responsibility for risk management which tends to emphasize identification, quantification and reduction of the potential for risk, in contrast to preparation for emergencies, an effort which is already ongoing. Of course, the concept that prevention can be cheaper than mitigation has been recognized belatedly for hazardous chemical waste disposal.

2.2 Philosophical Issues

MacLean (1982ab, 1983) discusses a range of philosophical ideas, some of which should be of interest to a risk manager in the early stages of a risk issue, before the battle lines have been drawn, as for asbestos-in-schools. His emphasis is on principles rather than on specifics. For example, he offers a perspective on how generally to factor effects on future generations into decision-making. However, pragmatic advice on how to cope with the issue of future generations is not available, nor is insight on why or whether risks to future generations from high level radioactive wastes and from hazardous chemical wastes should be managed on some comparable basis.

Of particular interest to the UCLA group are MacLean's ideas on risk and consent (MacLean, 1982a). He argues that the most promising way to justify centralized decisions that impose risks is in terms of consent of the people affected. Three models of consent are offered: implicit, hypothetical and nonconsent.

The concept of consent for centralized decisions is an important one. It has already been used in a pragmatic way, e.g., when the Advisory Committee on Reactor Safeguards (ACRS) in May, 1979 first recommended that the Nuclear Regulatory Commission develop quantitative safety goals, the

ACRS also recommended that these be provided to the US Congress for comment and consent (explicit or implicit).

The concept of hypothetical consent is probably most appealing to the technical community, and possibly to regulatory decision makers, since it suggests a high degree of "rationality" to the approach, which can include social values which have been identified and in some way, evaluated. As MacLean states: "A virtue of the hypothetical consent model is that it might justify decisions in areas where there is now controversy and where there will undoubtedly be winners and losers. It does this by dividing risk decisions into two stages. The first stage establishes procedures that are regarded as fair to all for making decisions, and in the second stage the argument shows that a given decision would result from these procedures; it is therefore rational for a person to accept decisions made in this way."

While this emphasis on consent and on process and procedures raises to attention these important basic concepts, what remains to be examined is how these can be actually achieved. This practical problem is clearly very complicated. Perhaps most important is the absence of good, let alone definitive, information on the potential risks. The uncertainties vary from large to huge. This is particularly the case for risks to future generations from disposal of hazardous chemical wastes. Also of interest is the extent to which society really benefits by political decisions that may, in considerable part, be molded by fears that do not relate well to best scientific knowledge.

In this context, it is of some interest to note that the recent CORADM report (NAS, 1982) on Risk and Decision Making, performed by a panel of the National Research Council at the request of NSF, begins with a discussion of the "apparent contradiction" between increased longevity and increased public concern about risk. CORADM summarizes the situation by noting that the type of risks has changed, e.g., from infectious diseases to chronic diseases, such as cancer; new sources of risk have emerged that the individual may fee powerless to prevent, e.g., nuclear war; there is a much intensified reporting of risks in the media, which, sometimes justified and sometimes not, magnifies people concerns; and there is a loss of faith in institutions created to deal with risk and a perception that individuals are unable to control events in a complex technological society.

2.3 The Role Of Quantitative Analysis

Slovic and MacLean have critiqued what they feel is an excessive tendency toward analysis and quantifications of the UCLA group. In a memorandum (1983), Slovic advances the idea that, "Risk analysis methods need to go beyond calculating costs in terms of probability x magnitude of loss computations to include the broader values and concerns that are important to people."

MacLean argues that the distinction between estimating and evaluating risks is often not made clear. While most risk analysts claim that their analyses are coherent presentations of relevant information, MacLean (1983) argues that" "Sometimes this is true, but often this modest stance is misleading. It is a little like describing an action as murder or lying but denying that one is evaluating it. The more that information relevant to a decision is brought into a risk assessment, the more one leaves only formal considerations, like consistency and transitivity of preferences, standing between the estimation and a decision. But if someone says he is only showing you what it would be rational to do, that he is not prescribing a choice, then he is disingenuous in claiming that his analysis is a modest one."

Similar concerns have been expressed by others. For example, in a study performed for EPA entitled "Decision Making for Regulatory Chemicals in the Environment", a National Academy panel said, concerning the proper use of analysis "Highly formalized methods of benefit-cost analysis can seldom be used for making decisions about regulatory chemicals in the environment. Thus, the development or such methods should not have high priority". They went on to add "However, the benefit-cost and decision frameworks described in this report can be useful in organizing and summarizing relevant data on regulatory alternatives which the decision maker must review". The panel further states "Value judgments about noncommensurate factors in a decision such as life, health, aesthetics and equity should be explicitly dealt with by the politically responsible decision makers and not hidden in purportedly objective data and analysis" (NAS, 1975).

Of course, there are those such as Crouch and Wilson (1982) who are strong advocates of tying to quantify what one can, and to do it thoroughly. Similarly, Lave (1981) says "A careful review of scientific evidence and a quantitative risk assessment should be the basis of regulatory decisions. Regulation without these elements is uninformed, arbitrary and unlikely to withstand litigation, induce cooperation from those being regulated, or produce the results desired". Lave goes on to say "The objective of risk assessment is not to eliminate judgment but to inform it". Lave, of course, is very conscious of the uncertainties in, and even impossibility of, quantification (1981). However, he concludes that analysis has the most to offer by clarifying issues for the most emotionally laden areas.

William Ruckelshaus, the EPA Administrator, has said (EPA, 1984) "Risk estimates are the only way we have of directing the attention of risk management agencies toward significant problems." In the same speech, he noted the difficulty EPA has in separating the assessment of risk from its management, in practice, although this is what the National Academy had recommended in a recent report (as well as back in 1975).

Merkhofer (1983) devotes an entire chapter reviewing criticisms of decision analysis and notes that many of the criticisms leveled at the use of formal decision-making approaches seem to assume that analysts view the approach as a rigid rule to supplant or replace the decision-making process. He says that this is not the case and that promoters of analysis view decision-making approaches as an aid to decision making.

The UCLA group has found the structured approach of decision analysis useful both in the organization of such an analysis and in calling forth in logical fashion the factors of significance. As such, it should be useful to risk managers in trying to apply MacLean's concept of hypothetical consent or otherwise attempting to assess a new risk situation. The method does not, of itself, provide a means of resolving the differing views of multiple decision-makers, nor does it provide the basis for making the ultimate judgment on how safe is safe enough for a risk under scrutiny. As Slovic emphasizes, quantitative analyses may frequently serve only as the first step in an iterative process, in which feedback from the various stakeholders is part of a deliberative political decision system.

As noted earlier, those recommending more use of quantitative analysis mostly apply the right qualifiers and caveats, e.g., that it be used to aid, not to substitute for, decision making. However, those concerned that quantitative analysis may be applied without proper qualification, or even be abused, are not without reason for such concern. Interestingly, while in this academic study on the one hand, Slovic (1984) and MacLean (1984) feel the UCLA group has been applying analysis too narrowly, in a real-life regulatory situation on the other hand, one member of the UCLA group (Okrent, 1984) has expressed concern with the way the Staff of the US Nuclear Regulatory Commission has been applying benefit-cost analysis to resolving nuclear safety issues. Finally, the frequent criticism by social scientists of quantitative methods as frequently being misused to produce the results one wants leads one to wonder if some of the social scientists have not been subject to the same human frailty, and if, on occasion, the psychological, sociological or philosophical points being argued are not, consciously or subconsciously, part of a hidden agenda.

2.4 Alternatives To Regulation

Baram (1983) and Baram and Miyares (to appear) provide an identification and clear description of a large number of alternatives to regulation. These include: Common Low (Negligence, Product Liability, Strict Liability, Nuisance, Trespass); Private Voluntary Standards (Industrial Self-Regulation, Licensure and Certification); Insurance and Compensation Plans (Workers Compensation, Government-Subsidized Insurance, Bonding Escrow and Restoration Fund); Contract Negotiation; Financial Incentives; and, Government Influence (Procurement, Information and Publicity, Advance Notice of Intent to Regulate).

Baram has examined these in detail for one of the issues, namely hazardous chemical waste disposal, and in summary fashion for asbestos in schools. Some of these alternatives have a clear potential for application to specific risk situations, for example common law suit for negligence or nuisance for an identifiable and attributable contamination of land or water as a result of highly inadequate disposal of a hazardous chemical. Baram has noted the pressure on industry which arises from laws and legal decisions involving criminal conviction for executors and a total financial liability for each contributor to a waste site. Another example of a promising alternative arises in the potential application of financial incentives for stakeholders living near a proposed waste site in the form of reduced tax rates. For naturally occurring hazards, such as radon, on the other hand, the application of these alternatives to regulation appears problematical, as is regulation itself. For uncertain, low-level, chronic, long-term, poorly attributable health effects, few, if any, of these alternatives seem to have much impact.

2.5 Public Involvement

Slovic, et al., (1980, 1981, 1982) offer not only insight into how some portions of society perceive risk, they also suggest an approach which involves early information to the public, the building of trust, and the achievement of a greater degree of consensus among stakeholders. Slovic (1983) strongly emphasizes that quantitative analysis can only be one input into decision-making and that there must be an iterative process which involves the values and opinions of stakeholders.

In a way, this process, coupled with the incentive alternatives of Baram, has been applied by Electricite de France. At one proposed nuclear plant site, they examined the concerns of different segments of the surrounding population. Where information was helpful (for example, the farmers wished to know how farms were doing near other nuclear plants) they made this available by arranging suitable tours. Furthermore, they reduced electricity rates for those living near nuclear power plants. However, for an issue like asbestos-in-schools, involving the fear of cancer, particularly for children, it is not clear whether Slovic's proposed approach can be effective, especially once the problem has received media attention or parents have otherwise become aroused.

Slovic, et al., point out that the problems in communicating risk information derive from human limitations and the complexity and uncertainties inherent in risk data, as well as from differences in values. People may be too much influenced by subtle differences in the form in which information is provided, a fact that creates moral and political issues.

The indoor radiological problem is an interesting case in this context. The lack of public information programs is one of Hurwitz's (1983) major complaints. He gives an informative example as follows: "A notable example of this attitude is contained in briefs submitted by the New York State Energy Office to the 1981 hearings on the State Energy Master Plan which reject mitigative action pertaining to indoor radon (even so much as warning the public of a possible hazard or including radon measurements in subsidized home energy audits) as "unnecessary encumbrances" to ongoing energy conservation programs." It is interesting to note that a public program has been established in Sweden (GAO, 1980), where the National Institute of Radiation Protection has warned the public about the risks of reduced ventilation and has offered advice on possible remedies.

MacLean's work (1982a) is particularly relevant at this point, especially the arguments for hypothetical consent, that is, the need to establish procedures that are regarded as fair to all for making decisions. Such a procedure is clearly lacking in the indoor air pollution problem and this lack is particularly irritating to Hurwitz.

How a fair procedure would be established is, of course, far from obvious. Building trust and mutual respect among the stakeholders, as Slovic recommends, is a positive suggestion, although actually doing it may be very difficult. In fact, two of the stakeholders, namely the DOE and EPA, have tried to resolve their differences without much success as of late August 1980 (GAO, 1980).

One can envision several levels of public information programs for indoor radon. A decision option to Congress, for instance, could be simply to inform the public (through the EPA, for example) about the risks from indoor air pollution, which is also one of Baram's alternatives. On a somewhat smaller scale, a public information program could be undertaken only in the context of energy conservation policies, where the risks from increased levels of pollutants would be identified among the many outcomes of the decision to reduce the air exchange rate. At an even smaller scale, such a program could be instituted only where the evidence shows that increased levels of radon and its daughters are likely, as is done in Sweden. The work of Slovic, Lichtenstein and Fischhoff would be useful in this effort.

As noted earlier, the argument has been made that such public information programs would be "unnecessary encumbrances" to energy conservation programs and that the risks are too uncertain in the sense that more information should be collected before any action is taken. The question is, of course, whether the need to remove uncertainties is so great as to warrant inaction. The kinds of uncertainties to be removed are also important. While the life-time risk given a level of exposure to radon and its daughters is reasonably well known, there are uncertainties about the radon source term (e.g., area to area dependence) and the relationship between the air exchange rate and indoor exposures which could be reduced by further research. It appears, however, that additional information at this level would affect the details of the public information program and not the larger decision of whether to inform the public at all.

3. RECOMMENDATIONS

3.1 Recommendations For Achieving Increased Use Of NSF/PRA Sponsored Studies By Risk Managers

1. NSF/PRA should encourage the inclusion of a chapter in the final report for all risk-policy-related research grants on how the results of that particular study can or should be used by decision makers, taking into account a wide range of risk situations.
2. Each NSF/PRA research grantee working in areas related to risk could be encouraged to include application of the research to at least one risk situation. This should include a critical assessment of the ideas evolving from the research as they relate to the risk situation.
3. Copies of the final reports for selected NSF/PRA sponsored studies should be sent to a wide range of decision makers in both the public and private sector, asking for opinions on whether and how such studies are useful or could be made more useful.
4. NSF/PRA should sponsor (and host) workshops for decision makers on risk from the public and the private sector at which the results of NSF/PRA sponsored studies are discussed and evaluated in terms of selected risk management situations.
5. NSF/PRA should sponsor and host small highly interdisciplinary workshops for research workers in its field of interest. Each workshop should have as a focus for discussion a few selected risk-management situations. The purposes of the workshops would include education, interchange of opinion and ideas and the broadening of perspective. The workshops should include a few participants actually engaged in real decision-making on the selected risk situations.
6. NSF/PRA should take steps to promote dissemination of the written results emanating from NSF/PRA grants to all NSF/PRA PIs active in the risk area.

3.2 Recommendations For Future Research

We present herein several recommendations for future research, which have evolved from recommendations of the four PIs, the CORADM report (NAS, 1982) and our own experience. It is interesting that the CORADM report presents a long list of widely diverse research efforts on risk and decision making and notes the near impossibility of establishing priorities for such research.

1. There is a need, recognized by several researchers, for multidisciplinary studies. Thus, CORADM says "There is also a need for expansion of interdisciplinary research projects on risk and decision making. The relative neglect of interdisciplinary research is understandable, given the inherent difficulties with such efforts: differences in problem definition, language barriers between participants, problems in finding sponsors who are sympathetic to joint projects, the complexities in establishing peer review mechanisms for interdisciplinary work, and the extra time and expense associated with combating all of

these difficulties. Despite these problems, the committee believe that an expanded interdisciplinary research program on risk and decision making should be undertaken".

Baram says that NSF/PRA should implement a three-pronged coordinated program, the second part of which involves "applied research which forces consideration of the generic principles and then selective application to specific real world problems". Merkhofer makes a related recommendation when he says "it would be very useful to conduct several showcase applications of decision analysis, cost-benefit analysis, applied social welfare theory, and so forth, for various risk situations".

The UCLA study group agrees with CORADM about all the difficulties of performing interdisciplinary research, having experienced many of these first hand. Nevertheless, the UCLA group recommends that a high priority be given to the support of systematic, high quality, interdisciplinary case studies of specific risk situations, studies which include social decision analysis and alternatives to regulation, studies which not only include sociological, psychological, economic, philosophic, political and real-life regulatory considerations, but also require specific proposals for resolution of the issues from the various disciplines as part of the overall task.

2. CORADM observes that "the whole field of conflict resolution - so integral to risk and decision making - is a fertile and valuable research domain for behavioral, social and management scientists. Research and experimentation with improved techniques for mediation and bargaining could be especially useful in resolving disputes about environmental policies that affect health, safety, and the environment".

The PIs and the UCLA group recommend increased research on option invention, the development of less adversarial or otherwise improved methods for controversy resolution, and the development of improved methods for structuring public participation in decision making.

3. Slovic and MacLean, among others, emphasize the importance of process and procedures in risk management. The UCLA group agrees and recommends that, using interdisciplinary case studies, more research be done on how to build a more nearly acceptable process and set of procedures and how to use these in arriving at decisions in the face of continuing conflict on values, objectives, and preferences among the stakeholders or multiple decision makers. In this regard, it is relevant to note the paper by Lave and Romer (1983), in which they discuss and elaborate upon the problems in making safety decisions that affect a large number of people who will not be able to control the outcome.

4. Slovic and CORADM both recommend research on risk perception. Concerning descriptive research on perceptions, CORADM lists several questions, including the following:

In perceptions about risks and in risk-taking behavior, what are the differences (if any) between cultures-between the United States and other countries, and, within the United States, between the sexes, races, religious affiliations, rural and city dwellers, socioeconomic groups, people with different levels of education, etc? What accounts for these differences? What is the role of the media in shaping and responding to people's perceptions about risks?

We might add a few questions to CORADM list such as:
How can the risk perceptions of the poor and the less literate, those

sectors of the public less likely to speak out or to have a truly representative spokesman, be ascertained and included in the decision process? Is it not important that this be done?

Can and should more be done to ascertain the relationship between knowledge of economic impact of alternative risks and risk perception?

Can and should more be done to ascertain the relationship between an improved knowledge of societal risks in general and risk perception?

5. CORADM recommends research to clarify the different responses to risks in different political, economic or national systems. The UCLA group recommends study of a specific subset of the CORADM recommendation as follows:

A changing regulatory system with regard to risk situations involving large hazards is emerging in the United Kingdom. We suggest an interdisciplinary study be made which includes the following:

a) What is the basis of the regulatory approach?
b) How has it been implemented?
c) What are its successes and failures?
d) How is it evolving?
e) How could it be improved (both socially as well as technically)?
f) How could it be adopted in the US?

6. A topic which has had the benefit of rather little research thus far relates to the development of threshold levels of risk for different kinds of actions:

- which preliminary level assessed by a state evaluation for some new risks, is sufficiently high that the legislative bodies and governor should be made aware of its potential, including the large uncertainties inherent in the assessment?
- what level assessed risk is sufficiently high that the public should be informed, even though there are large uncertainties and differences of opinion?
- what levels of assessed risk are sufficiently high that some form of regulatory action should be studied seriously?
- what levels of assessed risk are sufficiently high that actions to reduce or mitigate the risk should be taken, despite the uncertainties in the numbers?
- what is the effect on all of these threshold criteria, if any, of public perception of cost, etc.?

ACKNOWLEDGMENT

This research was supported in part by National Science Foundation Grant PRA 8305097.

The authors wish to thank Mssrs. Baram, MacLean, Merkhofer and Slovic, the four NSF PIs, for their important contributions to this effort. The authors also wish to acknowledge the advice of Professor Ward Edwards and of Dr. Patrick Johnson of NSF.

The authors also wish to thank members of the Safety Department of the Los Angeles School Board, especially Jack Waldron, for their cooperation, including participation in the workshop. They also wish to thank Mr. Thomas Bailey, Chief of Program Management Section, State of California, for his participation in the workshop.

REFERENCES

Baram, M., 1983, Memorandum to D. Okrent of November 14.

Baram, M. and J.R. Miyares, "Alternatives to Government Regulations for the Management of Technological Risks", to appear in: Contemporary Issues in Risk Analysis: The Behavioral and Social Sciences.

Crouch, E.A.C., and R. Wilson, 1982, Risk/Benefit Analysis, Ballinger Publishing Company, Cambridge, Mass.

EPA Journal, 1984, Risk in a Free Society.

General Accounting Office (GAO), 1980, Indoor Air Pollution: An Emerging Health Problem, CED-80-111.

Hurwitz, H. Jr., 1983, "The Indoor Radiological Problem in Perspective", Risk Analysis, 3: 63-67.

Lave, L.B., 1981, The Strategy of Social Regulation, The Brookings Institution, Washington, D.C.

Lave, L.B. and T. Romer, 1983, "Specifying Risk Goals: Inherent Problems with Democratic Institutions", Risk Analysis, 3: 217-227.

MacLean, D., 1982a, "Risk and Consent: Philosophical Issues for Centralized Decisions", Risk Analysis, 2: 59-67.

MacLean, D., 1982b, "Values at Risk", Prepared for Working Group on Risk and Consent, Draft.

MacLean, D., 1983, Memo to D. Okrent of November 8.

MacLean, D., 1984, Memo to D. Okrent of March 31.

Merkhofer, M.W., 1983, A Comparative Evaluation of Quantitative Decision-Making Approaches, SRI International.

NAS, 1975, Decision Making for Regulating Chemicals in the Environment, National Academy Press.

NAS, 1982, Risk and Decision Making: Perspectives and Research, National Academy Press.

Okrent, D., 1984, Additional comments in letter from J. Ebersole, Chairman, Advisory Committee on Reactor Safeguards to N.J. Palladino, Chairman, U.S. Nuclear Regulatory Commission. "ACRS Report on NUREG-1070", July 17, 1984.

Slovic, P., 1983, Memo to D. Okrent of October 17.

Slovic, P., 1983, Memo to D. Okrent of March 26.

Slovic, P., B. Fischhoff and S. Lichtenstein, 1980, "Facts and Fears Understanding Perceived Risk", Societal Risk Assessment - How Safe is Safe Enough? Plenum Press, New York, pp. 181-214.

Slovic, P., B. Fischhoff, and S. Lichtenstein, 1981, "Informing the Public About the Risks from Ionizing Radiation", Health Physics, 41: 589-598.

Slovic, P., B. Fischhoff and S. Lichtenstein, 1982, "Why Study Risk Perception"? Risk Analysis, 2: 83-93.

Spengler, J.D. and K. Sexton, 1983, "Indoor Air Pollution: A Public Health Perspective", Science, 221: 9-17.

DIVERGENT APPROACHES TO UNCERTAINTY IN RISK ASSESSMENT: MATHEMATICAL EXPRESSION COMPARED TO CIRCUMSTANTIAL EVIDENCE

Walter M. Gawlak* and Daniel M. Byrd**

American Petroleum Institute*
Washington, D.C. 20005
Science Advisory Board [A-101F]**
U.S. Environmental Protection Agency
Washington, D.C. 20460

ABSTRACT

Several cliches disguise a serious structural weakness in the process by which society makes regulatory decisions. First, regulatory decisions are multi-disciplinary. Science and law are critically involved. Next, these decisions are highly uncertain. This cliche has a different operational meaning for each of the two disciplines. Finally, dissatisfaction with the regulatory process raises a question of whether greater reliance upon analytical techniques can help.

In this paper we investigate the use of analytical techniques and suggest that an inherent divergence exists in the way scientific and legal minds approach uncertainty. This divergence may lead to regulatory conclusions assured to dissatisfy one or the other discipline. On one hand, a typical scientific assessor's approach to handling uncertainty, such as error propagation, compounds various weaknesses in the data base. Any final decision should comprehend this increased awareness of overall uncertainty. On the other hand, legal analysis , patterned on the judicial handling of circumstantial evidence, weaves together various uncertain elements to reach a determination from which overall uncertainty has been suppressed. Thus, a decision could be reached in which, from the perspective of the scientist, uncertainty apparently increases, whereas the legal profession apparently sees uncertainty as being removed. In short, each discipline has different structural concepts of uncertainty, and incommensurate ways to analyze and express uncertainty. One approach to resolving conflicting views of uncertainty lies in the area of risk management. The legalistic approach of imposing "convict/exonerate" outcomes on the regulatory process might give way to broadening the range of possible outcomes so that the degree of certainty can be related to the stringency of regulatory action.

KEY WORDS: Uncertainty; Inference; Circumstantial Evidence; Error Propagation; Risk Management

---

*This paper does not necessarily represent the positions of either the American Petroleum Institute or the U.S. Environmental Protection Agency.

1. INTRODUCTION

Most of us have heard the luncheon speaker at a professional conference begin his talk with a joke about: "a lawyer, a scientist and an economist found themselves on a desert island with a can of beans but no can opener..." These jokes invariably draw appreciative responses perhaps not so much for wit as for our appreciation for the joke's motivating premise -- that our profession's way of viewing the world is more enlightened or sounder or, at least, different than another's. This paper is concerned with a serious subject at the core of the foregoing joke; what we perceive to be a divergence between the way lawyers and scientists approach uncertainty in risk assessment. Our observations are based on our experience with the way society regulates potentially hazardous substances, a very serious issue to lawyers, scientists, economists and all of us collectively as citizens.

To develop these observations, like our hypothetical luncheon speaker, we have fabricated a simplistic universe for our lawyers and scientists, and generalized their behavior to the can-opener-and-beans level. We hope, nonetheless, that the anecdote tells us all something true and useful about structural problems inherent in our multidisciplinary approach to health risk regulation, problems which may limit the usefulness of tools such as quantitative risk assessment in achieving sound regulations.

2. ANALYTICAL FRAMEWORK FOR EXAMINING EVIDENCE

A common task set before both lawyers and scientists is the analysis and assessment of data. The scientist is more likely to perform the task mathematically, whereas the lawyer tends to do it verbally [1]. In order to draw comparisons, we have converted the task to a simple mathematical model and matched essential scientific and legal elements. For purposes of the relatively simple comparisons we hope to draw, most of us will be sufficiently satisfied with the surface parallels between legal and scientific causation and the assignment of approximate numerical values to such phrases as "beyond a reasonable doubt." We note that there have been ongoing efforts to apply more rigorous mathematical techniques -- probability theory, in particular -- to the presentation and weighing of legal evidence [2] [3] [4]. Although some legal scholars have strenuously and successfully led the resistance to these attempts so far, their reasoning has not persuasively demonstrated any mutual exclusivity between the legal and mathematical approaches [5] [6]. Furthermore, in the same regulatory system, lawyers and scientists both examine and judge evidence supporting the truth or falsehood of a proposition concerning the human hazard posed by a suspect substance. In short, scientists and lawyers are plausibly placed on the same desert island.

The model we propose is a simple one for illustrative purposes:

Probability that a proposition is true = $\pi_0^i p^i$

where i = number of essential elements which must be true.

and

$1 - p = q$ where

p = probability proposition is true.

q = probability proposition is false.

For purposes of illustrating this model, we have generated data for a few values of q for n=1 through 7. Because we are primarily interested in uncertainty, we have also generated data for uncertainty defined as:

$$1 - (p^n + q^n)$$

This is simply error propagation or enumeration of the region in which we do not know whether the final proposition is either totally true or false, because the argument contains elements that are a mixture of true and false. These numbers are tabulated in Tables I-X.

If a scientist were using this model, the several elements to be proven might be arranged hierarchically in the following scheme:

SCIENTIFIC UNCERTAINTY

(1)	CONCEPTUAL UNCERTAINTY	The Wrong Question.
(2)	MISSING THEORY	Assumptions; Descriptive Analysis of Qualitative Sources.
(3)	CHOICE OF MODEL	Equation Completeness; Moments of Uncertainty.
(4)	PROPAGATION OF ERROR	Sensitivity of Variables; Covarying Data
(5)	CHOICE OF DATA	Issues of Extrapolation, Interpolation and Interpretation.
(6)	DATA VARIANCE	Measurement Error.

Choosing n=7 for the number of steps in an argument does not seem far-fetched. Indeed, the National Academy of Sciences identified 33 inference guidelines where decisions had to be made on key elements in the assessment of risk posed to man by a potentially hazardous substance [7]. An inference guideline is defined as "an explicit statement of a predetermined choice among the options that arise in inferring human risk from data that are not fully adequate or not drawn directly from human experience." However, the most important and typical elements relied upon in the regulatory process amount to only a handful, for example, epidemiology, animal bioassay, extrapolation from animal to man and interpolation from high to low dose. A relative level of confidence can be assigned to each element. Because we rarely know the mechanism of action, we must often rely on assumption [8]. These assumptions can be subjected to sensitivity analysis and optional assumptions can be assessed in terms of their quantitative implications on the estimate of risk.

To apply the model to the legal system, we first must establish some basic definitions. For a positive verdict or decision to be reached in a case, factual proof must be adduced to support each of the essential elements of the case to a specified degree of proof [9] [10]. Both the elements of a case and the degree of proof have been established over time in the common law or codified in rules of evidence or procedure. Both elements and degree vary, of course, depending on the legal forum, kind of case, and so forth. In recent times, federal regulatory statutes have created new standards that must be met when the subject in contention is the "innocence" or "guilt" of a potential pollutant. These elements and degrees are intuitively familiar to us all and we have listed typical degrees of proof for criminal, civil and administrative cases in Figure 1.

BEYOND A REASONABLE DOUBT = 0.95 (0.90-0.99)
PREPONDERANCE OF THE EVIDENCE = 0.70 (0.51-0.75)
SUBSTANTIAL EVIDENCE IN THE RECORD = 0.50 (0.33-0.70)
ARBITRARY AND CAPRICIOUS = 0.30 (0.10-0.50)

PROBABILITY ESTIMATED FOR
NUMERICAL REPRESENTATIONS OF LEGAL
STANDARDS OF PROOF

Figure 1

For each we have assigned plausible numerical estimates of the probability values that correspond to these legal phrases.

The legal model also can be multiplicative; that is, usually each and every element of the case must be proven. The number of elements need not be fixed. Seven elements seem to be a reasonable number for discussion. For example, in a homicide case the prosecution might need to establish the corpus delicti (existence of a victim), weapon, identification of the defendant, the opportunity, and the mens rea (criminal intent). Under a regulatory statute, an Administrator might have the burden of demonstrating evidence in the record supporting the elements such as those identified by the NAS and referenced above. The number of elements may be greater or lesser depending on each individual statute which determines whether any element need be proven, can be taken as presumptively true, or is legislatively given as true [11]. Generally, each element of any case must be demonstrated at the appropriate degree of proof.

The demonstration is made by direct or circumstantial evidence. Direct evidence is simply direct knowledge as to the main fact to be proved. Circumstantial evidence is defined as:

> "the proof of various facts or circumstances which usually attend the main fact in dispute, and therefore tend to prove its existence, or to sustain, by their consistency, the hypothesis claimed. Or as otherwise defined, it consists in reasoning from facts which are known or proved to establish such as are conjectured to exist." [12]

We rely on circumstantial evidence in our study because it is analogous to the evidence examined by the scientist to determine a health effect, such as the potential human carcinogenicity of a substance. The legal direct evidence (symbolized by the "smoking gun") is as rarely available as, for example, the scientific direct evidence of a demonstrated mechanism of action. In both cases, reasonable inferences and links in the chain of circumstances are usually all we have.

Circumstantial evidence by definition often includes the linking of several subsidiary events (e.g., identification of the defendant by fingerprints, traces of blood on clothing and eyewitness account). Each of these sub-elements need not meet the "beyond a reasonable doubt" standard standing by itself so long as the necessary element in the chain of circumstances in which it fits passes the test [13]. The value of circumstantial evidence is based on the support each fact derives from others immediately connected therewith as to exclude all doubt of its existence as fact [14].

The Tables contain a column headed "Ratio," defined as q^2/p^2. We generated these calculations to examine how probability of truth (p) changes in relation to probability of untrue (q). If "true" is equated with one side of the argument before the court and "untrue" or "false" with the other, does this measure of "preponderance of the evidence" give us any insight into uncertainty? Note that the columns display the ratio of untrue to true rather than the reverse. This was done to display values comparable to a probability that was less than 1.0 in order to draw easier comparisons with p and uncertainty. As an examination of the Tables indicates, however, the ratio does not add anything to the analysis because the ratio decreases almost as rapidly as the uncertainty.

3. THE SCIENTIST'S APPROACH TO UNCERTAINTY

Examining the data in Tables I-X, what might a scientist conclude concerning the proposition in question? We suggest some observations. (1) It is only until the level of confidence in the evidence for a single step approaches about 0.95, that one sees the type of certainty for the overall proposition with which most scientists normally handle data and deal comfortably with the results. This value corresponds to the usual intuitive assumption for statistical significance. Even then, the probability that the proposition is true is only 70% at the seventh step. It is likely that a scientist might desire greater degrees of certainty before making a strong positive statement about the truth or falsity of a multi-element proposition or hypothesis.

(2) Uncertainty in the overall proposition increases rapidly until at least a 0.98 confidence level for a single step is achieved. AT the end of seven steps, our inability to be certain whether the proposition is either right or wrong is a manageable 13%. Even after only two stages at the 0.50 confidence level, our uncertainty is 50%.

(3) When p-values are close to 1.0 and constant, uncertainty rises slowly even over several steps. However, if any one p-value were low (<0.90), the uncertainty surrounding the proposition rises.

(4) Except for those cases where either the level of confidence is very high (>0.95) or few elements are involved, the scientist would probably forego drawing dispositive conclusions about the validity of the proposition but would instead describe the proposition in terms of probability and attendant uncertainty. This is a ramification of the typical scientific assessor's approach to handling uncertainty; displaying uncertainty by such a technique as error propagation clearly demonstrates the compounding of various weaknesses in the data base and tempers one's certainty. Thus, an exclusively scientific decision-making process comprehends an increased awareness of uncertainty in drawing conclusions.

4. THE LAWYER'S APPROACH TO UNCERTAINTY

(1) Apparently, at the level at which a scientist might barely entertain an argument, the lawyer will convict a criminal defendant beyond a reasonable doubt. The uncertainty when n=7 and p=0.95 is still reasonably high (30%); yet conviction is usually treated as a certainty. In other words, uncertainty tends to get suppressed even in the most serious cases treated by the legal system.

(2) The standards applicable to the regulatory system, even assuming the highest burden of proof (>0.50), do not apparently demand a high degree of confidence in the truth of the proposition that the substance is harmful to man [15]. The uncertainty need not be within any stringent bounds [16]. Of course, if only one or two elements were all that were required to meet the burden of proof, then setting n=2 and p=0.70 would

Table 1

$p = 0.5 \quad q = 0.5$

n	p^n	q^n	Remainder	Ratio
1	0.5	0.5	0	1
2	0.25	0.25	0.5	1
3	0.125	0.125	0.75	1
4	0.0625	0.0625	0.875	1
5	0.0313	0.0313	0.9375	1
6	0.0156	0.0156	0.9688	1
7	0.0078	0.0078	0.9844	1

Where: n = number of essential steps.
p^n = probability that the proposition is true.
q^n = probability that the proposition is false.
Remainder = uncertainty.
Ratio = relative falsity to truth.

Table 2

$p = 0.6 \quad q = 0.4$

n	p^n	q^n	Remainder	Ratio
1	0.6	0.4	0	0.667
2	0.36	0.16	0.48	0.444
3	0.216	0.064	0.72	0.296
4	0.1296	0.0256	0.8448	0.198
5	0.0778	0.0102	0.912	0.132
6	0.0467	0.0041	0.9492	0.088
7	0.0280	0.0016	0.9704	0.059

allow one a 50% chance of being correct, perhaps not an unreasonable situation, although the resulting uncertainty is a high 40%. The poor degree of certainty displayed above might not seem to weak a basis for action if one considers that science does not necessarily have strong data for such recurring issues as the mouse-to-man extrapolation or the interpolation of doses across wide ranges.

Table 3

$p = 0.7 \quad q = 0.3$

n	p^n	q^n	Remainder	Ratio
1	0.7	0.3	0	0.4286
2	0.49	0.09	0.42	0.1837
3	0.343	0.027	0.63	0.0787
4	0.2401	0.0081	0.7518	0.0337
5	0.1681	0.0024	0.8295	0.0145
6	0.1176	0.0007	0.8816	$6.1\text{x}10^{-3}$
7	0.0824	0.0002	0.9174	$2.6\text{x}10^{-3}$

Table 4

$p = 0.8 \quad q = 0.2$

n	p^n	q^n	Remainder	Ratio
1	0.8	0.2	0	0.25
2	0.64	0.04	0.32	0.0623
3	0.512	0.0008	0.48	0.0156
4	0.4096	0.0016	0.588	$3.9\text{x}10^{-3}$
5	0.3277	$3.2\text{x}10^{-4}$	0.672	$9.8\text{x}10^{-4}$
6	0.2621	$6.4\text{x}10^{-5}$	0.7378	$2.4\text{x}10^{-4}$
7	0.2097	$1.2\text{x}10^{-5}$	0.7903	$6.1\text{x}10^{-5}$

(3) Examination of the data suggests that uncertainty in scientific terms is being suppressed. The perception that regulatory cases are proven so convincingly might be a result of having the conclusions written by lawyers rather than by having inherently certain data. This result would be reached in a legal presentation to some extent merely by neglecting to treat uncertainty mathematically or at least with qualitative rigor. Indeed, if uncertainty analysis is replaced with a presentation that sums all of the evidence, the implied effect may be to give an impression that the sum of the evidence possesses a certainty not found in any one of the parts. This is the essence of the lawyer's trade; assigning high certainty to one's own argument while conceding none to the opponent, thereby suppressing uncertainty.

Table 5

$p = 0.9 \quad q = 0.1$

n	p^n	q^n	Remainder	Ratio
1	0.9	0.1	0	0.111
2	0.81	0.01	0.18	0.012
3	0.729	0.001	0.27	1.4×10^{-3}
4	0.656	0.0001	0.3438	1.5×10^{-4}
5	0.590	1.0×10^{-5}	0.4095	1.7×10^{-5}
6	0.531	1.0×10^{-6}	0.4686	1.9×10^{-6}
7	0.478	1.0×10^{-7}	0.5217	2.1×10^{-7}

Table 6

$p = 0.95 \quad q = 0.5$

n	p^n	q^n	Remainder	Ratio
1	0.95	0.05	0	0.0526
2	0.903	0.0025	0.095	2.8×10^{-3}
3	0.857	0.00013	0.143	1.5×10^{-4}
4	0.815	6.3×10^{-6}	0.185	7.7×10^{-6}
5	0.774	3.1×10^{-7}	0.226	4.0×10^{-7}
6	0.735	1.6×10^{-8}	0.265	2.1×10^{-8}
7	0.698	7.8×10^{-10}	0.302	1.1×10^{-9}

5. DISCUSSION OF DIVERGENCE OF APPROACHES

Like our colleagues on the desert island, we are attempting to find an insight where one might not expect one. The divergence in approaches to handling uncertainty described in our simple model of decision-making occurs most acutely in the regulatory process. This divergence, however, has probably been masked by a number of cliches that get applied to the complex social process. First, the decisions are multi-disciplinary. This cliche should be reassuring if the nature of the multi-disciplinary interaction is a synergistic effort which optimizes the outcome of the process. This synergism might, indeed, improve the confidence we had in

Tables 7

$p = 0.98 \quad q = 0.02$

n	p^n	q^n	Remainder	Ratio
1	0.98	0.02	0	0.0204
2	0.96	0.0004	0.039	4.2×10^{-4}
3	0.94	8.0×10^{-6}	0.059	8.5×10^{-6}
4	0.92	1.6×10^{-7}	0.078	1.7×10^{-7}
5	0.90	3.2×10^{-9}	0.096	3.5×10^{-9}
6	0.89	6.4×10^{-11}	0.114	7.2×10^{-11}
7	0.87	1.3×10^{-12}	0.132	1.5×10^{-12}

Table 8

$p = 0.99 \quad q = 0.01$

n	p^n	q^n	Remainder	Ratio
1	0.99	0.01	0	0.01
2	0.98	0.0001	0.020	1.0×10^{-4}
3	0.97	1.0×10^{-6}	0.030	1.0×10^{-6}
4	0.96	1.0×10^{-8}	0.039	1.0×10^{-8}
5	0.95	1.0×10^{-10}	0.049	1.0×10^{-10}
6	0.94	1.0×10^{-12}	0.059	1.0×10^{-12}
7	0.93	1.0×10^{-14}	0.068	1.0×10^{-14}

regulatory decisions in light of a second cliche: these decisions are highly uncertain. Unfortunately, this statement implies that the term uncertainty has the same meaning for everyone.

On the contrary, it may well be that an inherent divergence exists in the way two critical disciplines approach uncertainty. Thus, if science and law are inextricably wedded together, as they are in the process by which environmental regulations are promulgated, a decision could be reached in which, from the perspective of the scientist, uncertainty apparently increases, whereas the legal profession sees uncertainty as being removed. If so, this divergence may lead to regulatory conclusions assured to dissatisfy one or the other discipline. In turn, society will

Tables 9

$p = 0.998 \quad q = 0.002$

n	p^n	q^n	Remainder	Ratio
1	0.998	0.002	0	2.0×10^{-3}
2	0.996	4.0×10^{-6}	4.0×10^{-3}	4.0×10^{-6}
3	0.994	8.0×10^{-9}	6.0×10^{-3}	8.0×10^{-9}
4	0.992	1.6×10^{-11}	8.0×10^{-3}	1.6×10^{-11}
5	0.990	3.2×10^{-14}	0.01	3.2×10^{-14}
6	0.988	6.4×10^{-17}	0.012	6.4×10^{-17}
7	0.986	1.3×10^{-19}	0.014	1.2×10^{-19}

Table 10

$p = 0.999 \quad q = 0.001$

n	p^n	q^n	Remainder	Ratio
1	0.999	0.001	0	1.0×10^{-3}
2	0.998	1.0×10^{-6}	2.0×10^{-3}	1.0×10^{-6}
3	0.997	1.0×10^{-9}	3.0×10^{-3}	1.0×10^{-9}
4	0.996	1.0×10^{-12}	4.0×10^{-3}	1.0×10^{-12}
5	0.995	1.0×10^{-15}	5.0×10^{-3}	1.0×10^{-15}
6	0.994	1.0×10^{-18}	6.0×10^{-3}	1.0×10^{-18}
7	0.993	1.0×10^{-21}	7.0×10^{-3}	1.0×10^{-21}

have difficulty in determining in which direction to go in order to improve the regulatory system. Should greater reliance be placed on analytical techniques (grappling with more data and more clearly identified, higher uncertainty)? Or, should we move toward simpler, rapid response systems with fewer step and less analysis such as a Delaney Clause?

The problem of divergence and its ramifications will not be improved at all if one does not deal with uncertainty. At a minimum, uncertainty ought to be recognized and approached in some explicit manner. The first step in attempting to resolve the conflict is to examine further the reasons why uncertainty appears to be suppressed by the legal approach.

The need for binary resolution of a legal case has an influence on the characterization of evidence, which in turn magnifies the confidence in one argument and dismisses the other. As noted above, this tendency to overstatement may not be necessary if the standards of proof are meant to be low in the regulatory arena. A high degree of certainty may be subordinated to a desire to protect the public health: in the jargon, to avoid awaiting the body count before acting, to act on a presumption, or to accept "signal" as a substitute for replicated data. Recognition of the appropriate legal standard might take some pressure off lawyers attempting to prove that highly uncertain scientific evidence is really certain beyond a reasonable doubt.

The very nature of circumstantial evidence, also, may lead to paradoxical results depending on whether the legal or scientific system is in play. For example, the greater the number of clues surrounding a legal inference, the stronger the evidence (seven vague identifications may be better than three in terms of convincing the jury). In the case of science, a proposition with seven elements, each supported by poor quality data, is worse than one with three. Law and science appear to draw inferences differently. If this is so, some aspects of the divergence may not be resolvable and reducing the degree of conflict may be the best we can do.

It is important to note that the legal system does not assign numerical values to these standards, although a metric might be useful in some cases to instruct a jury or guide a reviewing court. One commentator has suggested that a simple explanation for not assigning a value such as 95% certainty to "proof beyond a reasonable doubt" might be that we prefer not to advertise the fact that we are willing to sacrifice one innocent person in order to secure the conviction of nineteen guilty ones [17]. Indeed, there has been a fierce debate waged in the legal literature over the past two decades over the use of probability theory to improve the operation and output of the legal system [2] [3] [4] [5] [6]. The stridency of the resistance to probability concepts by many noted legal scholars is itself circumstantial evidence for the proposition that the lawyer and the scientist have divergent views on causation and uncertainty.

Regardless of the causes for the suppression of uncertainty, in the hybrid legal/scientific regulatory arena overemphasis on the legal approach has the unfortunate side-effect of overselling the certainty of our scientific knowledge. The scientific approach obviously treats uncertainty more rigorously and openly. This is, of course, made possible by the scientific generation of data in numerical form. Although the scientist may be accustomed to higher quality data than is often available in the regulatory process, the techniques still have valuable application to the characterization of uncertainty in risk for the purpose of rulemaking. Quantitative risk assessment offers an opportunity for coming to grips with uncertainty even if this is bought at the expense of greater complexity and a higher cognizance of the uncertain nature of our actions [18].

One approach to resolving conflicting views of uncertainty lies in the area of risk management. The legalistic approach of imposing "convict/exonerate" outcomes on the regulatory process might give way to broadening the range of possible outcomes so that the degree of certainty can be related to the stringency of regulatory action [19]. Instead of burying uncertainty in the risk, the uncertainty should be articulated and used as a tool in the decision-making process. For example, the uncertainty could be used as a weighting factor in choosing among options.

6. CONCLUSIONS

Examination of a simple model of the legal system supports the observation that a divergence exists between the way the scientist and the lawyer approach uncertainty. This divergence is related to the systems of evidence and standards of proof with which each is accustomed to working. The result is that the typical scientific assessor's approach to handling uncertainty calculates and compounds various weaknesses in the data base, whereas the legal analysis, often using circumstantial evidence and aimed at reaching a binary judgment, tends to suppress overall uncertainty. When these two approaches must collaborate in a multi-disciplinary system, the outcome may be unsatisfactory unless both approaches are accommodated. We believe this conflict cannot be resolved unless uncertainty is explicitly addressed. One means of accommodation could be through an approach to risk management which broadens the range of possible outcomes so that the degree of certainty can be related to the stringency of regulatory action.

REFERENCES

1. D.C. Cox and P. Baybutt, Methods for Uncertainty Analysis: A Comparative Survey. Risk Analysis, 1: 251-258 (1981).
2. R. Cullison, Probability Analysis of Judicial Fact-Finding: A Preliminary Outline of the Subjective Approach, 19 U. Tol. L. Rev. 538 (1969).
3. L. Tribe, Trial by Mathematics: Precision and Ritual in the Legal Process, 84 Harv. L. Rev. 1329 (1971).
4. L. Brilmayer and L. Kornhauser, Review: Quantitative Methods and Legal Decisions, 46 U. Chi. L. Rev. 116 (1978).
5. D. Kay, The Laws of Probability and the Law of the Land, 47 U. Chi. L. Rev. 34 (1979).
6. M.G. Finkelstein, Quantitative Methods in Law. Free Press, New York (1978).
7. National Research Council, Risk Assessment in the Federal Government: Managing the Process. National Academy Press, Washington, D.C. (1983).
8. J. Jeffreys, Scientific Inference. Cambridge University Press, London (1973).
9. Brow v. West Riverside Coal Co., 143 Iowa 662; 120 NW 732.
10. Butt vs. State, 81 ARK 173, 98 SW 723; State vs. Alcorn, 7 Idaho 559, 64 P 1014.
11. J.V. Delong, Informal Rulemaking and the Integration of Law and Policy, 65 Va. L. Rev. 258 (1979).
12. Blacks Law Dictionary, (4th Edition) West Publishing Company, St. Paul, MI, (1968).
13. State v. Cohen, 168 Iowa 208, 78 NW 857, State v. Vestring, 144 KAN 167, 58 P. 2d 1087.
14. Commonwealth v. Polian, 288 Mass 494, 193 NE 68.
15. N.L.R.B. v. Lightener Pub. Corp. of Illinois, C.A. 7, 113 F. 2d 621.
16. Deena Products Company v. N.L.R.B., 73 S.Ct. 29, 344 U.S. 827, 97 L.Ed. 644.
17. Kay, Ibid., P. 40.
18. M. Henrion, The Value of Knowing How Little You Know: The Advantages of a Probabilistic Treatment of Uncertainty in Policy Analysis. Ph.D. Thesis, School of Urban and Public Affairs, Carnegie-Mellon University.

19. J.C. Davis, F. Irwin and S. Gusman, Determining Unreasonable Risk Under the Toxic Substances Control Act. Conservation Foundation, Washington, D.C. (1979).

MANAGING TECHNOLOGICAL RISKS: U.S. AND JAPANESE APPROACHES

Kazuhiko Kawamura*, Mark Boroush**, Saburo Ikeda***,
Paul Finger Lynes**** and Michael Minor*****,

Vanderbilt University*, Nashville, TN
The Futures Group** Glastonbury, CT
University of Tsukuba*** Sakura, JAPAN
Vanderbilt University**** Nashville, TN
Vanderbilt University***** Nashville, TN

ABSTRACT

Over the past two decades, concerns for the societal impacts- whether realized or potential- created by the adoption and application of technologies has both widened and deepened in most industrialized societies. This trend mirrors sharpening public concerns over the risks to human health and safety posed by technological hazards.

This paper is a preliminary report on a study sponsored by the National Science Foundation to conduct a comparative analysis of how technological risks are being managed in the United States and Japan. This paper examines one of the four case areas, airborne lead, which the study is currently investigating and outlines similarities and differences in the way both countries are managing risks.

KEY WORDS: Risk Management, Comparative Analysis, U.S., Japan, Airborne Lead

1. INTRODUCTION

In August, 1983 the National Science Foundation announced support for a 20-month exploratory study designed to conduct a comparative analysis of how technological risks are managed in the United States and Japan. The project is being conducted by Vanderbilt University, with cooperation from the University of Tsukuba, Tsukuba, Japan and The Futures Group, Glastonbury, CT.

The project is expected to accomplish the following objectives:

1) Conduct a systematic comparison of risk management in the U.S. and Japan.
2) Suggest possible cross-cultural learning, in areas related to methodologies, data collection, analysis, or decision processes, that might enhance the effectiveness of future technological risk management.(1)

In order to accomplish these objectives, the project will address several key issues:

1) What are the "structural" characteristics of the risk management processes in both countries? Who is involved? How are they involved? What is the character of the risk management decisions which are reached? How similar or dissimilar are the U.S. and Japan along these lines.
2) What do policy makers in both countries do as they reach for decisions on technological risk issues? How similar are the activities undertaken and the resources drawn upon?
3) What role(s) does risk assessment play in the formation of policy? What differences are there in the way each country uses such analysis?

2. APPROACH

We have approached the task of systematically comparing risk management in both countries through parallel comparison of several carefully selected cases of technological risk management. Five separate criteria were used to choose case study areas. (2) Included are:

1) Similarity of the Issue
2) Representativeness of Current Practices
3) Sufficient Issue Duration
4) Extensiveness
5) Data Availability

Based upon these criteria, we have chosen four case study topics through which to conduct the comparisons. These are:

1) risks from detergents;
2) seat belts;
3) risks from airborne lead; and
4) pesticides.

Each of these issues involves a different type of risk. (See Table One) Detergents involve longterm environmental damage risks. Seat belts address risks arising out of normal day to day human activities. The airborne lead and pesticide issues both involve human health risks. We are not attempting to find ideal types, in the Weberian sense, for our cases; we need cases which allow us to cut into the wide domain of technological risk issues from different angles. The case studies selected are intended to be heuristic: - that is, risk examples of a typical profile which can yield useful insights for an exploratory study such as this.

In addition to variance in the type of risk involved, these four cases differ in the nature of the task of inferring consequences. Both the detergent and seat belt cases originally allowed inferences to be made from experiential evidence, whether this be the levels of phosphorus on Lake Erie or the number of victims from car accidents. Over time, of course, the detergent issue has increased in complexity as the focus has moved from sudsing to phosphorus to other issues. Meanwhile, there are still the deaths from car accidents, so this issue has remained constant as to focus.

In the cases of both lead and pesticides, inferences of consequences generally have had to be drawn from laboratory experimentation rather than direct evidence. Both issues, in addition, have become more complex recently as improved measuring techniques indicate that smaller and smaller dose levels may have significant health impacts.

Preparation of case studies of these four risk areas has been ongoing since last fall in both the U.S. and Japan. In late October, 1984, some

Table 1

Case Study Topics

CASE STUDY	RISK TYPE	NATURE OF INFERENCE TASK
Detergents	Long-term environmental	Direct, originally from visual evidence
Seat Belts	Acute human	Direct, from experiental evidence
Lead	Chronic human, Point source	Indirect, from human measurement, monitoring
Pesticides	Acute human, Chronic human, Long-term environment, Non-point source	Indirect, from human and other measurements, monitoring, laboratory testing

20 U.S. scholars, government officials, and businessmen will meet with Japanese counterparts in Tsukuba, Japan to consider findings from both sides and discuss the nature of each country's risk management approaches.

For the rest of this paper, we report on one of the four topics addressed by the study, airborne lead. This is particularly instructive in describing the approaches to risk management used in the U.S. and Japan.

3. AIRBORNE LEAD IN THE U.S.

Tetraethyl lead was developed in 1921 by General Motors to increase the octane rating of gasoline and to prevent premature ignition in engine cylinders, a condition known as knock. The introduction of tetraethyl lead permitted the development of smaller, more powerful gasoline engines. Ethyl Crop. was founded three years later as a joint venture between General Motors and Standard Oil of New Jersey. (3) By the late 1960's over 400 million lbs. of tetraethyl lead were being produced per year. Present consumption, however, is only 120 million lbs. per year. (4)

If we ask "what events led to lead quality criteria," two incidents stand out. The first was a study of the lead levels in the blood of urban children. The study indicated that, of a total population of 10 million children, blood lead levels were as follows (5):

Blood lead level (μg/100ml)	No. of children (under age 6)
0-20	3,350,000
20-40	6,067,500
40-60	650,000
60+	32,500

Source: see J.F. Gilsinn, "Estimates of the Nature and Extent of Lead Paint Poisoning in the United States," National Bureau of Standards Technical Note 746, 1972.

A second incident, occurring in 1973, involved a baghouse from the

Bunker Hill Smelter in Kellogg, Idaho, which malfunctioned but continued to operate for a year. The baghouse discharged 35 metric tons per month of lead. The mean blood level of 170 children within one mile of the baghouse was 65.9 µg/100ml, and air levels measured 18 $\mu g/m^3$. A few children experienced blood levels over 100 µg/100ml and suffered permanent neurological damage(6).

3.1 U.S. Air Quality Regulations

In 1973 the Environmental Protection Agency (EPA) promulgated regulations for a phasedown of lead in gasoline as a result of a suit by the National Resource Defense Council (NRDC). (See Table Two) These limitations were initially struck down by judicial review but were reinstated in the landmark case of Ethyl Crop. v. EPA. (7) As a result, lead was listed under the Clean Air Act and air quality standards were required to be established. Regulations provided for the availability of leaded gasoline and required that new cars be equipped with catalytic converters.

In 1977, the proposed standard of 1.5 $\mu g/m^3$/calendar month average was issued, together with documents for Air Quality Criteria for Lead Air Emissions and the proposed regulations for State Implementation Plans. The final criteria document was issued in December, 1977. The following six items were cited as reasons for regulation:

1) Multiple sources of lead exposure: air, paint and ink, drinking water, pesticides, and fresh and processed foods.
2) Direct exposure from inhalation, or indirectly from the ingestion of food, water, or non-food items.
3) Significant individual variability in response to lead exposure, with young children representing the most vulnerable group.
4) Three systems within the human body were found most sensitive to lead levels. These were the blood-forming or hematopoietic system, the nervous system, and the renal system. Additional effects were found to the reproductive, endocrine, hepatic, cardiovascular, immunologic, and gastrointestinal systems.
5) The finding that blood level thresholds in children of 80-200 µg/100ml results in permanent, severe neurological damage or death: and that thresholds as low as 10 µg/100ml inhibited enzymes.
6) Finding lead to be a stable compound, ubiquitously distributed, which persists and accumulates in the environment and the human body.

This document identified leaded gasoline as the major contributing source of airborne lead representing 88.1% of the total. These findings were used to arrive at a standard of 1.5$\mu g/m^3$ for a calendar month average for airborne lead. The maximum safe blood level (geometric mean) for young children was set at 15 µg/100ml, and of this 12 µg/100ml should be from nonair sources. The difference, 3.0 µg/100ml, was estimated to be the safe allowable contribution from airborne lead. Epidemiological data indicated a general relationship of 1:2 between airborne lead ($\mu g/m^3$) and blood lead (µg/100ml). EPA determined that the level for the proposed standard should be 1.5 $\mu g/m^3$.

To summarize the anticipated impact of the ruling, it was estimated that the existing regulations for a phase-down of lead in gasoline, combined with the increasing use of unleaded gasoline in cars equipped with catalytic converters, would result in attainment of the standard in urban areas where automobile exhaust was the dominant source of lead in the air. No additional pollution controls were anticipated for these areas.(8)

Table 2

Ordered Phase Reduction of Lead in Gasoline by EPA, 11-28-73

	Grams/Gallon (Grams/Liter)	Date
	1.7 (0.45)	1/1/75
	1.4 (0.37)	1/1/76
	1.0 (0.26)	1/1/77
	0.8 (0.21)	1/1/78
	0.5 (0.13)	1/1/79
Comparison:	2.3 (0.61)	1970

(Source: D. R. Hattis, R. Goble, and N. Ashford, "Airborne Lead: A Clearcut Case of Differential Protection," ENVIRONMENT, 24:1, January/February 1982, pp. 41-42).

3.2 Recent EPA Actions

More recently, on July 30, 1981, EPA took action to reduce the amount of lead in gasoline 91% by 1986, to 0.1 gm/gallon (0.0264 gm/1) from the level of 1.1 gm/gallon (0.290 gm/1) in effect in 1981. EPA found that leaded gasoline still accounted for 45% of the motor fuel sold, and that 13.5% of the users of leaded gasoline were using this fuel in vehicles intended for unleaded gasoline use.

EPA announced that the number of children with elevated blood levels is expected to drop from 300,000 to 97,000 by 1986. The new restrictions are anticipated to further reduce the number by another 50,000. Additional data supported lowering the levels for concern in children.

4. AIRBORNE LEAD IN JAPAN

The number of vehicles in Japan grew from only 3.4 million in 1960 to 10 million in 1967 and 39 million in 1980. Particularly the number of passenger cars expanded very rapidly and consisted of more than 60% of the total vehicles registered in 1980. This rapid expansion of motor vehicles caused a severe air pollution problem by the late 1960s in various metropolitan areas. Thus a series of regulations controlling various air pollutants such as hydrocarbon and nitrogen oxide were established in early 1970s.

In 1970, a survey by a private medical group was released in Tokyo, stating that a number of residents in and near Yanagicho, Ushigome, Tokyo were suffering from poisoning caused by lead in automobile exhaust emissions. the results of the survey spurred social demand for a supply of unleaded gasoline.

Thus, in 1970 a public demand for unleaded gasoline began to

Table 3

Estimated Atmospheric Lead Emissions for the United States, 1975

SOURCE CATEGORY	ANNUAL EMISSIONS (metric tons/yr)	EMISSIONS AS % OF	
		Subtotal	Total
Mobile subtotal	142,000	100	--
Gasoline combustion	142,000	100	88.1
Stationary subtotal	19,225	100	--
Waste oil combustion	10,430	54.3	6.5
Solid waste incineration	1,630	8.5	1.0
Coal combustion	400	2.1	0.2
Oil combustion	100	0.5	0.1
Gray iron production	1,079	5.6	0.7
Iron and steel production	844	4.4	0.5
Secondary lead smelting	755	3.9	0.4
Primary copper smelting	619	3.2	0.4
Ore crushing and grinding	493	2.5	0.3
Primary lead smelting	400	2.1	0.2
Other metallurgical	272	1.4	0.2
Lead alkyl manufacture	1,014	5.3	0.6
Type metal	436	2.3	0.3
Portland cement production	313	1.6	0.2
Pigments	112	0.6	0.1
Miscellaneous	328	1.7	0.2
Total	161,225	--	100

Source: Environmental Protection Agency, Air Quality Criteria for Lead, EPA-600/8-77-017, December 1977.

appear. In the same year, the Ministry of International Trade and Industry (MITI) decided to promote the use of unleaded gasoline to purify exhaust emissions. MITI established the Industrial Structural Council (ISC) to study the matter. In July of 1970, the lead content allowed in premium gasoline was reduced to half of its original level under "administrative guidance" (gyosei shido) by MITI.

In August, 1970, the Industrial Structural Council submitted its recommendations to MITI. Recommendations included:

1) Both premium and regular gasoline should be lead-free in and after April, 1974.
2) The octane numbers of such gasoline should be 95 for premium and 88 for regular gasoline.
3) Production of vehicles powered by unleaded gasoline should begin.
4) Possible problems experienced as a result of lead elimination, such as valve recession, should be studied.

For the period 1971 through 1974, valve recession tests were conducted under MITI guidance. As a result of these studies it was found that:

1) The higher the vehicle speed and the greater the load being carried, the more often valve recession occurred:
2) As valve recession progressed there was increased chance of engine destruction.

Therefore, vehicles produced after April 1972 were to have valve seats of higher wear resistance.

Beginning in may 1973, installation of exhaust devices was required in order to reduce carbon monoxide, hydrocarbon and nitrogen oxide emissions. However, it was found that lead had an adverse effect on exhaust gas emission control system. While exhaust gas recirculation systems and catalyzers were required to reduce the levels of CO, HC and NO_X emissions, lead compounds deteriorated the catalyzers. Thus, measures were taken to reduce the quantity of lead additives in gasoline.

In March 1974, the ISC submitted another report recommending a revision of the total ban on leaded gasoline. Revisions were to accommodate cars produced before 1972. The ISC recommended that:

1) From October 1974, regular gasoline should be unleaded only.
2) To insure the safety of vehicles produced before 1972, premium leaded gasoline should be available for the time being.
3) Both premium and regular gasoline should be unleaded by April 1975.
4) An information drive should be mounted to inform both the general public and the refinery industry of these actions.

In September 1974, MITI issued its final notification on the unleaded gasoline program. MITI accepted the SCI recommendations almost totally. Specifically, MITI declared that:

1) From February 1975, regular gasoline should be unleaded only.
2) While enforcing this switch to unleaded regular, the public should be informed of these changes through a mass media campaign. Additionally, appropriate stickers should be placed on vehicles produced before 1972 to be eligible to purchase leaded gasoline. These stickers were to be of four types:
    a. Unleaded (blue), for cars with catalytic converters.
    b. Leaded for High speed, driving (orange) for cars which could use unleaded regular gasoline for driving on flat and smooth roads, but needed leaded premium for severe driving conditions.
    c. Mixed (green) for trucks and busses, where leaded gasoline should always be used with a mixing ratio of 1/3 or more.
    d. Leaded (red) for cars without catalytic converters.
3) Further studies should be done before a total ban on leaded gasoline was instituted.

As a result of a series of regulations controlling various air pollutants, the level of airborne lead decreased significantly.

In August 1976, the Central Council for Control of Environmental Pollution, an advisory organ to the Director-General of the Environmental Agency, made a recommendation that, at present, there was no necessity to establish environmental standards for lead in the atmosphere.(9) Thus Japan seems to have successfully eliminated the risks from airborne lead.

5. COMPARATIVE ANALYSIS

A comparison of the history of the airborne lead issue in U.S. and Japan draws attention to some key similarities:

- The largest contributor to the problem was automobile emissions. The mandated use of unleaded gasoline was the primary policy thrust to reduce the level of airborne lead in both countries.
- Both in the U.S. and Japan, public awareness and subsequent pressures influenced government policy to establish environmental quality standards in the early 1970s.
- Neither government has banned leaded gasoline totally.

At the same time, there are also some significant differences:

- One key difference in the implementation of risk management of airborne lead in the U.S. and Japan was the use of regular leaded gasoline. In Japan the use of regular leaded gasoline was stopped by the end of 1975. On the other hand, the U.S. has maintained the supply of regular leaded gasoline.
- In Japan, the use of unleaded gasoline resulted in:
  a) improved ambient air quality
  b) the ability to compete in the international automobile market-namely in response to emission controls in U.S. demonstrating a mutually beneficial arrangement between government, industry and national health interests. In the U.S., however, compliance with these standards has been contested repeatedly by automobile industry, demonstrating an adversarial role between industry and government.
- The individual response in the U.S. to efforts to reduce auto emissions has been perceived by many as a reduction in personal freedom and an infringement on personal liberties. The Japanese, on the other hand, seem to observe government regulations much more diligently.

6. CONCLUSIONS

In this paper, we reported on our effort to systematically compare risk management approaches in the U.S. and Japan through parallel analysis of several carefully selected cases. One such case, airborne lead, was described and tentative similarities and differences were shown. It seems ironic to note that the Clean Air Act Amendments of 1970 in the U.S. encouraged Japan to adopt stricter standards to control air pollution, which in turn, forced the Japanese automobile industry to develop much cleaner cars in order to export cars to the U.S. In the U.S., however, the implementation of the Clean Air Act Amendments has been delayed many times due to industry objections.

REFERENCES

1. Scoping Paper, "Managing Technological Risks: A Comparative Study of U.S. and Japanese Approaches," Vanderbilt University, January 5, 1984 (unpublished memo).
2. Mark Boroush, "Suggested Criteria for Case Study Selection," The Futures Group, September 12, 1983 (unpublished working paper).
3. New York Times, May 28, 1984, p. A15, and Wall Street Journal, May 16, 1984.
4. New York Times, May 28, 1984, p. A15.
5. J.F. Gilaim, "Estimates of the Nature & Extent of Lead Paint Poisoning in the United States," National Bureau of Standards Technical Note 746, 1972.
6. Environmental Protection Agency, Air Quality Criteria for Lead, EPA-600/8-77-017, pp. 12-15.

7. Environmental Review Cases, 8, 1975, p. 1353.
8. 43 Federal Register 46247, October 5, 1978.
9. Communication with Japan Automobile Manufacturers Association, Inc., Washington, D.C., April, 1984.

RISK MANAGEMENT, ASSESSMENT, AND ACCEPTABILITY

Chauncey Starr

Electric Power Research Institute
Palo Alto, CA 94303

Health and safety are moral sentiments generally given very high priority in most societies and by individuals. Like other moral sentiments - such as peace, freedom, and happiness - health and safety are not absolutes that can be defined in a quantitative sense or specified in detail. They are, in fact, intangibly measured by the absence of undesirable elements which tend to reduce these objectives. Good health is considered a state of physical and emotional well being which is achieved by the absence of detectable disease, physical malfunctions or early death. Safety is considered a circumstance of living in which physical injury or imminent threat of such injury is absent.

It is characteristic of such moral sentiments that each one is considered a social imperative justifying unlimited application of available resources. Obviously no society has enough resources to fully meet such open ended commitments. As a practical matter, therefore, national resources are allocated either by the operations of a free market, if one exists, or by governmental actions which mandate and regulate such allocations. In this presentation, I would like to discuss the interlocking roles of risk management, assessment and public acceptability in the formulation of social policy and government actions which lead to the allocation of our national resources for the improvement of public health and safety.

Analysis of the potential for harm from operational systems for delivering goods and services, and the quantitative assessment of risk probabilities and consequences, provides a very powerful tool for revealing the principle characteristics which can impair health and safety. I am an enthusiast for this process of quantitative analysis. It highlights the important risk-creating areas and thus encourages remedial changes. It encourages foresight rather than hindsight. It also provides a knowledge base for all subsequent actions taken by government or individuals to cope with these risks. And finally, the quantitative assessment provides a crude measure of the relative scale of the consequences of a specific risk in the total spectrum of all the other risks that we must live with. In combination with the corresponding spectrum of the benefits provided by various social activities, comparative risk assessment helps to provide a basis for the rational distribution of society's resources to improve public health and safety.

Nevertheless, the professional fascination with the methodology for the quantification of risk assessment, and public reactions to specific numerical findings, has tended to obscure the practical, functional end-objective which instigates assessment studies in the first place--namely, to assist society in setting criteria and allocating available funds for the management of risk.

Let me illustrate my point with a simple analogy. Almost every big city has a zoo, and in any popular zoo there is usually a tiger. It would be intriguing to make a public risk assessment of the consequences of the tiger escaping from the zoo. Perhaps we might crudely estimate that there is a probability of one in a thousand per day (or once in 3 years) of the tiger escaping and killing a nearby resident; or if we do a more sophisticated analysis, we may estimate the probability is one in a hundred thousand per day (once in 300 years) of such an occurrence. Regardless of probability, such an event can certainly occur. What are the options to protect the public? There are basically three -- political, technical, and managerial. A political solution is to order the elimination of all tigers from zoos. A technical solution is to declaw and defang the tiger. A managerial solution is to cage the tiger securely and provide alert zookeepers to keep the access gates closed. How has the public chosen among these three?

The public acceptability of zoos is obviously not influenced by risk assessments. Even though we know that wild animals occasionally escape from zoos, the public certainly considers them safe and acceptable enough to visit them with their children frequently. What is obviously acceptable to the public is the assurance by the zoo keeper that the tiger is securely caged. The point I wish to make is that it is not the risk assessment of the hazard from escaping tigers which is key to societal acceptance of zoos, but rather confidence that the management approach to coping with the risk is reasonably reliable.

My point is very simple. Public acceptance of any risk is more dependent on public confidence in risk management than on the quantitative estimates of risk consequences, probabilities and magnitudes. This, of course, shifts the important assessment from a frequency/consequence analysis to a determination of what is meant by "reasonably reliable" risk management. I am suggesting that the practical public answer to the question, "How safe is safe enough?", depends more heavily on the operations established for the management of risk than it does on the quantitative description derived from risk assessments. If this is indeed so, then we should recognize this in our public discussions of the societal acceptability of risks.

Much has been professionally written on the subject of risk assessment and its contribution to the decision process for choosing among alternatives for providing a specific service. For example, the electricity generation options of coal, nuclear power, and hydroelectricity have been compared as to benefits and risks, and have been persuasively defended by their proponents. In retrospect, the past decade has shown that the comparative risk perspective provided by such quantitative analysis has not been an important component of the past decisions to build any of these plants. Historically, initial choices have been made on the basis of performance economics and political feasibility, even in the nuclear power program. Pragmatic issues which permit a plant to be built and operated economically have guided the final choices. In every case, it has always been assumed by the decision makers that operation of any generating plant can be managed to meet reasonable objectives of public safety; i.e., that the "tiger" can be kept in its cage.

For example, let us consider the recent concern with bioengineering and recombinant DNA, and specifically, the proposed field experimentation using bacteria that have been genetically altered. As you may recall, these experiments were recently challenged by a few individuals fearful that bacteria genetically engineered for specific purposes might create unknown and substantial risks on a widespread scale. The court injunction halting these experiments was presumably based on the issue of whether the National Environmental Protection Act requirement for the filing of an Environmental Impact Statement had been properly followed. The opponents of such experimental activity ask for proof that atrocious effects will not occur, and focus on hypothetical worst case consequences. As a decision problem in a new scientific domain, the issue appears unsolvable. The uncertainties about the hypothetical risks can be made to appear very large, and the consequences can be imagined to be global.

However, in my opinion, the ability of the experimenters to manage or control the risk is the key issue. In the particular case of the DNA field experiment, the experiments were successfully carried out in greenhouses; the same experiment had been conducted in the field with bacteria produced by other than gene splicing; and a contingency plan to deal with the spread of the bacteria was part of the original proposal. However, these points were not emphasized in the public description of the issue. I would like to emphasize that those engaged in the experimental work were fully aware of the need to insure that the "tiger was caged." They recognized intuitively the need for risk management. The pragmatic issue here is not whether some unforeseeable and inadvertent manipulation can hypothetically destroy humanity, but rather whether the field experiments can be confined to specified areas. To ask the proposers of such work to file a ritualistic Environmental Impact Statement will miss this point. Such statements have become loaded with the "what if" and "assume that" issues and denigrate the pertinent questions of risk management.

Public fears can always be aroused by the concept of man's tempering with nature, creating global catastrophies. Such fears can easily be used to stop new scientific developments. The issue for new science is whether a national decision is made early, with little information, to deny or severely limit experimentation, or whether an approach is taken to cautiously manage the risk and expand the range of activity as more information is developed. Science cannot prove safety, only the degree of existing harm. Thus, new technologies cannot be proven safe before use. We must recognize that anticipatory arousal of public anxiety can inhibit the creation of new technology.

A comparable situation exists, of course, in the nuclear power field. The emphasis on Probablistic Risk Analysis (PRA), which has become a recent part of the safety review of the Nuclear Regulatory Commission, has focused professional concern on the frequency of core melt -- the most catastrophic failure that can occur with the internals of a nuclear power station. The arguments as to whether such core melt can occur with a projected probability of one in a thousand per year, or in a million per year, represent a misplaced emphasis on these quantitative outcomes. The virtue of these risk assessments is the disclosure of the system's casual relationships and feedback mechanisms, which might lead to technical improvements in the performance and reliability of the nuclear stations. When the probability of extreme events becomes as small as these analyses indicate, the practical operating issue is the ability to manage and stop the long sequence of events which could lead to extreme end results.

The historical nuclear regulatory approach is to assume as inflexible the end impacts of radiation release, and the thrust of regulatory prescriptions has been to reduce their probability rather than to reduce the consequences. This acceptance of end-of-spectrum consequences has perpetuated arguments about evacuation planning, distances, sirens, etc., which I believe are unnecessary and destructive of public confidence. I know this is both an ideological and emotional subject, but let me illustrate my point with a typical example. We know that a rain shower precipitates both dust and soluble gases from the air. In like manner, radioactive particles and gases (except the noble gases) can be precipitated by a man-made rain. The radioactive fallout during rain from the early atmospheric atomic bomb tests illustrated this effect. Thus, if a nuclear power plant was encircled with a man-made spray system, and with a standby pumped water supply, the public could be given reasonable assurance that "the tiger was caged." This has not been seriously considered because the nuclear industry doesn't believe the NRC can be diverted from its obsession with the "what if it didn't work" viewpoint -- no matter how feasible the management of the risk might appear. It depends on human intervention - like a fire department.

To repeat, my point is that public acceptability of nuclear power would be more likely to be achieved if it was assured of "reasonably reliable" risk management. This would mean that both the nuclear utilities and the NRC would need to place primary emphasis on the role of human intervention in the management of potential risks. It would also mean that the present NRC programs to establish safety goals based on quantitative risk assessments would become intellectual exercises useful for guideline engineering design, but not become the essence of regulatory policy.

We have a familiar technical system which clearly illustrates the difference between risk assessment and risk management. That is our prevalent form of transportation - the automobile. Every automobile is potentially a lethal device both to its operator and to the public at large. I leave it to you to imagine what an environmental impact statement of the automobile transportation system would reveal if it included the extreme scenarios of either driving with no hands on the steering wheel or with freedom of choice by each driver to go in any direction, at any time, in any manner he chose fit. Then I suggest you compare those end-of-spectrum consequences with the actual risk situation which, while large in the aggregate (50,000 deaths per year), is individually small enough that almost every one of us drives routinely on every errand. The answer, of course, is that a risk assessment which estimates the spectrum of frequency and consequences of auto accidents does not determine our social behavior in the use of the automobile. Obviously, we have learned how to manage the potential risks of the automobile by using both technical system design (roads and autos) and human intervention (good driving). While there is much that can be done to marginally improve the management of automobile risk, it is interesting that the public enthusiasm for such marginal improvements is not very strong. I am referring here to seat belt use, air bag proposals, and the attempts to reduce drunk driving.

In all these cases, there is a common thread. And that common thread is that real risk management involves capable human intervention, as well as the inherent characteristics of the technical system. Of course, it is easier to write a prescription for a technical design than it is to write a prescription for human behavior. We now use double insulated electrical appliances rather than teaching people to keep them out of the bathtub. It is also easier to regulate technology than to regulate people. We expect machines to be obedient and predictable -- and we realize that people are likely to be neither. For this reason, most regulatory agencies concerned with public health and safety find it most convenient to place the emphasis for acceptable performance on the shoulders of the technical system. There are systems where this is sufficient and preferable (e.g., the control of routine effluents). Unfortunately, we have found by experience that the machines occasionally fail and often in unpredictable ways. Happily, we have also found by experience that people can perform reliably, intelligently, and responsibly in novel situations - if the system is designed with positive interventions in mind.

Most every real operating system eventually involves man-machine collaboration, regardless of the regulatory approach. We attempt to design its technical characteristics to minimize potential risks, but we also depend upon human participation through operation and intervention to manage the situation. For example, our national policy is to develop drugs which are beneficial and yet minimize the risk to potential users. At the same time, because all drugs can be misused, we suggest or require medical supervision in their use. The public apparently finds this very acceptable. In like manner, the public accepts a myriad of daily risk exposures on the basis that these risks are reasonably managed by people.

Consider a few familiar additional examples of the dependence of risk management on human intervention. I call your attention to your home fire insurance. Based presumably on actuarial experience, there are several principal factors which determine your insurance premium. The materials used in the construction of your house, certainly are important -- these are the technical aspects. However, the location of the nearest fire department and hydrant are also very important -- these are the human intervention aspects. In urban areas where these are available, the insurance premiums are about half of those in rural areas where they are absent. The feasibility of human intervention makes the difference.

And another example, consider medical services in a community. The technical aspects encompass the quality of physician training and skills, and the quality of hospital facilities, particularly diagnostic and surgical equipment. Even with good doctors and good facilities, the data on recovery from nature's accidents, such as heart attacks and strokes, indicates that the availability of a quick response by an ambulance service with skilled paramedics can make a very large difference on the probability of recovery. We even try to teach lay persons the rudiments of CPR (Cardiopulmonary Resuscitation) in recognition of the value of a quick human response. Again, the importance of human intervention is recognized. When one compares the social consequences of natural disasters in underdeveloped countries with those in the U.S., enormous differences are apparent due to the rapidity and the scale of the response in the U.S. As we have little control over hurricanes or earthquakes, we have been forced to take a responsive approach based on human intervention to the management of these risks.

When the technical system performance is heavily dependent on human behavior, and so not easily managed by prescription, what does the regulator do? This is the case for autos, airplanes, ships, power stations, farm tractors, drugs, foods, manned military equipment -- you name it. People can cause accidents, and people can intercede to prevent them. So the regulator ordains the discipline of good habits. We license medical doctors, auto drivers, airplane pilots, nuclear plant operators, and hope for the best. Surprisingly, we have not as yet undertaken to license regulators -- but give us time.

Political and government agencies have little faith in human experience, judgment, or skills, and generally assume that these cannot be depended upon to provide effective accident intervention. However, regulators intuitively know that despite its uncertainty such intervention is essential for risk management, so they have undertaken to anticipate and prescribe the forms of human intervention where possible. We thus train operators on flight simulators and nuclear reactor simulators, both of which attempt to reproduce anticipated system failures and accidents. They certainly maintain the alertness of operators to off-design behavior of their systems, and perhaps contribute to their confidence when a real emergency arises. But the history of truly serious accidents indicate that they rarely duplicate each other, and limiting each accident will likely need individually tailored human intervention. One must conclude, therefore, that practical risk management must plan on human initiative to meet the special circumstances at the time of an accident. In fact, regulatory doctrine that discourages such initiative, and tries to prescribe ritualized response for accidents, may increase the danger of serious public consequences.

Returning to the importance of risk management, I should emphasize that it must be credible to lead to public acceptance of a technology. There are clearly conditions when it is not sufficient. You may recall our aborted civil defense program, with underground shelters and emergency supplies. It was unsuccessful in convincing the public that atomic bombs should be considered as merely an extension of our conventional military arsenal. Obviously, it appeared to the public that the civil defense program would only marginally reduce societal damage. While FEMA (the Federal Emergency Management Agency) continues to urge the civil defense concept, it is difficult to find much support for it. It is evident, then, that for risk management programs to create public confidence, they must have a reasonable plausibility that they will "keep the tiger caged."

In the early discussions of public acceptability of risk, I drew the distinction between voluntary and involuntary exposures. This distinction was based upon the degree of individual control on the management of the risk. My most elementary example is the use of the kitchen knife for slicing bread. If you hold both the knife and the loaf of bread, the distance between the knife blade and your hand may be quite small. If someone else holds the knife, you are likely to move your hand further away from the blade. The point of this simple illustration is that under those circumstances where the management of the risk is not under the control of the individual exposed to it, it takes a much higher degree of confidence to make the risk acceptable. It is that confidence which is at the root of public acceptability of a risk not directly under individual control.

You will quickly recognize that most of our major public debates on the acceptability of risk fall into this domain of involuntary exposure. We have involuntary public exposure in the domain of water pollution, air pollution, industrial effluents, food additives, drugs, and the like. In almost every one of these involuntary exposures, society has determined that the benefits of the activity involved are sufficiently large that the activity, whether it be food and water supply, the generation of electricity, or the production of an industrial product, should not be foreclosed. The public perception of benefits always precedes its concern with or the awareness of risks. In almost all existing cases, the perceived benefits are so large that the option of abandoning an activity is not seriously considered, and society's efforts are focused on reducing risks. In those few cases where it is uncertain that the benefits justify the risks, the issues have become idealogically politicized. (Handgun control in the U.S. is an example).

The political issue, when the benefits are clear, is how much of society's resources should be allocated to reduce a specific involuntary exposure to an acceptable level. The complexity of this problem is familiar to many of us. The benefits of most activities are not uniformly distributed, and the disbenefits in the form of involuntary risk exposures may be concentrated on a few. If, indeed, both the benefits and risks are focused uniformly on the same group, the question is usually resolved rapidly. A free choice of the group on the acceptable balance should certainly be feasible and forthcoming. Such simple cases must be very rare indeed, because I do not know of many -- outside of commercial and business decisions.

We are left then with the question of whether our national approach to regulatory methods and policies are generally capable of achieving the dual objectives of public acceptability and the equitable balance of benefit and risk which maximizes the total public welfare. It is generally assumed by the public that regulatory decisions optimize the public welfare. In fact, this is rarely the case, although public justification of a regulatory decision is often presented in this fashion. However, in almost every case, the regulatory decision is based on the pragmatic feasibility of risk management -- evaluated either implicitly or overtly. And this feasibility depends on the effectiveness of the triumvirate - technology, management, and politics.

This leads us to a major component of risk management -- the political aspects. In the broadest sense, this includes our societal values and public perceptions, and our institutions for implementing regulatory policies. In a democracy such as ours, with a very wide variety of special interest all trying to influence politicians concerned with re-election, public perceptions have a major influence on legislative decisions related to public health and safety.

In principle, legislators should be focusing on those major issues whose resolution can have the largest aggregate social benefit. In fact, however, most individuals view all health and safety risks as potentially capable of affecting them, and thus want them all addressed. Thus, little distinction is made politically between health risks that might affect a relative few or many. Special interest groups aggravate this situation by concentrating their political pressure on selected risks, as for example, radiation, or food additives, or waste storage. As might be expected, politicians also hesitate to irritate any large group by seeking reduction in risks that require a change in accepted public habits -- as for example, smoking, gun ownership, or drunk driving. Popular support of political decisions determines the legislative action, but not necessarily a societal optimum.

Such political factors determine our national priorities in risk management, usually without regard to risk magnitude, frequency, or quantitative importance. Political attitudes also determine the structure, policies, and methods of the regulatory agencies created by legislation. Legislative acts often prescribe in detail the administrative mechanisms and criteria to be used by the regulatory agencies, usually in simplistic and idealistic terms. The combination of legislative constraints and mandated idealism has created for our judicial system a major role in interpreting congressional intent, and thus the judiciary also has become an influential determiner of regulatory policy.

An element of our national political attitude on regulation is a basic distrust of the motivations and reliability of the private sector -- both corporate and individual -- to respond to emergencies. This distrust results in a heightened public belief on the necessity for regulation that will prescribe behavior or on technical fixes. It is usually unacceptable politically for a regulatory agency to openly place reliance for risk management on the personnel at an accident site. Such factors discourage regulatory acceptance of human intervention as a positive element in risk management.

For these reasons, we must consider the political aspects as perhaps the most important in determining the feasibility of risk management. One can only hope that the passage of time will permit the results of quantitative risk assessment, analysis of management options, and the comparison of benefits and risks, to eventually improve the perceptions of both the public, the politicians, and the courts.

Let me now repeat my main points that (1) public acceptance of a risk depends on public confidence in its effective management, and (2) that effective risk management requires an optimal combination of risk assessment, technical feasibility, human intervention, and political support. I suggest that our primary analytical emphasis in improving public health and safety should be on our ability to manage risks, rather than on the assessment of end-consequences arising from hypothetical situations. I am suggesting that an Environmental Impact Statement should concentrate on how an anticipated risk will be managed, contained or limited -- rather than on a frequency/consequence presentation of rare hypothetical scenarios. Further, that such a management analysis should include the role of plausible human interventions. I believe that such an approach would create the public confidence needed for public acceptance of new technologies with all their accompanying uncertainties.

I will finish my talk with a theme song tailored to my message (and with apologies to professional song writers):

"Hold that tiger, hold that tiger
We want him around;
Hold that tiger, hold that tiger
Keep him caged and bound."

RISK APPRAISAL - CAN IT BE IMPROVED BY FORMAL DECISION MODELS?

Hans Bohnenblust and Thomas Schneider
Ernst Basler & Partners
Zurich

ABSTRACT

The use of formal decision models can supposedly enhance the quality of safety decisions. It is discussed why there has been an increasing demand for such models in recent years. A short description of a safety model is presented. This model is based on the basic ideas of formal decision analysis and it has been applied in many different fields of technology.

The second part of the paper is concerned with a specific problem in the risk appraisal part of the model. In the real world it can be observed quite frequently that people do not base their safety decisions on expected damages. In this context the notion "risk aversion" is often brought into discussion. In a narrower sense "risk aversion" refers to the fact that a rare but catastrophic event seems to have more impact than many small accidents with the same expected number of fatalities. It is argued that an elaboration of this effect would be worthwhile. Some ideas are given on how the risk aversion effect can be modeled.

KEY WORDS: Decision Analysis, Risk Appraisal, Risk Assessment, Risk Aversion, Risk Judgment, Technological Risk, Safety Criteria

WHY DO WE SEARCH FOR FORMAL SAFETY CRITERIA?

In recent years formal risk analysis has increasingly been used to support safety decisions and so the consciousness grew that safety problems cannot be solved by the sole analysis of risks. Rather, there is a second part of the safety question which is even more difficult than the technical risk analysis. This part concerns the actual decision problem, i.e., the appraisal of risks which is a matter of value judgments (Figure 1).

In the traditional areas of technology these two parts usually have not been kept apart in a systematic manner. As a result, more than ninety out of a hundred safety decisions have been and still are being made by technical people. Yet it is clear that many non-technical aspects influence the question of what is safe.

Basically this is not satisfactory. One of the reasons for the difficulties we face today while dealing with large-scale technologies probably is that society is not used and not organized to make safety decisions in an interdisciplinary frame [Whipple, 1982]. Society left the

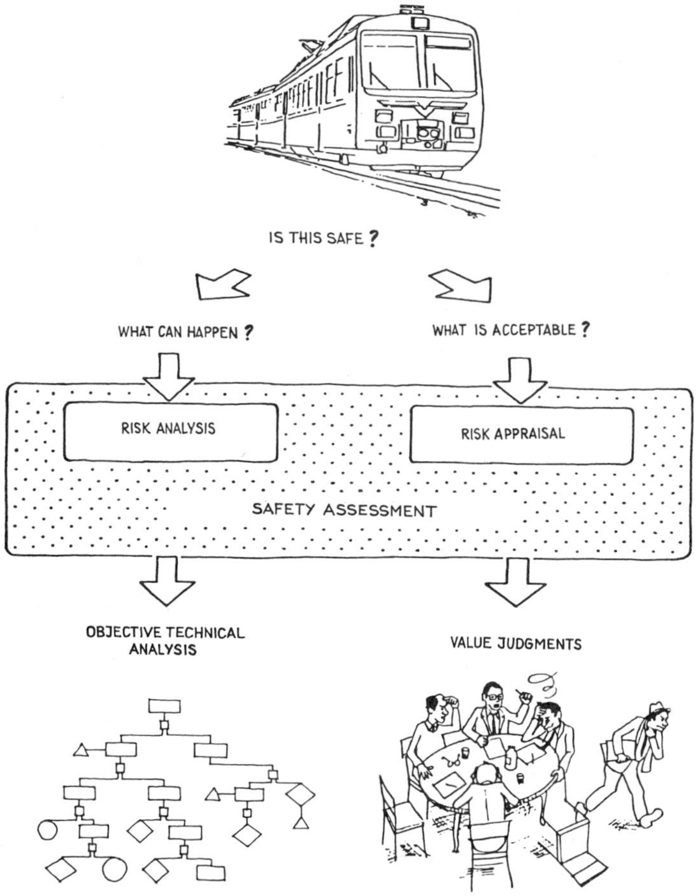

Figure 1

The Division of the Safety Problem into the Technical Risk Analysis and the Risk Appraisal

more simple decisions to engineers, and now it wants to tackle safety problems which are much more complex. But it is doubtful whether society will be able to solve extremely complicated problems before it has a thorough understanding of the simple ones.

Engineers really face a big problem today. If they act as before and make safety decisions on their own, they are called technocrats. If they ask what to do they do not get any practical answer. And still they are expected to do their daily work by designing all the things society needs for a comfortable life. [Bohnenblust and Schneider, 1983].

In their consulting work the authors already faced this problem about twenty years ago and had to look for a solution. Fortunately, we were not confronted with delicate technologies such as the nuclear industry. At the beginning we mainly studied problems in ammunition safety, later also in areas such as traffic safety and others. However, all of these problems more or less dealt with traditional technologies. We realize that many safety problems concerning modern technologies are subject to a much stronger influence of social aspects. Of course, this makes it more difficult to handle safety problems in a systematic way. This problem is illustrated in a graph by Grose, 1972 (Figure 2). It indicates that the applicability of formal models decreases with increasing importance of the social sciences.

Now, what was our issue? We were able to calculate risks. We had statistical data, we could simulate explosions and make full-scale tests and so on. In that sense we had a comparatively easy task. Nobody could seriously doubt our risk analysis, whereas in other fields even the analysis of the risks caused severe controversies [Huber, 1984]. But, what were we to do with our risk values? We faced the famous question: What risks are acceptable? And when we started to assail this question we soon were wondering how safety decisions could ever have been made.

But this is a crucial point: Safety decisions will be made anyway. As engineers we see that thousands of safety decisions are made day by day, but most of them lack in consistency (Figure 3). Moreover, in most cases there is no reasonable feed-back of the public. If an ammunition factory is designed poorly no one will take note. Especially, it will not be noticed if a lot of money has been spent uselessly. After an accident the design criteria may be changed, but this does not guarantee a solution which can be said to be basically better. Usually after an accident all safety efforts will be focussed on this specific accident type, eventually neglecting more important failure modes. However, a real opinion-forming process which is linked to a factual level in a comprehensive way will, as a rule, not take place. The basic problem is that we do not dispose of

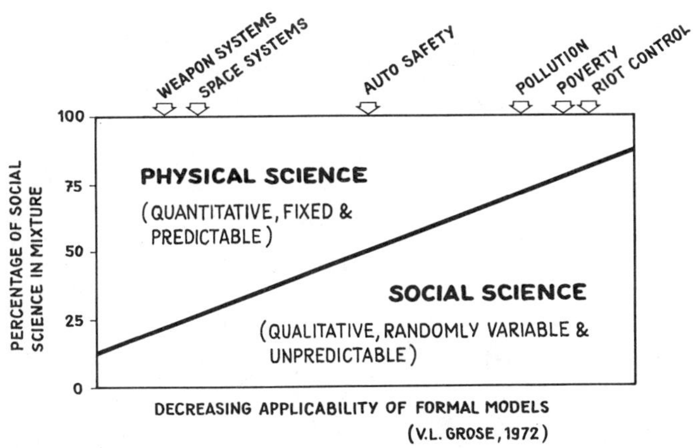

Figure 2

The Applicability of Formal Models Decreases with the Increasing Importance of Social Sciences

systematic approaches for the risk appraisal part as we do for the risk analysis part. It will surely be difficult to get elaborate models, but it is worthwhile to search for them. At the beginning, the models may be elementary though they help to address the relatively simple safety problems as they occur, for instance, in ammunition factories. More importantly, they allow a step forward and a gradual improvement towards sophistication. Also, they serve as reference point in evaluating our experience.

Why is this important? In a sense, engineers act as agents of society. However, society is not able to tell them in any single case what they are supposed to do. Still, there should be rules which show engineers in a general form how society's value system works. It could be said that there are such rules in the law. However, a statement such as "according to the state of the art" includes too many generalities. Society should express its value system in a more precise form, and it should not let this be the job of technical people alone. Else, any meaningful interaction between society and technology gets lost.

Now, the basic ideas of formal decision analysis are most appropriate for bringing some order into this situation. So we have tried to use decision analysis in searching for rules to make safety decisions since we were confronted with the ammunition safety problem. We have developed a safety model which is independent of a specific field of technology, but which includes precise rules. We have applied the model successfully in different though traditional fields, we have gained experience over many years and we have tried to improve the model on the way. However, we

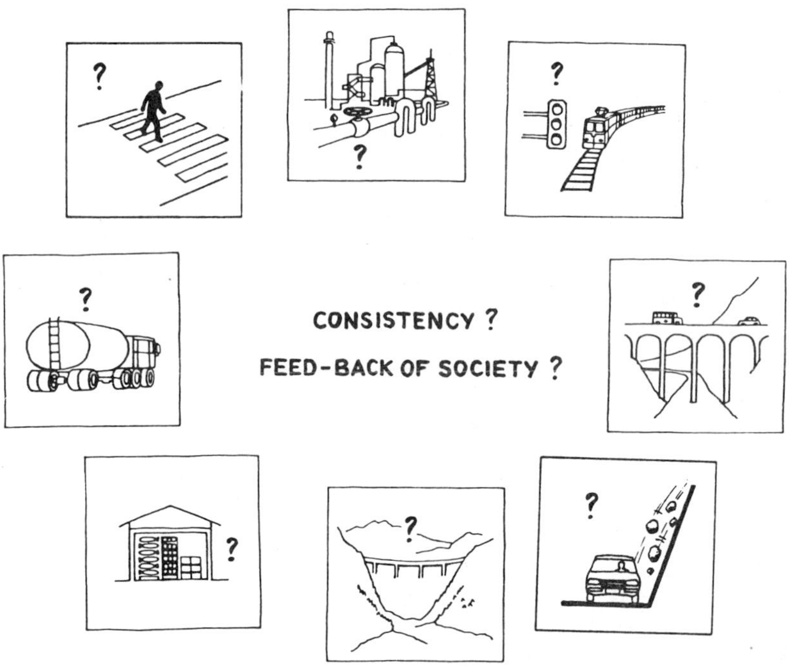

Figure 3

Many of Today's Safety Decisions Lack in Consistency and Get No Reasonable Feed-Back from Society

realize that the model is a simplification of reality and in its present state has many limitations. Still, we believe that it has helped to make better decisions than before. To incite discussions we want to present the conception of a safety model in the following:

SAFETY IS MORE THAN MINIMIZING LOSSES

It is generally recognized today that the risk assessment of a technical system requires that all accidental events are investigated with respect to the probability p of an event and its consequence C (Figure 4). For simplicity reasons let us assume here that we are only interested in fatalities. Now, when we look at a system in the long run and consider all possible events, the sum of the probabilities p_i times the consequences C_i will be the expected number of fatalities. Usually, this is called the collective risk R.

If we assess the value of R for any system it will, for sure, not be equal to zero. But, why don't we reduce it to zero? Of course, everybody dealing with safety knows that technical systems which are absolutely safe are not possible. But mostly it is technically feasible to reduce risks to a negligible value. However, this is rarely done due to the limitation of our resources. We can always reduce the risk of some facility, but it costs more and more money (Figure 5). The awareness of resource limitations is a fundamental aspect of all our technological activities, thus also of our safety efforts It simply means that what we overspend at one place will be lacking at another one.

Having realized this, it sounds rather reasonable that our resources, and herewith all safety measures, should be arranged in such a way that the expected number of fatalities will be minimized. This requirement is independent of how much money is spent. Well, a "naive" economist might expect us to behave in such a way. However, reality quickly teaches us that we are far away from such a simple goal. There clearly are a number of mechanisms which lead to a distortion from loss minimization (Figure 6). Some of these mechanisms are more evident than others. For instance, most people will prefer to reduce involuntary risks before voluntary risks are reduced [Starr, 1969; Slovic, 1980]. Somehow, this fact is appealing. However, it is less easy to understand why

POSSIBLE ACCIDENTS	PROBABILITY	CONSEQUENCES
E_1	p_1	C_1
E_2	p_2	C_2
⋮	⋮	⋮
E_n	p_n	C_n

STATISTICAL EXPECTATION OF LOSSES OR COLLECTIVE RISK

$$R = \sum_{i=1}^{n} p_i \cdot C_i$$

Figure 4

The Probabilities p_i and the Consequences C_i of all Possible Events are the Basis for Calculating the Statistical Damage Expectation which is Called the Collective Risk R

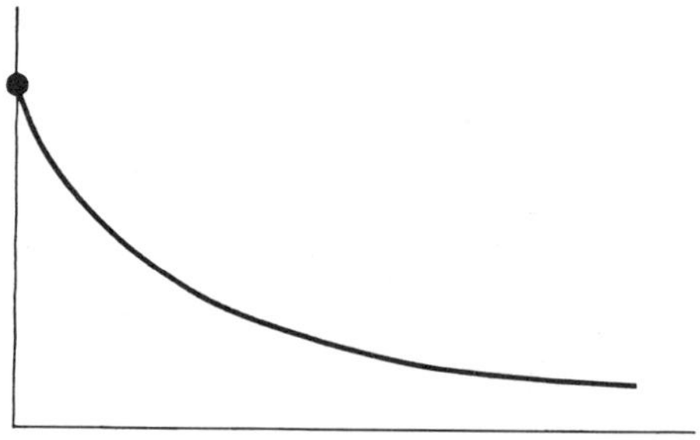

Figure 5

The Typical Shape of the Risk/Cost-Function

elaborate safety measures are taken in Swiss ammunition storages, though in the annual statistics there have not appeared any fatalities due to explosions in such facilities for about 40 years. At the same time, the reduction of speed limits still meets with opposition in Switzerland, though about 200 children die each year in traffic accidents. It is even less understandable if we compare the immense safety requirements at nuclear power plants with the moderate concern for the dying forests in Europe.

Obviously, safety is not just a matter of loss minimization. Rather something else is minimized. We want to discuss shortly how we did tackle this problem when developing a safety model.

SHORT DESCRIPTION OF A SAFETY MODEL

In judging safety it is important to realize that the notion safety covers different meanings. The question "Is this safe?" does not have a single answer. It rather depends on who is asking the question. We can distinguish between three different points of view, namely the individual who is exposed to a risk, the society and the company or agency which is responsible for a hazardous activity (Figure 7).

We can illustrate this with the example of a dam break risk (Figure 8). Each person staying below the dam has a certain annual probability of being killed by a breach in the dam. These probabilities, called the individual risks, are plotted in a decreasing order here.

Besides the individual risk, society is concerned with the expected number of victims among the entire downstream population. This is simply the sum of all individual risks and is represented by the area below the graph on the left of Figure 8. This risk measure is called the collective risk which represents the statistical damage expectation. Note that the shape of the graph and the area below it show two different aspects of a risk situation.

The collective risk can also be illustrated in a different way. In the second well-known graph of Figure 8 the size of the consequences of an

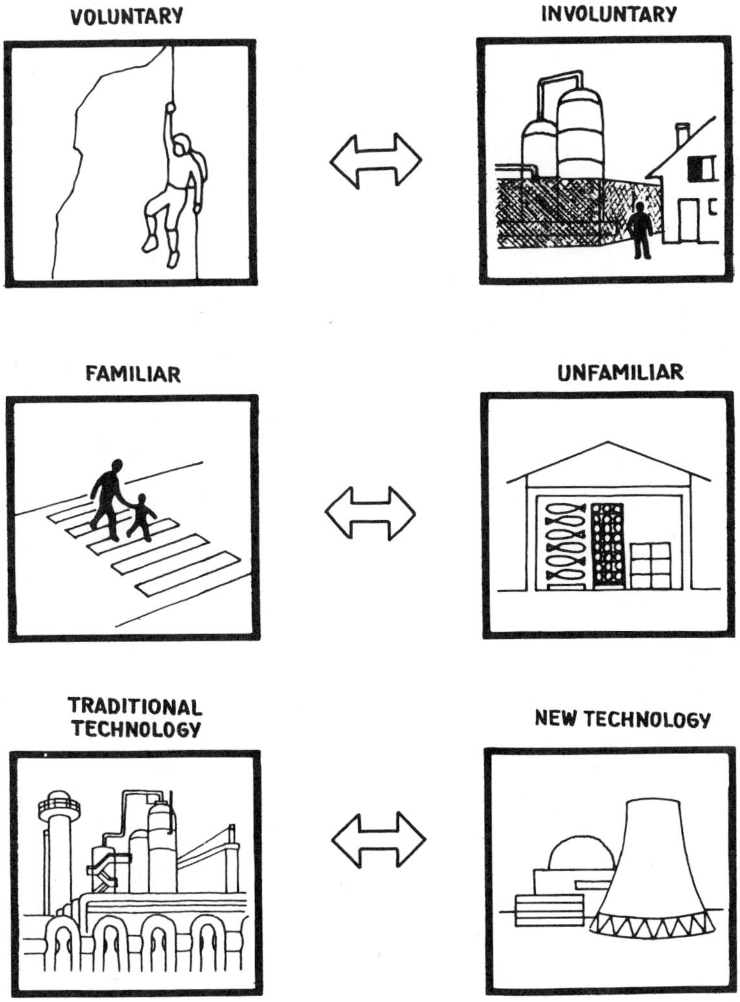

Figure 6

Some Examples of Observed Distortions from Loss Minimization

event are shown versus its exceedance probability. Here, the third point of view can be comprehended. It is represented by the company which is responsible for the dam and has an additional concern in the avoiding of catastrophic events with a large number of fatalities. Due to the enormous economic impact, but also due to the reaction of the public, such events are overproportionally more important to the prosperity of the company than small accidents. This is taken into account by extending the definition of the collective risk. The new risk measure gives a higher weight to high-consequences events. This effect is commonly called risk aversion, and we will shortly come back to it.

Now, what criteria do we have to judge these different risk measures? Let us come back to the question of loss minimization. Surely this leads to a first criterion which takes care of society's interest in

Figure 7

Three Different Points of View Can be Distinguished in Looking at the Safety Question

an efficient allocation of its resources. Well, the basic question in minimizing losses is: What is the appropriate effort we have to make in each specific case, still remembering that our funds are limited? This leads us straight to a classical optimization problem (Figure 9). From basic mathematics we know that the criterion for such an optimization is the equality of marginal costs. Hence, we have to stop our efforts at a certain slope of the risk-cost-diagram for any system we investigate. This slope has quite a real meaning. It actually is the price we pay for a marginal increase in safety and, if our risk unit consists of "fatalities", it is the marginal cost of saving life.

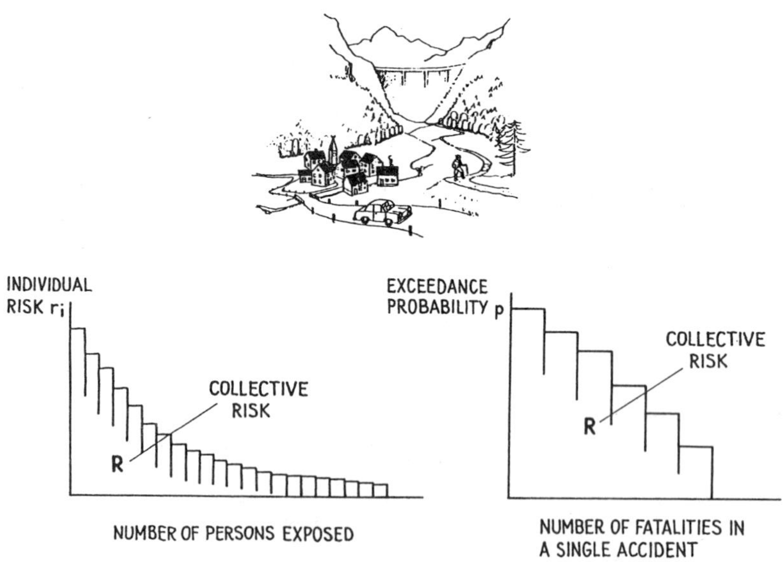

Figure 8

A Risk Situation can be Described by Two Diagrams Which Show the Risks to Individuals and Society and the Probabilities of Different Accident Sizes

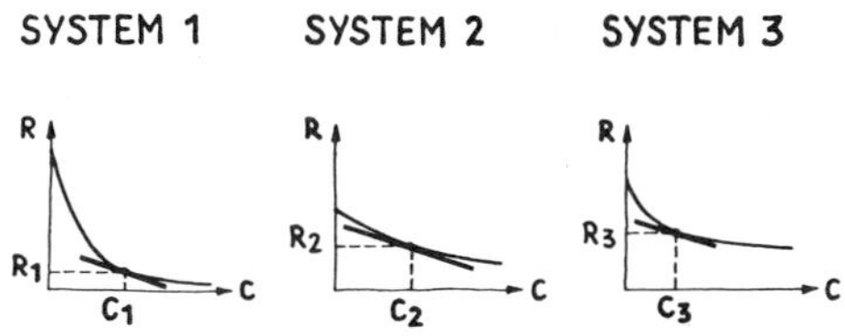

$$R = R_1 + R_2 + R_3 = MIN$$

$$C = C_1 + C_2 + C_3 = C_0$$

Figure 9

The Equality of the Marginal Cost for Saving Life is the Criterion for Loss Minimization. It is Indicated by a Certain Slope of the Curve in the Risk/Cost-Diagram

We already mentioned that loss minimization cannot be the only criterion. Rather there are some mechanisms which lead to distortions, as for instance the distinction between voluntary and involuntary risks. We have tried to combine some of these effects by introducing four different risk categories. To obtain an operational differentiation between the categories we consider the following factors (Figure 10): The ability to know about the risk, the ability to avoid the risk, the ability to influence the risk and the perception of the benefit resulting from the risky activity. With these factors we define four categories with continuous transitions. Depending on these categories, we suggest quantitative values for the marginal cost of saving life. It would need an extensive discussion to explain how the actual values have been chosen. We would like to emphasize that the quantitative values have to

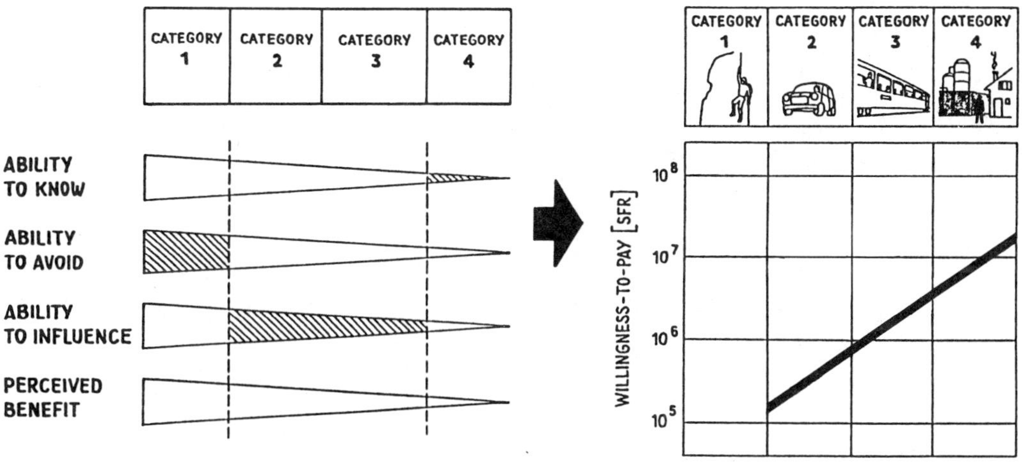

Figure 10

The Introduction of Risk Categories Leads to a First Distortion From Loss Minimization

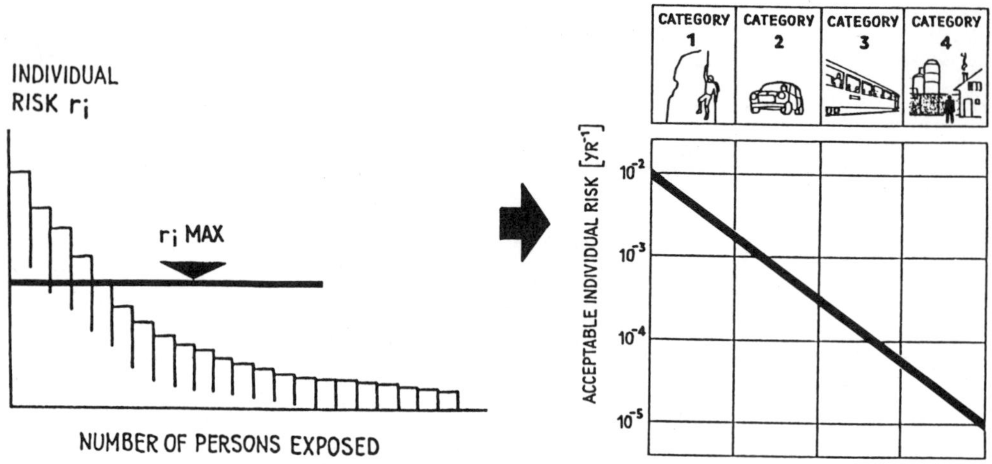

Figure 11

The Limitation of Individual Risks Leads to a
Second Distortion from Loss Minimization

be considered as suggestions which can be adjusted when more evidence is gained.

The second distortion stems from the claims of the individual. Though we did find an optimal solution with respect to the collective risk the individuals exposed still may face very different risks (Figure 11). From the point of view of equity the limitation of the individual risk will become essential if a situation is supposed to be acceptable. Also, such limits seem to be useful to restrict the individual risk due to a specific technical hazard in comparison to the total risk of an individual. Again we suggest a set of quantitative limits which depend on the risk categories defined above.

A last distortion addresses the so-called point of view of the company or responsible agency. Their concern for catastrophic events leads to an overproportional weight of these accidents. As a result we actually pay more to save a life in a catastrophic event than in a single-fatality-accident. For illustration the risk aversion functions of two different activities which we have used in our work are shown in Figure 12.

In the following, we will concentrate on the risk aversion effect. But first, let us give some general comments concerning the safety model. The model has been used for about 15 years now, and we have successfully applied it in various fields. However, we realize that the model is far from being complete. The problem of long-term damages, for instance, has not been included so far. Also, we feel that there is an urgent need for further discussions of any such model. The discussion

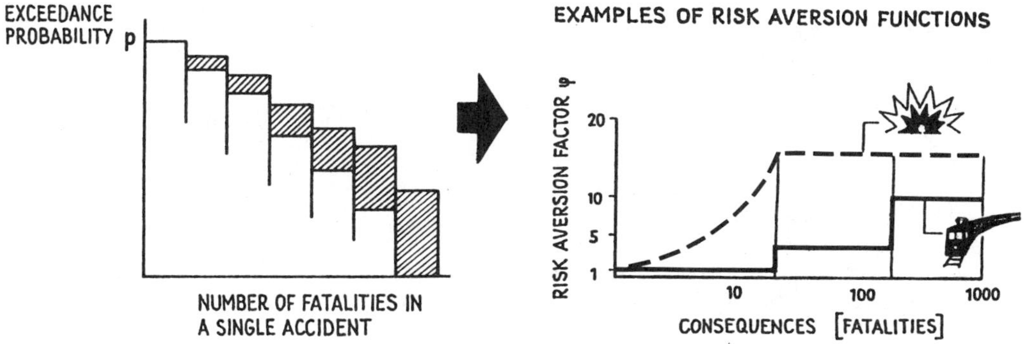

Figure 12

The Introduction of Risk Aversion Functions Leads to a Third Distribution from Loss Minimization

should lead (1) to a general review with emphasis on the limitations of the model, (2) a deeper understanding of the mechanism of risk appraisal and subsequently an improvement of the model and its reasoning, and (3) to an extension of the model. Still, even in its present shape the model is operable and it allows to make transparent and reproducible safety decisions. In that sense, it is an improvement compared to the large number of arbitrary decisions that have been made in the past. Also, it does help in gaining experience and in improving the decision makers' ability to make consistent safety decisions.

THE CONCEPT OF RISK AVERSION

One of the areas where we feel a further elaboration and a wider discussion to be most pertinent is the area of risk aversion. In many high technology fields, the so-called major hazards which can lead to

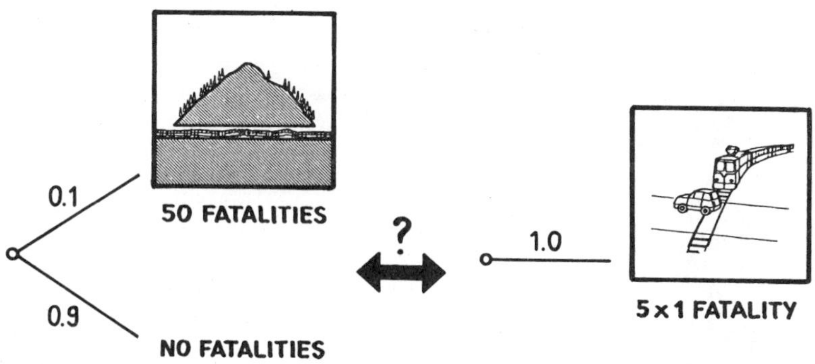

Figure 13

The Question of Certainty Versus Uncertainty and the Shape of the Value Function for Life Saving are Decisive for the Choice Between the Two Railway Systems

catastrophes are of growing importance. It could be observed in recent years that risk aversion is one of the major determinants of safety decisions in these fields.

In formal decision analysis the notion of risk aversion refers to the following fact: Consider, e.g., two hypothetical railway systems (Figure 13). You know that there is a 10 percent chance in a given time span that an accident with 50 fatalities occurs if you build the system on the left, which includes many tunnels. You could avoid the tunnels, but you would have grade crossings instead. If you choose this system, there will be five accidents with 1 fatality for sure. Now, if you prefer the system on the right, you behave risk averse. If you prefer the system on the left, you behave risk prone. Two basic things can be decisive for this choice. The first one is connected with the uncertainty in the system on the left. You may prefer the certain state of a comparatively small loss to the prospect of a major accident. Or if you are risk prone you may take the gamble of not having a loss at all. The second effect is connected with the possibly nonlinear shape of your value function. Here we understand the term 'value function' as the utility function under certainty [Kenney and Raiffa, 1976].

These two effects can be considered separately. In this paper we will concentrate on the second one, namely the nonlinear dependence of the value function on the number of fatalities. In short, the basic question is: Why do people prefer ten single-fatality-accidents to one accident with ten fatalities?

In our consulting practice we first encountered this effect in the field of ammunition storage. The responsible government safety committee realized that absolute safety is an unfeasible goal. So, it did accept the responsibility for accidents with limited consequences. However, it pronounced the strong will that catastrophic events should be avoided in any case. In this context, it was clear that the term "catastrophic" was referred to as an event with a large number of fatalities. The Swiss railway company seems to have very similar feelings. There are numerous grade crossings in Switzerland. Each year about 20 people die in car crashes at these crossings. But no one bothered very much until 39 people were killed in a bus crash in 1982. This accident had an extremely low probability and has no relevant importance in the long run statistics. All the same it made the railway company completely change their attitude towards grade crossings. A third example is the construction of a new high-speed railway net in Western Germany. The main characteristics of this net are the maximum speed limit of 250 km/h and the high percentage of tunnels which is about 30% of the total length. The safety in the tunnels was not a major concern in designing this system until someone sketched the scenario of a fire catastrophe in a tunnel which could result in as many as 500 fatalities. Subsequently, a detailed safety analysis was carried out. It turned out that the individual risks of the passengers were at a negligible level. Also the expected number of fatalities per year was of no concern. However, additional safety measures were planned in the tunnels due to the very small chance of a catastrophe.

In all these examples of catastrophic events the individual risks are negligible due to the low probability of the events. For the same reason these risks are not of great importance for society as a whole. Still, special measures are taken to avoid catastrophes. Well, it seems that the significance of an accident in a statistical sense, i.e. the expected value of the damages is not the decisive factor for a company or agency (Figure 14). For a company the so-called low-probability-high-

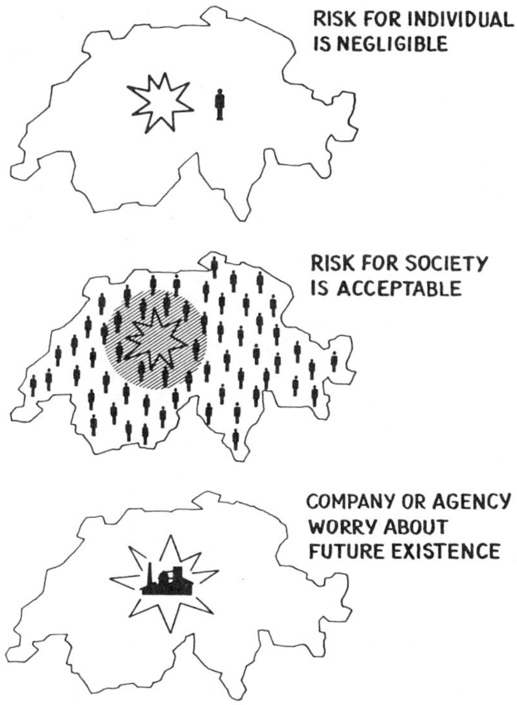

Figure 14

In Many High-Technology Fields Risk Values are Negligible from the Viewpoint of the Individual and Society, but not from the Viewpoint of the Company or Agency Responsible for the Hazardous Activity

consequences events may be much more threatening than for society because the company may not be able to bear these consequences. For example, society hardly takes any notice of the great number of fire blazes which occur each year. But, one out of three companies which went through a blaze has gone bankrupt within four years. There are two problems to this. (1) How can a company meet the direct costs and consequences of a catastrophic event? This is primarily a question of financing, and insurances may be of importance here. (2) There are the indirect consequences such as loss of image or the claim for exaggerated safety measures which can endanger a company's future existence. Though there may be several reasons for such effects, past experience shows that they can be caused by large number of fatalities.

Still, the question follows whether these indirect effects only are hypothetical and whether it would be possible to ignore them in a risk assessment. But let us recall the high-speed railway net. After a disaster some people simply might no longer use the high-speed trains. Even if this is only a small percentage it would have a real impact on the railway company. In the long run, it may look somewhat different. By educating the general public and explaining the underlying mechanism, a gradual reduction of the risk averse behavior eventually might be striven for.

In the short run we need to cope with the effect of risk aversion. The question remains whether it is possible to assess quantitative risk aversion functions and how we can go about it. It is quite obvious that the quantitative value of such functions finally is decisive for safety measures. Realizing this, one has to watch out that the introduction of risk aversion does not lead to a setback into arbitrariness. Rather, risk aversion functions need to be derived in a careful and consistent way.

Basically, there are two approaches to derive a risk aversion function: (1) questioning decision makers and the public about their attitude and (2) analyzing the mechanisms which underlie the risk aversion phenomenon and try to find a sound reasoning. A meaningful function probably has to be based on a combination of both approaches. The whole procedure can be seen as an iterating process. We have tried to make some initial steps towards such a procedure. We questioned the respective decision makers in two areas of our consulting work and tried to fit the answers into an analytical form. The results, e.g., are the two functions which are shown in Figure 12.

Now we feel the need for a further elaboration of the risk aversion phenomenon. In the next section we discuss some attempts at searching for explanations which should finally lead to an improved assessment.

THE UNDERLYING MECHANISMS OF RISK AVERSION

Let us assume that we did manage to assess a social value function which depends on the number of fatalities in a single accident, and that it shows a nonlinear shape. We do not want to exclude that part of the nonlinearity goes back to a real preference for say ten single-fatality-accidents to one accident with ten fatalities. But, we try to show that such an effect could also be explained by other reasons. Among those are certain misestimations of risks after severe accidents, the so-called community effect, the shake-up of the public, and others.

When people are questioned about their preferences they usually will not be able to distinguish between these mechanisms, but lump them together. However, the relative importance of the various mechanisms may differ from activity to activity. To learn more about this we need a thorough understanding of the mechanisms.

As an example let us consider what we mean by misestimation of risks. Risks can be misestimated with respect to the size of the consequences or to the probability of hazardous events. Figure 15 describes a system which has about one accident every one hundred years. Most of these accidents do have relatively small consequences. Once in a while, however, there may be a real catastrophic event. If a company happens to pick out this event, it is first of all bad luck. Moreover, subsequently the public may believe that all accidents occurring to that company do have catastrophic consequences. They will therefore claim for more safety measures than are justified by the actual damage expectation. As a consequence of this effect, the company might show a risk averse behavior. It would like to avoid such effects and is willing to provide some additional safety measures to avoid the catastrophic event. In the example shown the risk will be overestimated by a factor of five after an accident. Hence, the company might be forced to spend much more for safety measures than it spent before.

Figure 16 illustrates the misestimation of the probability of a serious accident. There may be one chance in ten thousand per year for a serious accident. However, after such an accident the public might not be

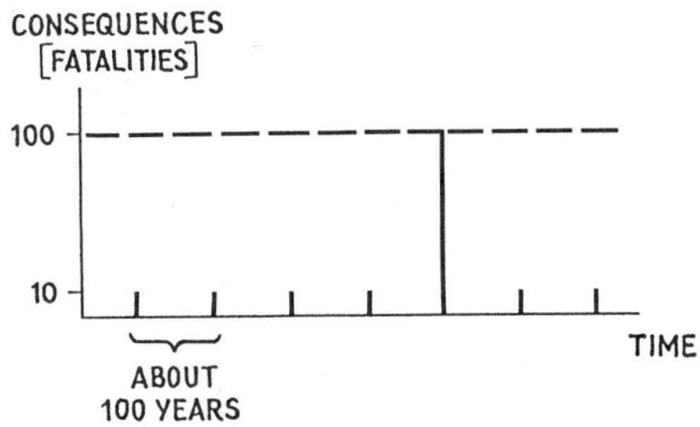

Figure 15

Due to a Misestimation of Consequences the Risk may be Perceived as R' = 1.0 Fatalities Per Year After a Catastrophe Occurred, though in a Statistical Sense the Real Risk is Only R = 0.19 Fatalities Per Year

able to imagine the long interval of ten thousand years. It might force the company to behave as if an accident occurred every thousand years. This means that the risk and therefore the safety costs are overestimated by a factor of ten.

Another example is the so-called community effect. Basically it says that there is a difference between an accident which kills say ten people out of a small community or a family, or ten people who are not related to each other at all. It is reasoned that in the first case the resilience of a community may be endangered. Basically, this has nothing to do with risk aversion. But, still the effect may overproportionally depend on the number of fatalities in a community. In addition, the strong perception of the community effect after being exposed to the aftermath of a catastrophe might influence risk aversion. For example, in a catastrophic railway accident it is highly probable that families or even communities

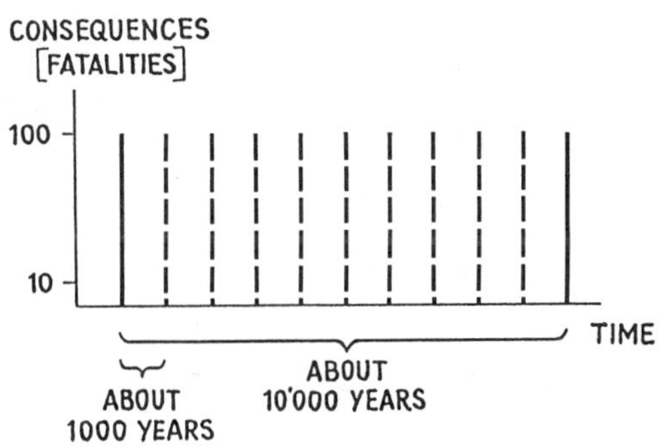

Figure 16

Due to a Misestimation of Probabilities the Risk May be Perceived as R' = 0.1 Fatalities Per Year After a Catastrophe Occurred, Though in a Statistical Sense the Real Risk is only R = 0.01 Fatalities Per Year

Figure 17

The Different Viewpoints of Safety are Interconnected with Each Other

are involved. Hence, part of risk aversion could stem from disapproving that communities as a whole are endangered in catastrophic accidents rather than from the threat of large numbers of fatalities.

A further effect is that the public simply is shaken up by a catastrophic event. People did not even take notice of small accidents that happened before. In that sense, a serious accident can bring a hazard into mind and cause all kinds of reactions which would not be provoked by the same number of victims in single-fatality accidents.

In our future work we will try to describe all the different mechanisms in a systematic and whenever possible quantitative way. A further complication to this stems from the fact that all occurrences in a hazardous activity need to be seen as a closed-loop system. It will not in every respect be possible to isolate a specific mechanism and to investigate it without considering the remaining mechanisms at the same time. Above all, the different viewpoints of the safety question are not independent from each other (Figure 17). Rather there are interconnections between a company or agency and society as well as individuals. Looking at this more closely a clear connection of risk aversion to a single viewpoint does not seem to exist.

Still, we try to strive for a better understanding of the risk aversion phenomenon. This may be the basis for creating questionnaires which can help to assess the quantitative shape of risk aversion functions in different fields. Moreover, they can be used to test the qualitative understanding of the phenomenon. Of course, the questioning will have a feedback on the analytical reasoning. The combination of both approaches should be seen as an iterative process.

We realize that we are far from being able to understand risk aversion. But we believe that it is possible to learn more and thus to improve our understanding by systematically analyzing the respective effects, by testing with questionnaires and by systematic discussions of the phenomenon. In the long run it may even be possible that some of the risk aversion effects could gradually be eliminated if the underlying mechanisms are understood. This could be done by a more adequate behavior of the responsible agencies and companies and a more appropriate strategy to inform the general public about hazardous activities.

CONCLUSIONS

In this paper we have presented a safety model which has been applied in our consulting work for many years. We also talked about the need for a general discussion of the risk appraisal part of such safety models. One area where this need is most urgent is the problem of risk aversion because this phenomenon is decisive for a large number of safety decisions.

We have seen that there is no physical law called "risk aversion". In this sense, it is useless to look for a universal rule by which the risk aversion phenomenon can be modelled. Risk aversion is a phenomenon created and formed by society and its individuals. The search for a risk aversion model can only be a search for a consensus as a basis for more consistent decisions. Similar remarks could be made to the safety problem as a whole.

Still, we are pleading for having the courage to develop models and rules as discussed in this paper for using them in practice. This will be the only way to gain experience and to enter the continuous process which may finally lead to a consensus concerning safety problem. We also need to promote or even provoke the interaction between engineers, politicians, the public, economists, lawyers, etc. As engineers we have to contribute to a public consensus by treating safety problems in more general terms and not just in terms of technical specifications which can only be understood by experts.

Well, we are still at the beginning, and today's safety models may be crude. But we think it is more important to capture the problem as a whole than to look for great accuracy in details. Or, as the saying goes: It is better to be roughly right than to be exactly wrong.

REFERENCES

Bohnenblust, H. and Schneider, Th.; 1983 "From Legal Principles to Engineering Decisions - Where do we Shift to Quantification?", Safety and Law, Proceedings of the V. International Summer Symposium on Safety Sciences.

Bohnenblust, H. and Schneider, Th.; 1984 "A Quantitative Safety Model for the High-Speed Railway Tunnels in Western Germany" (in German), Eisenbahntechnische Rundschau, 3, p. 193-201.

Grose, V.L.; 1972 "Applying System Safety in Non-Aerospace Situations", Space Technology - A Model for Safety Techniques and Accident Prevention? Proceedings of the IFU-Colloquium.

Huber, P.; 1984 "Discarding the Double Standard in Risk Regulation", Technology Review, p. 10-11, March.

Slovic, P., Fischhoff, B. and Lichtenstein, S.; 1980 "Facts and Fears: Understanding Perceived Risk", in Schwing, R. and Albers, W.A. (eds.), Societal Risk Assessment: How Safe is Safe Enough?, Plenum Press, New York.

Starr, C.; 1969 "Social Benefits vs. Technological Risks", Science, 165, p. 1232-1238.

Whipple, Ch.; 1983 "Opportunities for the Social Sciences in Risk Analysis - An Engineer's Viewpoint", presented at the NATO Advanced Study Institute in Les Arcs, France.

INSURANCE MARKET ASSESSMENT OF TECHNOLOGICAL RISKS

William B. Fairley, Michael B. Meyer and
Paul L. Chernick

Analysis & Inference, Inc.
Boston, Massachusetts 02109

ABSTRACT

This paper discusses three major problems with traditional risk assessments: risk assessments produced and used in circumstances that are adversarial in nature and often subject to adversarial bias; risk assessments tending to implicitly include cost-of-error and value-related biases; and risk assessments that cannot be performed, or, if performed, are in error when applied to hard-to-quantify risks.

This paper then discusses advantages of market risk assessments (MRAs, defined as probability and consequence products estimated by parties with substantial interests in the accuracy of the estimates, such as insurance companies) over traditional risk assessments: their potential for being unbiased; their combination of quantitative and qualitative information and assessments, which have tended to be difficult for traditional risk assessments; and their relatively wide availability and low cost. MRAs thus may be thought of as an untapped resource which might be used to supplement traditional risk assessments.

Four areas of future research are suggested: the characteristics and conditions required for the existence of MRAs; the methodologies necessary for extracting MRAs from insurance premiums; the structure of insurance programs, coverage designs, and insurable event definitions in order to maximize the value of MRAs; and how, if at all, MRAs can be appropriately used in public policy and regulatory dispute resolution.

KEY WORDS: Market Risk Assessment; Insurance; Technological Risks

1. INTRODUCTION

Public concern about scientific and technological risks is great and probably increasing. As a result, there is often a demand that risks be assessed or managed by public agencies. This demand has often been met in the past by regulatory agencies either producing or requiring risk assessments by traditional analytical methods which have focused on probability and consequence estimation. Insurance markets may provide an independent source of risk assessments, which may be extracted from coverage and premium information, and which may provide useful frequency (probability) and severity (consequence) estimates. This paper will discuss several aspects of market risk assessments (MRAs) stemming from insurance arrangements. This paper will focus on risks which might be termed "non-actuarial" in that actual accident experience data for the risk is sparse or non-existent. It is clear that MRAs work satisfactorily where data is plentiful and of high quality; life, health, auto and homeowner's insurance provide four example. Similarly, the state of the art is probably adequate for natural disasters of large consequence for which much data on events exists (such as hurricanes) or for any risks, artificial or natural in origin, which have been dealt with in a standard way by the traditional actuarial methods of life or causalty insurance ratemaking. It is by no means obvious, however, that MRAs would be useful or helpful in areas where data is sparse, non-existent, or of poor quality, such as technological risks which cause rare events with serious consequences, which are a subset of events often loosely referred to as "catastrophes" or "low-probability/high-consequence" (LP/HC) events. The remainder of this paper will therefore refer to "catastrophic" or "LP/HC" risks for which experience data is sparse or non-existent.

Three basic types of work remain to be done to validate the use of MRAs. First, some further analytical work is required to define the conditions required for the existence of MRAs, and to explore the expected characteristics of MRAs. Second, a model for insurance premium prices must be developed to predict premiums as a function of expected losses, expected expenses, required return on capital, and any compensation for risk aversion. Third, case studies must be conducted in each relevant application area to explore existing insurance arrangements (in terms of coverage design and extent, insurable event definition, premium design and amount, etc.) in order to discover what information is available for the extraction of risk assessments from existing arrangements. These last two tasks (constructing the premium model and conducting the case studies) will together form the basis for exploring three issues: (1) which methodologies are appropriate for extracting MRAs from existing insurance programs; (2) how future insurance programs could be designed to enhance their use as sources of MRAs; and (3) exactly how, if at all, MRAs can be appropriately used for regulatory and public policy purposes.

II. RISK RESOLUTION IN COURTS AND REGULATORY AGENCIES

A. Introduction

Courts and regulatory agencies face an increasingly heavy burden of resolving factual uncertainties and legal issues in the determination and evaluation of risks to health and safety from a variety of environmental, technological, or commercial sources. Green (1981) and Meyer (1982, 1983) cite several examples. Much of this burden constitutes the ordinary business of factfinding in legal and regulatory proceedings. For example, the treatment of uncertainties in the determination of events leading to an automobile injury or product liability dispute is the main concern of these legal proceedings.

Areas of risk determination in legal and regulatory proceedings that have met with less satisfactory treatment are risks in which large numbers of persons are put at risk but either (1) the risk of some future possible catastrophic event is in doubt, as for commercial nuclear power plants, or (2) where the identity of persons harmed by the specific cause is difficult to determine, but some harm is generally conceded, as in the health effects of air pollution, and in either case (3) where experience with the specific event or with the causation of harm is sparse or non-existent. See, for example, Fairley and Mosteller (1977) and Carndall and Lave (1981).

B. Major Problems of Factual Uncertainties

Three major problems confront courts and regulatory agencies which attempt to resolve factual uncertainties about risks of low-probability/high-consequence events with sparse or non-existent data. These problems are: (1) adversary bias; (2) cost-of-error or other value-related bias; and (3) difficulties surrounding the quantification of inherently hard-to-quantify risks. These three problems will be discussed in turn.

1. Adversary Bias. The parties to adversary proceedings put forward widely divergent risk estimates. The uncertainties in these estimates are rarely frankly described, because the parties advocating the risk estimates fear attacks on the soundness of their estimates and wish to convince the decision maker of the correctness of facts and analyses favorable to their side. Still worse, biased risk estimates are often explicitly produced solely for use in a particular regulatory proceeding. This problem is not often discussed openly in the professional literature, but it occasionally surfaces in print (U.S. Nuclear Regulatory Commission, 1980, p. 44) and is often discussed orally within the risk analysis profession (for example, there were numerous discussions on the problems of "analyst dependent" methodologies by conference participants at the 1982 Annual Meeting of the Society for Risk Analysis on Low-Probability/High-Consequence Risk Analysis). As a result, decision makers often have a difficult time in reconciling widely divergent risk estimates, and in discovering (even in a range) where the most reasonable value of various risk measures lie.

2. Cost-of-Error and Other Value-Related Biases. It is common for parties to cite "conservative" or "prudent" estimates of risk on the grounds that the cost of the error of under-estimating (or, alternatively, or over-estimating) the risk requires an estimate deliberately biased in a known direction. These deliberate biases are often not quantified, as in the possible bias resulting from linear extrapolations of cancer risk from high dose results in animals to low does results in humans, and these biases may cascade and compound if they appear at several steps of a multi-step process. The decision maker is left to imagine how much the estimated risk is over-stated and how to estimate the "true" risk. Furthermore, the decision maker's role may be usurped by the expert analyst, in the sense that the decision maker should be the source of society's cost-of-error or other value-related inputs, instead of the expert analyst providing these inputs sub silentio. The grounds on which expert analysts introduce these biases into risk estimates are admittedly value judgments; for example, where human life is at stake, it is often stated that an estimate of a risk should be higher than what one might otherwise believe was the expected value. Raiffa (1982) and National Research Council 91980) discuss "conservative" estimates of carcinogenesis.

3. Hard-to-Quantify Risks. Catastrophic risks of complex new technologies are inherently difficult to estimate. Historical data on the frequency of interest is generally quite limited, and may be non-existent. For example, Fairley (1982) shows how risk estimates from nuclear accident records yield only very imprecise upper bounds on risk that are far from the safety levels required by accepted standards. As an alternative to such direct data, risks can be estimated from strong detailed theories of event causations and the frequencies of various subsidiary or precursor events. Unfortunately, however, various limitations often preclude such strong models, either because the basic causation is not well understood (as in many biomedical issues), or

because it can not be determined that all event paths have been traced (as in commercial nuclear safety issues), or because the input traced (as in commercial nuclear safety issues), or because the input data necessary to drive the model does not exist (as in failure rates for new system components). In the absence of either direct data or strong models, a large measure of expert judgment must be used to determine and estimate risks, building upon whatever indirect data and weal models are available or on intuition and experience. Unfortunately, experts sometimes treat the best available data and models as if they provided precise risk estimates. For example, the results f complex quantitative models, such as Probabilistic Risk Assessments (PRAs) for commercial nuclear power plants, are often reported without any reference to such uncertainties as the quality of the data, the analyst's degree of confidence in the model, or the analyst's prior beliefs about the relevant probabilities.

III. INSURANCE MARKET RISK ASSESSMENTS

A. Definition

As part of their ordinary course of business, insurance companies make countless risk estimates every day. These estimates have not figured prominently in legal and regulatory risk determinations to date even though they appear to have several attractive features. For example, these MRAs are made independently of considerations of use in adversary proceedings, independently of cost-of-error or other value considerations (other than considerations concerning compensation necessary to induce insurance companies to sell the coverage and assume the risk), and they quantify risk in a particular way by putting a price on the assumption of risk. Fairley (1982) defines and discusses MRAs.

Insurance prices (premiums) must include as components estimates for several distinct types of costs and profits. Specifically, estimates or assumptions must be made about the frequency of accident events, about the severity (in dollar damages) of the accident events that do occur, and about the size of necessary loss adjustment expenses, general overhead expenses, and compensation for profit and risk assumption. Information on risk contained in insurance prices has probably not been prominently used to date because such information is contained in coverage prices that mix together different elements along at least three dimensions. First, insurance coverage prices mist together information on estimated frequency and severity, two quantities that are usually considered and estimated separately in more traditional risk analyses. Second, insurance coverage prices mix together loss estimates (from combined frequency and severity estimates) with expense, profit, and risk aversion estimates or loadings. Third, insurance coverage prices often mix together estimates for a wide variety and number of distinct types of accident events (in terms of their initiating chain of events) into single estimates for more aggregated types of outcome events useful for particular types of accident (insurable event) definitions. For example, insurance coverages tend to be for outcome events as generalized as fire, theft, collision, or property damage events, while more traditional risk analyses tend to focus more narrowly on particular types of events, usually defined as causation-outcome pairs. Finally, insurance coverage prices are the result of rather complicated actuarial methodologies, which may not be particularly accessible to non-actuaries, and are also at least partially the result of non-quantified business judgment.

The implicit frequency and severity estimates represent insurers' best available outcome estimates. Because insurers stand to lose money if their frequency and severity estimates turn out (after the fact) to be too low, these estimates might, if anything, be expected to be biased on the high side. On the other hand, to the extent that workable competition exists within the insurance industry and between the insurance industry and other providers of financial assurance (self-insurance pools, compensation pools, reinsurers, true self-insurers, etc.), these estimates could not become too biased on the high side as competing insurers should assume the risks at lower prices. Estimates of commercial nuclear power plant accident risks provide perhaps the best example of MRAs or "insurance market risk assessments," which merely means that insurance premiums may be interpreted (via a model and necessary collateral information) as risk estimates.

Although using insurance coverage prices as the sources of risk estimates involves some difficulties, discussed above, these difficulties can perhaps be overcome. To estimate certain types of nuclear power plant accident frequencies (probabilities), Denenberg (1973), Wood (1981), and Chernick et al. (1981) used collateral information on the estimated costs of payouts in the event of accidents, on estimated expenses, on return on equity, and on risk aversion, in conjunction with a standard analytical model for insurance pricing to estimate the expected frequency of accidents that underlies the premiums charged tot the plants' owner/operators.

B. Potential Advantages

Given that market risk assessments are made routinely by insurance companies, and that at least on example exists (commercial nuclear power plant accident risks) in which these market risk assessments provide quite different estimates from PRAs, see Denenberg (1973) and Wood (1981), are these compelling reasons for further study of their feasibility, the methodology for extracting them from premiums and coverage information, and their use in regulatory and policy decision making? We believe, in brief, that the potential advantages of MRAs, when used in conjunction with other types of risk assessments, are sufficiently attractive to warrant a closer look. We briefly discuss next four apparent or potential advantages of MRAs.

1. Potential Lack of Bias. Estimates of risk measures (probabilities and consequences) produced by special purpose studies to be used in agency decision making, are highly vulnerable to partisan influence. This partisan influence may come into play because of the high stakes involved in the outcome of the decision r because of the values which the various participants in the decision making process are attempting to pursue. For these reasons, risk estimates produced for or by regulatory proceedings are frequently either admittedly biased or subject to a strong suspicion of bias in known directions. MRAs, by contrast, should tend to offer independent, arms-length estimates that should not be subject to these kinds of influences.

2. Combination of Quantitative and Judgmental Estimates. MRAs combine both statistical or actuarial and other kinds of information on risk that are inputs to coverage and pricing decisions. This may be a particularly valuable quality for LP/HC risks for which inadequate data preclude complete reliance on actuarial methods. This problem (of actuarial methods being inadequate due to lack of relevant data) does not exist for some traditional life and casualty insurance lines, but appears in such casualty insurance coverages as hazardous waste disposal site insurance and commercial nuclear power plant insurance. MRAs thus provide a solution, albeit pragmatic, to the difficult problems of combining

information from several diverse sources and diverse methodologies, including purely intuitive judgments. By contrast, more traditional analytical disciplines have recognized the formidable obstacles in the way of formal combinations of such diverse evidence, including pure intuition or judgment, to reach conclusions about final risk estimates. Traditional analytical methodologies have, as a result, usually simply produced a model and then reported the model's results, but have not modified the model's results to take into account other data or opinions on the validity of the model. Further, traditional analytical methodologies are difficult to review and evaluate as to whether they form an adequate basis for decision making.

3. Interest Represented by MRAs. In one dimension, MRAs represent a polar case: parties who make MRAs are making serious bets, in that large financial gains or losses ride on the accuracy of the MRAs. Financial incentives thus appear to operate to favor the more unbiased and more precise estimates. Thus, there is an incentive mechanism, strongly rooted in general experience, which should operate to promote the accuracy of MRAs. Where the events insured are expected to occur with sufficient frequency, there will be actual periodic feedback that should operate to correct MRAs by experience.

By contrast, traditional quantitative risk assessments, such as PRAs, ostensibly represent the other polar case (along the interest dimension) in that they are produced by expert researchers who are assumed to be disinterested in the outcome of the regulatory decisions which will be made in partial reliance upon the PRA. In practice, of course, it may be difficult to separate advocates from truly disinterested researchers; in any case, neither truly disinterested experts nor experts who are acting as advocates have the strong financial incentives provided by the risk market.

4. MRAs as an Untapped Resource. Various regulatory agencies, and regulated industries or parties, have committed very substantial resources towards producing and improving quantitative, analytic risk assessments. This commitment of resources is laudable, and almost certainly represents a wise investment of research resources. The source of this activity is perhaps well represented by the Nuclear Regulatory Commission, which has recently sponsored or supervised major research efforts n quantitative safety goals (U.S. Nuclear Regulatory Commission (1980, 1982b) and Strip (1982b)), on PRAs (U.S. Nuclear Regulatory Commission (1982a)), and on quantitative consequence estimates (Strip (1982a)), all relating to commercial nuclear power risks. Market risk assessments, by comparison, are an untapped resource based upon information already collected and analyzed, which should be subject to extraction from existing premium and coverage data with relatively modest effort and expense, compared to PRAs and quantitative consequence estimates. As is true of any source of information about risk, if risk assessments derived from other methodologies differed significantly from MRAs, decision makers should want to note these differences, understand the reasons for these differences, and weigh the reliability of the alternate sources before forming opinions about risks. In short, MRAs may provide a relatively inexpensive and relatively available alternative to traditional, analytic risk assessments, which can be used to supplement, but not necessarily to replace, the analytic risk assessments.

IV. REQUIREMENTS FOR FUTURE RESEARCH

Four areas of interest concerning MRAs need to be explored. These four areas of interest are: (1) what are the characteristics of MRAs, and what conditions will allow the determination of useful MRAs; (2) what methodologies are appropriate for extracting MRAs from existing insurance programs and premiums; (3) how can insurance programs, coverage designs, and insurable event definitions be best structured so as to make the resulting insurance premiums interpretable as MRAs; and (4) how, if at all, are MRAs appropriately used for public policy purposes. These four research areas will be discussed in turn in this section.

A. What are the Characteristics of MRAs, and What Conditions are Required for the Existence of MRAs?

Five characteristics of MRAs need to be investigated: (1) whether, and under what conditions, MRAs are unbiased; (2) what precision MRAs embody; (3) whether MRAs can be used for risk classification (as well as total risk assessment) purposes; (4) how, if at all, MRAs combine quantitative and qualitative risk assessments; and (5) whether, and to what degree, MRAs separate factual judgments from value judgments.

Each of these five characteristics might be investigated in light of formal models for insurance premiums as these are developed; of the theory of risk and insurance developed in the literature; of portions of modern financial theory, such as portfolio theory and the capital asset pricing model; and of insurance industry practice.

With respect to bias, an obvious attraction of MRAs is their potential for supplying statistically unbiased estimators of risk parameters. A stochastic specification for the insurance-premium model appears appropriate for determining when and under what conditions data on premiums and collateral information can be used to define unbiased, or nearly unbiased, estimates of risk parameters. For example, the simplest theoretical condition for obtaining unbiased estimators for risk parameters will probably result from conditions of pure competition in the insurance industry. Under pure competition conditions, premiums should be bid to a level that has a simple structure. In the real world, of course, things may not be so simple. More rough-hewn operational approaches may be necessary for studying whether conditions in existing insurance markets provide a basis for some notion of workable competition and for reasonably unbiased estimators. For example, there may be attributes of the markets in which arms-length, bargained-for transactions for premiums are believed to occur; under some conditions such transactions may define unbiased or nearly unbiased estimators. Alternatively, examination of the attributes of insurance markets may allow analysts to adjust for or bound the size of the bias which is present in the risk parameter estimates in such premiums. The theory of contestable markets may provide a helpful theoretical framework for considering these questions (see Bailey and Friedlaender (1982)). The most extensive recent treatments of the economic organization of the insurance industry are Hill (1978), Joskow (1973), and National Association of Insurance Commissioners (1974).

With respect to precision, the quality, character, and the size of the data base upon which insurance premiums are based will obviously affect the precision of the premium's implied risk parameter estimates. The insurance premium model(s), as they are developed, will form the basis for exploring ways of measuring the precision of risk parameter estimates in MRAs primarily as a function of the size of the data base and of other features of the data. Precision is, of course, a central concern when making risk estimates where actual accident experience is sparse or non-existent.

With respect to risk classification purposes, there are several relevant questions. Existing insurance markets supply risk estimates only for those classifications of risks that are separately priced. That is, insurance premiums reflect the structure of insurance companies' adopted risk classification schemes. The insurance premium model(s) should indicate how the cells of the existing risk classification schemes are priced currently. In addition, it is necessary to determine how fine the risk classification process should be, how the fineness relates to other qualities, such as bias and precision, and how the fineness of the classification scheme in terms of number of rating variables employed or number of classification cells may be balanced off against rating methods which combine information from other cells of the classification scheme, such as empirical Bayes techniques for rating; see Morris and Van Slyke (1978) and Tomberlin (1982).

With respect to combining statistical with other information, there is a major unexamined question of how insurance premiums combine statistical (or other narrowly actuarial) data with "pure intuition" or other types of qualitative information. It appears that insurance ratemakers in fact use heuristic analogs to formal Bayesian techniques to perform this combining function. Credibility theory is explicitly used by casualty actuaries now for estimating individual risk classification cell parameters where adequate data exists for the population as a whole but not for the risk classification cell alone. Whether some analog of credibility theory is used to mix quantitative with qualitative risk assessments remains an open question. This is largely an empirical issue, to be answered by determining what major types of information are used in insurance price setting, and how these disparate types of information are combined. This question should probably be addressed in connection with case studies on MRA applicability.

With respect to separating factual judgments from value judgments, it must be determined whether risk parameter estimates used in MRAs are or can be constructed to be independent of value considerations. Both insurance premium models and case studies will be helpful in examining whether elements of value such as the costs of errors of estimation or the degree of risk aversion of the insurer have been separated from other risk parameter estimates (empirically, in the real world) and whether one should expect them to be so separated under our model (as a matter of theory).

Beyond these five characteristics of MRAs which need to be investigated at some level, five possible pre-conditions, or requirements for practically feasible determination of MRAs need to be examined: (1) the existence of some notion of workable competition, contestable markets, etc. among providers of financial assurance (insurance companies, compensation pools, reinsurers, self-insurance pools, true self-insurers, etc.); (2) the discount rate and the degree of risk aversion for providers of financial assurance should be roughly equivalent to, or should at least have some known and acceptable relation to, society's discount rate and degree of risk aversion; (3) the party providing the financial assurance should have a true financial interest in the vent's outcome; (4) information on the probabilities and the consequences of the event should exceed some minimal threshold level to permit intelligent risk assessment by any party, whether in an MRA or in a PRA; and (5) the existence of sufficient total insurance capacity (including reinsurance and self-insurance) to permit sufficient "laying off" of the assumed risk to keep any risk aversion term in the insurance premium model tolerably small. (Two pertinent discussions of insurance capacity in relation to large risks are contained in Solomon, Whipple, and Okrent (1978) and Stone

(1973).) For each of these five possible conditions, two questions must be answered: (1) to what extent, if any, are they necessary prerequisites for MRAs; and (2) are they largely met in the real world.

B. What Methodologies are Appropriate for Extracting MRAs from Existing Insurance Premiums and Programs?

Insurance premiums clearly depend upon estimates of the expected frequency of insured-against events, the estimated cost of these events, the expenses of writing insurance and adjusting losses, and the profit required on the insurance business for the particular line of insurance, including compensation (if any) for risk aversion resulting from risks of losses over and above expected losses. A number of models of insurance premiums have been proposed, in insurance ratemaking regulatory hearings, in insurance ratemaking calculations performed within insurance companies, or in academic journal literature. See, for example, Chang and Fairley (1979), Fairley (1979), Hill (1979), Joskow (1973), Seal (1969), Tomberlin (1982), and Weisberg and Tomberlin (1982). These listed models have been applied to date, however, only in "actuarial" situations where adequate data existed. They differ in the precise theories by which the elements are determined and in how they combine various elements to reach an overall premium.

The various models differ in whether or not, and how well, they specify parameters for the measures of risk that were previously selected as being of interest for the purposes of decision making, that is, probability and consequence estimates. Models for premium estimation may or may not correlate well with implicit models used for underwriting decisions made by insurers on how risks should be classified. In short, insurers may not act in the real world exactly in the way the model might show that they "should" act.

We believe that models for insurance premiums can be specified which meet two criteria simultaneously. The preferred model(s) must: (1) be consistent with currently accepted economic, financial, and risk theory; and (2) be capable of specifying estimable parameters for the chosen measures of risk. This includes, for example, components representing returns on equity and risk aversion compensation amounts, if any. The specification of this general model for insurance premiums (which will then be used to extract probability/consequence estimates from premiums) has been stated in our previous work for the U.S. Nuclear Regulatory Commission; see Chernick et al. (1981, pp. 63-84), but much work remains to be done.

C. How Can Insurance Programs, Coverage Designs, and Insurable Event Definitions Be Structured so MRAs Can Be Best Extracted From Premiums?

As previously discussed, one of the problems that makes it difficult to extract consequence and probability estimates from insurance premiums stems from the fact that coverages often include relatively broad groups of particular accident sequences within the defined insurable event. One does not wish to design insurance coverages solely for the ease of the analyst who wishes to use the resulting premiums as sources of MRAs, to the detriment of the appropriateness of the coverages and the coverage prices. This would truly be an example of the tail wagging the dog. However, it is possible that insurance coverage definitions (which include dollar amounts of coverage, and as applicable, deductible amounts, co-insurance percentages, coordination of benefits provisions, and the like) and the insurable event definitions can be modified to improve the accuracy with which the resulting premiums can be interpreted as MRAs, without causing any corresponding detriments.

We expect that this issue can be studied most effectively through two or three case studies of potential applications of MRAs, such as commercial nuclear power, hazardous waste storage, liquefied energy gas transport, and biomedical engineering. Especially as the issues are clarified by the work described above of defining the characteristics of and prerequisites for MRAs, and of modelling insurance premiums, we believe that it will be fruitful to examine the existing insurable event definitions, coverage definitions, and premium structures available in each application area to determine whether or not improvements are possible to permit more accurate extractions of MRAs from premiums. The insurance premium models may indicate that certain modifications to the coverage definitions and the insurable event definitions would result in more accurate MRAs without otherwise degrading the efficiency of, or the financial assurance provided by, the insurance program.

D. How, If At All, Are MRAs Appropriately Used in Public Policy and Regulatory Dispute Resolution?

The ultimate purpose of studying MRAs is to draw appropriate public policy conclusions which may improve the public risk assessment and risk management processes. This section will discuss a few areas in which public policy conclusions may be appropriate.

First, decision makers will need an outline of potential application areas, either in terms of types of activities or types of existing economic arrangements, in which MRAs can be used to estimate the sizes of loss risk externalities (which can be internalized if an insurance program is in fact put into operation, and charged to the activity generating the loss risk) and the other areas in which MRAs can be used as sources of probability/conseqeunce estimates of loss risks (whether or not the coverage ends up being put into place and the loss risks internalized). On the other hand, there may be very substantial areas in which MRAs can not be put to either one of these two possible uses.

Second, the accuracy of MRAs must be compared to that of traditional analytic probability and consequence estimation methods, to determine whether MRAs are credible sources of information in various application areas. It must be remembered, of course, that experience data in these areas may be quite sparse, and thus may preclude all but the weakest inferences from the data. Despite this serious handicap, this analysis can determine whether MRAs and analytic risk estimates in the various study areas are "within each others' confidence intervals" or are more discrepant, and similarly, whether either the MRAs or the analytic risk estimates are inconsistent with experience data. Some very general conclusions about the credibility or the usefulness of MRAs in risk assessment and risk management may be possible.

Third, some highly tentative conclusions may be possible concerning the desirability or possibility of combining MRA results with other data, including analytic risk assessment results, in order to capture the information contained in both methods. One of the subsidiary questions to be resolved in this regard is whether or not MRAs somehow already contain within them the results of prior, analytic risk assessments.

V. THE CURRENT STATE OF MRA RESEARCH

There is a limited level of activity in the areas discussed in this paper, or in closely related areas. This section will briefly describe the work that is being done currently and that has been recently completed.

Investigators at Analysis and Inference, Inc. have completed a research project entitled "The Application of Modern Statistical Principles to Automobile Insurance Risk Classifications," supported by the Division of Applied Research of the National Science Foundation. The purpose of this project was to study traditional actuarial approaches to the classification and premium rating of risks in automobile insurance, and to propose modern statistical models and estimation methods to improve actuarial practice. This project has determined that the currently-used additive and multiplicative models (which are employed to combine relativities derived from univariate analyses of data cross-classified in several dimensions) can be improved upon by mixed additive and multiplicative models, which in turn can be improved upon by empirical Bayes estimators. See Tomberlin (1982) and Weisberg and Tomberlin (1982). This work forms an important part of the theoretical underpinning for public policy applications for MRAs. The performance of insurance classifications and premium rating is critically important to MRAs; if insurers do not assess risks well, their assessments are not very useful to the public. Improvements in technical understanding of the strengths and weaknesses of the risk classification and rating process will allow for better estimation of MRA accuracy. One of the goals, in fact, of the project on statistical theory in actuarial practice is to develop methods that are broadly applicable in risk assessment. The approaches used by casualty actuaries are similar in many respects across different lines of insurance. There are, for example, commonalities in risk classification and rating approaches between the automobile insurance line and the commercial general liability (CGL) line is the insurance line most relevant to most MRAs; see Williams, Head, and Glendenning (1978).

The present authors and colleagues at Analysis and Inference, Inc. completed a project for the U.S. Nuclear Regulatory Commission which analyzed self-insurance pools as financial assurance mechanisms for assuring the adequacy of funds for nuclear power plant decommissioning expense. This project developed the most detailed MRA methodology in existence to date, building upon earlier work by Denenberg (1973) and Wood (1981), for extracting catastrophic probability and consequence estimates from insurance premiums. This work was published by the NRC as NUREG/CR-2370; see Chernick et al. (1981).

On the policy side of the present proposal, Michael Baram has recently completed two projects, "Alternatives to Government Regulation" and "An Assessment of the Use of Insurance for Managing Risks to Health, Safety, and Environment," both funded by the National Science Foundation. The first study was a broad review from a public policy perspective of alternatives to direct governmental regulation for managing technological risks. It reviewed a wide variety of proposed alternatives, including common law actions for negligence and strict liability, industrial and professional self-regulation, insurance mechanisms, other compensation mechanisms, efficient charges, and tax systems. The second study was a general investigation of the qualities of insurance as a regulatory mechanism, through four case studies.

The project at Clark University and Decision Research, Inc. on "Improving the Social Management of Technological Hazards" has worked on exploration of: (1) the size and trends in technological hazards; (2) taxonomies of technological hazards; (3) causal models for risks as a basis for planning managerial strategies of intervention; (4) judgmental basis in risk perception; and (5) generic approaches to "acceptability of risk" decision making.

Similarly, the International Institute for Applied Systems Analysis (IIASA) in Vienna has sponsored research for several years on the general subject of risk management. This work has recently focused on decision making for LNG safety.

The Division of Risk Analysis of the U.S. Nuclear Regulatory Commission has sponsored or directed work on three related topics: (1) quantifying safety goals for commercial nuclear power plant regulation, see U.S. Nuclear Regulatory Commission (1980, 1982b) and Strip (1982b); (2) improving probabilistic risk assessments, the traditional fault-tree methodology, see U.S. Nuclear Regulatory Commission (1982a); and (3) improving consequence of accident estimates, see Strip (1982a) and Ritchie et al. (1982).

Several other U.S. governmental agencies have sponsored work on risk assessment. For example, the Environmental Protection Agency has conducted a study of private insurance availability for "non-sudden" accidents resulting from hazardous waste disposal. The Department of Commerce has conducted a study on availability of insurance, and on competition in the insurance markets, with respect to oil pollution insurance for oil spills from off-shore oil drilling activity. The Department of energy's Office of Pipeline Safety has conducted a study of LNG and LPG production, transmission, and storage risk assessment, and on the various methods of assuring financial responsibility of parties engaging in LNG and LPG operations.

VI. CONCLUSIONS

Market risk assessments, extracted from information contained in insurance premiums, may provide a valuable source of risk assessments which may be used to supplement more traditional types of quantitative risk assessments. Although substantial work remains to be done on the qualities, the interpretation, and the use of MRAs, there is reason to be optimistic that these problems can be overcome to a great enough extent to permit MRAs to become a valuable supplement to more traditional quantitative risk assessments.

REFERENCES

1. Buhlmann, H. (1970), Mathematical Methods in Risk Theory, Springer-Verlag, 1970.
2. Bailey, Elizabeth E. and Friedlaender, Ann F. (1982), "Market Structure and Multiproduct Industries," Journal of Economic Literature, September, 1982.
3. Chang, Lena and Fairley, William B. (1979), "Pricing Automobile Insurance Under a Multivariate Classification," Journal of Risk and Insurance, March, 1979.
4. Chernick, Paul L., Fairley, William B., Meyer, Michael B., and Scharff, Linda C. (1981), "Design, Costs, and Acceptability of an Electric Utility Self-Insurance Pool for Assuring the Adequacy of Funds for Nuclear Power Plant Decommissioning Expense," NUREG/CR-2370, U.S. Nuclear Regulatory Commission, December, 1981.
5. Crandall, Robert W. and Lave, Lester B., Editors (1981), The Scientific Basis of Health and Safety Regulation, The Brookings Institution, 1981.

6. Denenberg, Herbert S. (1973), "Testimony Before the Atomic Safety and Licensing Board, Licensing Hearings for Three Mile Island Nuclear Power Plant," November, 1973.
7. Fairley, William B. (1979), "Investment Income and Profit Margins in Property-Liability Insurance: Theory and Results," Bell Journal of Economics, Spring, 1979.
8. Fairley, William B. (1981), "Assessment for Catastrophic Risks," Risk Analysis, Vol. 1, No. 3, 1981.
9. Fairley, William B. (1981), "Market Risk Assessment of Catastrophic Risk," in Kunreuther, Howard, and Ley, Eryl, Editors, The Risk Analysis Controversy, Springer-Verlag, 1982.
10. Green, Hardol P. (1981), "The Role of Law in Determining Acceptability of Risk," in Nicholson, William J., Editor, Management of Assessed Risk for Carcinogens, Annals of the New York Academy of Sciences, Vol. 363, 1981.
11. Hill, Raymond (1978), Capital Market Equilibrium and the Regulation of Property-Liability Insurance, Ph.D. Dissertation, Massachusetts Institute of Technology, 1978.
12. Hill, Raymond (1979), "Profit Regulation in Property-Liability Insurance," Bell Journal of Economics, Spring, 1979.
13. Joskow, Paul (1973), "Controls, Competition and Regulation in the Property-Liability Insurance Industry," Bell Journal of Economics and Management Science, Autumn, 1973.
14. Meyer, Michael B. (1982), "Catastrophic Loss Risks: An Economic and Legal Analysis, and a Model State Statute," June 1982, in Low-Probability/High-Consequence Risk Analysis: Issues, Methods, and Case Studies, Plenum Press, 1984.
15. Meyer, Michael B. (1983), "Regulating Catastrophes through Financial Responsibility Requirements: A Model State Statute," Harvard Journal of Legislation, Summer, 1983.
16. Morris, Carl and Van Slyke, L. (1978), "Empirical Bayes Methods for Pricing Insurance Classes," Proceedings of the Business and Economics Section, American Statistical Association, Annual Meeting, San Diego, 1978.
17. National Association of Insurance Commissioners (1974), Monitoring Competition: A Means of Regulating the Property and Liability Insurance Business, Volumes 1 and 2, May, 1974.
18. National Research Council (1980), Regulating Pesticides, A Report Prepared by the Committee on Natural Resources, National research Council - National Academy of Sciences, 1980.
19. Raiffa, Howard (1982), "Science and Policy: Their Separation and Integration in Risk Analysis," in Proceedings of the Sixth Symposium on Statistics and the Environment, American Statistician, August, 1982.
20. Ritchie, L, Johnson, J., and Blond, R. (1982), "Calculations of Reactor Accident Accident Consequences, Version 2: User's Guide," NUREG/CR-2326, SAND 81-1994, Sandia National Laboratories, 1982.
21. Seal, Hilary (1969), Stochastic Theory of a Risk Business, Wiley, 1969.
22. Solomon, Kenneth A., Whipple, Chris, and Okrent, David (1978), "More on Insurance and Catastrophic Events: Can We Expect De Facto Limits or Liability Recoveries," Rand Corporation, P-5940, March, 1978.
23. Stone, James M. (1973), "A Theory of Capacity and the Insurance of Catastrophic Risks," Parts I and II, Journal of Risk and Insurance, 1973.
24. Strip, D.R. (1982a), "Estimates of the Financial Consequences of Nuclear Power Reactor Accidents," NUREG/CR-2723, SAND 82-1110, Sandia National Laboratories, September, 1982.
25. Strip, D.R. (1982b), "Analysis of a Proposed One Thousand Dollar Per Man-Rem Cost Effectiveness Criterion," NUREG/CR-2899, SAND 82-1870, Sandia National Laboratories, October, 1982.

26. Tomberlin, Thomas J. (1982), A Statistical Perspective on Prediction Losses in Automobile Insurance, Ph.D. Thesis, Statistics Department, Harvard University, September, 1982.
27. U.S. Nuclear Regulatory Commission, Advisory Committee on Reactor Safeguards (1980), "An Approach to Quantitative Safety Goals for Nuclear Power Plants," NUREG-0739, 1980.
28. U.S. Nuclear Regulatory Commission, American Nuclear Society, and Institute of Electrical and Electronics Engineers (1982a), "PRA Procedure Guide: A Guide to the Performance of Probabilistic Risk Assessments for Nuclear Power Plants," NUREG/CR-2300, Rev. 1, april, 1982.
29. U.S. Nuclear Regulatory Commission (1982b), "Safety Goals for Nuclear Power Plants: A Discussion Paper," NUREG-0880 (for comment), February, 1982.
30. Weisberg, Herbert I. and Tomberlin, Thomas J. (1982), "A Statistical Perspective on Actuarial Methods for Estimating Pure Premiums from Cross-Classified Data," Analysis and Inference, Inc., Journal of Risk and Insurance, 1982.
31. Williams, C.A., Head, G.L., and Glendenning, G.W. (1978) Principles of Risk Management and Insurance, American Institute for Property and Liability Underwriters, 1978.
32. Wood, William C. (1981), "Nuclear Liability after Three Mile Island," Journal of Risk and Insurance, September, 1981.

PUBLIC RECOGNITION OF HAZARD[1]

George Oliver Rogers

University of Pittsburgh
Pittsburgh, PA 15260

ABSTRACT

Like hazard identification, the public recognition of hazard reduces uncertainty by delineating the relatively risky from the less risky. For centuries humans have relied on actuarial experience in recognizing hazardous situations. However, the risks associated with more complex technological systems often veil the nature, likelihood, and severity of the risk. Moreover, many risks are obscured by relatively delayed, and long-term effects, sometimes associated with very small "doses." While it is clear that these cases require scientific intervention in the identification of hazard, little is known about the social process involved in the use of experiential data in the recognition of hazard. This paper examines the relationship between prior experience and risk estimates for a variety of hazards.

KEY WORDS: Risk Recognition, Hazard Experience

INTRODUCTION

The public recognition of hazard rests on a foundation of experience and values. Like hazard identification, the public recognition of hazard reduces uncertainty by delineating the relatively risky and potentially risky, from the less risky. People have historically relied on actual experience with hazards as the primary mechanism for recognizing hazardous situations. Complex technological systems often veil the nature, likelihood and severity of risk. Many risks are obscured by delayed or long-term effects. Such risks clearly require scientific intervention in the identification of hazard. But for hazards, where actualized hazard experience is available, the public seems to recognize the potential for harm. Little is known, however, about how such experience is used in the public recognition of hazard. The public recognition of hazard often forms the larger context for the assessment of risk by risk analysts. Their credibility is frequently jeopardized by failing to adequately

---

[1]This research was partially supported by the Federal Emergency Management Agency (Cooperative Agreement No. EMW-K-1024). This report has not been reviewed by FEMA and in no way reflects the view and policies of the Agency. The author accepts full responsibility for the contents herein and gratefully acknowledges the support, comments and criticisms offered by colleagues at the University of Pittsburgh and the Annual Meeting of the Society for Risk Analysis.

understand this context of the public recognition of hazard. This paper examines the relationship between experience and hazard by the general public.

Potential hazards are actualized in terms of initiating event(s), their outcome(s), exposure(s) and consequence(s). The risk event initiates the actualization of any hazard. It may be the result of a social action or an environmental process. The risk event is characterized by an unknown probability of occurrence, which implies that "...risks can never be measured precisely" (Whyte and Burton 1980:3), because "...it is always a question of estimation." However, the estimates of the likelihood of event occurrence, with varying degrees of precision, must inherently be used in forming attitudes about the hazard and, of course, in determining potential actions for dealing with the event, both ex ante and post hoc. For example, the probability of a flood is based on the weather patterns of the season, the river system capacity, the snow pack, ground water capacity, and the history of the particular river system for flooding under comparable circumstances, but the probability of flooding cannot be precisely determined. "An event is defined by its complete description and the probability of its occurrence" (Rowe 1977:99). To varying degrees, both the true probability and the estimates are temporal and spatial. The true probability of some hazards (e.g., weather-related hazards such as snowstorms, tornadoes, and hurricanes) are certainly seasonal and regional. Other hazards may pose more or less constant threats to specific areas (e.g., earthquakes, chemical hazards, nuclear war, and nuclear power plants). The true probability and the associated estimates are based on varying degrees of human experience and knowledge, and form probability distributions that may or may not vary in accord with one another.

THE RECOGNITION OF RISK

Hazard recognition comprises an awareness of the conditions and circumstances that constitute the potential for harm. It is the discovery, identification or recognition of potentially hazardous situations. Uncertainty is reduced through recognition of potentially hazardous situation. "For most of human history, the identification of...risk arose from direct human experience" (Kates 1978:14). However, in modern times society seems to be increasingly dependent upon science for identification of risk (cf. Nehnevajsa 1981) in part because science and technology have created more effective ways of observing and predicting hazardous situations, and in part because hazards are more likely to be detectable only by sophisticated, scientifically developed technology. While scientists sometimes support conflicting knowledge-claims and thereby produce an aura of questionable credibility, the experience of everyday living or ordinary knowledge remains a fundamental component of the knowledge-base on which people recognize hazardous situations (Nehnevajsa 1981). The knowledge-base on which hazard recognition rests is in part ordinary or working knowledge, based primarily on experience, and in part scientific knowledge, based primarily on systematic study. Hence, a significant portion of the recognition of hazard is based on previous experience.

Hazard perception is contingent on its recognition. For example, non-recognized hazards may exist, even for some time, before they are recognized (cf. Lawless 1977). The causal links between the various components of hazard actualization is fundamentally based on knowledge. For example, if a particular consequence is attributed to an event, say event X, but is actually the result of some other risk event, say event Y, the second event (Y) may or may not be recognized as hazardous. It is when knowledge of the linkage between the event (Y) and the consequence exists that situations leading to event Y may be recognized as being hazardous.[2] The immediacy of the effect and the familiarity associated with a particular hazard are fundamental elements of the knowledge-base and determine to some extent its use. In the sense that non-recognized hazard cannot be estimated, the perception of hazard is dependent on the recognition of the potentially harmful situation. Like perceived hazard, the acceptability of risk is contingent upon the recognition of hazard. In essence, a hazard cannot be deemed either acceptable or unacceptable when it remains non-recognized. The foundation of our fear, anxiety and concern, which is associated with the acceptability of risk, is dependent on: (a) the commonly understood nature of the risk in terms of peoples' familiarity with it, (b) immediacy of effect, (c) the general knowledge concerning its riskiness, and (d) potential and actual benefits. These dependencies are summarized in Figure 1.

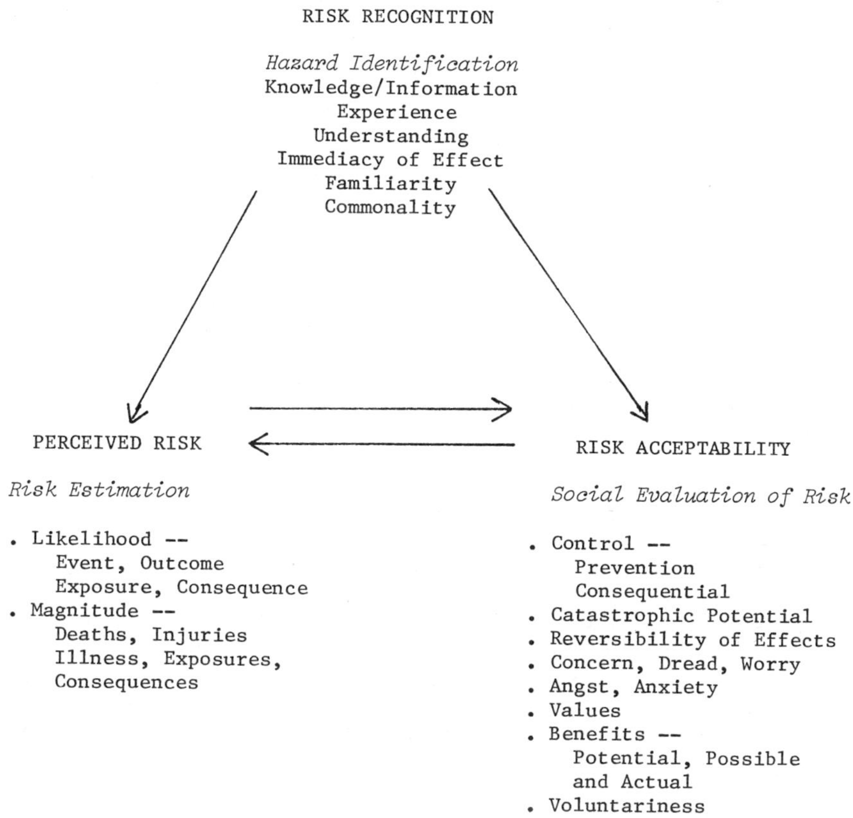

Figure 1

Risk Recognition, Perceived Risk and Risk Acceptability

---

[2]Event Y could have other recognized consequences which would make it recognizable as hazardous, but only when knowledge of the link exists may Y be recognized as hazardous in terms of this consequence.

The recognition of hazard is addressed in terms of estimated likelihood of various risks by geographic region of the United States. The distribution of perceived hazard across the various geographic regions of our country serves to underscore the idea that people tend to know, if only in the ordinary sense, the hazards they face. This does not mean, nor should it be inferred, that residents of an area, know in a comprehensive sense, the hazards they potentially face, but rather that there is a sensitivity to the major hazards recognized as part of the area's ambience.

Conceptually, life-experience is inherently related to hazard recognition because "ordinary knowledge" principally comprises experiences of various kinds, and the perception of hazard rests firmly on this data-bank of knowledge. The cognitive processes, through the use of heuristics (cf. Tversky and Kahneman 1973, 1974 and 1981; Tversky 1972; and Kahneman and Tversky 1972) utilize this inventory of experience in estimating the likelihood of a hazard actualizing through its event-outcome(s)-exposure-consequence(s) chain. Even identical experiences are likely to have different meanings among people of various social positions. The relationship between perceived hazard and experience for specific hazards are examined in this section.

People of different social structural locations are likely to be characterized by different experiences. Human beliefs, attitudes, and behaviors are, in general terms, shaped by self-image. The conception of "who we are" is recognized as a product of the social structure, our integration into the social system, the values associated with that, and significant life-experiences (cf. Goffman 1970; Mannheim 1936; Merton 1968; and Holzner 1968). The frequency of life events during a given time period, the nature of the events, and even their order have been shown to be associated with significant effect on attitudes, beliefs, and behaviors (cf. Thoits 1981; Coates, Moyer and Wellman 1969; Dohrenwend 1973; Dohrenwend and Dohrenwend 1974; and Hogan 1978). These life changes have even been associated with illness and susceptibility to illness (cf. Rahe 1968, 1972 and 1974; Hinkle, et al. 1958; Holmes and Rahe 1967; Hong and Holmes 1973; Rahe, et al. 1964; and Rahe, Mahan and Arthur 1970). While typical life events include marriage, separation, reconciliation, divorce, remarriage, death of a spouse, birth of a child, beginning and ending employment, schooling, etc., life histories of severe events (e.g., the Great Depression, World Wars) have been shown to have long-term effects (cf. Elder and Liker 1982).

Life events usually represent fundamental social-system, social-structural, or social-position changes. These shifts are often associated with the stage of life and life cycle (cf. Rogers 1980 and 1983; Linton 1942; and Parsons 1942), but may also be associated with career or economic changes. While life-experiences may have fundamental social-system, social-structural, role, or position changes associated with them, they are not conceptually dependent on the occurrence of the life event. Life events may be thought of as altering fundamentally the roles played by an individual. Life-experiences, on the other hand, represent personal life histories and their associated special meanings for the individual. Life-experience may be thought of as a segment of personal life history with special meaning for the individual. These experiences are sufficiently unique, prominent, or important in the individual's data-bank of experience to fundamentally alter common or ordinary knowledge. While some of these life-experiences may represent temporary or relatively brief historical or personal events (e.g., severe economic conditions, periods of unemployment or extreme poverty, personal injury or disease, etc.), others may represent more cumulative experiences (e.g., child abuse, living in extreme or under special conditions). While life-experiences, by virtue of their unique, prominent, and important position in the personal life history, have special meaning for the individual, these experiences, even if only of a similar nature, have a shared group meaning for those who hold the life-experience in common. Furthermore, by their very nature, for example, severe economic conditions and periods of war are simultaneously experienced by groups. While these experiences are not uniformly experienced by all, they do have a shared meaning--a cohort effect.

Prior disaster experience is a shared experience. It is shared with those other individuals, families, groups, or communities affected by the same disaster. Prior disaster experience may be thought of as historical, in the sense of occurring in the past; it is more periodic than on-going, and is not necessarily cumulative. Because the individual is likely to have a clearer image of self-efficacy if a prior disaster of any kind has been experienced, prior disaster experience itself has meaning beyond the experience with the particular hazard. Previous experience with a particular hazard has been found to enhance the understanding of the process by which risk is actualized (cf. White and Haas 1975). One relatively well-documented finding suggests that prior disaster experience is directly related to warning belief with prior disaster experience being associated with higher warning credibility, and no experience with less warning credibility (cf. Mileti, Drabek and Haas 1975; Williams 1957; Mack and Baker 1961; and Drabek and Boggs 1968). Hence, whether prior disaster situations become actualized or not affects the credibility of the warning message directly. Direct experience, even though it may not be actualized, works as a partial reinforcement schedule in the perception of the hazard. Experiences with other disasters are expected to affect the perception of likelihood in the realization that "bad things can happen," and risk evaluation by clarifying our image of self-efficacy in dealing with extreme situations.

Only residential proximity with nuclear power plants seems to affect the perception and acceptability of the associated risks. This casts the relationship between hazard experience and perceived and acceptable risks in terms of experience of various kinds, and the risks associated with nuclear power plants. Rogers (1983:149) speculates, based on this finding, that "...only direct experience with the particular hazard in question affects the perception and acceptability of that risk." For the case of nuclear energy, the history has been rather accident-free, but the "...recent referenda in the Three Mile Island area concerning the potential restart of the undamaged reactor..." points out that when experience includes accidents, the effect is likely to enhance the magnitude of perceived hazard and decrease any associated acceptability of the risks. This seems to suggest that people recognize the hazards they face. The geographically diffuse nature of the operational definitions, of perceived and acceptable risk of nuclear power plant accident, and the relatively accident-free history, precluded the direct observation of experience and perceived hazard.

METHODOLOGY AND MEASUREMENT

This analysis of hazard experience and perception is based on one of three "piggyback" surveys sponsored by the Federal Emergency Management Agency and conducted in 1982 by the Gallup Organization. In the second survey conducted during the summer of 1982, 1030 adult Americans within the 48 contiguous United States were interviewed. The Gallup Organization employees a post-sample weighting procedure to the survey to represent

1500 respondents. Because we found little difference in the weighted and unweighted sample, this analysis is based on the unweighted sample. The interpretation of these data remains conceptually clear and the sampling error remains at approximately the ±5 percent level with a 95 percent confidence level for the conservative 50-50 distribution.

Substantively, for each of a series of possible hazards, people were asked if they had experienced the specific disaster. The operational variable for disaster experience is a binary variable where the value one represents the existence of an experience, and zero represents no reported disaster experience of this type. This represents only the individual's reported experience with the given hazard and not the actual occurrence of the hazard in their area. That is, it cannot be assumed that the individual experienced the given disaster in the present residence or even the more general geographic location. For example, a respondent could have experienced a hurricane at a previous residence in a Gulf of Mexico state even though the current residence is in a Northeastern or Western state. For that matter, the experience could represent an experience while on vacation in an area affected by such a disastrous event. In addition, the concept of experience is somewhat unclear. Does this refer to direct personal experiences? Disaster experience of one's immediate family? Experiences that one's friends may have incurred? Experiences of one's community? or Disaster experiences that have been heard about, via social contacts or even the media? What role do experiences of fiction play (e.g., "Silent Spring," "The China Syndrome," "The Day After," or "Countdown to Looking Glass")? The temporal dimension of such reported experiences is not operationally clear. These disaster experiences could be associated with occurrences last week, last year, several years ago or even childhood experiences. These disaster experiences could be relatively traumatic in terms of a loved one's death or injuries, property damage and the like, or have relatively little associated stress. The reported disaster experiences consist of experiences significant enough to become mentioned in a telephone interview.

In addition, the respondents were asked to report any evacuation circumstances they may have experienced. These experiences are conceptually limited by the occurrence in the respondent's past. While these occurrences are again spacially limited in the sense of not being limited to evacuation experiences in their present residential location, as with the disaster experiences, they are more specific for personal, family or household experiences. In contrast to the general disaster experience variable, the evacuation experience variable is limited temporally to a five year time-frame. It is also somewhat more specific in terms of stress inducement. The evacuation itself, at a minimum, requires crisis response action, which inherently has some heightened anxiety level associated with it. It also connotes a degree of direct impact, contingent only upon the occurrence of the hazard itself, and the resulting hazard outcomes, and consequences. The evacuation variable is operationally a binary variable where one represents an evacuation experience in the previous five years, and zero represents no such experience.

The estimated likelihood of occurrence of the given disaster is represented as "very likely," "somewhat likely," "about a 50-50 chance," "somewhat unlikely," "very unlikely" or "never." In this sense, the operational definition of the likelihood estimate may be derived by setting the "very likely" response equal to one, the "somewhat likely" response to .75, the 50-50 chance response equal to .5, the "somewhat unlikely" response to .25, and the "very unlikely" and "never" responses equal to zero. The operational definition has the added feature of making the respondents estimate "mirror" the actual probabilities in the sense of having an appropriate range, and the 50-50 mid-point of the scale reflects a degree of specification to the likely and unlikely ends of the likelihood estimates. However, it is not conceptually clear that "very likely" is adequately represented by the likelihood estimate of one, or put another way, "almost certain to occur." The unlikely extreme is conceptually somewhat clearer in that the "very unlikely" response is joined with the "never" response as the zero likelihood estimate of the hazard occurrence. These estimates are spacially and temporally specific: "in your area" and "in the next five years." Furthermore, these estimates do not include any hazard consequence(s), such as personal damage or destruction of property, so the operational specification of the estimated likelihood of these specific disasters is conceptually clear.

Regional Distribution of Hazard Recognition and Experience

The geographic distribution of recognized hazards reflects, to some extent, the distribution of hazards themselves. The distribution of hazard experiences and exposure rates (Rossi, et al, 1983) among regions of the United States indicates the relative accuracy of the hazards being recognized by the general public. Natural hazards are presented in Table 1 in terms of exposure rate, reported experience, and likelihood estimates associated with natural hazards of relatively extended onslaught, including floods and hurricanes, and natural hazards of relatively rapid onslaught, including tornadoes and earthquakes. Technological hazards, including gas line breaks, chemical accidents and dam failures, are presented in Table 2. This cross-cutting dimension of hazard recognition by geographical distribution implies that people generally recognize the hazards they face, if only in a common sense way. Some (e.g. flood and dam failure) are rather diffuse among all the regions, while others (e.g. hurricanes, tornadoes and earthquakes) are relatively concentrated in one or more regions. Rogers (1984) describes these distributions in detail and discusses how they relate to a common-sense understanding of the hazards. In general, the regional correlations are quite low and substantively insignificant. The methodological problems with these rather crude measures may explain many of the non-significant relationships described.

HAZARD EXPERIENCE AND RECOGNITION

The correlations between reported experiences with hazard and the estimated likelihood of common hazards are presented in Table 3. Of the 64 correlations presented, 28 are not significant at the .10 level and 7 are marginally significant at the .10 level but not the .05 level. The five largest correlations between experiences and perceived likelihood of hazard actualization are between experience and estimated likelihoods of the same hazard. The largest correlation is between experience and estimated likelihood of the somewhat recurring flood (.343). This certainly reflects the kind of hazard recognition associated with living near a river system, such as in conjunction with emerging disaster subcultures. The second largest correlation reflects the relative predictability and seasonal nature of hurricanes. The correlation between experience and estimated likelihood (.320) clearly shows that people recognize the hazards they have faced in the past and that they expect to face similar hazards in the future. Tornadoes are less predictable in the sense of where and when they will touch down even after a tornado producing weather system is detected -- a fact that scientists through technology are working to remedy. Yet the correlation between experiences and estimated likelihood of tornadoes is third highest (.266). The next

Table 1

Natural Hazard Likelihoods and Experiences by Region of the Country

Region	N	Flood	Hurricane	Tornado	Earthquake
New England					
1)	1032	18.4	18.4	31.0	1.0
2)	65	18.5	3.1	4.6	0.0
3)	65	37.7	52.0	29.1	23.0
Middle Atlantic					
1)	2113	31.7	22.2	41.2	4.3
2)	167	10.8	4.8	1.2	0.0
3)	167	31.1	29.8	19.4	12.0
East N. Central					
1)	2825	13.8	10.6	98.8	8.1
2)	210	14.3	1.0	12.9	0.0
3)	210	35.0	11.2	73.3	15.9
West N. Central					
1)	1188	21.0	8.4	89.2	4.2
2)	83	26.5	1.2	16.9	0.0
3)	83	40.4	11.0	75.6	2.5
South Atlantic					
1)	2384	13.4	36.5	42.4	2.5
2)	186	13.4	16.1	3.2	0.0
3)	186	34.2	47.2	49.4	12.9
East S. Central					
1)	942	19.1	62.6	41.7	7.4
2)	61	32.8	1.6	19.7	0.0
3)	61	44.7	19.5	71.7	25.0
West S. Central					
1)	1265	19.8	34.8	78.3	3.2
2)	83	21.7	18.1	7.2	0.0
3)	83	54.8	42.0	70.4	9.7
Mountain					
1)	405	14.8	4.9	42.0	17.3
2)	36	36.1	2.8	8.3	5.6
3)	36	53.5	1.4	39.6	37.5
Pacific					
1)	838	10.7	6.6	26.3	84.7
2)	139	16.6	0.7	0.7	9.4
3)	139	39.1	6.3	9.2	79.0

1) Hazard Victimizations (1970-1980) per 1000 Households Reported by Peter Rossi et al in VICTIMS OF THE ENVIRONMENT Plenum Press New York, 1983.

2) Percent Reporting Hazard Experience in 1982 Gallup Survey.

3) Estimated likelihood of Hazard Actualization in respondent's own area and within five years, from 1982 Gallup Survey. Responses are "very likely" (scored 100), "somewhat likely" (scored 75), "50-50 chance" (scored 50), "somewhat unlikely" (scored 25), "very unlikely" and "never" (scored 0).

highest correlation is between the experience and estimated likelihood associated with the less predictable earthquake (.183). The fifth largest correlation occurs between the estimated likelihood and prior experience with chemical accidents (.173). The magnitude of the relationship seems

Table 2

Technological Hazard Likelihoods and Experiences by Region of the Country

Region	Gas Line Break	Chemical Accident	Dam Failure
New England			
1)	41.0	45.6	7.9
2)	0.0	1.5	0.0
Middle Atlantic			
1)	59.3	44.8	10.2
2)	2.4	1.8	0.6
East N. Central			
1)	58.3	48.8	9.6
2)	1.4	4.8	0.0
West N. Central			
1)	51.2	44.8	9.0
2)	0.0	4.8	1.2
South Atlantic			
1)	46.6	42.9	9.6
2)	1.6	6.5	0.0
East S. Central			
1)	39.2	43.4	17.6
2)	1.6	13.1	0.0
West S. Central			
1)	55.6	47.5	13.5
2)	4.8	8.4	1.2
Mountain			
1)	56.4	53.5	31.3
2)	11.1	0.0	0.0
Pacific			
1)	56.6	53.5	25.2
2)	1.4	2.2	7.2

1) Likelihood estimate consists of Respondent's estimate of the occurrence of the disaster in their own area and within five years. Responses are "very likely" (scored 100), "somewhat likely" (scored 75), "50-50 chance" (scored 50), "somewhat unlikely" (scored 25), "very unlikely" and "never" (scored zero).
2) Experience variable equals one if Respondent reports having experience with disaster, zero, otherwise.

to be a function of the predictable nature of the hazard, with the highest correlation being between experience and perception of hazards of a highly predictable nature, and the significant, but smaller correlation, being among hazards of recognized areas but of little predictability. Hence, it appears that people recognize the hazards they face, and estimates of the likelihood of future occurrences are based on prior experiences with these same hazards.

Comparing the even more abstract recognition of being in potential risk areas associated with nuclear attack confirms this trend. Each

Table 3

Correlations Between Disaster Experiences and Likelihood Estimates
1982 Gallup

Experiences**	Likelihood Estimates 1.	2.	3.	4.	5.	6.	7.	Nuclear Attack
1. HURRICANE	.320	NS	.057	-.071	-.109	NS	-.109	NS
2. TORNADO	-.080	.266	.079	NS	NS	NS	NS	NS
3. FLOOD	NS	.055	.343	.056	NS	NS	-.043	.041
4. DAM FAILURE	NS	NS	NS	NS	NS	-.041*	NS	-.056
5. EARTHQUAKE	-.063	-.123	NS	.046*	.183	.055	NS	.057
6. CHEMICAL ACCIDENT	NS	.081	NS	.046*	NS	.173	.062	NS
7. GAS LINE BREAK	NS	NS	.045*	.042*	NS	.089	.109	NS
SOME DISASTER EXPERIENCE	.078	.097	.279	.051*	.046*	.120	NS	.051

NS - not significant at .10 level

*Not significant at .05 level

**Experiences are binary variables representing either expressed experience or evacuation due to specific disaster

respondent is associated with a particular county based on the reported Postal Service Zip-code and then related to nuclear attack risk areas as specified by the Department of Defense (TR-82). In comparing this objective risk area estimate with the respondent's perception of the likelihood of the local area being a target area, we find that those respondents objectively at risk place their likelihood at .608, while those objectively not in target areas place their target area likelihood at only .395. Hence, people tend to recognize that, while hazards affect all people "equally" in the sense of status, they also recognize their own susceptibility to specific hazards, including the most abstract and never actualized nuclear attack.

LINKING HAZARDS

This analysis examines direct experience with, and likelihood estimates of specific hazards. Hazard experience is operationalized as a general or evacuation event of a specific hazard. Table 4 presents the correlation among eight specific and two more general experience indicators. Of the 45 correlations among these variables, about half (21) are significant at the .10 level. Of these correlations, only the

Table 4

Correlations Among Disaster Experiences**
1982 Gallup

	2.	3.	4.	5.	6.	7.	8.	9.	10.
1. HURRICANE	NS	.097	NS	NS	NS	.065	NS	NS	.309
2. TORNADO		.149	NS	NS	.080	NS	.057	-.052	.334
3. FLOOD			.133	NS	NS	NS	.045*	NS	.507
4. DAM FAILURE				NS	NS	NS	NS	NS	.080
5. EARTHQUAKE					NS	.192	.150	NS	.139
6. CHEMICAL ACCIDENT						.043	.087	NS	.244
7. FIRE							.122	NS	.331
8 GAS LINE BREAK								NS	.169
9. OTHER									.267
10. SOME									

NS - not significant at .10 level

*Not significant at .05 level

**Experiences are binary, variables representing either expressed experience or evacuation due to specific disaster

correlations between the individual hazard and having some disaster experience have significant magnitudes. These correlations range from .080 to .507 and may be thought of as measuring the contribution of each individual hazard experience to the overall disaster experience profile. The highest correlation among hazard experiences with having some experience is between earthquakes and fires at .192. The correlation between earthquakes and gas-line breaks (.150), tornadoes and floods (.149), fire and gas-line break (.122) comprise the most significant relationships. These correlations represent logical connections among hazards of various kinds; a further example of this is the .133 correlation between flooding and dam failure.

The correlations among the estimated likelihoods for the various hazards, presented in Table 5, are on the whole more significant and of greater magnitude than those among the actual experiences. The largest correlation reflects the relationship between estimated likelihood of a chemical accident and a gas-line break (.456). The correlation of nuclear and chemical accident (.443) represents the second largest correlation. While these two relationships seem to suggest the recognition of the similar technological context of these hazards, the correlation between the likelihood estimates for a potential nuclear accident and gas-line

Table 5

Correlations Among Likelihood Estimates
1982 Gallup

	2.	3.	4.	5.	6.	7.	8.	9.
1. HURRICANE	.082	.156	NS	-.089	.065	NS	.065	.055
2. TORNADO		.158	NS	-.139	.137	.130	.057	.106
3. FLOOD			.263	.099	.174	.160	.148	.088
4. DAM FAILURE				.234	.185	.126	.214	.098
5. EARTHQUAKE					.213	.182	.145	.065
6. CHEMICAL ACCIDENT						.456	.443	.174
7. GAS LINE BREAK							.160	-.109
8. NUCLEAR ACCIDENT								.141
9. NUCLEAR WAR								

break is only (.160). The estimates for possible dam failure and potential nuclear accident seem to represent the overall relationship of hazards involving large-scale technology, but the correlation of .214 may only reflect the physical relationship of many nuclear power plants that are located near river systems that have upstream dams. The completely logical relationships between dam failure and earthquakes (.234), and dam failure and flooding (.263) are reflected. Furthermore a relation seems to be reflected between chemical accidents and earthquakes (.213). A considerable structure, that was only modestly reflected among experience indicators, is apparent in the likelihood estimates.

CONCLUSIONS

Lay-people reduce risk uncertainty by delineating relatively risky situations from those deemed less risky. Drawing on centuries of human experience, lay-people use actualized hazards in the recognition of risky and potentially risky situations. Hence, public recognition of hazard rests on a foundation of social values and experience. This research examines the general relationship between prior disaster experience and recognition of hazard. While recognizing that lay-people cannot use actual experience with hazard in identifying many technological risks, which tend to veil the nature, severity and likelihood of the associated risks, the public seems to recognize globally the hazards they face.

This research takes an important first step in understanding which experiences, and for what hazards, public recognition of risk is linked to actual hazard experience--particularly those that are relatively common, predictable or familiar. While the public recognition and estimation of risk is less than precise, in terms of expert risk assessments, the resulting distributions of recognized hazard bears a marked resemblance to patterns of hazard exposure rates. It is in this sense that we conclude that people tend to recognize, at least in common terms, the hazards they face. In addition, the pattern of linking between hazard likelihoods and experiences suggests that people understand, at least in common-sense terms, the nature of potential threat.

While this research has foccussed on direct experience with similar and related hazards, much remains in the examination of more vicarious experience and the recognition of hazard. Many interesting questions remain: How severe does an experience need to be in order to affect public recognition and perception or risk? Do these experiences need to be personal? In direct familial relationships? Within the community? Or within society at large? How do severity and directness interplay to impact on recognition and perception of risk? In essence, under what conditions does the public use their experience, and that of others, in recognizing potentially risky circumstances?

REFERENCES

Coates, D., S. Moyer and B. Wellman, 1969, "Yorklea Study: Symptoms, Problems and Life Events," Canadian Journal of Public Health, 69, 471-481.

Dohrenwend, Barbara S., 1973, "Life Events as Stressors: A Methodological Inquiry," Journal of Health and Social Behavior, 14, 167-175.

Dohrenwend, Barbara S. and Bruce Dohrenwend, 1974, Stressful Life Events: Their Nature and Effects, John Wiley and Sons, New York.

Drabek, Thomas and Keith Boggs, 1968, "Families in Disaster: Reactions and Relatives," Journal of Marriage and the Family, 30, 443-451.

Elder, Glen H. Jr. and Jeffrey K. Liker, 1982, "Hard Times in Women's Lives: Historical Influences Across Forty Years," American Journal of Sociology, 88, 241-269.

Goffman, Erving, 1970, "The Presentation of Self in Everyday Life," Sociological Essays and Research, Charles Anderson (ed.), The Dorsey Press, Homewood, Illinois.

Hinkle, L.E. et al. 1958, "An Investigation of the Relation Between Life Experience, Personality Characteristics, and General Susceptibility Sceptability to Illness," Psychosomatic Medicine, 20, 278-295.

Hogan, Dennis P., 1978, "The Variable Order of Events in the Life Course," American Sociological Review, 43, 573-586.

Holmes, Thomas H. and Richard H. Rahe, 1967, "The Social Readjustment Rating Scale", J. of Psychosomatic Research, 11, 213-218.

Holzner, Burkart, 1968, Reality Construction in Society, Schenkman Publishing Co., Inc., Cambridge.

Hong, E. Michael and Thomas Holmes, 1973, "Transient Diabetes Mellitus Associated With Culture Change", Arch. Gen. Psychiatry, 29, 683-687.

Kahneman, Daniel and Amos Tversky, 1972, "Subjective Probability: A Judgment of Representatives," Cognitive Psychology, 3, 430-4544 June.

Kates, Robert W., 1978, Risk Assessment of Environmental Hazard: Scope 8, John Wiley and Sons, New York.

Lawless, Edward W., 1977, Technology and Social Shock, University of Rutgers Press, New Brunswick, NJ.

Linton, Ralph, 1942, "Age and Sex Categories," American Sociological Review, 7, 589-603.

Mack, Raymond W. and George W. Baker, 1961, "The Occasion Instant," National Academy of Sciences/National Research Council Disaster Study #15, Washington, D.C..

Mannheim, K.,1936, Ideology and Utopia, London.

Merton, Robert K., 1968, Social Theory and Social Structure, The Free Press, New York.
Mileti, Dennis, Thomas E. Drabek and J. Eugene Haas, 1975, "Human Systems in Extreme Environments: A Sociological Perspective," Institute of Behavioral Science, University of Colorado.
Nehnevajsa, Jiri, 1981, "Knowledge as Uncertainty: Some Implications for Knowledge Utilization," Conference on Knowledge Use, Pittsburgh, PA March.
Rahe, R.H., 1968, "Life-Change Measurement as a Predictor of Illness", Proc. of the Royal Society of Medicine, 61, 1124-1126.
Rahe, R.H., 1972, "Subjects' Recent Life Changes and Their Near-Future Illness Susceptability", Adv. Psychosom. Med., 8, 2-19.
Rahe, R.H., 1974, "The Pathway Between Subjects' Recent Life Changes and Their Near-Future Illness Reports: Representative Results and Methodological Issues", in Stressful Life Events, B. Dohrenwend and B. Dohrenwend (eds.), Wiley, NY.
Rahe, R.H., J.L. Mahan and R.J. Arthur, 1970, "Prediction of Near-Future Health Change from Subjects' Preceeding Life Change", J. of Psychosomatic Research, 14, 401-406.
Rahe, R.H., et al., 1964, "Social Stress and Illness Onset", J. of Psychosomatic Research, 8, 35-44.
Rogers, George O., 1984, "Hazard Perception and Crisis Behavior", University Center for Social and Urban Research, University of Pittsburgh, Pittsburgh, PA May 15.
Rogers, George O., 1983, "Toward A Sociology of Risk: Values, Experience, Perceived and Acceptable Risk," Ph.D. Thesis, University of Pittsburgh.
Rogers, George O., 1980, "Presidentially Directed Relocation: Compliance Attitudes", University Center for Social and Urban Research, University of Pittsburgh, Pittsburgh, PA May.
Rossi, Peter, et al, 1983, Victims of the Environment: Loss from Natural Hazards in the United States, 1970-1980, Plenum Press, New York.
Rowe, William D., 1977, An Anatomy of Risk, John Wiley and Sons, New York.
Thoits, Peggy A., 1981, "Undesirable Life Events and Psychophysiological Distress: A Problem of Operational Confounding," American Sociological Review, 46, 97-109.
Tversky, Amos, 1972, Imagination by Aspects: A Theory of Choice," Psychological Review, 79, 281-299.
Tversky, Amos and Daniel Kahneman, 1982, "The Framing of Decisions and the Psychology of Choice," Science, 211, 453-458.
Tversky, Amos and Daniel Kahneman, 1974, "Judgment Under Uncertainty: Heuristics and Biases," Science, 185, 1124-1131 September.
Tversky, Amos and Daniel Kahneman, 1973, "Availability: A Heuristic for Judging Frequency and Probability," Cognitive Psychology, 15, 207-232.
White, Gilbert F. and Eugene J. Haas, 1975, Assessment of Research on National Hazards, The M.I.T. Press, Cambridge, MA.
Whyte, Anne V. and Ian Burton (eds.), 1980, Environmental Risk Assessment, John Wiley and Sons, New York.
Williams, Harry B. 1957, "Some Functions of Communication in Crisis Behavior," Human Organization, 16, 15-19.

MANAGING DIFFERENCES IN INDIVIDUAL RISK PERCEPTIONS:

A PILOT EXPERIMENT TO INTEGRATE INDIVIDUAL PERCEPTIONS

W. Larry Petcovic

Advanced Communication Techniques
Columbia, Maryland 21045

ABSTRACT

There are two ways of dealing with risk perception and biases: attempt conformity to logical, quantitative manners; develop an awarenesses and recognition of risk preferences and biases with the intent to team the individual results so as to balance perceptions as a group. A pilot learning experiment, using an interactive computer based program, was created with the intent to provide participants feedback concerning their cognitive style preference, risk aversion or risk seeking preference for small quantities in the 10E-4 to 10E-6 range, and to provide indirect experience with feedback on decision biases concerning availability, overconfidence, weighting, and framing. No judgments were made concerning the correctness of any input or feedback. Instead, comments were made within the context of the individual determining relevance.

KEY WORDS: Process consultation; Computer based training; Heiristics and Biases; Risk Perception; Availability; Overconfidence; Framing; Weighting; Cognitive Style.

1. INTRODUCTION

Health issues today are a common happening. Any daily newspaper will carry the continuing story of "Agent Orange", "Hazardous Waste Dump", "Nuclear Winter", "Caffine", "Lead", "Microwaves" ... and the list goes on. Health issues range from global impact to personal in terms of effects and in terms of decision making. With limited resources in this world, how do we manage our decision processes about risk, both at the personal and group level?

If the world were totally logical and quantitative, then risk could be ranked in order of greatest harm and actions taken to control effects. In the works of Daniel Kahneman et al. (1982) on judgment of uncertainty, he paints a picture of perceptions of risk by individuals that are more subjective, qualitative and unpredictable. Peter Moore (1984) extends these findings into many other risk decision areas beside health issues. Moore includes financial and economic uncertainty as well. Individual risk perceptions and decision making as expressed in group activities are not subject only to logical process. Rather, perceptions and decision styles are varied and decisions at the group level will inevitably result in polarizied issues, especially in health subjects. there is little hope for unity of decisions and for logical control winning over individualized perceived risk.

2. THE POINT

The management of risk perceptions is a fundamental question of "process vs prescription". Is there a correct perception of risk at the individual and collective levels? How much do we really know about group risk decision making anyway? Is being prescriptive the same as being an advocate for a specified perception? How are perceptions changed? What is the role of a communicator? Is the design of a training program about risk prescriptive? Process, as used by Edgar Schein (1969), could involve the sharing of insights about risk decision making without a prescriptive agenda. The management of risk perceptions in this case qould be to allow individuals to determine their own biases used in decision making and to experience those biases in different situations. The individual could then determine the significants of their findings and set their own agenda for changes in how they would prefer to personally make decisions. Prescriptive, as used by Schein, would be any design of a program that identifies an individuals biases and then attempts to change biases so as to reinforce desirable traits as defined by the program designer.

It is the goal of this research to maintain a "Process" style of program design and implementation. My own biases for a "thinking and intuitive" style probably serve as a shadow in the design of this program. My personal experience also indicates that I have a difficult time designing for a "feeling" style. It will be necessary for many individuals of varying styles to share their thoughts on the design if a process oriented program is to be fully achieved.

3. THE DESIGN

With the fundamental structure of the program "process" oriented, adult learning techniques that optimize the process format are needed. The works of Malcolm Knowles (1978) would suggest the following structural design considerations:

- an accepted model of human differences that could serve as a bases for accepting individual perception differences. This model should be individualized and easily internalized. Human integrity is the value. human differences at a collective level should be seen as adding value to the system. This program will use the Myers-Briggs Type Indicator (1980) as a model.

- an identifiable model of some aspect of differences in individual perception that relate to integrity at the collective level also needs to be constructed. This program uses the "risk-averse and risk-seeking" concepts of Kahneman (1982) within the context of a risk range of 1 in 10,000. Integrity is maintained by demonstrating how other perceptions (opposite of the participant) are of value.

- an exercise for the learner to experience different situations with the requirement to exercise judgments with feedback. The interest is not to judge the response, but to structure the situation so as to evoke typical biases used in decision making. The learner will determine the value of the response. This program exercises the biases of availability, weighting, framing and overconfidence as defined by Kahneman (1982).

- feedback as much as possible to keep the learner involved. This program will use an IBM interactive computer based system so as to provide interactive feedback specific to learner responses.

Fundamental to this program is the individual awareness of preferred risk attitudes that serve as "risk interpreters" and "decision boundries" which result in behavioral outcomes. Managing change in individual perceptions in a process consultation format requires that individuals gain insight into their own biases that shape their perceptions, to value the perceptions of others as serving some role at the collective level, and to become self-aware of decision processes in future risk decisions. Collective decisions about risk could then be managed in a less polarized environment.

4. PROGRAM PRODUCTION CONSIDERATIONS

The development of the interactive programs has been performed using the "InterAct" authoring software. The design required a total of 7 lessons. Only three lessons are available at the main menu. This was necessary so that the learner would first be profiled as "thinking of feeling", then as "risk-averse or risk seeking", and then have access to the exercise lesson which acts as a sub-menu for the four lessons of availability, weighting, framing, and overconfidence. Only one password is made available in a feedback loop of the risk lesson.

A brief description of this program would be as follows: Individual preferences for thinking styles have been known for sometime. Individual preferences in decision making concerning probabilistic risk have recently been shown to be either of a risk aversion or risk seeking nature. Individual biases brought to the decision process can also be identified. If the individual is aware of these preferences and biases, perhaps a more considerate decision will be made concerning risk. An interactive computer based program has been developed to assess and to make the participant aware of results in regards to thinking style, risk preference, and probability biases. These individual results may then be structured for small group interactions as they relate to perceptions.

5. MAJOR COMPONENTS

Thinking Style

The purpose of assessing thinking style is to provide the participant data in a communications style that is most acceptable to the life experience of the individual. This should provide the participant a richer meaning in evaluating the program questions. The basis of the thinking style is the MBTI of Isabel Briggs Myers (1980). The "thinking" and "Feeling" styles will be used. This could also be thought of as left brain or right brain style.

This lesson is scored such that a 100% score in total feeling responses and a 25% score is total thinking responses. Scores that are close to the 60% level would indicate a balance response of 7 or 8 questions of the 15 questions as thinking or feeling.

Risk Preference

This section is based on the work of Daniel Kahneman and Amos Tversky. An article titled "The Psychology of Preferences", Scientific American, Vol 246, Number 1 best illustrates the preference. Through a series of situations and choices of comparative risk in the 10E-5 range, the participant will be assessed as risk-averse or risk-seeking. The participant will be made aware of this preference.

This section is scored such that 100% would be all risk-seeking responses and a score of 25% would be all risk-averse responses. Scores of about 62% would indicate a 5 or 6 question response of the 11 questions as seeking or adverse.

Risk Biases

Availability - this bias is described as our judging a probable occurrence based on our ability to recall or imagine the event. This bias will be explored by presenting choices to evoke the expected bias. The participant will be given feedback demonstrating the bias. Additional choices will be presented to permit practice in dealing with the bias.

Overconfidence - this bias is described as our belief in our judgments which are typically excessive in confidence. This bias will be explored with choices that evoke the overconfidence. The participant will be made aware of the overconfidence and given additional choices to practice responses. This lesson is not scored.

Weighting - this bias relates to our assigning value to the likely occurrence of a probability. Several situations will be presented for the participant to evoke the biases. Additional situations will be presented for the participant to exercise control of the bias. This lesson is not scored.

Framing - this bias relates to our selecting a base reference from which our decisions are made. The frame of reference can vary for most decisions but can greatly affect the outcome. The participant will be provided several situations to evoke this bias and given the same date with a different frame for comparison. This lesson is not scored.

Student Records and Feedback

All responses are recorded on the second disk drive under the last name of the student. All responses are also printed out if a printer is attached. A cumulative score is also given. Musical sounds provide feedback as do comments for each response. Hints have not bee provided for any questions. Time on and off station are also recorded.

6. THE BOTTOM LINE

The original purpose of this entire lesson series was to develop a computer based product that would play a role in managing change. Managing change, as it relates to probabilistic risk decision making at the individual and collective level, has now been shown to have many more considerations than the original design included. Allow me to share the bottom line.

Prescriptive vs Process Instructional Design

It became apparent in the scoring of the Decision and Risk lessons that both possible responses of "thinking, feeling" or "risk adverse, risk seeking" were important as a balance to collective decision making. Was the purpose of this series to PRESCRIBE a desired outcome - a selected style as being correct? In this case, since each lesson was designed to identify a style of decision making or risk taking, the judgment was only about an identify and not a response to a preferred style. However, the original design was going to attempt to predict the learner response in each Exercise lesson (the four biases lessons) taken by the learner based on their identity in the decision and risk lessons. How well could I predict responses in the different biases lessons if I knew the learner's preferences for decision and risk style? I found this to place me in the position of judging the learner response as correct or incorrect according to my concepts. I found it to say that I knew the correct response to any question, regardless of the perception and interpretation of the learner. How would the learner self-discover the interplay of the biases if I play "expert" and judge responses for them? How will the desired learning style of "internalizing" responses be facilitated?

A non-advocate, PROCESS form of instructional design would allow the learner to explore different biases without my judging and scoring. Preferences for decision and risk style are still developed, but for the purpose of creating an "integrity" for different perceptions. The learner can develop the meaning of responses to the lessons on biases. Experience to date indicates that most people find it interesting, but also want to take the test again to experiment with their own biases. In the field of artificial intelligence and expert systems, prescription vs process will be an issue.

As it turns out to date, learners seem to enjoy the identification of style in the decision and risk preferences and in general, do see value in preferences different from their own. They also enjoy "playing" the biases and experimenting. For example, one physicist answered all the questions the way he thought Jane Fonda would respond. He developed an opposite preference style, but indicated that there was some value in Jane Fonda's choices. The choice of PROCESS as a program design guide allows the user to explore the management of different preferences as a goal, rather than prescription of one preference. It also places the instructional designer, THE CHANGE AGENT, in a non-advocate position for dealing with the audience on polarized issues.

7. CALL FOR FURTHER EVALUATION

The author welcomes any organization interested in exploring this form of communications and its possible utility to risk perception and decision making. To use this software, one needs the IBM PC with 128K, color graphics card and a color monitor. Other options as a second disk drive will permit electronic recording keeping and a printer will also document learner results. The author anticipates the program being modified as users share their feedback. Information on the authoring system is also available from the author.

REFERENCES

Briggs Meyers, Isabel and Peter Briggs Meyer, "Gifts Differing", Consulting Psychologist Press, 1980.

Douglas, Mary and Aaron Wildavsky, "Risk and Culture", University of California Press, 1982.

Kahneman, Daniel, Paul Slovic and Amos Tversky, "Judgment Under Uncertainty" Heuristics and Biases", Cambridge University Press, 1982.

Kahneman, Daniel, and Amos Tversky, "The Psychology of Preferences", Scientific America, Vol 246, Number 1.

Knowles, Malcolm, "The Adult Learner: A Neglected Species", Guld Publishing, 1978.

Moore, Peter, G. "The Business of Risk", Cambridge University Press, 1983.

Schein, Edgar H., "Process Consultation: Its Role in Organization Development", Addison-Wesley Publishing Company, 1969.

STRESS FROM RISK UNCERTAINTIES

D.S. Mileti

Colorado State University
Fort Collins, Colorado 80523

ABSTRACT

The phenomenon of stress in disasters and emergencies has been examined in a variety of ways over several decades. Results are not cumulative, and a wide range of opinion exists as to the stress impacts of disasters and emergencies on psychological stress. At the same time, stress does have a variety of meanings. Stress--as fear and anxiety versus stress as resulting in illness--does manifest itself in disasters and emergencies. One key reason is the use of uncertain information in presentations to the public. Certain information--even about uncertain risk--can reduce stress. Stress due to the misuse of public information in emergencies can be reduced, and not all stress should be viewed as bad.

KEY WORDS: Stress, Disasters, Emergencies, Emergency Information, Illness

1. INTRODUCTION

Human reaction to events that involve risk and risk uncertainties has long been a topic for investigation. Although events examined have been of varied sorts, research into natural and technological disasters and emergencies provide a rich and varied history. Early efforts set an interdisciplinary tone and included work that had sociological (cf., Fritz, 1954; Killian, 1952; Carr, 1932; Prince, 1920), psychological (cf., Menninger, 1952; Prasad, 1935) and social-psychological (cf., Quarantelli, 1954; Eliot, 1932) orientations. Subsequent research continued to push in different theoretical directions (cf., Dynes, 1970, 1986; Dacy and Kunreuther, 1969; Crawshaw, 1963); and, currently, the diversity of research into disasters and emergencies flourishes. Despite this long tradition, some aspects of human reaction to risk and risk uncertainties in disasters are better understood and documented than are others. The effect of risk uncertainties on stress is positioned, perhaps, somewhere inbetween theoretical closure and ignorance. It is the purpose of this paper to attempt to sate the character of this relationship and offer some hypothesized directions for future research; attention is limited to disasters and emergencies. Moreover, focus is placed n the linkage between the sociological characteristics or risk which may be prime determinants of psychological effects, if and when the latter occur in disasters and emergencies.

2. THE PHENOMENON OF STRESS IN EMERGENCIES

A review of the findings and conclusions from work to investigate stress in disasters and emergencies reveals a wide range of differences of opinion about stress impacts. These can readily be grouped into three classes. First, there is a wide body of research that concludes that disasters and emergencies do not produce negative stress impacts; in fact, it is concluded that disasters and their associated risks serve to enhance group solidarity and individual stability. This conclusion was reached early-on by those who investigated emergencies and disasters (Fritz and Marks, 1954; Janis, 1951), continued as the field matured (Coleman, 1966; Wilson, 1962; Fritz, 1961) and still prevails into the more recent disaster research literature (Quarantelli, 1979; Quarantelli and Dynes, 1973). Second, there is an equally well-established set of findings which reach the opposite conclusion; namely, that disasters and emergencies tear the social and psychological fabrics of the human condition, elicit psychological stress in general, and can produce continuing long-term stress states in some members of a population leading to stress-induced disorders. This viewpoint was also reached early in investigations of disasters and emergencies (Wolfenstein, 1957; Wallace, 1956; Rosenman, 1956; Menninger, 1952; Tyhurst, 1951), continued as the field grew (Krystal, 1986; Lifton 1967; Farber, 1967; Crawshaw, 1963), and remains a conclusion for some who have done more recent investigations (Kasl et al., 1981a, 1981b; Houts et al., 1980a, 1980b; Gleser et al., 1981; Titchener and Kapp, 1976; Rengell, 1976; Lifton and Olson, 1976; Erikson, 1976). Third, there have also been some who have reached conclusions in the middle-ground. Disasters and emergencies do elicit stress in some people who experience them, but this stress is short-lived and does not result in psychological stress-induced disorders--although there are on occasion some rare exceptions. This conclusion is also to be found in the early literature on disaster and emergency research (Moore et al., 1963; Bates et al., 1963; Form and Nosow, 1958; Fritz and Marks, 1954; Janis, 1951), in the second generation of disaster research (Hall and Landreth, 1975; Drabek et al., 1973; Drabek and stephenson, 1971) and in more contemporary studies (Mileti et al., 1984; Bromet et al., 1982; Bromet, 1981, 1980; Dohrenwend et al., 1981, 1979). Obviously, there is a wide range of difference of opinion in the research literature on the stress impacts of disasters and emergencies.

There is a good reason why these different conclusion exist. Put simply, different researchers have gone about defining, operationalizing and measuring the concepts of disaster and stress very differently from each other. Disaster has meant atomic bomb explosions to some, while to others it was taken as a localized tornado disaster. Few researchers have sought to decompose "disaster" into a theoretically meaningful typology of elements, for example, level of physical destruction, uncertainty in emergency information, and so on. At the same time, researchers have measured stress in different ways; some have equated it with evacuation, others asked people to estimate it themselves, still others used behavioral measures of stress manifestations. An obvious need is to seek theoretical refinement before research findings can be rendered consistent.

3. A WORKING DEFINITION OF STRESS

In general, and for the purposes of this paper, stress can be conceptualized as some altered state in a person stemming from changes in the environment (social, economic, cultural, physical, and so forth) which, if not adjusted for, can lead to negative health (physical or mental) consequences. Cannon (1928) first explored the relationship between psychological states, for example, anxiety, fear and anger, and illness; and Meyer (1951) elaborated that work to illustrate the relationship between life events and mental health. Work by Selye (1956, 1950) and Wolff (1950), in addition to Hinkle (1974, 1957) led to understanding of the links between stress, physiological adaptations to stress and illness. These works and others, suggest that stress can be adapted to, mitigated, or lead to illness. Others (Brown, 1973; Dohrenwend, 1973, 1970; Myers et al., 1972; Paykel et al., 1972, 1971; Rahe and Holmes, 1968, 1967) have also sought to specify the link between stressful life events and both physical and mental health. The conclusion of a reading of these works, and others, suggests that stress is an altered state in an individual which need not be negative, but which could be if continued for long periods of time. As such, it seems only useful to view stress in two different but related ways. First, stress can be seen as an altered emotional state manifest in factors such as fear, anxiety, concern, worry, excitement and the like. Viewed in this way, stress is not necessarily a negative concept in emergencies and disasters. States like those catalogued above are, perhaps, necessary to elicit some emergency actions, for example, evacuation. Second, however, is an alternative view of stress. This view would see the consequences of stress for some in the long-term as negative health effects. It is beyond the scope of this paper to address how or why the former type of stress may result in the latter. What is relevant here, however, is to be reminded that stress in emergencies can be viewed in two ways. These are negative health effects, and fear which may be either negative or positive depending on how it is viewed and the consequences that it may have for the person who possesses it.

4. RISK UNCERTAINTY AND OTHER DETERMINANTS OF STRESS IN EMERGENCIES AND DISASTERS

Although research in this area is far from theoretical closure in an academic sense, a good deal of evidence has been accumulated to suggest that stress--defined as anxiety, fear, motivation and so forth--in disasters and emergencies is largely the result of disaster or emergency-specific situational factors. Of these, the aspect of the situation with the greatest impact on stress is the emergency information which is made public (cf., Mileti, 1975a, 1975b). This is the case because information is the key determinant of how people form situation perceptions of the emergency.

Several aspects of emergency information that effect the formation of situation perceptions, and which can result in stress, follow. These are: (1) consistency of emergency information--inconsistency in the tone or content of emergency information creates confusion, uncertainty and stress (Mileti, 1975a, 1975b; Fritz, 1957; Clifford, 1956); (2) accuracy of emergency information--information must be presented in such a way that people do not suspect that they are not receiving the whole truth (Mileti, 1975a); (3) certainty of emergency information--information must be certain as to the events that are taking place and what people should do; even if there is a low-probability or ambiguous situation, the emergency information should be stated with certainty--even about ambiguity (Mileti et al, 1981); (4) sufficiency of information--enough must be said so that the public has a good idea of what is happening; not knowing creates anxiety (Petty and Greene, 1983; Leik et al., 1981); (5) frequency of information--information that is confirmed can reduce anxiety (Mileti, et al., 1975); (6) specificity of location in information--public emergency information should clearly state the areas at risk to avoid confusion and anxiety (Perry and Greene, 1983; Diggory, 1956).

The over-riding conclusion of research into disasters and emergencies is that uncertainty about risk does produce stress in the population at risk. At the same time, objective uncertainties are often high in emergencies; however, what is important is not objective uncertainty but rather how uncertainty is perceived. Emergency information which is certain about uncertainty, who should do what, and confirmatory will likely elicit less negative psychological stress than information which is of a different character. Risk uncertainty, therefore, can be a source of stress in emergencies if it is not well managed in terms of how it is presented to the public.

Obviously, other factors can lead to stress in emergencies beyond risk uncertainty. Stress resulting from risk uncertainty can be mitigated through how public information abut the risk is presented to the public. Risk uncertainty from emergencies and disasters when it results in stress, likely does not lead to illness; however, risk uncertainty that continues across time for prolonged periods could, for some unique individuals, result in medically diagnosible illness.

REFERENCES

Bates, F.L., et al., 1963. The Social and Psychological Consequences of a Natural Disaster. Washington D.C.: National Academy of Sciences, National Research Council.

Bromet, E., 1980. Three Mile Island: Mental Health Findings. Pittsburgh: University of Pittsburgh and Western Psychiatric Institute.

Bromet, E., and L. Dunn, 1981. "Mental Health of Mothers Nine Months after the Three Mile Island Accident," Urban and Social Change Review, 14: 12-14.

Bromet, E., H. Schulberg and L. Dunn, 1981. "Reactions of Psychiatric Patients to the Three Mile Island Nuclear Accident," Archives of General Psychiatry, 39 (June): 725-730.

Brown, G.W., T.O. Harris, and J. Peto, 1983. "Life Events and Psychiatric Disorders. Part 2: Nature of the Causal Link," Psychological Medicine, 3: 159-176.

Cannon, W.B., 1928. "The Mechanism of Emotional Disturbance of Body Function," New England Journal of Medicine, 198: 877-884.

Carr, L.J., 1932. "Disaster and the Sequence-pattern Concept of Social Change," The American Journal of Sociology, 38: 207-218.

Clifford, R.A., 1956. The Rio Grand Flood: A Comparative Study of Border Communities. Washington, D.C.: National Academy of Sciences, National Research Council.

Coleman, J, 1966. "Community Disorganization." In R. Merton and R. Nisbet (Eds.) Contemporary Social Problems. New York: Harcourt, Brace and World.

Crawshaw, R., 1963. "Reactions to Disaster," Archives of General Psychiatry, 9: 157-162.

Dacy, D.C., and H. Kunreuther, 1969. The Economics of Natural Disaster. New York: The Free Press.

Diggory, J.C., 1956. "Some Consequences of Proximity to a Disease Threat," Sociometry, 19 (March): 47-53.

Dohrenwend, B.S., 1973. "Social Status and Stressful Life Events," Journal of Personality and Social Psychology, 28: 225-235.

Dohrenwend, B.S., and B.P. Dohrenwend, 1970. "Class and Race as Status-related Sources of Stress." In S. Levine and N. Scotch (Eds.) Social Stress. Chicago: Aldine.

Dohrenwend, B.P. et al., 1981. "Stress in the Community: A Report to the President's Commission on the Accident at Three Mile Island," Annals of the New York Academy of Sciences, 365 (April): 159-174.

Dohrenwend, B.P., et al., 1979. Report of the Task Force on Behavioral

Effects of the President's Commission on the Accident at Three Mile Island.
Drabek, T.E., et al., 1973. Longitudinal Impact of Disaster on Family Functioning. Denver: University of Denver.
Drabek, T.E., and J. Stephenson, 1971. "When Disaster Strikes," Journal of Applied Social Psychology, 1: 187-203.
Dynes, R., 1970. Organized Behavior in Disaster. Lexington, Mass.: D.C. Heath.
Dynes, R., and E.L. Quarantelli, 1968. "Group Behavior under Stress," Sociology and Social Research 52 (July): 416-429.
Eliot, T.D., 1932. "The Bereaved Family," The Annals of the American Academy of Political and Social Science, (March): 1-7.
Erikson, K., 1976. "Loss of Community at Buffalo Creek," American Journal of Psychiatry, 133 (March): 302-305.
Farber, I., 1967. "Psychological Aspects of Mass Disasters," Journal of the National Medical Association, 59: 340-345.
Form, W. and S. Nosow, 1958. Community in Disaster. New York: Harper and Row.
Fritz, C., 1961. "Disasters". In R. Merton and R. Nisbet (Eds.), Contemporary Social Problems. New York: Harcourt, Brace and World.
Fritz, C., 1957. "Disasters Compared in Six American Communities," Human Organization, 16 (Summer): 6-9.
Fritz, C., 1954. "The NORC Studies of Human Behavior in Disaster," The Journal of Social Issues, 10(3): 26-41.
Gleser, G., B. Green and C. Winget, 1981. Prolonged Psychosocial Effects of Disaster. New York: Academic Press.
Hall, P. and P. Landreth, 1975. "Assessing some Long-term Consequences of a Natural Disaster," Mass Emergencies, 1: 55-61.
Hinkle, L.E., 1974. "The Concept of Stress in the Biological and Social Sciences," International Journal of Psychiatry in Medicine, 15: 335-357.
Hinkle, L.E., and H.G. Wolff, 1957. "Health and the Social Environment." In Leighton, Clausen and Wilson (Eds.) Experimental Investigations: Explorations in Social Psychiatry. New York: Basic Books.
Houts, P., et al., 1980a. Health-related Behavioral Impact of the Three Mile Island Nuclear Incident. Part I. Hershey, Pennsylvania: Report Submitted to the TMI Advisory Panel on Health Research Studies of the Pennsylvania Department of Health.
Houts, P., et al., 1980b. Health-related Behavioral Impact of the Three Mile Island Nuclear Incident, Part II. Hershey, Pennsylvania: Report Submitted to the TMI Advisory Panel on Health Research Studies of the Pennsylvania Department of Health.
Kasl, S.V., et al., 1981a. "The Impact of the Accident at the Three Mile Island on the Behavior and Well-being of Nuclear Workers: Part I: Perceptions and evaluations, Behavioral Response and Work Related Attitudes and Feelings," American Journal of Public Health, 71(5): 472-483.
Kasl, S.V., et al., 1981b. "The Impact of the Accident at the Three Mile Island on the Behavior and Well-being of Nuclear Workers: Part II: Job Tension, Psychophysiological Symptoms, and Indices of Distress," American Journal of Public Health 71(5): 484-495.
Killian, L.M., 1952. "The Significance of Multi-Group Membership in Disaster," American Journal of sociology, LVII(4): 309-314.
Krystal, H. (Ed.), 1968. Massive Psychic Trauma. New York: International University Press.
Leik, R., et al., 1982. "Community Response to Natural Hazard Warnings," Unpublished Report. Department of Sociology, University of Minnesota.
Lifton, R., 1967. Death in Life: Survivors of Hiroshima. New York: Random House.

Lifton, R., and E. Olson, 1976. "The Human Meaning of Total Disaster: The Buffalo Creek Experience," Psychiatry, 39: 1-18.
Menninger, W.C., 1952. "Psychological Reactions in an Emergency Flood," American journal of Psychiatry, 109: 128-130.
Meyer, A., 1951. "The Life Chart and the Obligation of Specifying Positive Data in Psychopathological Diagnosis." In E.E. Winter (Ed.) The Collected Papers of Adolf Meyer, Volume III: Medical Teaching. Baltimore: Johns Hopkins Press.
Mileti, D.S., et al., 1984. "The Three Mile Island Incident: A Study of Behavioral Indicators of Human Stress," Mass Emergencies and Disasters 2(1).
Mileti, D.S., J. Hutton and J. Sorensen, 1981. Earthquake Prediction Response and Options for Public Policy. Boulder, Colorado: Institute of Behavioral Science.
Mileti, D.S., T.E. Drabek, and J.E. Haas, 1975. Human Systems in Extreme Environments. Boulder, Colorado: Institute of Behavioral Science.
Mileti, D.S., 1975. Natural Hazard Warning Systems in the United States. Boulder, Colorado: Institute of Behavioral Science.
Moore, H.E., et al., 1963. Before the Wind: A Study of Response to Hurrican Carla. Washington, D.C.,: National Academy of Science, National Research Council.
Paykel, E.S., and Uhlenbuth, E.H., 1972. "Rating and Magnitude of Life Stress," Canadian Psychiatric Association Journal, special supplement II: 93-100.
Paykel, E.S., et al., 1971. "Scaling of Life Events," Archives of General Psychiatry, 25: 340-347.
Perry, R., and M. Greene, 1983. Citizen Response to Volcanic Eruptions. New York: Irvington Press.
Prasad, J., 1935. "The Psychology of Rumor," British Journal of Psychology, 26 (July).
Prince, S.H., 1920. "Catastrophe and Social Change, Based Upon a Sociological Study of the Halifax Disaster." Ph.D. Thesis. New York: Department of Sociology, Columbia University.
Quarantelli, E.L., 1979. The Consequences of disasters for Mental Health: Conflicting Views. Columbus, Ohio: Preliminary Paper 62, the Disaster Research Center, the Ohio State University.
Quarantelli, E.L., 1954. "The Nature and Conditions of Panic," American Journal of Sociology, 60: 267-275.
Quarantelli, E.L., and R. Dynes, 1973. "When Disaster Strikes," New Society, 1-4:5-9.
Rengell, L., 1976. "Discussion of the Buffalo Creek Disaster: The Course of Psychic Trauma," American Journa of Psychiatry, 133 (March): 313-317.
Rosenman, S., 1956. "The Paradox of Guilt in Disaster Victim Populations," Psychiatric Quarterly Supplement, 30: 181-221.
Selye, H., 1956. The Stress of Life. New York: McGraw-Hill.
Selye, H., 1950. The Physiology and Pathology of Exposure to Stress. Montreal: Acta.
Titchener, J., and F. Kapp, 1976. "Family and Character Change at Buffalo Creek," American Journal of Psychiatry, 133 (March): 295-299.
Tyhurst, J., 1951. "Individual Reactions to a Community Disaster," American Journal of Psychiatry, 107: 764-769.
Wallace, A., 1956. Tornado in Worcester. Washington, D.C.: National Academy of Sciences, National Research Council.
Wilson, R., 1962. "Disaster and Mental Health." In G. Baker and D. Chapman (Eds.) Man and Society in Disaster. New York: Basic Books.
Wolfenstein, M., 1957. Disaster: A Psychological Essay. Glencoe, Illinois: The Free Press.

THE INDUSTRY - REGULATORY INTERACTIVE PROCESS FOR RESOLUTION OF SEVERE ACCIDENT QUESTIONS FOR NUCLEAR POWER

Anthony R. Buhl

International Technology Corporation
Knoxville, TN 37922

ABSTRACT

This presentation briefly reviews the historical views of the industry and regulators toward the risks associated with potential severe accidents in nuclear power plants and focuses on actions and changes since the accident at Three Mile Island in 1979. Primary emphasis is placed on the intense interactions between the Nuclear Regulatory Commission and the Industry Degraded Core Rulemaking (IDCOR) Program since 1981.

KEY WORDS: Industry Degraded Core Rulemaking (IDCOR) Program; Severe Accidents; Nuclear Safety

> Lightning Rods:
> "The more points of iron are erected around the earth, to draw electrical substance out of the air, the more the earth needs must be charged with it. And therefore, it seems worthy of consideration, whether any part of the earth being fuller of this terrible substance may not be more exposed to more shocking earthquakes. In Boston are more erected than anywhere else in New England; and Boston seems to be more dreadfully shaken. Oh! There is no getting out of the mighty hand of God! If we think to avoid it in the air, we cannot in the earth: Yet it may grow more fatal: and there is no safety anywhere..."

This quote is from a sermon the right Reverend Thomas Prince delivered in the old South Boston church in 1751 against Benjamin Franklin's use of the lightning rod - the first intervenor against electricity! We have always found energy debates wrought with extreme views. It is easy to be emotional or angry about energy issues, particularly nuclear power. What we need is technical truth - not emotion.

NUCLEAR POWERED ELECTRICITY GENERATION PLANTS IN THE UNITED STATES ARE SAFE!

This major milestone in understanding nuclear safety has been achieved by the IDCOR program. IDCOR--an acronym for Industry Degraded Core Rulemaking program--is sponsored by over 60 electric utility companies, major manufacturers and engineering firms in the United States as well as by Germany, Finland, Japan and Sweden.

IDCOR has conpleted the most intensive evaluation of nuclear power safety ever attempted in a single program. Spanning more than 3 years and supported by 23 contractors and dozens of scientists and engineers, IDCOR's findings are being released to the U.S. Nuclear Regulatory Commission (NRC) five years after the accident at Three Mile Island and a decade after publication of WASH-1400, the Reactor Safety Study. WASH-1400 was a landmark publication of the first integrated quantitative risk assessment of operating nuclear power plants. IDCOR has found that nuclear power is much safer than experts believed five years ago.

The exciting results of IDCOR are best understood in the context of history. Nuclear safety grew up in the U.S. under an umbrella called "Design basis accidents" (DBA). The introduction of enlightened thinking through risk analysis 10 years ago was received by regulators as a Samaritan in Judea. The watershed event at TMI in 1979 caused a rethinking of safety views and lead to considerable efforts to understand nuclear safety, including the IDCOR program. IDCOR was developed to provide a clear technical understanding for all of the technical issues related to potential severe accidents in nuclear power plants. IDCOR was the nuclear industries' response to an official call by the U.S. NRC for a rule making on severe accidents.

This paper proceeds as follows:

1. History - the umbrella that leaked;
2. Risk Analysis - a technology mistrusted;
3. TMI - a plant that broke;
4. IDCOR - the new approach; and
5. NRC - severe accident policy improves

HISTORY - THE UMBRELLA THAT LEAKED

Following World War II, fission power was first used for ship propulsion and then to provide steam for producing electricity. A troubling preoccupation with nuclear safety developed. In 1956, William F. Libby, the acting Chairman of the U.S. Atomic Energy Commission (AEC) said, "Regardless of location and isolation, there is no such thing as an absolutely safe reactor--just as there is no such thing as an absolutely safe chemical plant or refinery."

Early attempts to define the risks of nuclear power failed. WASH-740 did more to confuse reactor safety understanding and mislead regulation than perhaps any other single document. WASH-740 drew national attention to the possibility of thousands of deaths, injuries and billions of dollars lost from a single accident. The study was based on small reactors at that, and both probabilities and consequences of accidents were misstated.

The excitement caused by WASH-740 and other pronouncements led to the design basis accident approach to place a ceiling over commercial reactor accidents. The idea was to select big accidents which would "UMBRELLA" the small ones and use these to design the plants.

A reasonable philosophy of "prevent the accidents but if they occur contain them" evolved using DBAs as the test for the latter. The DBA was defined in 1961 in 10CFR100 as a double-ended rupture of the large primary coolant piping--the umbrella approach lasted until the TMI accident in 1979.

RISK ANALYSIS - A TECHNOLOGY MISTRUSTED

The "rush to megawatts" in the 1960s became the albatross of the nuclear industry-generating both reliability and safety problems: (a) the residual "decay" heat after reactor shutdown became too great for passive containments to handle, (b) the Ergen report gave us the "China Syndrome" fear, and (c) emergency core cooling systems were litigated at great length in public hearings.

About this time in 1973, the U.S. AEC commissioned the Reactor Safety Study--WASH-1400. WASH-1400 calculated the expected frequency of severe accidents and levels of consequences. This work included cataloging accident sequences, analyzing physical processes which could lead to radiological releases and determining potential offsite consequences.

WASH-1400 said:

-- pre-occupation with large pipe breaks is ill-advised--the umbrella regulatory approach should be examined.
-- 98% of all coremelt accidents will result in little or no injury to the public.
-- transient events like loss of offsite electrical power and small pipe breaks dominate public risk along with human error.
-- the tight containment prescribed by the umbrella regulations can be completely bypassed in several potential accidents.

An international controversy over nuclear safety followed WASH-1400 publication and lead to nothing--the essential insights from WASH-1400 were largely ignored from 1974 until the accident at TMI. The NRC itself generally misunderstood the value of risk analysis and WASH-1400. The Commissioners under Congressional pressure appointed a "panel of experts" to study the matter--the so-called "Lewis Committee".

In 1977, this committee suggested that the NRC redirect its regulatory emphasis away from the umbrella approach and look at the true contributors to risk

--transient events,
--small pipe break,
--human error.

The message was never understood by the Commissioners until TMI rehabilitated risk assessment.

Indeed the Commissioners responded in 1978 and 1979 by withdrawing the Executive Summary of WASH-1400 and misdirecting the NRC staff to be wary of risk evaluations.

TMI - A Plant That Broke

On March 28, 1979, nuclear safety entered a new era--after the accident at TMI, technical understanding became more important than gamesmanship or politics. The advantages of risk analysis were finally realized by parts of the NRC. The need for a more realistic understanding of accident phenomena and plant behavior under severe accident conditions was clear.

IDCOR - The New Approach

In light of the TMI-2 accident, the NRC issued an advanced notice of rulemaking (ANR) on severe accidents in October 1980. The thesis of this pronouncement was to determine whether additional large design changes or regulation changes were needed to protect the public should another severe accident occur. For example, the ANR focused on strange devices like "CORE CATCHERS". IDCOR was the industry's response to develop a technically sound, well-documented basis for industry participation in the decision process. Technology for Energy Corporation was selected to manage the IDCOR program.

The potential threat to the public or risk is dominated by four factors:

1. The probability of reactor core damage.
2. The capability of the containment.
3. The radioactive fission products which could be released should a severe accident occur.
4. The consequences of the loss of containment in case of an accident.

IDCOR found that the probability of core damage is very low--in agreement with WASH-1400 and several other more recent studies. The major enhancement in understanding reactor safety by IDCOR results came from discovering that containments are far more effective than previously thought to be and from a deeper understanding of the chemistry and behavior of fission products.

Further, IDCOR added several new insights about the physics and chemistry of accident progression and subsequent effects. Perhaps the most important factor is that the amount of radioactive material available for release during severe accidents is much smaller than earlier research had suggested.

Accident and Core Damage Prevention

IDCOR found that additional hardware would do little to reduce the frequency of accidents. IDCOR did find that alternative uses of existing in-plant systems could be beneficial.

Severe Accident Phenomenology

IDCOR evaluated existing technical data and work from industry and government in the U.S. and several other countries. IDCOR developed additional data and computational models for understanding the physical and chemical processes leading to a fission product source term. The important conclusions are:

1. The necessary conditions do not exist to allow steam explosions to cause primary system or containment failure.
2. Hydrogen detonation on a scale which challenges containment functions is not likely (Negligible in probability).
3. Containment will function effectively for many hours following most accidents and release very little radioactivity.
4. The China Syndrome is a myth--core debris, even if outside the reactor, will not penetrate the floor of containment if water is present, as is usually the case.

Real Plant Evaluations

IDCOR used the new phenomena insights, new data and computational tools to evaluate real plants representative of the existing reactor and containment types.

--Zion - a Westinghouse PWR with a large dry containment
--Sequoyah - a Westinghouse PWR with an ice condenser containment
--Peach Bottom 2 - a General Electric BWR with a Mark I pressure suppression containment

--Grand Gulf - a General Electric BWR with a Mark III pressure suppression containment

IDCOR also developed these tools for Oconee, a B&W plant; Susquehanna, a GE Mark II plant and Calvert Cliffs, a Combustion Engineering plant.

IDCOR included evaluations of operation actions to terminate accidents as well as human errors which could worsen accidents.

CONCLUSIONS

IDCOR showed that major new design features are not needed to assure public safety. Equally important, IDCOR found that earlier estimates of fission product release to the environment had been greatly over stated by at least a factor of 10 and, in many accidents, much more.

NRC - Severe Accident Improves

These new technical results are being analyzed by the NRC in a careful and intensive interaction process. The data suggests that not only are existing plants safe enough but that, indeed, many current licensing requirements could be reduced.

The adverse effects of nuclear accidents on the public are greatly reduced from what the technical community and regulators thought they were a few years ago.

There is presently a severe accident policy paper before the U.S. NRC Commissioners which has been reviewed by the public and which has passed muster of both the NRC staff and the Advisory Committee on Reactor Safeguards. The existing NRC ruling paper states:

> On the basis of currently available information, the Commission concludes that existing plants pose no undue risk to public health, safety, and property and dismisses therefore the need for prompt action on generic rulemaking or other regulatory changes for these plants. However, the experience of NRC and the nuclear industry with plant-specific probabilistic risk assessments is that each of these analyses, which provide a more detailed assessment of possible accident scenarios, has exposed relatively unique vulnerabilities to severe accidents. Generally, the undesirable risk from these unique features has been reduced to an acceptable level by low-cost changes in procedures or minor design modifications. Accordingly, when NRC and industry interactions on severe accident issues have progressed sufficiently to define the methods of analysis, the Commission plans to formulate an integrated systematic approach to an examination of each nuclear power plant now operating or under construction for possible significant risk contributors (sometimes called "outliers") that might be plant specific and might be missed absent a systematic search. Following the development of such an approach, an analysis will be made of any plant that has not yet undergone an appropriate examination. The examination will include specific attention to containment performance in striking a balance between accident prevention and consequence mitigation.

THE BUREAUCRATIC DIMENSION TO RISK ANALYSIS: THE ULTIMATE UNCERTAINTY

J. Ward Wright

Eastern Kentucky University
Richmond, Kentucky 40475

ABSTRACT

This paper examines how bureaucracies respond to hazardous waste emergencies. In particular, the research has focused on the use of risk analysis techniques in the decision-making process as these techniques are affected by emotional public reactions to these emergencies. The study shows there is a direct relationship between the uncertain seriousness of the hazardous waste threat, the level of public emotional reaction, and the lack of rationality in the bureaucratic response. Hazardous waste threats are a special problem for state and local governments because of the difficult technical analytical problems they pose, the unknown (if not unknowable) nature and extent of the danger threatened, the willingness of the media to highlight governmental indecision and scientific speculation, and the extreme fear of the public ostensibly threatened. This study is based on the results of a 2-year study of hazardous waste decision-making processes in the state governments of Taxes and Maryland, and preliminary findings are presented.

KEY WORDS: Hazardous Wastes; Bureaucracy; Risk Assessment; Risk Management; Decision Making; Dumps.

INTRODUCTION

The risk analysis literature is strangely silent on one of the most important contextual factors in decision making: the bureaucratic setting in which these activities take place. Virtually all phases of risk management are undertaken by bureaucrats, and even the "scientists" engaged in risk assessment are not immune from the influence of bureaucratic rewards and penalties. My remarks are concerned with how these bureaucratic factors affect the manner in which risk assessment and management are undertaken.

The materials discussed here are based on a two-year study, funded by the National Science Foundation, of how decisions are made in state governments in dealing with hazardous wastes. The research base was four case studies, two each in siting hazardous waste facilities and coping with dumps. Two of the studies were undertaken in Texas and two in Maryland. In each case, an effort was made to trace the decision making process from the first person involved with the matter (e.g., a field inspector discovering the dump site) to the ultimate resolution of the problem, to the extent that there was a resolution. Throughout this process, the emphasis was on identifying the principal factors influencing the decision making process.

THE BUREAUCRATIC FACTOR

At the bureaucratic levels, both Texas and Maryland can accurately be described as progressive and professional operations. Nonetheless, they are traditional bureaucratic operations, though neither is as administratively hidebound as federal agencies.

While anyone who has worked extensively with a large, complex organization, public or private, has experience with what I refer to as the "bureaucratic factor," he or she does not always reflect on its characteristics and its implications for the manner in which people in these organizations undertake their work. We must ask, "What is meant by bureaucracy?"

Bureaucracies have been with humanity since the dawn of recorded history; in fact, they were largely responsible for much of what was recorded. However, in their modern forms, Max Weber's description of more than sixty years ago still serves us well. I will summarize briefly the pertinent points from his classic essay.

Bureaucracies are created to "provide specialized administrative functions according to objective considerations." Business is undertaken according to calculable rules "without regard for persons." Within the organization, there is functional specialization of work, and every effort is undertaken to make each employee the master of "single yet methodocially integrated functions." To attain this, bureaucratic officeholders are selected on the basis of appropriate training. The expectations of the bureaucrat include lifetime tenure within clearly delineated career paths, together with "mechanical fixing of conditions of promotion."

From an overall organizational standpoint, there is a "specific development of administrative tasks" within a hierarchical, top-down system of control and direction. Strict adherence to prescribed procedures is sought of all participants. Authority is "strictly delimited" within the narrowly defined rules which articulate each specialized job. This is referred to by Weber as the "reduction of modern office management to rules..."[1]

The descriptive literature in this field can, in part, be summarized by these four characterizations: (a) anonymity, Weber refers to the "principled separation of the private sphere of the official from that of the office;" accordingly, the appointed bureaucratic official is expected to abjure any personal proclivities for public recognition; (b) discipline, Weber refers to the "official's integration into the given functional conditions of a fixed mechanism based upon discipline;" (c) records, bureaucracies undertake management by written documents, and Weber refers to this process in terms of everything going to "the files;" and (d) tradition, the conservative, relatively changeless nature of bureaucratic activity is well recognized.

---

[1]Gerth and Mills, eds., From Max Weber: Essays in Sociology (Oxford University Press, New York, 1966), pp. 196-244.

Two of these features deserve special comment. Victor Thompson addresses the following especially pertinent observation with respect to "records": "[F]ailure for the modern bureaucrat means the loss of part of his identify. A report of his failure goes into his file -- his paper identify, a paper alter ego that follows him inescapably through life -- and alters his identify unfavorably. Innovation is more risky for the bureaucrat than for the entrepreneur."[2] The threatening nature of innovative change also relates to the foregoing comment on "tradition." Here again, Thompson's observations are enlightening: "The bureaucratic orientation is conservative. Novel solutions using resources in a new way, are likely to appear threatening. The bureaucratic orientation is politically minded. It is more concerned with the internal distribution of power and status than with the accomplishment of the organization's goals."[3]

These are the generally recognized characteristics of bureaucracies which I found in practice to be most relevant to decision making in the hazardous waste field. The results of these characteristics of bureaucracy are organizations which emphasize the routine performance of carefully prescribed functions, and each agency has its specialized role. It is an arrangement in which interdisciplinary or interfunctional responses to problems either do not occur or are matters of high level negotiation. By contrast, the industries producing hazardous wastes seize any opportunity to dispose of their wastes as cheaply as possible. As a consequence, tens of thousands of toxic dumps have been created, each of which potentially poses health and safety threats to the public which require the coordinated attention of a broad range of specialists. The need for biologists, chemists, hydrologists, geologists, epidemiologists, and a variety of engineering specialists by no means exhausts the possibilities that could be required to analyze even one dump. These needs are obvious when one refers to EPA's eight general categories of "perilous residues": radioactivity, infectiousness, phytotoxicity (plant life), teratogenicity-mutagenicity (genetic abberations), ignitability, corrosivity, reactivity, and toxicity. Clearly, few modern threats to public health and safety require more tightly integrated, interdisciplinary responses by highly qualified specialists if they are to be assessed properly and cleaned up effectively.

Does this mean that traditional, specialized bureaucratic agencies cannot cope with this problem? It does not. however, fundamentally new institutional arrangements are required. One such new arrangement will be discussed as the conclusion to this presentation.

RISK MANAGEMENT

While the new discipline of risk analysis has been applied principally to the development and application of federal regulatory policies, especially with respect to the assessment of proposed new drugs, it is a methodology which is also relevant to hazardous waste decision making. Faced as they are with more existing sites than they have resources to handle, many industrial states desperately need effective

---

[2]Victor A. Thompson, Bureaucracy and Innovation (University of Ala. Press, University, Alabama, 1976), pp. 4-5.

[3]*Ibid.*, p. 22.

methods for the identification of the most serious situations, the evaluation of the degrees and types of threats (e.g., environmental, personal) each dump poses, and the delineation of possible ameliorative actions that could be taken with the limited resources available. These are the types of activities commonly associated with the term "risk management."

Writers in the field of risk analysis call for a careful delineation of the following roles: (a) scientific, the process of undertaking a rigorous analysis and assessment of the dangers posed by a particular situation; and (b) political, the process of evaluating the results of this analysis and exercising the regulatory and corrective options available and affordable. For the purposes of this presentation, the first of these steps is referred to as "risk assessment," and the second as "risk management."

A model for interrelating these functions has been set forth by Doderlein: the policy makers set the risk objectives; the scientists gather facts, identify risks, or the spectra of risks, and develop methodologies for contending with these risks; and the policy makers weigh the results of this assessment and decide on the means for implementation of a response.[4] This logical approach is basically the model suggested by this presentation; however, its application is complicated at the outset in the public sector by problems associated with establishing "risk objectives." The objective is safety. Lowrance makes the pertinent observation that one cannot measure "safety." A situation can be deemed "safe" only if the associated risks are acceptable.[5] Anyone who has participated in public meetings concerned with hazardous wastes, or who has merely read about them, is well aware that the only "acceptable" condition for the public is "zero risk." since this level of risk is an impossibility, public policy makers are faced with the type of quandary that often, if not usually, leads to administrative torpor. This becomes obvious as we look at some of the results of the case studies.

CASE STUDY FINDINGS

In both of the states studied, one case concerned the authorization of a hazardous waste site and the other concerned coping with a hazardous waste dump. With respect to the siting cases, like virtually all other hazardous waste situations, the problem was one of finding a location with respect to which public opposition would not be effectively overwhelming. The NIMBY (not-in-my-backyard) syndrome was abundantly present in both; however, in both cases, for different reasons, the states were able to establish the sites.

---

[4]Jan M. Doderlein, "Understanding Risk Management," Risk Analysis (March, 1983), p. 2.

[5]William W. Lowrance, of Acceptable Risk, Science and the Determination of Safety (William Kaufman, Inc., Los Angeles, California, 1976), p. 8.

Three major problems arise here that are also present, though in somewhat different form, in the emergency response cases. In the first place, the presence of an authorized dump almost inevitably does, in fact, impact seriously and adversely on property values. A second problem concerns the much less explored result of emotional and psychological stress on residents living in the vicinity of such dumps. Both of these problems stem, however, from a third problem; namely, the paucity of reliable scientific evidence that can guarantee that any hazardous wastes in the ground can be deemed safe in the long run regardless of the techniques utilized.

Throughout the study, the shortcomings in available scientific methods and knowledge plagued the policy makers seeking to do the right thing. At permit hearings, neither those in favor of the site nor those opposed seemed to have difficulty in producing "experts" testifying to the safety or lack of safety of the proposed facility. Without belaboring this point, it is clear that safety cannot be guaranteed, and a public that seeks zero risk finds this situation unacceptable.

Nonetheless, both states proceeded with these disposal facilities. In Texas, a deep well injection system was proposed. Texas uses these wells extensively to deal with its huge waste problem. The technology is well-established, and the permitting process is a quasi-judicial procedure that extends over along period of time and is costly to contend with. After initial public arousal, the procedure ground on, and the well was approved. In Maryland, the State decided to acquire and operate a facility extending out into the middle of Baltimore Harbor, a site distant from any effective residential protest. This too succeeded.

The stories with respect to emergency responses, however, provided a different picture of bureaucratic complexity. When considering these types of cases, it is vitally important to recognize that state legislatures have not been willing to appropriate the funds required simply to move in and clean up dumps as fast as they are found. In fact, they are not even willing to allocate the funds required for detailed, in-depth assessments of suspected sites. Nor will they provide the funds for the types of laboratories and technicians that would be required by a comprehensive program. In both cases studied, until EPA authorized "Superfund" money to analyze and clean up the dumps, the government's efforts to contend with the situations were not effective.

The Resource Conservation and Recovery Act (RCRA) of 1976 provided federal assistance to states willing to adopt legislation conforming to RCRA provisions and associated EPA regulations. This law had a very interesting twist which has stirred most states to act in response to it. It provides that if a state government does not enforce RCRA, EPA will. Faced with this alternative, it appears that most hazardous waste generators, for whatever reasons, would prefer to deal with their state agencies rather than the minions of EPA. Accordingly, an increasing number of states have responded with new programs tailored to RCRA requirements. Both Texas and Maryland were among the early leaders in this respect, with Texas, in fact, being the first state fully to conform.

The result of this is that many states now have comprehensive, RCRA-like legislation and regulations, but little or no money for their agencies to use in the enforcement process. These have come to be referred to as "paper programs."[6] Under these straitened circumstances, bureaucrats have come to believe that either a given dump qualifies for Superfund or there is little that they can do.

In facing hazardous waste dumps, bureaucrats are often damned if they do and damned if they don't. These are exactly the types of situations calculated to sully otherwise impeccable personnel records. In our cases, even years after a threatening site was discovered, there was little awareness of the true extent of the problem. In one case, preliminary investigation over a period of more than a year convinced officials they were dealing with a situation in which up to 45 drums, containing about a dozen chemicals, were buried in a remote location. Ultimately, after being approved for Superfund and after excavation got underway, 1,200 drums were buried in the pit and 23 different chemicals were discovered. In the other case, after two and a half years of investigation of a five-acre industrial site, state inspectors assumed the area was contaminated with arsenic to depths ranging from a few inches to three feet. After the site qualified for Superfund, and more than one million dollars was spent on assessment, heavy contamination was found nearly 40 feet below the surface throughout the five acres, and unacceptable (i.e., unsafe) levels of arsenic were found to depths of 300 feet even beyond the five acres of the site itself.

While state programs have a number of administrative weak points, one of the most serious is with field inspectors. Typically, these "front line troops" are poorly paid and undermanned. For example, a report to the Texas legislature noted that the entire RCRA program there included no more than 50 field inspectors for the entire state. The district office for the Houston area, in which region it is said up to 60 percent of the nation's petrochemicals are manufactured, the inspection staff is never more than ten. Whereas EPA would like to see 1,600 inspections annually, the state agency does well to undertake somewhat more than 600.

These small staffs scour the state looking for dangerous sites and, when suspicious circumstances are observed, they dip vials of water from ponds or scrape samples of dirt from the surface to be examined in the limited available laboratories. State bureaucratic officials are well aware of these limitations; however, all requests for additional personnel or funds for scientific analyses meet the stiff opposition of the petrochemical lobby. The results of these conditions are hundreds of suspect sites which have been only superficially studied. Prosecution of suspected polluters, where they are known at all, is often hamstrung by the paucity of hard evidence the agencies can put into the hands of the prosecutors.

Faced with this overall situation, state officials are loath to reveal new sites to the public or expound on the dangers such dumps portend. Where industrial concerns are involved and the evidence of contamination is slight, vigorous prosecution is regarded by the business community as an attack on the industrial base of the state. It is often difficult to make these legal actions stick. On the other hand, once a suspected serious situation is publicized, the public reaction is

---

[6] James P. Lester, James L. Franke, Ann O'M. Bowman, and Kenneth W. Kramer, "A Comparative Perspective on State Hazardous Waste Regulation," The Politics of Hazardous Waste Management (Duke University Press, Durham, North Carolina, 1983), p. 230.

visceral, if not hysterical. Fanned by media reminders of Love Canal and Times Beach, responsible officials with no financial resources find themselves unable to promise effective clean-ups nor even able to delineate accurately the characteristics of the situation. While Weber stressed that "every bureaucracy seeks to increase the superiority of the professionally informed by keeping their knowledge and intentions secret," in the case of hazardous waste dumps, secrecy is often a matter of survival. The resulting delays in taking any action are interpreted by the zealots of the media as official callousness and disregard for pregnant mothers and their unborn children.

No agency or bureaucrat wants to face these types of situations, and every effort will be made to shift responsibility elsewhere. Thus, the matter is handed off from the field office to the central office, from administrative enforcement officials to prosecutors, and from prosecutors to the courts. And if these shifts do not succeed in ridding them of the problem, the state will look to local authorities or federal agencies. And as soon as the case is transferred to someone else, the relieved agency will withdraw from the scene as a matter of professional courtesy to the newly burdened operatives, who themselves soon will seek a new victim on to whom they may foist this unwelcome conundrum. Even if the case ends back where it started, responsible officials can say they have been actively dealing with the problem by passing it around. The seriousness of the situation is emphasized by the fact that throughout the nation there are tens of thousands of suspected dump sites about which little is known or can be determined with the bureaucratic arrangements now in existence. Meanwhile, every year, of nearly one-half billion metric tons of hazardous waste being produced, more than 130 million are disposed of in ways for which no one can account. Until something can be done to provide for proper disposal, this flow of contaminants will continue unabated with or without the awareness of the affected state governments.

When a course of action is clear, one can predict the policies of bureaucracies with some accuracy. Given an understood problem and the means for dealing with it, every state agency will proceed with remedial action. On the other hand, when they are faced with situations which are dangerous, both institutionally and personally, and no funds are available, there is no way to predict what bureaucracies will do. There is no rational course of action available, as they see it. Under such circumstances, bureaucrats are likely to suffer from what Katz and Kahn refer to as "cognitive myopia" where the easiest way to avoid seemingly intractable problems is to respond to "the immediate, the visible, and the palpable and neglect the dimensions of the problem which are more remote in time and place."[7] Discussing bureaucratic decision making, Herbert Simon notes that "behavior is determined by the irrational and nonrational elements that bound the area of rationality."[8] Given the circumstances which prevail in the field of hazardous waste, it is easy to see why bureaucratic behavior is the "ultimate uncertainty."

---

[7]Daniel Katz and Robert L. Kahn, The Social Psychology of Organizations (John Wiley and Sons, Inc., New York, 1966), p. 289.

[8]Herbert A. Simon, Administrative Behavior, A Study of Decision-making Processes in Administrative Organization (Free Press, New York, 1957), p. 241.

WHAT CAN BE DONE?

Unless a hazardous waste site qualifies for Superfund treatment, publicity about it inevitably leads to a public admission of bureaucratic ignorance and impotence. Not surprisingly, the public response is uncomplimentary both to the responsible agency and to the poor bureaucratic victim unlucky enough to have to deliver this message. Hence, the real problem is with the tens of thousands of sites which have been identified but which have not, and most likely never will be, certified for Superfund treatment. So these sites, for the most part, remain as they are, largely unattended, unknown threats to the population in the vicinity. Each of them poses a potentially dangerous, if not lethal, threat to public safety and health.

During the course of the case studies, the research team noted that there was little conscious awareness on the part of state officials of the field of risk analysis, much less its methodologies. Their approach to each site was to find the means, or some organization that had the means, to cart the hazardous materials away. It is readily recognized that sufficient amounts of money would go a long way toward solving hazardous waste problems; however, the following comments are based on the assumption that such funds will remain in very short supply for the foreseeable future. This means that alternatives short of total clean-up must be found that are not costly but which will ameliorate the more serious consequences of each situation. Risk assessment could play an important role in these processes.

There is a major need for the application of risk assessment in determining the safest location and proper construction of new hazardous waste sites and in dealing with dumps. The emphasis in this presentation is on how this discipline may assist in dealing with dumps. The principal problem is that all unanalyzed dumps are potentially serious threats to the public, and delays in dealing with them simply exacerbate the problem. In one of the cases studied, the polluting company was allowed to continue to contaminate the site for four and half years after it was discovered because the state agency did not understand the true seriousness of the problem and was not in a position to develop an adequate solution. It is simply intolerable to accept the proposition (which seems to characterize many bureaucracies) that without sufficient funds to remove all contaminants from a dump, nothing at all can be done. A program must be developed in each state that does everything possible to (a) aggressively locate all existing dumps that contain hazardous substances; (b) analyze each site as rigorously and thoroughly as possible with the limited means available; and (c) develop acceptable standards for assessing threats to safety and health in terms of the types and intensities of contaminative materials, the types of affected media (e.g., air, water, soil), and the proximity of the threatened populations.

Organized as they are in specialized functional agencies, each state bureaucracy approaches each problem with the tools of its dominant discipline -- engineering and public health being two of the most prominent. As has been emphasized, an effective risk assessment should be undertaken by a multidisciplinary team. With few exceptions, state governments cannot put such teams together from the ranks of their fulltime employees. In fact, in most instances, the relatively few scientifically trained people on their staffs have largely administrative tasks that require their fulltime attention. This means that risk assessment teams will have to be drawn from public and private research resources (e.g., universities, corporate laboratories, foundations). From these organizations and institutions, panels of scientists should be drawn up from which appropriate individuals would be selected to act as risk assessment boards with respect to one or more dumps under scrutiny at any particular time.

This plan is proposed in the face of much critical comment on the scientific board concept. Wynne has written of "the myth of natural consensus," and he emphasizes that scientists are people subject to the same pressures as others. There is no sole, objective truth of which they are privy. They have their own biases, cognitive limitations, and similar frailties.[9] Lowrance too warns against the "arrogation of wisdom" by the scientific community.[10] Nonetheless, there does not seem to be any reasonable alternative available to state governments.

Such risk assessment boards would provide the following advantages: (a) they would operate outside the fragmented, specialized agencies which are legally responsible for solving hazardous waste problems; (b) they would provide the coordinated combination of scientific and technical skills not routinely available to these agencies; (c) the principal careers of board members would not be dependent on the bureaucracies they would be assisting; (d) they would be able to work together in an environment and under circumstances that would foster a consensus; (e) they would provide the public with factual information and thus force state officials to take action that would at least ameliorate conditions to the extent legal and administrative resources permit; (f) they would provide state officials with a much needed information basis for taking legal action or seeking financial support from the legislature; and (g) they would provide state officials with a much firmer information base for establishing rational priorities for addressing the many situations the larger industrial states now face.

The creation and use of such scientific assessment groups is only the centerpiece of the broader risk assessment program being proposed in the complete report to the National Science Foundation. It should be emphasized that we are under no illusions that these boards, or the complete program for that matter will be a sufficient palliative in the long run. However, such a step would be a major improvement in the present haphazard situation, would be within the means of virtually any state government, and would lay a very valuable experiential foundation for the application of multidisciplinary techniques to these extremely dangerous threats to the health and safety of millions of Americans.

---

[9]Brian Wynne, "Institutional Mythologies and Dual Societies in the Management of Risk," The Risk Analysis Controversy (Springer-Verlag, New York, 1982), pp. 129, 133.

[10]Lowrance, op cit., p. 110.

RISK ASSESSMENT AND THE LAW: EVOLVING CRITERIA BY WHICH CARCINOGENICITY RISK ASSESSMENTS ARE EVALUATED IN THE LEGAL COMMUNITY

R. Bruce Dickson

Paul, Hastings, Janofsky & Walker
Washington, D.C. 20036

ABSTRACT

Where scientific uncertainty precludes the definitive resolution of factual issues, there has been a tendency among decision-makers in the regulatory and judicial arenas to use policy-making as a substitute for fact-finding. This tendency led several regulatory agencies in the 1970s to limit the use of risk assessment because of the complexity of the risk assessment process and because of their concern that methodologies were not sufficiently developed to remove uncertainty. Several judicial decisions have since reversed the trend. Agencies are now moving in the direction of scientific risk assessment, although there is much ground yet to be covered. Simultaneously, trial courts appear to be replacing scientific fact-finding with policy-making when confronted with the uncertainty of cancer causation. This paper reviews these two developments and provides a contrast between the criteria currently used by courts to evaluate carcinogenicity risk assessments by regulatory agencies and the criteria used by a recent trial court to evaluate evidence of cancer causation.

KEY WORDS: Regulation of Carcinogens; Carcinogenicity Risk Assessment; Toxic Tort; Significant Risk

OVERVIEW: Fitting An Uncertain Science Into Our Fact Finding Process

The question of cancer causation is a fairly recent entrant into the realm of governmental and judicial decision-making. During the 1970s governmental agencies began to face the many difficult issues involved in carcinogenicity determinations as they initiated the task of implementing the many environmental and safety laws enacted during that time period. More recently, the same issues have been brought into the trial courtroom as civil tort litigation based upon claims of liability for cancer causation have become increasingly common.

Each of these decision-makers -- the regulatory agency deciding whether and to what extent a particular substance should be controlled because of its potential to cause cancer and the court or jury deciding whether cancer was caused by exposure to a substance -- must make a determinative ruling on a complex and uncertain scientific question. As the complexity of that scientific question first became apparent to government decision-makers, their response was to retreat from the challenge of making scientific decisions and to place most decisions in the realm of policy-making. The result was to shift the normal burden agencies bear to show the need for regulation. As a result of policy decisions, agencies ruled that the burden lay with others to show that there is no need for regulation. A similar shift can now be seen in the context of civil litigation - a shift apparently designed to provide a remedy for injured parties, despite the uncertainties inherent in proving the cause of a chronic illness.

Government agencies have followed a circuitous route to risk assessment. Early efforts to avoid risk assessment and to substitute policy decisions for scientific analysis have failed. Most agencies have now come to recognize the need to incorporate risk assessment methodology into their decision-making processes. A number of reviewing court decisions, including decisions of the Supreme Court, have established the need for quantitative risk assessments in agency decision-making. The current case law indicates that agency risk assessments must be based upon scientifically sound experimental data and a reasonable exposure analysis. While policy decisions do play a role in resolving issues, their role is being rapidly replaced with science.

This paper reviews several agencies' early attempts to avoid scientific risk assessments. It also reviews relevant judicial decisions that require that agencies assess risks and that certain minimal criteria for risk assessments be satisfied. It also reviews the rejection of risk assessment methodology in a recent federal court decision involving liability for radiation-induced cancer.

1. THE EVOLUTION OF CARCINOGENICITY RISK ASSESSMENT IN THE FEDERAL REGULATORY PROCESS

Most federal authority to regulate hazardous materials has evolved from legislation written within the past 10 to 15 years. Table 1 lists federal laws granting regulatory authority over the hazards posed by toxic substances and identifies the agencies exercising such authority.

a. Efforts to Simplify the Risk Assessment Process

As experimental data began to emerge to show that particular chemicals induced tumors in rats and mice, several agencies began the process of limiting their use. The laws described above call for a certain quantum of evidence regarding the hazard to be regulated, and in the absence of such evidence authorize reviewing courts to overturn the agency decision.

Agencies were faced with the prospect of converting an experimental data base into a regulatory decision. The process was an ardous one, requiring a large investment of agency resources and time, and producing a result that might well be overturned. Indeed, the first major regulatory effort to limit exposure to alleged carcinogens was vacated on judicial review. See, Dry Color Manufacturers Association v. Department of Labor, 486 F.2d (3rd Cir. 1973), in which the Occupational Safety and Health Administration's ("OSHA's") Emergency Temporary Standards for certain supposed carcinogens were rejected by a reviewing court. While OSHA's efforts were ultimately successful (see, Synthetic Organic Chemical Manufacturers Association v. Brennan, 503 F. 2d 1155 (3rd Cir. 1974)), the large amount of time consumed in the process led agencies to seek means to simplify the process.

At the same time that OSHA was regulating workplace exposure to potentially carcinogenic substances, EPA was beginning to regulate potentially carcinogenic pesticides under the Federal Insecticide

Table 1

Federal Laws Pertaining to Hazards Posed by Toxic Substances[1]

STATUTE	YEAR ENACTED	RESPONSIBLE AGENCY	SUBSTANCES COVERED
Toxic Substances Control Act	1976, amended 1981	EPA	All new chemicals (other than food additives, drugs, pesticides, alcohol, and tobacco) and existing chemicals not covered by other toxic substances control laws
Clean Air Act	1970, amended 1977, 1980, 1981	EPA	Hazardous air pollutants
Federal Water Pollution Control Act (now Clean Water Act)	1972, amended 1977, 1978, 1979, 1980, 1981, 1982, 1983	EPA	Toxic water pollutants
Safe Drinking Water Act	1974, amended 1977, 1977, 1980	EPA	Drinking water contaminants
Federal Insecticide, Fungicide, and Rodenticide Act	1947, amended 1972, 1975, 1978, 1980, 1983	EPA	Pesticides
Act of July 22, 1954, (codified as Section 346(a) of the Food, Drug & Cosmetic Act)	1954, amended 1982	EPA	Tolerances for pesticide residues in food
Resource Conservation & Recovery Act	1976, amended 1978, 1980, 1982, 1983	EPA	Hazardous wastes
Marine Protection, Research and Sanctuaries Act	1972, amended 1974, 1975, 1976, 1977 1980, 1981, 1983	EPA	Ocean dumping
Comprehensive Environmental Response, Compensation, and Liability Act	1980, amended 1982, 1983	EPA	Hazardous substances released into environment
Food, Drug & Cosmetic Act	1938	FDA	Basic coverage of of food, drugs, and cosmetics
Food additives amendment	1958	FDA	Food additives
Color additives amendments	1960	FDA	Color additives
New drug amendments	1962	FDA	Drugs
New animal drug amendments	1968	FDA	Animal drugs and feed additives
Medical device amendments	1976	FDA	Medical devices
Federal Meat Inspection Act	1967	USDA	Food, feed, and color additives; pesticide residues in meat and poultry products
Poultry Products Inspection Act	1957	USDA	
Egg Products Inspection Act	1970	USDA	
Fair Packaging and Labeling Act	1976	FDA	Packaging and labeling of food and drugs for man or animals, cosmetics, and medical devices
Public Health Service Act	1944	FDA	Sections relating to biological products
Occupational Safety & Health Act	1970, amended 1974, 1978	OSHA, NIOSH	Workplace toxic chemicals
Federal Hazardous Substances Act	1960	CPSC	Hazardous (including toxic) household products
Consumer Product Safety Act	1972	CPSC	Hazardous consumer products
Poison Prevention Packaging Act	1970	CPSC	Packaging of hazardous household products
Lead-Based Paint Poisoning Prevention Act	1973, amended 1976	CPSC, HEW, HUD	Use of lead paint on toys or furniture, on cooking, drinking, and eating utensils, in federally-assisted housing
Hazardous Materials Transportation Act	1975, amended 1976	DOT (Materials Transportation Bureau)	Transportation of toxic substances generally
Federal Railroad Safety Act	1970	DOT (Federal Railroad Admin.)	Railroad safety
Ports & Waterways Safety Act	1972	DOT (Coast Guard)	Shipment of toxic materials by water
Dangerous Cargo Act	1952		
Federal Mine Safety and Health Act	1977	Labor (Mine Safety and Health Admin.), NIOSH	Toxic substances and other harmful physical agents in coal or other mines

[1]See, Environmental Law Institute, An Analysis of Past Federal Efforts to Control Toxic Substances, Washington, D.C., 1978.

Fungicide and Rodenticide Act ("FIFRA"). Unlike the issuance of workplace exposure standards under the Occupational Safety and Health Act, which are the result of informal rulemaking proceedings, pesticide decisions are the subject of adjudicatory proceedings, in which holders of pesticide registrations are entitled to present evidence and to cross-examine experts presented by the Agency Staff.

After several carcinogenicity adjudications, in which the Agency litigated the scientific issues on which a finding of carcinogenicity must be based, EPA attempted to avoid the need to re-try the same scientific issues in case after case. EPA had completed lengthy fact-finding proceedings regarding Aldrin/Dieldrin and DDT and was attempting to cancel or suspend Mirex, Heptachlor and Chlordane. In the Mirex proceeding, the Agency attempted to avoid the prospect of re-trying the scientific principles of carcinogenicity upon which the earlier decisions were based -- and upon which the pending Mirex cancellation depended -- by asking the Administrative Law Judge in FIFRA Docket No. 293 to take official notice of seventeen "well established recognized facts" concerning carcinogenicity.

The taking of official notice would have had the effect of removing the issues noticed from the adjudicatory proceeding.[1] As a result, the Agency could have avoided submitting evidence to prove the principles, and the pesticide registrant would not have had an opportunity to refute the principles or to cross-examine the experts used to establish them as fact.

The seventeen principles which EPA sought to establish included uncontroverted facts, as well as propositions lying at the heart of scientific controversy. Among the principles were the following:

> Principle 4 - A chemical carcinogen is any agent which has been shown to cause benign or malignant tumors in adequately conducted studies in man or in animals. Critical quantitative and qualitative evaluation is necessary to establish this causal effect which may be evidenced by significantly increased tumors or by shortening the latency period between the exposure and the development of tumors, or by the development of unusual tumor types.
>
> Principle 5 - Any substance which has been shown to cause tumors in any mammalian species in adequately conducted studies at any dose level must be considered a carcinogenic hazard to humans, unless proven otherwise.
>
> Principle 9 - The long latency period of cancer, its frequency in the population and the difficulty in isolating control groups which are not exposed to certain chemicals hinders epidemiological research. With chemicals to which exposure is widespread in the population, it is extremely difficult if not impossible to detect a specific carcinogenic effect in the human population.
>
> Principle 16 - There is no method for establishing a "no-effect" level for human exposure to carcinogens.
>
> Principle 17 - The implication of carcinogenicity should be drawn both from tests resulting in tumors diagnosed as

---

[1]The Agency amended its regulations just before requesting official notice, to permit Administrative Law Judges to take notice of "facts within the specialized knowledge and experience of the Agency," 40 Fed. Reg. 25815 (June 19, 1975).

benign and from tests resulting in tumors which are more obviously malignant....

Had the Agency been successful in its attempt to have these "facts" made the subject of official notice, the fact-finding process would literally have been replaced by the policy-making process. However, the Administrator of EPA rejected the narrow approach embodied in the seventeen principles and instead adopted Interim Procedures and Guidelines for Health Risk and Economic Impact Assessments of Suspected Carcinogens. 41 Fed. Reg. 21402 (May 25, 1976). The Interim Guideline for Carcinogen Risk Assessment did not resolve issues, as the official notice would have, but categorized the evidence of carcinogenicity into various levels of substantiality, with the "best evidence" coming from positive epidemiology studies, "substantial evidence" from animal studies producing malignant or pre-malignant tumors, and "suggestive evidence" from animal studies producing benign tumors and from mutagenicity, cell-transformation and initiation-promotion skin tests. Structural relationships would be considered as "ancillary evidence." Under the guidelines, EPA would use a variety of extrapolation models, would describe the degree of uncertainty and would evaluate the relevance of the mode of exposure if an animal study is used to support a finding of carcinogenicity.

The next effort to remove areas of uncertainty from the regulatory risk assessment process came from the Consumer Product Safety Commission ("CPSC"). On June 13, 1978 the CPSC published as a final regulation its Interim Policy and Procedure for Classifying, Evaluating, and Regulating Carcinogens in Consumer Products, 43 Fed. Reg. 25658. The Interim Policy created a categorization scheme whereby substances that are shown to be positive in epidemiology or animal carcinogenicity studies would be banned from consumer products. CPSC concluded that thresholds had not been established and therefore any carcinogen would be considered to pose some risk. As a result of its Policy, there would be no debate over scientific principles, no evidence of pharmacokinetic differences between animal and man, no evidence of a threshold and no argument that a carcinogenic risk was insignificant or insufficient to justify a ban.

The CPSC Interim Policy was short-lived, however. It was vacated upon legal challenge on the ground it was adopted without compliance with the requirements of the Administrative Procedures Act, which calls for notice and an opportunity to comment on substantive rules. See, Dow Chemical, USA v. Consumer Product Safety Commission, 459 F. Supp. 378, 395 (W.D. La. 1978).

On October 10, 1979 EPA proposed a generic carcinogenic risk assessment method for promulgating national emission standards for hazardous air pollutants. The proposal, entitled National Emission Standards for Identifying, Assessing and Regulating Airborne Substances Posing a Risk of Cancer, 44 Fed. Reg. 58642 (October 10, 1979), would have defined the Agency response to supposed carcinogenic air pollutants on the basis of the type of evidence of carcinogenicity that was available. Such evidence was to be classified in accordance with the 1976 EPA Interim Guidelines discussed above. As a result, a substance with either "best evidence" or "substantial evidence" would probably be listed as a hazardous air pollutant on the ground that it was a "high probability" human carcinogen. This generic approach was thus intended to eliminate disputes over the significance of the "principal" data, arguments over basic scientific issues and evidence that the risk presented by a particular pollutant was insignificant.

The ultimate generic regulation seeking to eliminate debate over the meaning and significance of carcinogenicity data was adopted by OSHA in January 1980. In a regulation entitled Identification, Classification and Regulation of Potential Occupational Carcinogens, 45 Fed. Reg. 5002 (January 22, 1980), OSHA sought to establish -- and thereby preclude all future debate regarding -- several principles of carcinogenicity. Among the issues over which debate was barred in the OSHA Cancer Policy were the following:

- that substances that produce only benign tumors pose a carcinogenic hazard;
- that positive results at any dose level establish a carcinogenic hazard;
- that threshold levels of exposure will not be considered;
- that all risks must be controlled to the greatest degree possible, regardless of their magnitude.

In promulgating its Policy, OSHA made clear its intent to avoid repetitious debate over these and other scientific issues and to eliminate quantitative risk assessment from the rulemaking process. In perhaps the most direct statement of the underlying policy that lay behind all of these efforts to simplify the fact-finding process for identifying carcinogens and evaluating their risks, OSHA stated:

> "The existing case-by-case approach, with its constant re-examination of already resolved scientific and policy issues, does not permit regulation of such substances in a timely and efficient fashion. Moreover, OSHA believes this approach is needed to overcome the administrative difficulties created by the case-by-case approach." 45 Fed. Reg. 5005.

In a separate but related vein, Congress took action in 1958 to provide that no food additive be considered safe if it is found to induce cancer in man or animals. 21 U.S.C. §§ 348 (c)(3), 360(d)(1)(H), 376(b)(5)(B). The so-called Delaney clause was used by FDA during the 1970s to reject several food additives on the ground that, while the additive itself was not shown to be carcinogenic, it contained an impurity or constituent which was. Thus FDA terminated the provisional listing of carbon black and graphite because they contained carcinogenic polynuclear aromatic hydrocarbons. 41 Fed. Reg. 41857 (September 23, 1976); 42 Fed. Reg. 60734 (November 29, 1977). The agency removed several other dyes from its provisional list solely on the basis of their carcinogenic constituents. In so doing, the agency did not attempt to estimate the magnitude of the risk presented by each constituent, but merely cited the Delaney clause requirement. Quantitative risk assessments were considered irrelevant to such a determination.

As a result of these efforts to simplify the process, the role of risk assessment -- particularly the quantitative element of risk assessment -- was in doubt at the end of the 1970s. Agencies that had begun to regulate carcinogens in the early part of the decade had turned their attentions to creating an administrative system in which risk assessment followed certain rigid, predetermined courses to identify carcinogens and then stopped. Quantitative risk assessment was believed to be too uncertain a science to be considered in the regulatory process.

b. The Move Toward Quantitative Risk Assessment

The agency efforts to establish generic principles of carcinogenicity and to avoid quantitative risk assessment were short-lived. EPA's 17 principles were never fully accepted by the Agency. CPSC's Interim Policy and Procedure was overturned in the courts for procedural flaws in its promulgation. EPA's generic carcinogenic air pollutant proposal was never issued as a final rule. And OSHA's generic cancer policy was subsequently stayed, pending a reevaluation by the Reagan Administration. As a result, courts have not had an opportunity to determine whether the generic approach to scientific issues arising in informal rulemaking satisfies the requirements of procedural due process found in the Administrative Procedures Act and in the United States Constitution.

Several courts have, however, reviewed specific regulatory decisions which were, in turn, based upon risk assessments. While this area of law is by no means settled, several principles have emerged regarding the need for a quantitative assessment of risk and the quality of the risk assessment itself.

First, it is becoming increasingly clear that if an agency is to take action to remedy a carcinogenic risk, it must determine whether the risk is significant. The significance of the risk is viewed by courts as a quantitative measure. The generic agency determination in the 1970's that all carcinogenic risks must be eliminated or reduced to the maximum extent feasible would no longer be acceptable under current case law.

When OSHA adopted a workplace exposure standard for benzene which limited exposure to 1 ppm, the Court of Appeals for the Fifth Circuit and ultimately the Supreme Court ruled that the new standard had not been supported by necessary findings and was therefore invalid. In Industrial Union Department, AFL-CIO v. American Petroleum Institute, 448 U.S. 607 (1980), the Supreme Court ruled that the Occupational Safety and Health Act requires that the Secretary of Labor find "as a threshold matter, that the toxic substance in question poses a significant health risk in the workplace and that a new, lower standard is therefore 'reasonably necessary or appropriate'...." 448 U.S. at 614-615. The record before the Secretary showed only that exposure to benzene at high concentration levels produced adverse effects, but did not support a finding that exposure at 10 ppm (the existing permissible exposure level) posed a significant risk. OSHA was unable to extrapolate to low exposure levels by constructing a dose-response curve. It relied, however, on its policy -- one of the principles developed during the mid-1970s discussed above -- that it must be assumed that there is no safe level of exposure to a carcinogen unless proven otherwise. The Court ruled that the burden was on the agency to show, on the basis of substantial evidence, that it is at least more likely than not that long-term exposure to 10 ppm of benzene "presents a significant risk of material health impairment." The agency must also show that its new standard will eliminate or reduce the risk. Such a finding need not be based upon scientific certainty, but must be suppored by the best available evidence. These principles were reaffirmed in American Textile Manufacturers Institute, Inc. v. Donovan, 452 U.S. 488, 505 (1981).

Several other courts have ruled that, when agencies evaluate the need for regulatory controls, they should distinguish between significant risks and insignificant or de minimis risks. In 1979, when FDA determined that acrylonitrile copolymers used in beverage containers should be considered food additives, the reviewing court found in the definition of "food additive" an implied limitation. "Congress must have intended," the court said, "the Commissioner to determine with a fair degree of confidence that a substance migrates into food in more than insignificant amounts." Monsanto Co. v. Kennedy, 613 F.2d 947 (D.C. Cir. 1979), is a landmark case that has substantially expanded FDA's authority to rely on carcinogenicity risk assessments. The court said:

> "[T]here is latitude inherent in the statutory scheme to avoid literal application of the statutory definition of 'food additive' in those de minimis situations that, in the informed judgment of the Commissioner, clearly present no public health or safety concerns," 613 F.2d at 954.

This decision has provided justification for FDA's use of quantitative risk assessments for carcinogenic impurities in food additives despite the Delaney clause prohibition. Thus, in its final rule listing D&C Green No. 5 for use in drugs and cosmetics, the agency used two extrapolation models to estimate the upper limit individual lifetime risk of cancer resulting from exposure to a carcinogenic constituent in D&C Green No. 5 47 Fed. Reg. 24278 (June 4, 1982). The resulting risk estimates -- 1 in 30 billion and 1 in 300 million -- led FDA to conclude that there is a reasonable certainty that no harm will occur as a result of the use of D&C Green No. 5. Citing Monsanto v. Kennedy, supra, the agency ruled that it had authority, despite the Delaney clause, to list a food additive that contained a carcinogenic constituent. 47 Fed. Reg. 24280.

FDA's ruling was upheld on judicial review in a recent decision. Scott v. Food and Drug Administration 728 F.2d 322 (6th Cir. 1984), upheld FDA's application of the de minimis principle of Monsanto v. Kennedy, supra, to find that the level of carcinogenic risk posed by a constituent was so low as not to render the food additive unsafe. 728 F.2d at 325.

The concept of de minimis or insignificant need for regulatory action is not new. It has historically been applied to ensure that agencies have the authority to refrain from acting if the risk to be regulated is too slight to warrant regulation. See Alabama Power Co. v. Costle, 636 F.2d 323, 360 (D.C. Cir. 1979); Ethyl Corp. v. EPA, 541 F.2d 1, 18 (D.C. Cir. 1976), cert. denied, 426 U.S. 941 (1976).

Not all agency risk assessments are as successful upon judicial review as FDA's assessment of D&C Green No. 5. In a decision that is already having a major impact on agency risk assessments, the Fifth Circuit overturned the decision of the CPSC banning area-formaldyhyde ("UF") foam insulation. Gulf South Insulation v. United States Consumer Product Safety Commission, 701 F.2d 1137 (5th Cir. 1983).

CPSC had banned UF insulation on the basis of a quantitative risk assessment which estimated an excess cancer risk of between 0 and 51 per million. The court ruled

- that it was unreasonable for the Commission to base a complex risk assessment on a single study, while disregarding other studies;
- that the CPSC may not make a finding that a product presents an unreasonable risk of injury on the basis of an upper confidence limit, "up to" estimate of risk;
- that the agency may not apply an exacting risk assessment model to unrepresentative and imprecise exposure data.

On the basis of this decision, agencies whose statutory mandates are comparable to that of CPSC might be required to design their risk assessments to yield the most likely or most probable risk estimation. Upper confidence limits may not provide the substantial evidence of actual risk on which most agency risk findings must be based.

In estimating risk, agencies might also be precluded from relying on a single set of data or a single extrapolation model. Agencies should use all appropriate models that fit the available data. And finally, agencies must use valid exposure data -- not worst case assumptions -- in the risk assessment process.

In part because of the increasing acceptance of risk assessment among the regulatory agencies, there have been several attempts recently to outline consistent principles to be used in conducting risk assessments. These efforts can be distinguished from the efforts of the 1970s to make generic policy decisions on scientific issues arising in agency risk assessments. The recent report prepared by the Office of Science and Technology Policy ("OSTP"), entitled Chemical Carcinogens: Review of the Science and Its Associated Principles, provides an overview of the science, while limiting the issues for which policy decisions displace scientific analysis.

There remain important issues for which agencies continue to rely on policy decisions. One important question pertains to the existence of thresholds and the issue of low-dose linearity. The principles of carcinogenicity risk assessment issued by OSTP state:

> No single mathematical procedure is recognized as the most appropriate for low dose extrapolation in carcinogenesis. When relevant biological evidence on mechanism of action (e.g., pharmacokinetics) exists, the models or procedures employed should be consistent with the evidence. However, when data and information are limited, as is the usual case, and when much uncertainty exists regarding the mechanism of carcinogenic action, models or procedures which incorporate low-dose linearity are preferred.

This expression of preference for low-dose linearity when the mechanism is uncertain represents the continued use of policy choices in lieu of scientific choices in areas of uncertainty. Where evidence of mechanism exists, however, under the OSTP principles, agencies would select models for low-dose extrapolation that are consistent with available data.

Under this view, evidence would be permitted to show that tumors are produced in experimental systems as a secondary effect of tissue damage, which is itself the direct effect only of high dose exposure to the substance of concern. Except for a decision by FDA in 1974, agencies have generally either refused to consider the possibility of a threshold for a carcinogen or have concluded that a non-genotoxic mechanism, for which a threshold might be found, has not been adequately proven in any given case.[2] FDA itself has not used this rationale to find a threshold for any other potential carcinogen.

Thus uncertainty continues to generate the substitution of policy decisions for scientific decisions. The public will see shortly how EPA intends to address these areas of uncertainty in its carcinogenicity and mutagenicity risk assessment process. The agency is preparing a new set of carcinogenicity and mutagenicity risk assessment guidelines, to be released within the next month.[3] In addition to EPA, several agencies have joined forces to form an Interagency Risk Management Council, which is expected to coordinate agency policy on the many difficult issues that

---

[2]In the 1974 FDA decision, the Commissioner of FDA stated:

> "While selenium at high dietary levels (in excess of 2 parts per million in experimental animals) has been proved to be a hepatotoxic agent in these and other studies, <u>the Commissioner concludes that its capacity to induce liver damage when consumed in excessive amounts, which is possibly associated with a higher incidence of liver cancer, does not warrant its classification as a carcinogen</u> (see the preamble to the proposal published in the FEDERAL REGISTER of April 27, 1973 (38 FR 10460))."

See also, Hearings Before a Subcommittee of the Committee on Appropriations, House of Representatives, 93rd Cong., 2d Sess., Part 8, at 210 (1974), in which FDA stated:

> "The issue has arisen whether the substance tested, or some other intervening factor, has caused the cancer. If a food additive causes a pathologic change at a particular level (*e.g.*, liver damage), and it is that pathologic change (rather than the additive directly) which in turn leads to cancer, the anticancer clause does not preclude approval of that additive for use at levels which provide an adequate margin of safety below the point at which it causes the pathologic change involved."

[3]EPA has, since this talk was presented, proposed a new set of guidelines for carcinogen risk assessment. See 49 *Fed. Red.* 46293 (November 23, 1984). The new proposed guidelines provide a more detailed discussion of various factors to be considered in a risk assessment. EPA has also established a classification system for evaluating evidence of carcinogenicity. See 49 *Fed. Red.* 46299 (November 23, 1984). Although the proposed guidelines appear to allow a thorough evaluation of potential risk, the classification system might impose a rigid scheme for categorizing chemicals based upon only certain types of data. Under this system, there is "sufficient evidence" of carcinogenicity if a substance is positive in multiple species or strains, in multiple experiments or to an unusual degree. There is "limited evidence" if studies involve a single species or strain, are restricted by inadequate dose, duration or follow-up or involve benign tumors only. There is "inadequate evidence" if the studies cannot be interpreted regarding carcinogenic effect. There is "no evidence" if at least two well-designed and conducted studies in different species are negative. Under some circumstances (which are not fully specified) EPA will decrease the evidence from sufficient to limited. The only example given pertains to the "widely diverging scientific views" regarding the significance of mouse liver tumors. It is too early to determine whether the level of evidence will be reduced for other endpoints of questionable relevance to humans.

arise now that risk assessments are firmly entrenched in our regulatory process.

2. THE TREATMENT OF RISK ASSESSMENT IN RECENT CIVIL TORT CASES

The movement in the mid-1970s toward replacing the risk assessment fact finding process with precautionary policy decisions may now have found its place in civil tort litigation.

The central issue in tort liability actions differs significantly from the issue before regulatory agencies in most risk assessment matters. Regulatory agencies must make findings of fact regarding the risk ("unreasonable risk," "significant risk," etc.), which generally must be supported by substantial evidence. In the context of tort liability, courts must generally find that a duty of care involving the type of harm that has occurred ran from the defendant to the plaintiff and that a violation of that duty was the proximate cause of the injury

Proof of causation is a particularly difficult hurdle for a plaintiff to overcome in the context of a chronic health effect. The long latency period and the presence of other potentially causative factors are two additional elements of uncertainty that arise when a carcinogen risk assessment plays a role in civil litigation.

These issues were recently discussed at length in a comprehensive decision holding the United States liable for cancer allegedly resulting from nuclear weapons testing between 1951 and 1963. In Allen v. United States of America, Civil No. 79-0515-J (D. Ut., May 9, 1984), the court fashioned an approach to the issue of proximate causation that is comparable to the approach of courts facing claims of DES-related injury.

Faced with a large number of claims of radiation-induced cancer injuries in which plaintiffs were unable to prove an actual link between exposure and cancer, the court created a new test for proximate causation in the context of chronic health effects. Under that test, a plaintiff must prove the following:

- that a defendant acted negligently;
- that that negligent act created a hazard;
- that a member of that group developed a biological condition which is "consistent with" having been caused by the hazard, "such consistency having been demonstrated by substantial, appropriate, persuasive and connecting factors."

If each of these elements has been established, a fact finder may conclude that the hazard caused the condition, although the defendant might offer proof to the contrary. Slip Opinion at 343-44.

The action was a consolidation of the individual claims of 1,192 named plaintiffs. The court found there to be no question whether ionizing radiation is a carcinogenic hazard. With regard to the question of a threshold, however, the court rejected arguments that at least some of the plaintiffs were exposed at levels too low to produce an effect. In a vein similar to the regulatory approach described above, the court ruled that in the absence of proof of the mechanism of cancer induction, it would be presumed that any level of exposure increased the risk of cancer.

This did not end the inquiry, however. The increase in risk of cancer cannot be equated with causation, and in a tort litigation proximate causation must be proved. But, the court held, proximate causation need not necessarily be shown by proof of a cause-in-fact connection between the action and the injury. Such proof would probably be impossible in the context of an injury that followed a long latency period. The court ruled that proximate causation can also be shown by establishing "the most exclusive factual connection" possible between the action and the injury. Slip opinion at 323. If a sufficiently exclusive factual connection can be established, proximate cause can be found. The court found support for this ruling in a line of cases originally developed to help a plaintiff who could not identify the individual whose action caused the injury.

Summers v. Tice, 33 Cal. 2d 80, 199 P.2d 1 (1948), in which two defendants had each fired a gun in the direction of plaintiff, upheld liability despite plaintiff's inability to prove causation with regard to either defendant. Sindell v. Abbott Laboratories, 26 Ca. 3d 588, 607 P.2d 924 (1980), ruled that proof of a cause-in-fact connection between an individual defendant's action and plaintiff's injury was not necessary to establish liability for a DES-related injury. Also cited was Basco v. Sterling Drug, Inc., 416 F.2d 417 (2d Cir. 1960), in which a plaintiff was allegedly blinded by one or both of two drugs. A sufficient factual connection was demonstrated simply on the basis of the fact that the drug was used and the fact that plaintiff's injury was consistent with observed side effects of the drug.

The court also found support in a number of other cancer causation cases, including Krumbach v. Dow Chemical Co., 676 P.2d 1215 (Colo. App. 1983); McAllister v. Workmen's Compensation Appeals Board, 69 Cal. 2d 408, 445 P.2d 313 (1968); and Besner v. Walter Kidde Nuclear Labs, 24 App. Div. 2d 1045, 265 N.Y.S. 2d 312, 313 (1965), in which cases the causal connection between exposure and injury was found to be sufficient, despite potentially confounding factors. In the Krumbach v. Dow Chemicals Co. case the burden of proof to show that ionizing radiation exposure had caused cancer was ultimately satisfied by a showing of a "reasonable probability" of causation.

The court noted Restatement (Second) of Torts, Section 432(2), which states:

> "If two forces are actively operating, one because of the actor's negligence, the other not because of any misconduct on his part, and each is of itself sufficient to bring about harm to another, the actor's negligence may be found to be a substantial factor in bringing it about."

Where a "substantial factor" is found -- that is, where a detailed factual connection is shown -- the cases shift the burden to the defendant to refute causation.

The court rejected arguments based on the lack of statistical probability that defendant's actions caused plaintiffs' injuries. As noted above, it also rejected arguments that plaintiffs were exposed to radiation at a level too low to have been a proximate cause, noting that the relationship between exposure and risk is largely unknown.

Under the Allen decision, risk assessment plays an extremely limited role in determining causation. The test prescribed by the court for determining proximate causation appears to limit the issues in the same fashion as was attempted by regulatory agencies in the 1970s. If a hazard is identified and an identifiable population is put at an increased risk, the party whose negligence created the hazard may be liable for injuries which are consistent with having been caused by the hazard. Under this decision, the size of the risk at actual exposure levels and the presence of other potential causes are not relevant to a finding of causation.

It is too early to know whether the Allen ruling on proximate cause for a carcinogenic risk will be adopted generally in cases of exposure to substances shown to be carcinogenic. The decision does fit into a growing trend among courts to fashion a theory of recovery despite a plaintiff's inability to prove all traditional elements of his case because of scientific or factual uncertainty. Whether that trend continues as chemical carcinogenicity risk assessments are developed -- or whether the uncertainties of the risk assessment process and the ability to quantify risk assessments will cause other courts to take a different course -- must await future decisions. The Supreme Court's recent refusal to review a state court DES decision which allowed a plaintiff to proceed to trial despite her inability to identify the defendant who supplied the drug suggests that state courts will be free to continue to fashion a remedy in these circumstances for plaintiffs who are unable to prove all elements of their claim. See E.R. Squibb & Sons v. Collins, 53 L.W. 3226, No. 83-1932 (October 2, 1984).

CONCLUSIONS

As carcinogen risk assessments moved into the regulatory arena, regulatory agencies initially sought to convert the risk assessment fact-finding process into a policy-making process. However, courts soon rejected these efforts as inconsistent with the underlying nature of the fact-finding process as it pertained to carcinogens. Agencies were told that they must assess risks and that risk assessments must satisfy minimal criteria. The regulatory risk assessment process is now undergoing changes which may bring it into compliance with judicial requirements.

Carcinogenicity has more recently moved from the agency to the trial courtroom. The recent Allen v. United States decision appears to reflect a tendency similar to that found in regulatory agencies in the 1970s. Scientific uncertainty has led to judicial policy-making to enable plaintiffs to remain in court despite their inability to prove a causal relationship between exposure and illness.

These two developments are of tremendous significance to the future of carcinogenicity risk assessment and warrant careful scrutiny in the years ahead.

ESTIMATING THE RISK FROM FLOODING AND EVALUATING WORRY

C.H. Green, E.C. Penning-Rowsell and D.J. Parker

Flood Hazard Research Centre
Enfield Middx EN3 4SF

ABSTRACT

Benefit-cost analyses are required for proposed flood alleviation and sea defense schemes in the United Kingdom. However, generally flood impacts such as stress, worry and evacuation have been left as 'intangibles'. Economic theory predicts that the avoidance of such impacts is a real benefit of flood alleviation. But only those impacts for which there exist a value concensus as to the bases of their measurement can be included within the framework of economic efficiency benefit-cost analyses.

In recent benefit-cost analyses we have, therefore, employed exploratory procedures to measure worry, health damage and other traditionally 'intangible' flood losses. A method of deriving the monetary equivalents of these impacts has also proved both feasible and to have face validity. The magnitude of these losses generally exceeds the physical damage caused by flooding.

Since worry has been shown to correlate with perceived risk, such evaluations include willingness-to-pay to reduce risk to health and safety. Estimating the risk to life in a particular location is problematic since the risk is almost wholly a function of the behavioural choices made by those at risk. Results of several studies of behavioural response to warnings and to discovering water are discussed.

KEY WORDS: Floods; worry; intangibles; benefit-cost analysis; risk estimation

INTRODUCTION

As compared to most hazards that will be discussed here, the risk of flooding is high. In the UK, flood protection is not usually contemplated to a standard of protection in excess of the probability of a flood per year of 1 x 10(-2) for urban schemes, and agricultural land to not more than to a standard of 0.1 or 0.2. The London Thames Tidal Barrier is an exception since its design standard is to protect up to the flood whose annual probability is 1 x 10(-3).

Risks to life through drowning, exposure and injury are, however, low in most instances. The risk of death from all natural disasters in the UK is 4.3 x 10(-8) (Grist 1978). However, there is evidence that there are premature deaths as a consequence of flooding particularly amongst the elderly (Bennet 1970), and this is certainly believed to be the case amongst those who have been flooded (Parker et al, 1983).

In addition, the traditional masonry construction for houses is being to a degree superceded by timber framed construction in the UK. The low death rate in the UK has almost certainly been partly due to the predominance of two story masonry construction. Effective warning of flooding has seldom been possible in the UK, and the predominant response of the affected residents has been to sit out the flood in the upper floor of their house. Since only very rarely have masonry structures collapsed during flooding in the UK this has been a successful strategy. It is not clear that the new forms of timber framed construction will show the same resistance to collapse and consequently we may experience a higher death rate in the future in the UK.

Most flood mitigation schemes are promoted by the Water Authorities which are autonomous tax raising watershed authorities who have a multifunction purpose. The schemes are, however, partly funded out of central government funds, and any scheme with a capital cost in excess of 500,000 pounds must be supported by an economic efficiency benefit-cost analysis. In most cases, these BCA's are based on the principles outlined in Penning-Rowsell and Chatterton (1977), and the damage data contained in that Manual and in subsequent publications by the Flood Hazard Research Centre (Penning-Rowsell, Chatterton and Witts, 1983).

The Treasury requires the use of economic efficiency benefit-cost analysis: the demonstration that aggregate willingness-to-pay to avoid future flood losses exceeds all scheme costs, both sides of the equation being discounted to net present values, and the distribution of benefits and costs being ignored (HM Treasury, 1981).

As a procedure economic efficiency BCA has a number of limitations: it rests upon very strong assumptions of concensus. Where there is no concensus either as to the magnitude of the impact, the rate of trade-off between impacts or against the numeraire money, then distributional considerations arise which exclude the inclusion of that impact in an economic efficiency BCA. Such considerations can only be resolved over and above the results of the BCA. BCA is in no sense a mechanism for resolving conflict in which instances the procedures being applied by Schmidtke (1984; Bayerisches Landesamt 1983) are more appropriate.

Secondly, to evaluate an impact the change in the parameter variable must take place over an area where that variable is locally continuous. Charges across discontinuities cannot be evaluated: hence the recognition that concepts of lump sum values of life are incompatible with the logic of benefit-cost analysis (Dreze 1962, Jones-Lee 1969, Mishan 1971).

Thirdly, in flood alleviation benefit-cost analysis we do not derive willingness-to-pay directly but infer willingness-to-pay on the basis of the rational man model. That is the rational man will be willing to pay the expected value of future losses averted, discounted to present value. This requires that an exhaustive and mutually exclusive listing of the impacts of floods be undertaken and moreover that none of the impacts interact synergistically (Green et al 1983).

Flood impacts are normally defined by antithesis: direct impacts are damages done to physical structures and contents whereas indirect losses are everything else. Similarly, whereas tangible losses are those which can be measured and evaluated, intangible losses are those which, although they may have no other common property, cannot be evaluated at the present time. We have drawn a distinction between intangible losses which cannot be evaluated in principle in economic efficiency terms, unquantifiable losses, and those for which there is merely a technical problem in evaluating them: unquantified losses.

Exclusion of some impacts as intangibles necessarily means that a benefit-cost analysis will underestimate willingness-to-pay. More seriously perhaps because some impacts are left unevaluated it is impossible to know whether the analysis captures the majority benefits and costs or only a small fraction. Similarly the degree to which impacts are captured may vary between different options and hence mean that the results of the analyses are not strictly comparable.

For urban flooding, the great majority of properties flooded are dwellings. Hence the importance of evaluating the intangible or unquantified impacts of flooding on households, given the relatively complete assessment of impacts on other land users (Green et al 1983).

In discussing these impacts, reference will be made to the results from interview studies of residents of a number of flood plains. Table 1 summarizes the principal characteristics of these areas and the associated flood hazard.

The Risk To Safety From Flooding

The requirement is to determine the risk from flooding in particular locations. In turn at such locations the risk to life safety is a function of the characteristics of the event into which we may subsume local topography; the characteristics of the dwellings; and the characteristics and the behavioral responses of the population. Those floods in Britain which have resulted in high death tolls have been characterized by rapid onset, high velocities and/or deep water (Robinson 1979; Delderfield 1976). Conversely, the significance of water velocity and depth is the ability of the dwellings to remain intact as places of refuge. The characteristics of the population influence the behavioral options open to them and the appropriateness of the behavioral response of the individual that individual's conditional survival probability. Failure to choose the optimum behavioral options has lead to high death tolls in some floods (Gruntfest, Downing and White 1979=8).

If we now replace the generic term 'risk' by the more appropriate term of 'detriment' (Royal Society 1983), a further problem is to establish the appropriate measure of detriment in order to assess the detriment from flooding in a particular location.

Any measure of detriment involves value judgments and hence can only be established on a concensual basis (Green 1980). We have previously shown (Green, Brown and Goodsman 1983) that the public's cognition of detriment are organized as shown in Figure 1 and that these measures are underlain by the required concensus.

Restricting ourselves to the immediate risk of death or injury in a flood, we can then compute the detriment from flooding in a particular location as follows. If we can compute the conditional probability of death given a flood event, then we can compute the risk of death from flooding standardized to a given unit of time. There is however a problem in determining the relevant unit of time period: the risk of flooding is not constant over time, the risk of the sea overtopping sea walls obviously being zero at low tide. Equally, people do not regard the risk as constant but as varying according to conditions: they are only

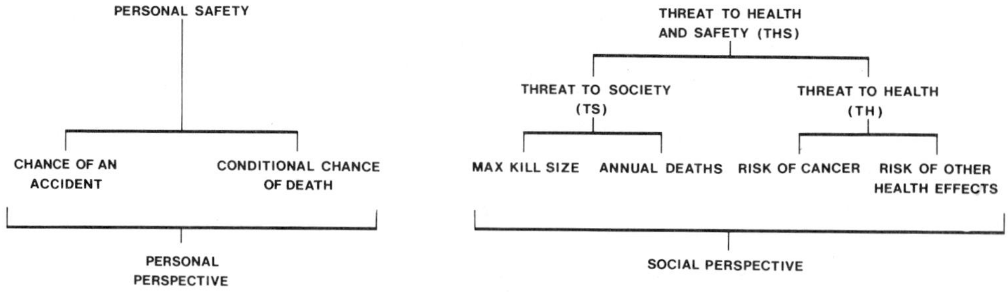

Figure 1

(Each Item is Determined by the Items Below it.)
(For Example, "Threat to society" is determined by annual deaths andmaximum kill size.)

concerned about the risk of flooding when there is a risk of flooding. In short whilst standardizing on the risk of death per year, thereby including in the duration of exposure to the risk, may be acceptable, it is more appropriate to keep the return period, which is the inverse of annual exceedance probability of the flood event and the condition probability of death distinct. This then gives the equivalent of the "personal safety" measure of detriment.

On the components of the "threat to society" measure of detriment, the detriment from a particular event is then estimated first by multiplying the population at risk times the conditional probability of death to obtain the expected kill size. Multiplied by the event probability gives the predicted average annual deaths (Green et al 1983).

However that in any one location, the risk of flooding is characterized by floods of different return periods each affecting different populations and possibly possessing different event characteristics adds complications. So do the likely variations in the conditional probabilities of survival across both events and different sub groups in the population. All these estimates of detriment on the relevant measures depend upon the estimated conditional probability of death given a specified flood event. This then is the key variable. This conditional probability is a function of the behavioral options open to residents, their choice amongst those options and the likelihood that each option will result in survival.

Similar considerations apply to damage reduction which in the UK has been the principal reason for the implementation of flood warning systems (Chatterton, Pirt and Wood 1979). In the UK we do not yet have a sufficient data base to separate out these various factors which may influence both behavioral choice and conditional survival probability.

The available behavioral repertoire is clearly influenced by the characteristics of the individual household at risk: the elderly and disabled clearly having a more restricted behavioral repertoire than the able-bodied. The characteristics of the flood also influence both the range of options and the likely success of each: speed of onset, water velocity and depth are all likely to be important.

Table 1

Background to Interview Studies

Location	Date of most recent flood	Characteristics of location and flood hazard	Number of house/holds interviewed
Loughton	June 1983	London commuter belt. Flash thunderstorm flooding of a stream through two housing estates. The new estate suffered severe flooding whilst minor flooding was experienced on the other estate. Return period of event estimated as 50 to 75 years. No warning.	131
Southgate	June 1982	Outer London suburb. Mixture of sewer back up plus flooding from stream both caused by severe thunderstorms. Flooding mainly confined to gardens, garages and outbuildings. No warning. Flooding occurs on average once every 1/2 years.	58
Swalecliffe	February '78	Coastal village. Overtopping of existing sea wall. Severe flooding at night of mainly elderly population many living in single storey dwellings. Warning was not effective. Previous major flood occurred in 1953.	48
Uphill	December '81	Coastal village. Breach of sea defences. Severe flooding of mainly elderly population many living in single storey dwellings. No warning. Previous flood occurred in 1929. New defences completed prior to interviews.	101
Whitstable	No recent experience	Small town on coast. Risk of overtopping or collapse of existing sea defences; possible severe damage by wave action in some areas, and widespread inundation lasting for some time.	110

Given the absence in the UK of pre-flood advice as to the actions that should be taken, residents' knowledge of the likely success of different behavioral options can only be gained by experience. Consequently it is of questionable validity to transfer the results of studies in one flood plain to another. Figure 2 illustrates the kind of event tree which it is necessary to construct. Residents' choice of behavioral actions is initially likely to be determined by their expectation as to the nature of flooding. In the areas studied this expectation has been of severe disruption to life and damage to property but of only a small risk to life and health (Table 2 and 3).

Moreover there is a difficulty in determining how people respond to warnings since it is quite difficult to find floods where warnings have been effective. Furthermore, warnings may be most successful in those instances where there is a recent history of flooding. In these cases, it is necessary to separate out the effect of experience from that of warning. From work on behavior in fire (Canter et al 1980) it is also

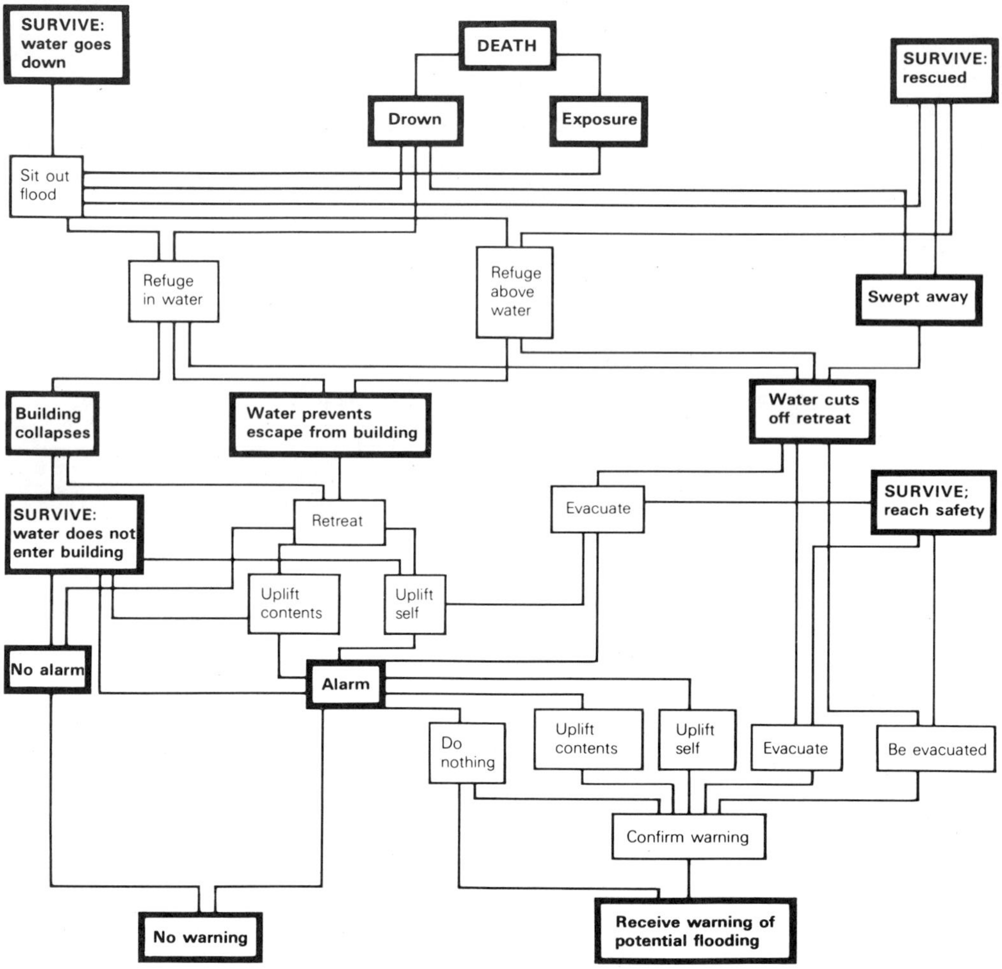

Figure 2

Event tree-Deaths from floods

known that people generally seek to confirm a warning by personal observation. Those who are at risk from flooding, and those in particular who have experienced flooding, have their own observational warning system (Table 4). The degree to which a formal warning system will supercede such informal warning systems is consequently problematic.

Consequently we have a long way to go before the kind of event tree illustrated in Figure 2 can be completed.

From Table 5 it can be seen, however, that the majority of residents of this particular floodplain, one with a long history of flooding and a well developed flood warning system, would anticipate sitting out any future flood after taking what action they could to minimize their losses.

Table 6 on the other hand records what happened in a severe flood which occurred when a sea wall was breached during a winter storm and up

Table 2

Relative Degree of Distress and Disruption Assessed to Result from Different Events (Medians)

House burgled	8.0
Car stolen	4.0
Home and garden flooded	8.0
Burst water pipe in house	5.5
Discovery of dry rot in house	5.0
House fire - fire brigade called	10.0

Sample: Southgate

Scale : 11 point category scale anchored at 0 'least' and 10 'most severe'

Table 3

Worry and Beliefs as to the Risk (Sample Medians)

	Southgate	Loughton		Swalecliffe	Uphill	Whitstable
		households whose homes have been flooded	households whose gardens have been flooded			
risk to health and safety from flooding	3.0	2.0	1.0	4.0	0	N/A
worry about the possibility of flooding	6.0	4.0	1.0	6.0	0	2.0
likelihood of future flooding	6.0	3.0	2.0	5.0	0	2.0

All questions used a 7 point category scale anchored at '0' and '6', higher scores refer to greater levels of worry or beliefs of greater likelihood or risk.

to five feet of water swept into houses. As part of a detailed interview schedule residents were asked to report the sequence of actions they carried out when they found water entering their houses. This discovery occurred in a number of ways from the front door breaking open under the pressure of water to a spreading damp patch on the living room carpet initially being blamed on an overflowing bath or burst boiler. Elderly households characteristically went upstairs, looked out the window to find out what was happening and then waited for the water to go down. The

Table 4

Behavioural Adaptions to Worry

	Loughton			Southgate
	Households who have been flooded (% Yes)	Households whose gardens have been flooded (% Yes)	Households potentially at risk (% Yes)	(% Yes)
We stay up all night when it rains	31	6	6	31
If we go away we arrange with neighbours how they can contact us in case of a flood	68	46	15	74
When it rains we check the level of the water in stream*	59	60	18	81
We are afraid to go out when it rains heavily	24	9	2	52
When we go away on holiday or a visit we move important things above the possible water level	22	6	2	57
We are too worried to sleep at night when it rains heavily	35	6	7	52
When it rains heavily we move the car(s) to a safe area*	31	8	5	46
We are afraid to go out when a thunderstorm is forecast	27	9	2	38
We would move to another area if we could	26	0	3	40
We are worried every time it rains	59	15	11	81
We keep a stock of candles etc.	39	16	8	45
We cannot use the cellar*	0	0	0	24
Given up trying with garden	17	7	3	26
Spent money on private flood prevention	21	17	0	22
We move important items above possible water level	N/A	N/A	0	39

* of households to whom behaviour is relevant

elderly and households without a male at home were those whose first action was ring for help or advice from family or the emergency services: telephones then went dead. The violence of this flood is shown by the rapid abandonment of attempts to keep the water out of the house and the general, where possible, retreat upstairs. At that point, many people went to bed which, if interpreted as an attempt to keep warm, was a sensible thing to do.

In so far as it is legitimate to draw any general conclusions from the data to date, it raises doubts as to the likely effectiveness of any plans for evacuation on warning. Even assuming that the warning is transmitted successfully, evacuation is likely to contradict the public's expectations as to the appropriate action to adopt. Consequently, the proportion of households who would evacuate is likely to be quite low.

Table 5

Behavioural Intentions of Floodplain Occupants in Shrewsbury

Do nothing	17
Evacuate to upstairs	75
Evacuate premises	16
Move furniture	48
Flood proofing (contingency)	5
Flood proofing (emergency)	20
Prepare to move	10
Buy provisions	14
Listen to weather forecast	6
Ring Fire Brigade, Police	2

Source: Harding and Parker 1974

Evaluating Intangibles

For some years there has been concern that estimating willingness-to-pay for flood mitigation in terms of the net present value (NPV) of the expected value of future flood losses leads to gross underestimation of willingness-to-pay. Sugden and Williams (1978) argued that differences in the market price of houses due to differences in their flood risk should be equivalent to the NPV of the expected uninsurable losses. In the UK context, however, the housing market is generally believed to be insensitive to environmental bads (Pearce 1978). In the case of flooding, anecdotal and interview data suggests that house purchases are generally unaware of the flood risk at the time of purchase. Consequently no price differences can be expected to be observed. Such differences have, however, been reported in the United States (Antle 1977).

An interview study conducted by Sterland (1973) indicated that the compensation required by households for uninsurable and non-monetary losses from flooding would be substantial.

Consequently we have conducted a series of studies to assess the relative importance to flooded households of the different impacts, both direct and intangible, of floods. The major limitation of these contracts is that they have been conducted as part of consultancy contracts to assess the benefits of flood alleviation for particular locations. As such we have been carrying out essentially basic research in a consultancy context and operating in a very short time frame. Hence the results have to be read as exploratory.

For each of the areas for which results are described extensive interviews were carried out with victims of past floods (Parker et al

Table 6

Uphill: Behaviour on Discovering Water Entering House

	(% of actions)		
	1st action	2nd action	3rd action
Tried to mop up water	2	6	3
Went upstairs	18	21	19
Tried to find out what was happening	6	13	10
Turned off electricity	7	13	0
Awaited rescue	1	2	9
Warned neighbours	4	0	0
Tried to keep water from coming into house	19	6	3
Tried to lift furniture etc. above water level	12	8	12
Left the house	0	6	3
Waited for the water to go down	0	7	30
Trapped by water	2	2	0
Rang for help/advice	12	0	1
Moved dependents/ pets upstairs	8	5	4
Moved necessities upstairs	1	1	3
Other action	4	7	0

1983; Green and Penning-Rowsell, 1983; Green et al 1984; Penning-Rowsell et al 1984). The interviewers were instructed that we placed prime importance on their verbatim record of significant comments by the flooded household, but the interview schedule contained a number of close-ended questions. The interviews took from one to three hours to complete.

In order to assess the relative significance of the different impacts of the flood, respondents were asked to rate on an eleven point category scale each of the different impacts of the flood in terms of the 'relative severity of affect on the household's life' - if they had experienced that impact. The scale was anchored at 0 'no effect' and 10 'most serious effect'. The scale and the set of impacts were listed on a card given to the respondents.

Table 7

Judged Relative Severity of Impacts Over Those Who Suffered Each Impact

	Swalecliffe	Uphill	Southgate	Loughton
Affect on health	7.5	5.0	2.0	0
Having to leave home (evacuation)	10.0	6.0	0	8.0
Damage to replaceable furniture and contents	9.0	7.0	0	5.5
Worry about future flooding	10.0	2.0	8.0	6.0
Loss of irreplaceable objects (memorabilia)	10.0	7.0	0	6.0
All the problems and discomforts of trying to get house back to normal (disruption)	10.0	10.0	6.0	8.0
Damage to the house itself	5.0	5.0	3.0	5.0
Stress	10.0	na	6.5	6.0

The principal areas where we have conducted studies, and the main characteristics of both the flood events and local population, were described in Table 1. The different populations assessments of the severities of the different impacts are given in Table 7. It is notable that both forms of direct damage - that to the structure of the house and to furniture and furnishings - are in each case rated amongst the least significant impacts. So too is health damage to assessed even though in the case of Uphill, 75% of households believed that the health of one or other members of households had been affected by the flood (Green et al 1984).

It could, however, be argued that the relative low significance attached to direct damages compared to impacts like disruption and stress results from recovery of those losses from insurance. In general differences in households' assessments of the relative severity of each form of direct damage are partly explicable in terms of the differences in the degree to which they recovered their losses from insurances as well as the differences in the value of the loss (Green et al 1984). However, whilst households who achieve only a poor recovery of their losses from insurance judge the severity of their direct losses greater than do those households who make a fuller recovery, the former do not judge the severity of their direct losses to have exceeded the severity of their non-monetary losses. The differences in the different households' assessments of severity of affect of the direct damages they have suffered is however principally explicable in terms of the gross value of the loss and, in some cases, whether or not they suffered financial problems as a result of the flood (Green et al 1984).

That there be a correlation between the different households' subjective assessments of the severity of direct damage and the monetary value of their losses is a precondition of the procedure we have termed 'bootstrapping' (Green et al 1983) to enable the non-monetary losses to be evaluated. Given this relationship and assuming, because respondents were asked to compare the relative severities of the different impacts, that the severity scale was used in the same way for each impact, we argue that a judged severity of say '5' for damage to contents is identical to a judged severity of '5' for loss of memorabilia. If from the regression equation relating households' judgments of the severity of direct loss to the monetary value of their direct loss, results in the prediction that a monetary loss of 500 pounds would be expected to correspond to judged severity of 5, we then argue that the monetary equivalent of the loss of memorabilia is 500 pounds.

Although we have to date carried out a number of benefit assessments on these principles, we do not here propose to itemize the results of such assessments given their exploratory nature. A considerable amount of work is required to test the validity and reliability of the different measures and to check the specification of the regression equations before the procedure is generally applicable. Equally such generalization of the procedure requires a data base ranging across a variety of flood events and populations affected. Whilst the magnitude of some impacts on households can be predicted from knowledge of the predicted flood depth (Penning-Rowsell et al 1984) this specification is almost certainly incomplete. However, it may be judged from Table 7 that the monetary equivalents of the non-money losses generally exceeds that from direct damages. Perhaps of more general interest are other lessons from this work.

Measuring Worry

It is generally desirable to develop multiple methods of measuring each of the different impacts. Equally the costs of the impacts do not stop at the households but spillover on to others: health damage results not merely in a loss of utility to those suffering the damage but also may result in additional costs to the health and social services, a loss of productivity and so on (Green et al 1983). Hence we have adopted techniques such as the Nottingham Health Profile (McKenna, Hunt and McEwen 1981) to measure the subjective health status of flood victims compared to that of control samples (Green et al 1984; Emery 1984; Parker, Penning-Rowsell and Green 1984). Event stress is being assessed by Holmes and Rahe's (1967) Social Readjustment Rating Scale (Emery 1984).

Worry about the risk future flooding was not an impact of flooding, or more correctly a response to the risk of flooding, to which we had given a great deal of thought until we undertook the study at Southgate. There during pretesting of the interview schedule residents complained that we asked too few questions about their worry. Rather rapidly we had to decide what additional questions might usefully be asked. We already included in the interview schedule three 7 point category scales on worry about the risk of future flooding, how likely the respondent believed future floods to be, and the threat to health and safety posed by such flooding.

For the Southgate study and subsequently we therefore have included a series of questions on behavioral responses to worry. That is, given that worry is a cognition we argued that it is only useful analytically if it results in some changes in behavior. This is a rather similar approach to the way Fishbein and Ajzen (1975) have related attitude to behavioral intentions. The initial set of questions are given in Table 4 together with the proportions of the sample who agreed with each statement, as are the responses of three subsequent sample at Loughton.

Three points are relevant here. First, at least crudely the questions distinguish between the degree of flood experience and risk of flooding to which the groups are exposed. Secondly, those residents who adopt each of the behavioral options are incurring some cost: in terms of opportunities foregone, as are those who do not go out when a thunderstorm is forecast; additional resource costs, as have those who have invested in private flood protection; or in foregoing uses of their dwelling - those for example who have given up with their gardens. Hence merely being worried involves in these instances a real economic loss. Thirdly, it is difficult to find any set of behavioral options which is applicable across the populations in all situations of flood risk. At Southgate for example only those who experienced flooding through overtopping by the stream checked the level in the stream: those who attributed the flooding to the backing up of sewers found the behavior irrelevant. Equally, from the interviews at Uphill and Swalecliffe, installation of a loft ladder looks as though it might be a useful question - but only for those living in single-story dwellings. Purchase of a tide table also looks useful as a test item for those at risk from tidal flooding though obviously not to those at risk from fluvial flooding. In general then not all behavioral options are relevant or available in all circumstances.

In a large scale study of the costs of the sewer flooding, collapse and renovation to be begun ext year, we intend to develop this measure of worry. Ideally we would wish to derive a Guttman scale of worry by this process.

Worrying About What?

We had, however, until very recently simply considered worry to be either parallel to, or a response to, beliefs as to the risk of future flooding. In part this was because in the Swalecliffe study, respondents had answered when asked about the risk of future flooding that they had already answered that when asked about worry. In all the studies undertaken, including some not discussed here, worry and beliefs as to the risk of flooding were highly correlated. Thus we assumed people said they were worried because they believed the probability of a flood to be high.

But, whilst in the initial benefits studies incorporating the evaluation of the unquantified impacts of flooding on households we derived monetary equivalents of worry, this has progressively appeared to be an incorrect procedure (Parker et al 1984, Green 1984). It seems logically unsound to treat worry, which is some belief about the future, as an impact of some past event. Whilst the experience of the event may have changed the degree of worry felt about potential flooding this change might be described as a change in expectations. This could be characterized as a rational model of worry: that is that worry is caused by consideration of what may happen rather than being a neurotic fear resulting from past experiences. If such a rational model of worry is correct, then it would be double counting to derive an expectation value of the unquantified losses from future flooding by the inclusion of worry since worry itself would be an expectation value for those losses.

Hence the relevant question is why are people worried or what are they worried about? In particular is worry merely a response to beliefs as to the probability of flooding in the future or is it an expectation value?

Worry as noted earlier is found to be strongly correlated to beliefs as to the likelihood of future flooding and also beliefs as to the threat to health and safety posed by future flooding. If it is an expectation value then it should also be correlated to some aggregate measure of the likely impact of future flooding. We have not asked respondents to assess the impact of such a future flood. However, we have included in the interview schedules a measure of aggregate impact of the most recent flood.

Respondents were asked to summarize the overall severity of the flood as it affected their household. The scale used was a seven point category scale anchored at 0 'no affect' and 6 'still have not recovered'. As a test of whether this scale captured the impacts of flooding, we first carried out a factor analysis on the judgments on the severity of the individual impacts for the Uphill sample. The Pearson's product moment coefficients between the households' judgments of the relative severity of the different impacts were taken as the initial matrix. A two factor solution was obtained for which the loadings after varimax rotation are given in Table 8.

Again using the households as the sample, the overall severity of the flood was then regressed upon the household factor scores on these two factors. We could explain the majority of the differences in individual household's assessments of the severity of the flood in this way (Green et al 1984):

Table 8

Uphill: Severity of Impacts, Rotated Factor Matrix

Overall severity = 5.03 + 0.59 Factor 1 + 0.93 Factor 2
($R2 = 0.56$; $F = 36.9$; $< .0001$)

	Factor 1	Factor 2
Health affects	0.77	0.19
Evacuation	0.50	0.38
Damage to contents	0.26	0.74
Worry about future flooding	0.79	0.09
Loss of memorabilia and irreplaceable contents	0.72	0.23
Disruption to household's life	0.26	0.68
Structural damage	0.07	0.86
Eigen value	3.081	1.057
cumulative %	44	59

Source: Green et al 1984

Thus we can explain the majority of the variation in the judgments of severity without knowing anything about the depth of water to which a household was flooded, the extent of financial losses or many other variables.

At this stage we are only interested in factor analysis of the correlations between the individual impacts of the flood as a statistically efficient procedure for summarizing these correlations. In particular we do not believe it useful to name each factor nor do we wish necessarily to ascribe a common underlying causality to the impacts loading on a particular factor. Thus what is of interest is that in this instance at least disruption is associated with the two direct loss variables and not with the other unquantified impacts. We do not wish however to draw any conclusions as to why this pattern of association occurs.

For immediate purposes it is also more important that in spite of its simple form, the overall severity question captures quite well the differences in the various households' experiences of the flood. If then worry is an expectation value we would expect the degree to which a household expresses worry to be a function not only of their belief as to the likelihood of future flooding but also their judgment of the overall severity of the flooding they will experience. The problem with the Uphill data is that interviewing was undertaken after the completion of a new sea wall. And, whilst worry was low and the believed likelihood of flooding in the future was low, the first test of the new scheme neatly divided high worry, belief about high risk respondents from those who reported low worry, and beliefs of a small risk. until this test we understand that there were considerable doubts in the village whether the flood gates could be shut or whether they would be jammed open by accumulated stones. So the data is not ideal for a text of the expectation value hypothesis, nevertheless:

$$\text{worry} = -0.144 + -0.198 \text{ likelihood} + 0.798 \text{ overall severity}$$
$$\qquad\qquad (0.093) \qquad\qquad (0.151)$$
$$R2 = 0.32;\ F = 19.08;\ n = 85;\ < .0001)$$

As a better test of the hypothesis we have, therefore, also used the Loughton data. Here we found that:

$$\text{worry} = 0.326 + 0.421 \text{ likelihood} + 0.568 \text{ overall severity}$$
$$\qquad\qquad (0.083) \qquad\qquad (0.086)$$
$$(R2 = 0.67;\ F = 85.1;\ n = 85;\ < .0001)$$

In neither instance did the inclusion of the variable risk to life and health significantly increase the explained variance.

Consequently worry looks as though it is best considered as an expectation value.

In that we have not applied personal inventory tests to respondents we cannot however wholly rule out the hypothesis that personal characteristics influence the degree to which people worry. But, analysis of survey data for Whitstable, an area which has not experienced severe flooding for thirty years, showed that the degree of worry experienced was not associated with age, sex or social class (Parker, Penning-Rowsell and Green 1984).

Worry, Perceived Risk and Statistically Estimated Risk

To date we have only been able to make extremely crude comparisons between beliefs as to the risk of flooding and some statistical estimate of the probability of flooding. In none of four cases examined in detail has there been sufficient hydrological or hydrographic data to allow the engineers to supply us with precise estimates of the return periods of flooding to different magnitudes.

We can compare the beliefs expressed by households as to the likelihood of future flooding to the depth of water experienced during the last flood but it is not clear whether such a comparison is between perceived risk and a surrogate for statistical probability, or between the former and flood severity. In the crude terms of Table 3, it does appear that those at greater risk believe that they are at a higher risk than those who are less likely to be flooded.

CONCLUSIONS

Worry appears to be interpretable as the subjective expectation value of the future detriment from flooding. This conclusion mirrors the earlier findings that differences in affective beliefs explained differences in willingness-to-pay to reduce the detriment across a range of hazards; between, for example, rock-climbing and living near a nuclear power station (Green, Brown and Goodsman 1983). Similarly that interpersonal differences in affective beliefs provide the best explanation for differences in individuals' willingness to pay to reduce the 'threat to society' from flooding (Green 1984).

Thus the evaluation of worry is potentially a method of determining directly the benefits of flood alleviation to households. The alternative approach of extending the rational man model by way of the bootstrapping procedure similarly looks sufficiently promising for it to be worth undertaking the detailed methodological work necessary to validate the procedure.

More generally relevant to risk research is the strong indication that affective beliefs are essentially rational responses to threat, rather than being the vapouring of feeble minds, and that worry can be comparatively easily measured. There is a danger, which is expressed by that inelegant tautology "psychological fear", of making the double assumption that emotional reactions are not based on any real appraisal of the threat from a hazard and can, in consequence, be massaged away. Instead it may well be that to explain worry is to account more completely for those characteristics of a hazard which are significant to the public. As a research paradigm perhaps we should adopt the parody of Kipling's "If": "if you can keep your head when all about you are losing theirs, then maybe there is something you don't know about the situation".

ACKNOWLEDGMENTS

The work reported here was in part conducted under contracts from: Canterbury City Council; Posford, Pavry and Partners; thames Water Authority and Wessex Water Authority.

REFERENCES

Antle, G., 1977. "Analysis of property value to determine the impact of flooding" in Proceedings of the Social Scientists Conference, Memphis, US Army Engineer Institute for Water Resources: Fort Belvoir, Virginia.

Bennet, G., 1978. "Bristol Floods 1968, Controlled survey of effects on health of local community disaster", British Medical Journal, 3, 454-458.

Bayrisches, Landesamt fur Wasserwirtschaft, 1983. Nutzen-Kusten-Untersuchung zur Teilruckleitung der Oberen isar, Munich.

Canter, D., Breaux, J. and Sime, J., 1980. "Domestic, Multiple Occupancy and Hospital Fires" in Canter, D. (ed.), Fires and Human Behaviour, John Wiley" Chichester.

Chatterton, J.B., Pirt, J. and Wood, T.R., 1979. "The benefits of flood forecasting", Journal of Institute of Water Engineers and Scientists, 33 (3), 237-52.

Dalderfield, E.R., 1976. The Lynmouth Flood Disaster, ERD: Exmouth.

Dreze, J., 1962. "L'utilite sociale d'une vie humane", Rev. Fr. Rech. Oper., 6 (23), 93.

Emery, P.J., 1984. "Health affects of flooding" in Handmer, J. (ed.) Proceedings of an Invitational Workshop, International Flood Research: Lessons for the UK, Geo Brooks: Norwich (in press).

Fishbein, M. and Ajzen, I., 1975. Belief, Attitude and Behavior, Addison-Wesley: Reading, Mass.

Green, C.H., 1980. "Revealed preference theory: assumptions and presumptions" in Conrad, J. (ed.), Society Technology and Risk Assessment, Academic Press: London.

Green, C.h., 1984. "Not looking under lampposts: benefit-cost analysis and social values" in Handmer, J. (ed.), Proceedings of an Invitational Workshop, International Flood Research: Lessons for the UK, Geo books: Norwich (in press).

Green, C.H., Brown, R.A., and Goodsman, R.W., 1983. The Perception and Acceptability of Risks, Building Research Establishment Note N7/83, Fire Research Station: Borehamwood.

Green, C.H., Emery, P.J., Penning-Rowsell, E.C. and Parker, D.J., 1984. The health effects of flooding: a survey at Uphill, Avon, Flood Hazard Research Centre, Middlesex Polytechnic.

Green, C.H., Parker, D.J. and Emery, P.J., 1983. The Real Cost of Flooding to Households: Intangible Costs, Geography and Planning Paper No. 12, Middlesex Polytechnic.

Green, C.H., Parker, D.J., Thompson, P.M. and Penning-Rowsell, E.C., 1982. Indirect losses from urban flooding: an analytical framework, Geography and Planning Working Paper No. 6, Middlesex Polytechnic.

Green, C.H. and Penning-Rowsell, E.C., 1983. Western Enfield Flood Relief Scheme: An extended assessment of selected benefits, Flood Hazard Research Centre, Middlesex Polytechnic.

Grist, D.E., 1978. Individual risk - a compilation of recent British data, SRD, R125, London: HMSO.

Gruntfest, E.C., Downing, T. and White, G.F., 1978. "Big Thompson flood exposes need for better flood reaction system to save lives", Civil Engineering, February, 72-3.

Harding, D.M. and Parker, D.J., 1974. "Flood hazard in Shrewsburg, UK" in White, G.F. (ed.) Natural hazards local, national and global, Oxford: New York.

Holmes, T.H. and Rahe, P.H., 1968. "The Social Readjustment Rating Scale", Journal of Psychosomatic Research, 11, 213-218.

Jones-Lee, M., 1969. "Valuation of reductions in probability of death by road accident", Journal of Trans. Eco. Policy, 3, 37.

McKenna, S.P., Hunt, S.M. and McEwen, J., 1981. "Weighting the seriousness of perceived health problems using Thurston's method of paired comparisons", International Journal of Epidemiology, 10 (1), 93-97.
Mishan, E.J., 1971. "Evaluation of Life and Limb: a Theoretical approach", Journal of Political Economy, 79, (4), 687.
Parker, D.J., Green, C.H. and Penning-Rowsell, E.C., 1983. Swalecliffe Coast Protection Proposals: Evaluation of Potential Benefits, Flood Hazard Research Centre, Middlesex Polytechnic.
Parker, D.J., Penning-Rowsell, E.C. and Green, C.H., 1984. Whitstable Central Area Sea Defense Scheme: evaluation of potential benefits, Flood Hazard Research Centre: Middlesex Polytechnic.
Pearce, D.W., 1978. "Noise Nuisance" in Pearce, D.W. (ed.) The Valuation of Social Cost, Allen and Unwin: London.
Penning-Rowsell, E.C., Parker, D.J. and Green, C.H., 1984. Loughton Brook Improvement Scheme: Assessment of potential benefits, Flood Hazard Research Centre: Middlesex Polytechnic.
Penning-Rowsell, E.C. and Chatterton, J.B., 1977. The benefits of flood alleviation: a manual of assessment techniques, Saxon House: Farnborough.
Penning-Rowsell, E.C., Chatterton, J.B., and Witts, R.C., 1983. An "intermediate" updating of the Middlesex Polytechnic standard potential flood damage data, Flood Hazard Research Centre, Middlesex Polytechnic.
Robinson, D.N., 1971. The Book of Louth, Barracuda: Buckingham.
Royal Society, 1983. Risk Assessment - A Study Group Report, Royal Society: London.
Schmidtke, R., 1984. "Project Appraisal, Resource Allocation and Public Involvement" in Handmer, J. (ed.) Proceedings of an invitational Workshop, International Flood Research: Lessons for the UK, Geo Books: Norwich (in press).
Sterland, F.K., 1973. "An evaluation of personal annoyance caused by flooding" in Penning-Rowsell, E.C. and Parker, D.J. (eds.) Proceedings of a Symposium on the Economic Aspects of Floods, Middlesex Polytechnic.
Sugden, R. and Williams, A., 1978. The Principles of Practical Cost-Benefit Analysis, University Press: Oxford.
H.M. Treasury, 1982. Investment Appraisal in the Public Sector, HM Treasury: London.
Wood, D.G., 1972. The Behaviour of People in Fires, Fire Research Note No. 953, Fire Research Station: Borehamwood.

COMMUNICATING UNCERTAINTY FOR REGULATORY DECISIONS[1]

Rex V. Brown and Jacob W. Ulvila

Decision Science Consortium, Inc.
Falls Church, Virginia 22043

ABSTRACT

To compare immediate regulatory options, the decision maker needs a current assessment of outcomes (e.g., health impact of a proposed standard), including, say, a mean and a range of uncertainty. To evaluate an option to delay the decision, he needs to appreciate how that current assessment might change, introducing a higher order uncertainty. This paper explores graphic and other devices for communicating this elusive distinction, with a view to clarity, action orientation and logical soundness. Other common interpretations of uncertainty are treated as secondary for regulatory purposes. A nuclear backfit decision is used as an example.

KEY WORDS: Uncertainty, Regulation; Decision Analysis; Communication.

1. INTRODUCTION

This paper reports on an effort to develop aids for communicating uncertainty in information developed for regulatory and other risk management decisions. It is part of a larger, ongoing effort on communication methods sponsored by the National Science Foundation, Technology Assessment and Risk Analysis Group, and by the Nuclear Regulatory Commission, Division of Risk Analysis and Operations (Brown and Ulvila, 1984; U.S. Nuclear Regulatory Commission, 1984).

The argument is illustrated in the context of a nuclear regulatory decision: whether to require a certain reactor to "backfit" a costly safety feature, either a vented containment or an alternate decay heat removal (ADHR) system. The information to be communicated is contained in several extensive studies, and synthesized in Burke, et al., 1984.

The word "uncertainty" covers several important issues which need to be kept apart and communicated distinctively. There are no well-accepted vocabulary and graphical conventions for conveying these distinctions, and so we will propose some candidates for consideration.

---

[1]This material is based upon work supported by the National Science Foundation, under Grant No. PRA-8413114, and by the Nuclear Regulatory Commission, Division of Risk Analysis and Operations.

The primary distinction in kinds of "uncertainty" is between what we will call "outcome uncertainty" and assessment uncertainty" or "firmness."

2. OUTCOME UNCERTAINTY AS CURRENTLY ASSESSED

Outcome uncertainty addresses the question: "Based on current information and analysis, what might actually happen and with what probability?" This is a major--perhaps the sole--uncertainty of concern to the user (decision maker) who is prepared to make a decision now, one way or another, without delay. It can be argued that, in this case, only central measures like means are important if the user is risk neutral--i.e., prepared to "play the averages."

For an event, like a core-melt in a nuclear power plant, this assessment is a single number--the probability of the event. For a quantity, like the number of fatalities, it is a distribution. Two important quantity cases need to be distinguished: continuous and discontinuous.

Continuous Uncertainty

In the continuous case, typified by "number of cancers, given a core-melt," the probability changes smoothly in a familiar tapered contribution, like a normal or a lognormal.

Pictorially, we can either represent the complete continuous distribution, say in the density curve form of Figure 1 (many people like cumulative versions, such as complementary cumulative distribution functions, CCDFs--see Figure 6 below--more difficult to interpret); or summarize it as an outcome estimate (say, mean) and a range (say, 90%) as shown in Figure 2. The summary seems adequate for most purposes.

Normally, we believe that some kind of range bar with a cross hatch for the mean estimate, is sufficient to represent the spread of the quantity distribution, but we use the convention of a dashed line for outcome uncertainty, to distinguish it from assessment uncertainty to be discussed later.

It is important that the interpretation of the range stays the same. It would be seriously misleading, for example, if the bar sometimes means a 99% probability and sometimes 80% probability. This inconsistency is often found in current risk analysis reports, when ranges are carelessly propagated by combining upper and lower bounds, which tends to make the bounds progressively more extreme as more and more distributions are combined. What that interpretation numerically should be, is less important. In this paper, we are using, more or less arbitrarily, a 90% range. Possible arguments can be made for anywhere between 80% and 99%. What is important, we repeat, is consistency.

Similarly, the point estimate should have a consistent interpretation--say, a mean as proposed here. Other central measures, such as median, may be markedly different in the skewed distribution commonly found in risk analysis. A more seriously flawed (but, infortunately, common) practice is to use "conservative" (i.e., deliberately biased) estimates in an inconsistent and undeclared way.

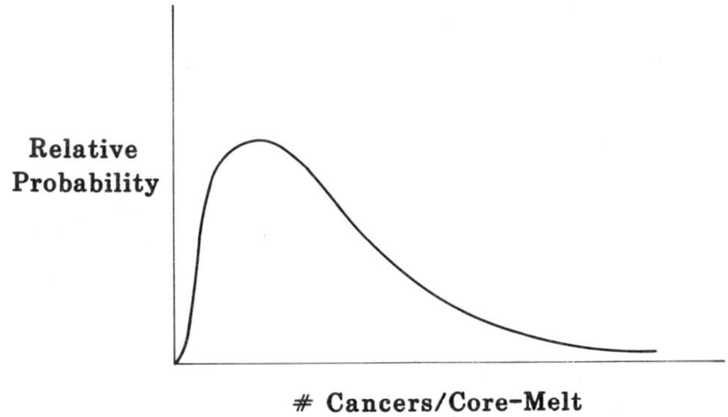

Figure 1

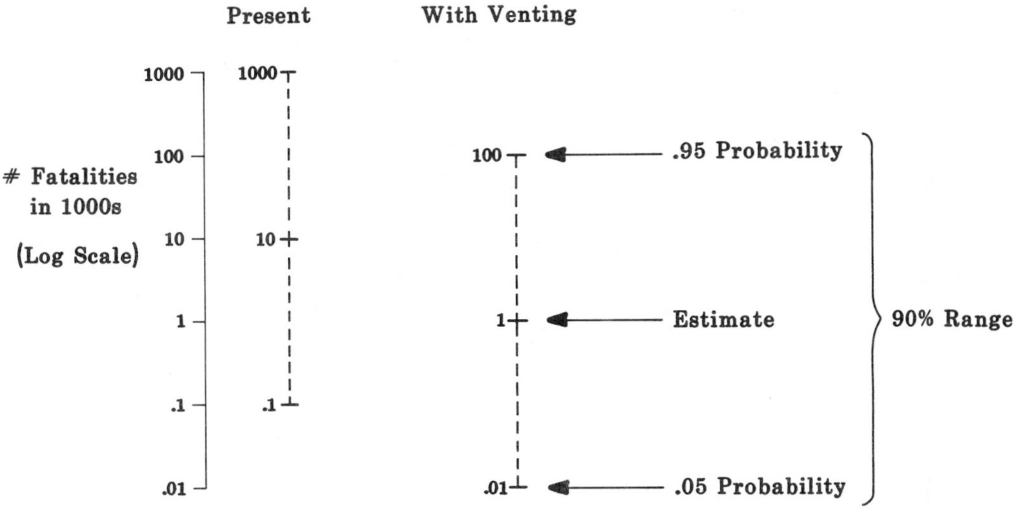

Figure 2

Discontinuous Distributions

By "discontinuous" we mean a distribution with a large mass at zero (corresponding to the "no accident" case) plus continuous segment. This is typified by the assessment of "number of cancers, unconditional on accident."

The best communication of such a discontinuous distribution is less clear than in the continuous case. The full distribution is awkward to represent precisely as a density curve (as in Figure 3) because of the spike at zero and the very small probability of a non-zero quantity. Probability here cannot be conveniently captured by area. A complementary cumulative distribution function (CDDF)is more informative but difficult

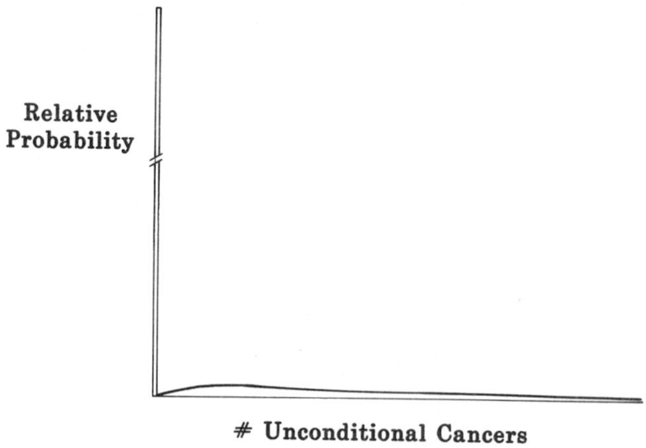

Figure 3

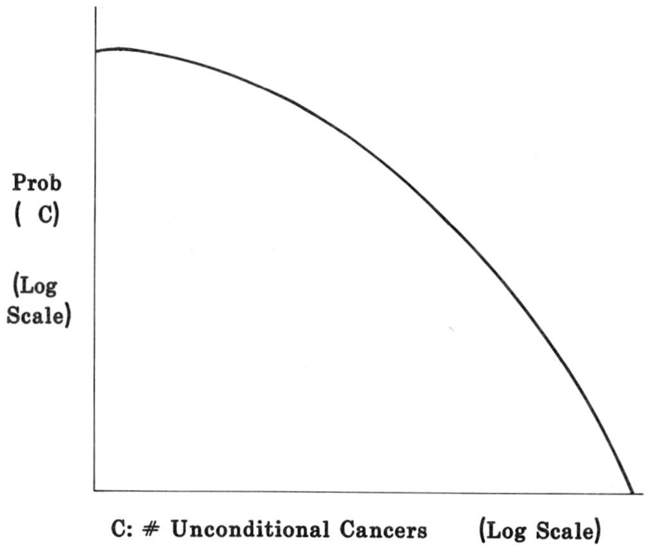

Figure 4

to interpret (see Figure 4). In any case, it is not usually necessary to convey the full distribution.

However, the appropriate summary measures are not as clear as for continuous distribution. A 90% range would be seriously misleading about the shape of the distribution and the mean will usually give an implausible value between the two "lumps" (i.e., one with negligible density). These measures do not, therefore, communicate the essence of the outcome uncertainty. They may seem unreal to the user.

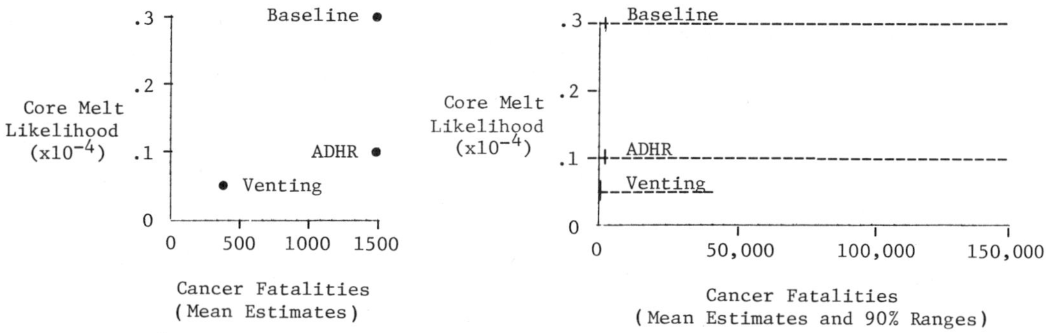

Figure 5

Backfit Impact on Core Melt Likelihood and Resulting Cancer Fatalities

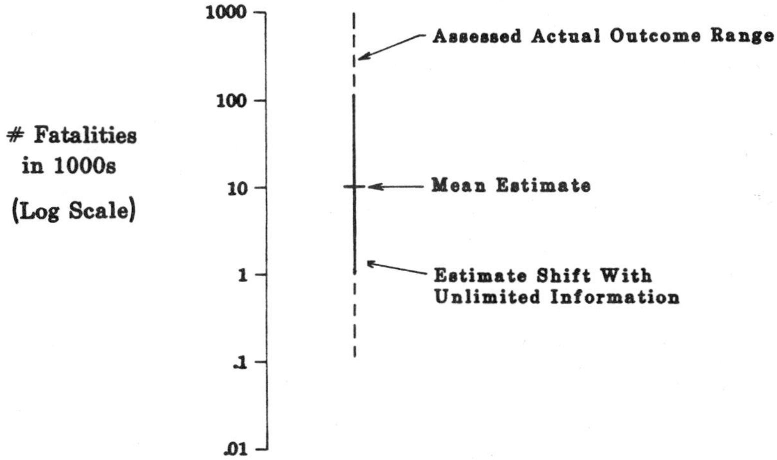

Figure 6

Outcome Range Vs. Assessment Range with Unlimited Information
(Cancer Fatalities Given Core-Melt)

The mean of the distribution does have some significance, in that it is the basis of conventional dollar evaluation of risk in terms of the statistically "expected" value of a life or man-rem. For such purposes, we suggest using the term "statistical expectation" for this mean (where we used "estimate" for the mean of a continuous distribution). The term "mean" can be used more generally to cover both continuous and discontinuous case. (It can be replaced by a "certain equivalent" which will be larger than the statistical expectation, if the assessor is risk averse.)

A more meaningful representation of the essence of the discontinuous distribution might be as two numbers: the mean and the probability of the continuous segment. This description can be enriched (with corresponding loss of conciseness) by including a conditional range for the continuous segments.

Figure 5 shows a variant of this approach, where the conditioning is on a particular, major event--in this case, a core-melt. It is easier to understand, bit it loses information associated with significant events not involving core-melt. This can be handled (again at the cost of conciseness) by adding other conditioning events, such as less serious core damage and by distinguishing core-melt, with and without containment failure (giving, for each event, its probability and conditional deaths, etc.).

A novel, but as yet untried, divice for collapsing a discontinuous distribution into a single number, which might seem less "unreal" than the familiar "statistical expectation" would be to express it as a "standard catastrophe probability." The assessed distribution would be equated to a risk where a standard catastrophe is defined (e.g., 10,000 cancers) and only its probability varies. The measure is logically equivalent to statistical expectation (or a "certain equivalent" in the risk averse case), but whether it communicates better remains to be shown.

3. ASSESSMENT UNCERTAINTY AND FIRMNESS

The notion of firmness. We considered above how to characterize uncertainty within a current assessment about what the outcome of the present or a proposed--state of affairs might be. We now address the higher order question of "firmness:" how might that current assessment change with new information or analysis? In many ways, this is a more important issue--because it has more serious action implications--and one that is certainly more difficult to communicate clearly and concisely. The approach we outline below can be considered as no more than a first pass at a solution.

There are actually three quite distinct types of messages to be communicated in the area of firmness: assessment shift with unlimited information; with achievable information; and prospective reduction in uncertainty. They are difficult concepts and require distinctive language and graphical devices to keep them apart, since they have quite distinct action implications for the user.

The following discussion is intended to provide guidance on how to communicate this assessment uncertainty, through terms and graphic devices. However, any communication of this idea to users will need to be accompanied by some kind of expository text on the concepts themselves, which will be unfamiliar to many users.

These action implications could be analyzed within the framework of personalized decision analysis, using formal values of information concepts. We do not address that issue here, and our approach here does not depend on it.

Assessment Shift with Unlimited Information

The above outcome assessments, including event probability, estimates and statistical expectations, may shift with new information analysis or reflection. How far could they conceivably shift? In particular, could they shift enough to indicate a different decision? (The answer to this question is relevant to help the decision maker second guess the findings themselves, wait for more information, or anticipate controversy.) This shift can be itself assessed probabilistically and we will refer to this as "assessment uncertainty" or "firmness." This uncertainty is sometimes referred to as uncertainty about the "true risk." However, we can think of no meaningful interpretation of what that might mean, other than in the limiting information sense we are using. It has been referred to as the "phlogiston" theory of risk, on the grounds that it treats risk as something akin to a substance contained within a risky situation.

The means of such distributions will always be the same as of the corresponding current outcome assessment, and so, probably, do not require new terms. The distributions about these means, however, will be quite different. In particular, they will be tighter and almost invariably continuous (as defined above). We can again represent them by 90% ranges, but to avoid confusion with outcome ranges, we need a distinctive term, say, "assessment range."

Pictorially, we can use a solid bar for the range (contrasted with the broken line for outcome uncertainty) as in Figure 6.

<u>Assessment Shift After Achievable Research</u>

A related, but separate issue is how the outcome assessment (event probability, estimate, or statistical expectation) might shift with information or analysis that might actually become available, say, by waiting a few years. This is directly relevant to the decision maker in deciding whether in fact to wait for such information. (If the potential shift in assessment is unlikely to change the decision, this would argue against waiting.) The shift can clearly be expected to be less than with the unlimited information on which the assessment uncertainty discussed above is based.

Again, we have a distribution centered on the same outcome assessment mean, but with a still tighter distribution (how tight depends on actually how much new information is to be gathered and this must be specified). As before, it can be summarized by a 90% range, but there is no obvious clear and concise term to refer to it by. It can cumbersomely be described as "range on assessment shift with information achievable by Study A." (Contrasted with: "range on assessment shift with unlimited information.")

Pictorially, the range can be represented by a thicker bar (see Figure 7). There may be situations where we want to distinguish shifts due to <u>different amounts</u> of achievable information. (Perhaps this can be handled by having lines of varying thickness.)

<u>Reduction in Outcome Uncertainty</u>

After new information will have been obtained (whether achievable or unlimited) the new outcome assessment can be expected to have less uncertainty in it. This is the flip side of assessment uncertainty and is related to it, and to the original outcome uncertainty, by well know statistical formulas. Essentially, as the assessment range gets larger, the new outcome range gets smaller, but not in a simple additive way. (In technical terms, the prior variance is the sum of the prior expectation of the posterior variance and the prior variance of the posterior expectation.)

We can call this future assessment "achievable outcome uncertainty" (contrasted with "current outcome uncertainty"), and it can also be summarized by a 90% range. (Technically, an expected range, but the distinction is fine.) Pictorially, we might simply show the current and achievable uncertainties as adjacent dashed lines as in Figure 8.

An Alternative Pictorial Representation

There is serious perceptual problems with the pictorial representation of assessment uncertainty we have used in Figures 7 and 8. Uncertainty about shift in assessment (which gets longer with increasing information) may be confused with uncertainty about outcome (which gets smaller, and is addressed in Figure 9).

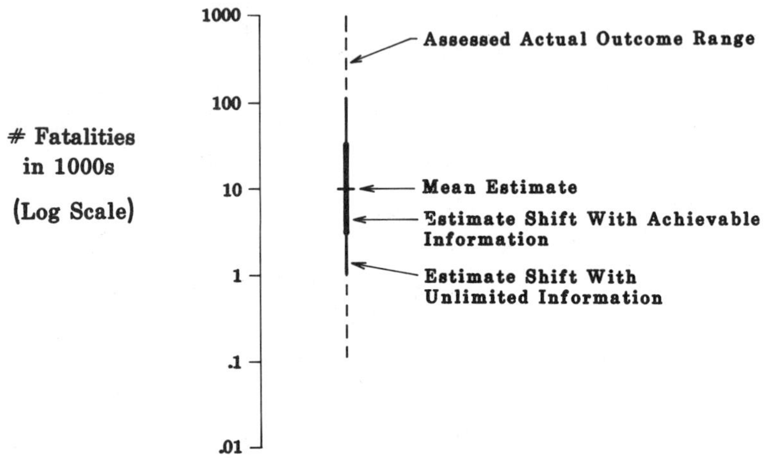

Figure 7

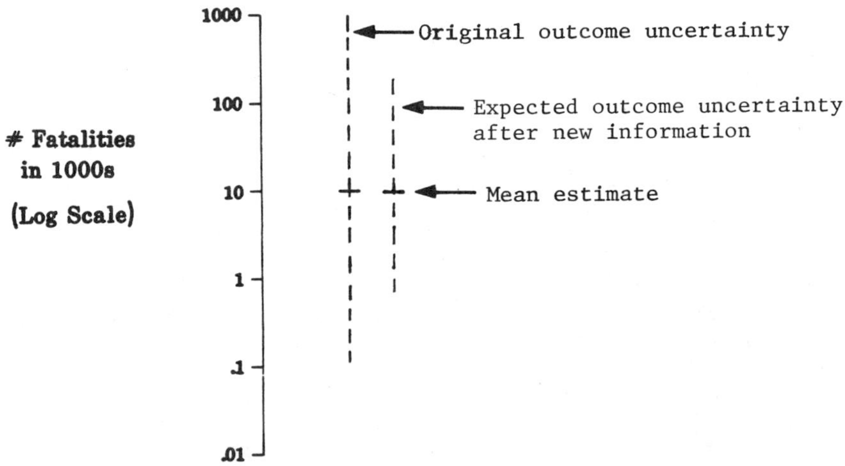

Figure 8

Reduction in Outcome Uncertainty

Figure 9 attempts to clarify the distinction by showing the assessment shift as a fanning out from the current estimate. It also highlights the action implications of the shift by displaying a decision threshold. In this example, it shows that neither achievable nor unlimited information is likely to change the decision, which would argue against delaying a decision. (However, the threshold is liable to change as a result of other information which may come to light).

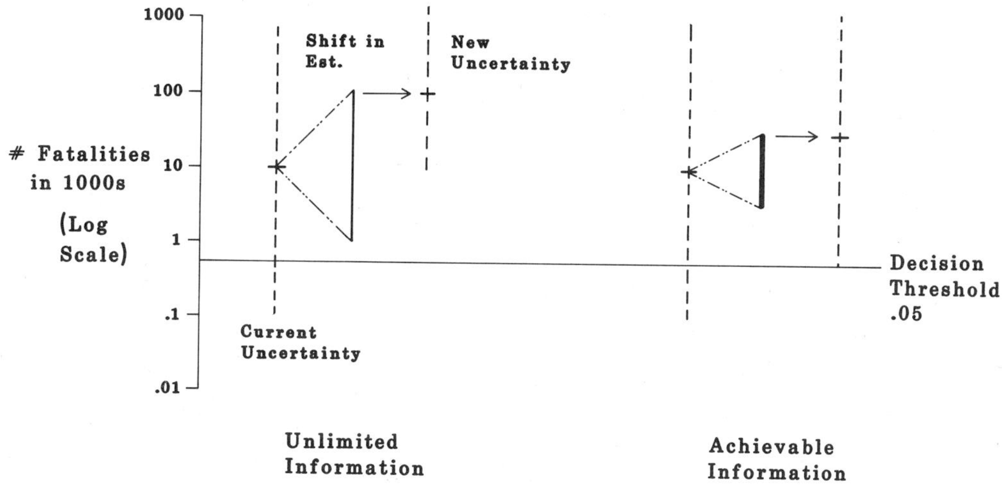

Figure 9

Assessment Uncertainty for Events

Uncertainty about event assessments is somewhat simpler to characterize than for quantities, since current assessment involves only one probability number (not a range). The corresponding concepts, using the same conventions as we have used for quantities, are illustrated in Figure 10. Note that the decision threshold of a probability of 10^{-4} is fixed, by published NRC design objectives, and so is more meaningful than in the "cancer given core-melt" case used above. In particular, this display indicates that, while unlimited information might show the reactor to be unacceptably risky (with or without the venting backfit), achievable information is unlikely to show this, in either case.

4. SOURCES OF UNCERTAINTY

It is often important for a user to understand what accounts for assessment uncertainty (however defined) about a quantity or an event of real interest--let us call it a "target variable." This may help him to validate or second guess uncertainty in the target variable; or give him guidance on where it might be reduced before a decision is made.

The sources of uncertainty can be expressed (and displayed) in terms of the corresponding uncertainty about the component variables from which a target estimate was derived. This is true whether we are dealing with outcome uncertainty or one formulation or another of assessment uncertainty.

In the case of assessment uncertainty, it is normally insufficient to display range bars for the components--at whatever level of disaggregation is of interest. In a decision package, it may be enough to decompose uncertainty in the target evaluation of a regulatory option, say the net-benefit of a vented containment option, into its two first-level components--benefits and costs. Figure 11 displays this for the unlimited information case. Finer disaggregation of the sources of uncertainty may, of course, be added as needed.

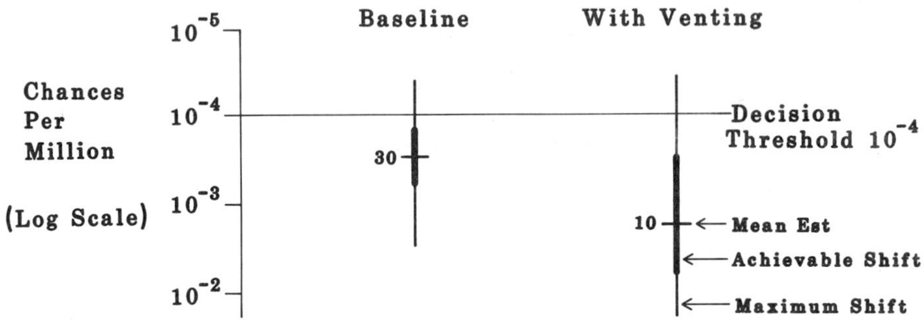

Figure 10

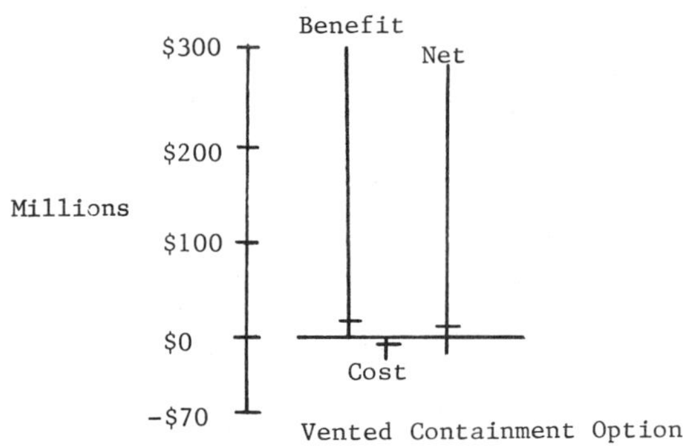

Figure 11

Decomposition of Uncertainty in Net Benefit Assessment
(Estimates and 90% Ranges)

Note, however, that if we are dealing with outcome uncertainty, the target or component distributions may be discontinuous and not usefully represented by range bars (as noted earlier). This would be the case for Figure 11, since the outcome benefit is most likely to be zero (e.g., no deaths or damage). Other display devices would need to be used, perhaps building on the discontinuous distribution ideas suggested earlier.

CONCLUSION

The above uncertainty messages, especially those having to do with prospective new information, are extremely difficult to communicate. However, they are questions which regulatory decision makers ask, and need to ask, and we see no alternative to conveying them as accurately as we can, however difficult they may be to grasp. (In time we may find simpler ways to do it and would appreciate suggestions from readers.)

Past attempts by others to communicate uncertainty that we have seen, seem either logically confused (for example, in not making the above logical distinctions), or their action implications are not clear (e.g., to wait for more information), or they are impenetrably mathematical.

REFERENCES

Brown, R.V. and Ulvila, J.W. Communicating information and uncertainty in the regulatory process (Draft Report). Falls Church, VA: Decision Science Consortium, Inc., October 1984.

Burke, R.P., Strip, D.R., Aldrich, D.C. Regulatory analysis for severe accident issues: An example (Draft Report). Albuquerque, NM: Sandia National Laboratories, August 1984. (NUREG/CR-3976; SAND84-1727).

Trial decision support summary (Draft working paper). Washington, D.C.: U.S. Nuclear Regulatory Commission, Divisions of Risk Analysis and Operations, November 1984.

THE RELIABILITY OF INDIVIDUAL AND POOLED EXPERT JUDGMENTS

Sheldon J. Reaven

327 S. Stadium Hall
Knoxville, Tennessee 37996-0710

ABSTRACT

Can the opinions of individual experts be amalgamated into a collective 'best estimate' that is more likely to be correct than the individual estimates? This question is of interest to risk analysts in taking account of differences among experts within probabilistic (and other) risk studies, and to decision makers in judging among the competing conclusions of the risk studies themselves. Several mathematically and conceptually ingenious methods, Bayesian and otherwise, have been proposed for synthesizing expert opinions, especially when expressed as probability estimates. I argue that (except in special circumstances where a large body of experience with phenomena known to be similar to those under study in virtually all relevant aspects makes the effort hardly worthwhile) no such method can be shown to be more successful than method R: adopt the estimate of a randomly chosen expert. This result is reached by criticizing a supposed parallel (in the Lewis panel's evaluation of the Rasmussen reactor safety report's methodology) between the track records of scientists and bookies in making judgments. The analogy wrongly presupposes that a trait of 'reliability' can be ascribed, in one degree or another, to both a scientist and a measuring instrument. I also argue that expert opinion pooling depends for its justification on some very debatable assumptions about the nature of probability and of science. The fundamental issue is whether mistaken agreement and thoroughgoing disagreement are the rule or the exception in science, especially in areas of importance to public policy.

KEY WORDS: Opinion Pools; Expert Opinions; Nature of Science; Probability; Methodology of Risk Assessment

1. INTRODUCTION

Americans have a healthy distrust of experts, or so we fancy. By this I mean both that we like to think we do in fact bring a skeptical, country shrewdness about believing everything the experts say, and that we like to think that this "show me" attitude is healthy in a democracy. Surely there is every reason to take so Socratic a stance. Thoroughgoing disagreement among the experts, over technical "facts," scientific theories, and personal and community values, characterizes most contemporary issues, from acid rain, nuclear power, and toxic wastes, to the relative merits of supermarket food and health food. And democracy relies upon the free exchange of independently formed opinions in any case. We are nudged to think for ourselves whenever we learn that, in their complacency, "they" were wrong all along about, say, continental drift or the effect of ice cream on diabetics.[1,2]

Still, our claim to healthy wariness may only be so much pretense and self-congratulation. Americans are as quick and uncritical as any people in flocking to accept as gospel the latest (if ever-changing) expert pronouncements as to the real or imagined dangers of caffeine, jogging, cholesterol, termite gas, smoking, suntans, you-name-it. Indeed, too automatic and unwillingness to consider and act upon advice from experts can be too "healthy" for our own good. Who would not defer to a civil engineer in fixing a bridge, or to an M.D. in curing an infection (albeit some would count acupuncturists and herbalists along with the M.D.'s, by way of a perhaps double-edged compliment)? Finally, a sometimes intimidating popular conception of science (even among many scientists) as what Mr. Spock does on Star Trek -- as the apotheosis of objectivity and ironclad logicality -- can over come any proclivity to question what experts say. The lamentable record of the "media" in explaining, cross-examining, and comparing expert claims (as with the risks just mentioned) is one symptom of this conception.

These ambivalent traditions concerning expertise form part of the cultural backdrop, as it were, to our problem, that of justifying methods of combining the opinions of individual experts on a particular question into a single, consensual judgment. For it is precisely in order to deal with the diversity of opinion on questions having significant scientific and technical dimensions that many practitioners of (experts in?) risk analysis and allied subjects have sought to develop ways to synthesize expert judgments into a "best bet" resultant. In the hope that n heads are better than one (or none), several government departments (e.g., the Department of Energy, Nuclear Regulatory Commission, Environmental Protection Agency, and National Science Foundation), sometimes prodded by the requirements of legislation and the need to make and justify some very hard decisions, have appealed for studies of proposed amalgamation methods.

These methods seek to transform the often unyielding problem of examining several expert judgments in depth with a view to determining which one is true, into a problem of processing these judgments according to some widely applicable technique that (typically) assigns differing weights to each judgment. Now no-one has a recipe for constructing sound methods, or for guaranteeing the soundness of a method once it has been invented. So it should not come as a surprise that there is plenty of room for argument on the validity and potential usefulness of amalgamation methods.

What should we expect of such methods? Do any fill the bill? I shall try to clarify these questions and the issues underlying them. My approach will consider what arguments might legitimately persuade a public official or other decision maker that an opinion pooling scheme will improve his or her deliberations, that it works -- not in every case, but more often (or more accurately) than other available procedures. The overriding consideration seems to be this: is the collective judgment more likely to be correct than the individual ones?

The answer turns on assumptions about the nature of probability and of science, I argue. In particular, the answer depends on whether considering more points of view in science is tantamount to increasing the information at our disposal in answering a question. I conclude that amalgamation methods can be expected to work in narrow applications in which the expert/scientist functions essentially as a recording instrument, but that these conditions are not met in the cases for which help from pooling methods is being sought, such as estimating human error probabilities in nuclear power plants, dose-response relationships in toxicology, or the effects of acid rain. I also argue that the very notion of what would count as empirical confirmation of an analgamation method is highly problematic, so that such methods could have a chaotic track record if science is an enterprise that upsets consenses as often as it fosters them.

2. SYNTHESIS METHODS

Genest and Zidek[3] provide a comprehensive survey of proposed techniques for synthesizing expert opinions. Some of these methods simply average the experts' individual numerical estimates;[4] others apply sophisticated (possibly iterated, usually linear) averaging techniques that weight each estimate according to the "reliability" or "predictive ability" of the expert making it.[5] Some methods treat the single judgments as observed data for transformation in accordance with a subjectivist Bayesian analysis of how knowledge rationally is to be updated;[6] still others derive optimal syntheses by axiomatizing the fundamental properties that a pooled judgment might reasonably be expected to exhibit.[7] The latter approach has led to some powerful and surprising[1] theorems restricting the possible forms of synthesis functions.[8] Some methods allow or require the experts to interact, or to be given information on one another's interim estimates, before formulating the final synthesis.[4,9] Some methods can produce counterintuitive[2] surprises.[10]

Most importantly, the methods differ in their requirements as to the "mode" for <u>expressing</u> the expert judgments themselves: in terms of simple probabilities, probability measures, distribution functions, or probability densities; in tems of estimates of non-probabilistic quantities (cost, utility, bounds of uncertainty, or number of options); in terms of Yes-or-No answers to "Will event X occur?"; even in terms of ethical judgments.[3-13] While most methods are designed to handle simple probabilities (for simplicity) or probability measures (for mathematical generality), many modes can be accommodated mathematically to other modes. For instance, the Yes-or-No verdicts to "Will event X occur?" can be regarded as assigning a probability of more than 1/2 or less than 1/2 (as the case may be) to event X. For convenience in presentation, we shall examine the case where the synthesized judgment is to be an estimate P(E) of the actual probability P'(E) of an event E, given individual estimates

---

[1] See Ref. 3. A major issue is whether it matters that linear amalgamation methods do not preserve the independence of events, even if all experts agree on such independence in a given case. I agree with Lehrer and Wagner[46] that independence is not an appropriate requirement. However, hardly anyone questions the "zero preservation property," which has the pooled judgment assign zero probability to an event whenever every expert does so. This requirement is not as reasonable as it may seem. As noted below, some writers would view a unanimous verdict that an event is impossible with skepticism, as a sign of unexamined assumptions and a lack of alternative theories.[14]

[2] For instance, Morris[10] says that a Bayesian decision maker who initially thinks the chance of rain is 1/2 should revise his estimate to 81/82 upon learning that two well-regarded weathermen estimate the same probability as 3/4.

$P_i(E)$, i = 1,2,...,n, from a group of n experts.[3] However, the arguments can be adapted to apply to other modes for expressing expert judgments.

3. TESTING THE METHODS

One can spend a great deal of time discussing the relative merits of the several pooling methods. However, I think it more fruitful to consider what it would mean for any one of them -- call it method M -- to be successful. Proponents of synthesis need to be able to say something like "M improves the chance of making accurate estimates, compared to non-pooling methods of estimation," or "M gives estimates that tend to be more accurate than the non-pooling counterparts." After all, the decision maker needs assurance that M yields dividends in practice. If any theoretical arguments for M's efficacy are valid, this fact ought to be revealed in practice.

We can view the problem as involving a hierarchy of possible ways of estimating, say, the probability P'(E) that a high-level radioactive waste repository in a certain geological formation will not leak for 1,000 years. Lacking well-established, mathematically tractable applicable scientific theories, and lacking a significant body of direct test results in these or very similar conditions, we could (a) select P(E) at random, (b) let a randomly chosen layperson select P(E), (c) let a randomly chosen expert in one of the disciplines bearing on radioactive waste management select P(E), or (d) use M with a group of experts from the various relevant fields.[4] I think most people would hold that (c) is more reasonable than (b), and that (b) is more reasonable than (1) -- although even these claims are open to challenge.(14-16) However, the relevant issue here is whether (d) is to be preferred to (c).

Consider a thought experiment: as opportunities for application become available, we subject problems in a wide range of fields to amalgamation method M. Each application is a trial; for each trial k we record the group estimate g_k and the estimate e_k of a randomly chosen expert. Refer to the latter technique as method R. Now, in many cases, often over a period of decades or more, we shall come to know the true probability t_k with some significant degree of accuracy. Thus an experimental evidence accumulates, we may become convinced beyond what we regard as any reasonable doubt that t_k in the case of the radioactive waste repository leakage has a particular value. In such a case, we will say that M or R turned out to be successful in trial k depending on whether g_k or e_k is closest to t_k. (Geometric or logarithmic measures of closeness may be desirable for low-probability events). Ties are credited to R, since the decision maker's interest is in M's out-performing R, which is always easier and cheaper to execute. We gradually accumulate a running track record for M and R:

---

[3] The questions "who is and isn't an expert?" and "which experts should be consulted?" are very important. I bypass them here because the arguments concerning the relative advantage of the pooled opinion over a randomly chosen individual opinion apply no matter how they are answered. However, I urge the widest tolerance in deciding whose opinions are worth considering.

[4] Other options include letting an expert the decision maker trusts or likes estimate P(E).

$$F(M) = \frac{\text{number of times M turns out to be successful}}{\text{number of trials}},$$

and similarly for F(R). The question, then, is whether the success rate F(M) exceeds F(R) in the long run.[5]

Note that I am not claiming that our coming to believe that t_k is the true probability disproves the view that some other number is, nevertheless, the true probability. Probability claims are in general not subject to deductive proof or disproof on the basis of samples. Nor do I assume that we won't change our mind abut t_k as our knowledge evolves. Even the most celebrated examples of scientific knowledge are tentative and subject to replacement.(17,18) It is just that we use our total system of belief at later times to evaluate our best guesses at earlier times. Finally, in many applications of M, the verdict as to the value of t_k may not be forthcoming for very long, historical time spans, if ever, either because we never happen to acquire the required experimental evidence or theoretical certitude, or because the probability is indeterminate or meaningless in principle. For example, many proponents of the frequency theory of probability would argue that expressions like "the probability that Shakespeare really wrote 'his' plays," "the probability that another oil embargo will occur by 1990," and "the probability that the theory of relativity is true" fall into the indeterminate or meaningless category.(19) Any cases in which t_k is undecided or undecidable accrue to the credit of neither M nor R (regardless of the theory of probability one adopts).

4. DOES F(M) EXCEED F(R)?

Let us examine three possible reasons for thinking amalgamation methods exist for which F(M) exceeds F(R) in the long run. We shall take up the considerations that underly these reasons in the next section.

Empirical Support

In a few studies, the amalgamated opinion of a group of experts, and of a group of laypersons in some instances, provided a better estimate than the opinions of any participant (and a fortiori than R in an overwhelming majority of possible sets of successively and randomly chosen experts).(7,20-22) These studies, while certainly not on the scale of our thought-experiment, involved predictions of weather, football point spreads, Canadian wheat production, and similar items where accuracy was easy to ascertain. The advantage of pooling was slight. However, several other studies of collective and individual expert reliability, while not specifically designed to test amalgamation methods, strongly suggest that pooling has no advantage in several fields.(23,24) Empirical support is

---

5 Most amalgamation methods, particularly ones using weighted averages, aim for maximal specificity compatible with the individuals' judgments. (This notion, like its relative, Popper's idea of corroborability,(47) is hard to pin down). So we can exclude from consideration trivial M's that perform no synthesis: e.g., M = "the true probability lies somewhere between the lowest and highest individual estimate, inclusive," Such lame conclusions tell the decision maker nothing new, and succeed by default if at all. In any case, interval estimates can be handled more naturally within the distribution function mode of encoding expert judgments.

thus mixed; I shall suggest in the next section that the difference between applications where pooling has some success and those where it does not correspond to the difference between occasions when the scientist/expert can be assimilated to a recording device, and occasions when he or she cannot.

Mathematical Support

A very few statistical arguments attribute an advantage to pooling. Foremost among these is the classical result that given a group of unbiased, independent statistical estimators (with known variances) of a quantity Q, there is a unique weighted arithmetic mean of the estimators that minimizes the total variance.[(8)] In order for this result to carry over to the competition between M and R, one would have to regard each expert as a random variable whose values are his or her probability estimates, and say that the expert is an unbiased estimator, that one expert's estimates are stochastically independent of another's, and that the assignment of variances is meaningful. Then the unique weighted mean would have a greater success rate than R. But these assumptions as to the modelability of expertise on an instrumental analogy are highly debatable individually, and beg the question at issue.[6]

Philosophical Support

Some philosophers think it constitutive of rationality that our beliefs and actions take into account the "total information" available to us at any moment. The idea is simply that the more information we consider, the more correct our views become. K. Lehrer, for instance, argues that the opinions of other observers comprise a portion of the total information, and that these opinions perforce must be considered in forming our own views.[(5,25)] He develop an ingenious, if unwieldy, iterated weighted average technique for doing so in which the weights reflect each expert's successively revised ratings of one another's reliability. Lehrer gives an example in which one wants to estimate the probability that the office Xerox will be working on a particular day. It is reasonable to ask the secretaries who work with the machine for their estimates of, say, how many days per week the machine operates. If their estimates differ sufficiently, one's information as to the attentiveness, memory, and honesty of the secretaries will be used to weight each secretary's estimate. Such an approach is more reasonable than making several randomly timed visits to sample the machine's reliability oneself.

I agree with Lehrer's analysis of this case, but it does not clinch the case for amalgamation. The information experts provide often is not the kind of fairly direct observational experience that the secretaries is essence recorded mentally. We turn to experts for new ideas, for creative analysis, for critical thought, as much as for their experience. Thus when we consult another doctor for a second opinion on the chances of recovering from a disease, we benefit not only from the doctor's prior experience with recovery rates -- experience which surely increases the size of the sample of cases over that comprised by the first doctor's experience -- but also (very possibly) from a new theory of the nature, development, and causal mechanism of the disease. Lehrer groups information that is not derived from sampling in the category of

---

[6] Particularly with respect to the variances of the individual expert's estimates (see the discussion of reliability below).

"unregimented information." But it is not at all clear in what way a new idea (say) may be called "information" at all, or whether it is commensurable, integrable, with other modes of information. Nor is it clear that the way to "process" unregimented information is by amalgamating probability judgments. These points become clearer when we examine the assumption that experts have enduring (or temporary) traits called 'reliabilities' that license the assignment of weights to their estimates.

5. THE RECORDER ANALOGY

We have seen that amalgamation methods seem to work in certain special circumstances, but that we have yet to find a good reason to think that such methods tend to be more successful in general, or across wider ranges of applications, than other ways of forming opinions. Now it seems to me that all of the arguments for the superiority of synthesis methods, and the examples offered to explain and justify these methods, depend crucially on two features, the recorder analogy and the tally analogy. It is striking that the problem areas that seek the most help from pooling methods in making judgments are the very areas where these analogies appear to break down.

The recorder analogy explicitly or implicitly regards the expert/scientist primarily as an observation instrument possessing the trait of reliability in similar situations to one degree or another. Thus Lindley, Tversky, and Brown claim "the human mind is used as an instrument for the assessment of uncertainty, much as the ruler, the pan balance, and the pendulum are used as instruments for the measurement of length, weight, and time."[26] Lehrer calls the scientist a "finely tooled...instrument for finding truth;" the weight given to a scientist's opinion is to reflect the degree to which his or her other opinions are thought to exhibit the trait of "trustworthiness".[25,5] Similarly, Winkler's weights are obtained by assessing the "predictive ability" of each expert.[6] Furthermore, the recorder analogy surfaces in talk of "calibrating" experts to ensure that, say, 80% of an expert's claims of the form "the probability of event X is .8" are true -- i.e., that the hit rate matches the expectation.[26]

Let us see what this reification of an "ability to get it right" entails. When we characterize the level of reliability of a measuring device, the reliability figures reflect real and enduring attributes of the instrument. For example, an ordinary thermometer exhibits a certain degree of reliability that reflects factors like the purity of the mercury, the smoothness of the capillary tube, the accuracy of the marking lines, and the rate of expansion of the glass. This reliability may conveniently be summarized in terms of the variance of readouts made when the actual temperature is, say, 60°, and by similar statistics.

Now it is crucial to the case for opinion pooling that it make sense to talk in a similar vein of the reliability of experts. This is because the philosophical and mathematical arguments presented above can be invoked with much greater authority if the recorder analogy holds. The philosophical argument assumes that opinions as to the reliability of individual experts can be construed as information (however tentative or mistaken) about a real property the experts possess in varying measure. The amalgamated opinion would in effect maximize this property by discounting the judgments of the "worst" experts and emphasizing those of the "best" ones. And the mathematical argument would uniquely minimize the variance for the group judgment by in effect constructing it from the enlarged sample of single instrument (each expert) readings, adjusted in proportion to the variance typical of each instrument. So the recorder analogy, where valid, promises to explain not only why the judgment of an individual (well-calibrated) expert can be reliable, but also why the group opinion can be expected to be more reliable than those of its various members. In our terms, F(m) can be expected to exceed F(F) when the recorder analogy holds.

It appears that for some tasks experts can indeed be viewed as measuring instruments (if never entirely so). To be sure, the expert's performance may be very unsteady. Studies in psychology and in the history and philosophy of science have uncovered many sources of "instrumental error": unreliability of eyewitness courtroom testimony, experts' overconfidence in their own judgments, a tendency to pessimism and to extreme views in groups, a tendency to ignore or discount counterexamples, illusory correlations, and the coloring of observations and assumptions by ethical and/or scientific theories.[27-34,24] The fact remains that despite such flaws (if flaws they are), experts do pretty well singly (and better jointly) in a number of areas, such as forecasting local weather.

It was this fact that led the Lewis panel to argue in its review of the methods of the Rasmussen nuclear power safety study[35] that subjective probability judgments are acceptable in probabilistic risk analysis:

> subjective judgments as to the probability of an event can also be reliable. When the judgment is made by a person or group highly experienced with the system in question, the predicted probability is sometimes quite accurate. A case in point is a horse race. The post-time odds are precisely the subjective probabilities of winning estimated for each horse by an experienced group -- the handicappers and to some extent the betting public -- and one finds empirically that the post-time odds and the race outcomes agree with one another to a remarkable degree over the long haul.[36]

6. WHERE THE RECORDER ANALOGY BREAKS DOWN

The Lewis panel's argument is instructive in understanding the difference between areas where experts do well and areas where they don't. It seems to me that three factors explain the successful track record of bookies, meteorologists, and other experts in fields where the recorder analogy has some merit:

Well-defined populations available for study. We have a clear enough sense of where to look for evidence for and against the claim that the oddsmaker's success rate is high. For example, we might examine the population of announced odds at licensed American racetracks since 1960. We also know which situations sufficiently resemble one another to be called horse races: no one thinks a dog race is a horse race.

Clear standards for counting successful and unsuccessful estimates. We know how to tell where one horse leaves off and another begins, which horses win and lose, whether odds were announced at post-time, etc.

Estimates reflecting a sizable body of experience and a limited set of clearly identified variables of unquestioned relevance. This means that the probability estimates reflect knowledge of the ages and past records of the horses, track conditions, how well rested each horse is, and so on. The relevance of such factors to the outcome of the race is clear. Other factors affect the results, but such confounding is not statistically overwhelming.

Are these elements present in the fields where government agencies (and other parties) look to opinion pooling methods for help in "dealing with uncertainty" -- e.g., human error probabilities in nuclear power plants, relative significance of unresolved generic nuclear safety issues, probability of reductions in acid rain resulting from alternative policies, probability of toxic effects as a function of similarity of chemical structure, probability of another oil embargo by 1990, choice among alternative models of dose-response relationships underlying molecular mechanisms in carcinogenesis, probability of leaks from alternative designs for packaging radioactive wastes? I think the three factors mentioned above are not present in these cases, and in other cases where theoretical uncertainty plays a predominant role, and that accordingly talk of expert reliability just does not carry over to such areas.

For consider just how and where we might seek evidence of an expert's (high or low) reliability in such a problem area. Our "knowledge" in such fields is characterized by the presence of many competing theories[7] and mathematical models -- or sometimes by none at all (indeed, dreaming up new theories is the goal of much of the research).[14,30,31,34] Any estimate of probabilities such as those mentioned above depends crucially on the choice of a theoretical framework (whether explicitly formulated, implicit in a mathematical model, or partially specified).[8]

Are we then to suppose that a particular, ostensibly very reliable expert has an uncanny gift, a knack, for picking the right theory or model (much as one might have a knack for accurately estimating distances)? This would be a fundamental mistake, for several reasons.

In choosing between theories A and B, we are in no position to claim that (say) theories "like A" in some respect stand a greater chance of being correct than theories "like B". No metric of similarity among theories, or among occasions of theory selection, is available. So an expert could not legitimately ground an estimate made on the basis of A

---

7 This is by no means a failing. See note 12.

8 See Refs. 33 and 37 for examples in toxicology and radioactive waste management.

(or B) on some characteristic of A (or B) _qua_ theory or model[9]. Nor is the expert's past performance a valid indicator of his or her "trustworthiness." For one thing, the careers of many of even the most acclaimed scientists are marked by a flash or two of brilliance followed by a long twilight of negative results and unfruitful hunches. But this most definitely does not mean that their views deserve less weight than those of the scientist who has had a string of successes. Science profits from "negative" results as much as from positive ones. For another thing, last year's successful judgment and the unsuccessful judgment of the year before may involve different detailed problem areas; which estimate(s) should be counted in assessing reliability? Indeed, the very similarity of occasions over which reliability is to be assessed is often a matter settled only by the advent of a powerful theory.

Another complication is that we often are not in a position to _tell_ whether or not an expert's past judgments were successful, either because the verdict is still out on the phenomena in question, or because the judgment is such that it is unclear whether or not a particular verdict supports it. For example, what do we say about a correct probability estimate that is based on a false theory?

The more closely we examine the _mechanics_ of attempting to study an expert's reliability, the worse the difficulties become. Do we count only the expert's published judgments? This rule would be doubly biased, by the expert's self-selection in submitting materials for publication, and by the filtering of the peer review system. The expert deemed to possess the trait of low reliability may simply be someone who likes to stick his or her neck out with bold ideas that stimulate discussion; an expert with high reliability may reflect a personality that likes to play it safe. And just how far back do we sample opinions? For example, do the estimates rendered by a famous geologist 30 years ago, when continental motion was heresy and plate tectonics had not even been imagined, count against his or her reliability today? Perhaps not -- but it is hard to pin down just _why not_. Moreover, do ten successes in one area count as much as one success in each of ten areas? Does one big success (e.g., cracking the

---

9 One might reply that the experts assess the characteristic of "degree of support by the evidence," and that the better supported theory _ipso facto_ has more chance of being true. This reply does not stand up. (a) "Degree of support" lies in the beholder's eye. Two equally competent experts given identical information can and often do reach opposite verdicts as to which theory best fits "the evidence". Theories in science float amid a complex web of strengths and difficulties; few come out clean from the "ocean of anomalies" they face. So perceptions of degree of support depend on what pros and cons are examined or ignored. (b) External factors too numerous and varied to list influence both the experts' "degree of support judgments" and the estimates of how skillful an expert is in making these judgments. (c) The concept of "degree of support" collapses because "the evidence" is in general not independent of the theory of theories on which it is supposed to serve as a check. (d) The program of some philosophers of science to operationalize the trait of degree of support has led nowhere -- except to counterexamples. At best the trait is highly multidimensional and hence subject to intransitivity problems (cases in which A is better supported than B, B is better supported than C, and C is better supported than A). See the discussions by Lakatos, Kuhn, and others in the book cited in Ref. 30 for more on these points.

genetic code) weigh equally with a run-of-the-mill success (e.g., estimating where Venus will be at this time next Tuesday)? Perhaps not -- but whatever weights we employ can be vitiated by loading the expert's record with many small successes (e.g., estimating Venus' position for Wednesday, for Thursday, and so on).

Some opinion pooling methods(6,7) would bypass these issues by relying upon the expert's subjective judgments as to one another's reliability. This maneuver merely sweeps the difficulties under the rug, as if the issue were just a matter of difficulty in accurately getting a fix on a real trait, the expert's reliability. What evidence can the subjective estimates be based upon that is not itself affected by the difficulties I have outlined? Our reliability assessments are in any case strongly influenced by extraneous characteristics of the expert being assessed: prominence, luck, friendship or rivalry, school ties, research funds, persuasiveness, number of laboratory assistants, and ability to use computers. And the peer review system is not necessarily the best judge, even in the long run, of who is and isn't an expert in the first place.(15,34,38)

In view of the foregoing considerations, I think it is fairly clear that (a) the population of opinion-rendering occasions that would have to form the basis for reliability assessments is highly amorphous and elusive, (b) the standards for what counts as a successful and unsuccessful estimate are also amorphous and elusive, and (c) the estimates themselves mirror basic choices of theory and model, often involving a very large number of known and unknown variables; since past experience in choosing theories does not "project" into current choices, no-one could have a knack for making such choices. This is because nature is more surprising than horse racing.

My quarrel is not with the fact that estimates of reliability may be very arbitrary, a point that many might acknowledge. Rather my objection is to the idea that anything is there to be estimated in the first place. This inappropriate reification is the underlying reason for the breakdown of the recorder analogy in fields where theory is crucially important but far from settled (again, in the very areas where help is being sought from opinion amalgamation methods). The factors that permit the racetrack oddsmaker to make a living are missing in situations in which the expert does not function like anything remotely resembling a measuring instrument: inventing and evaluating theories, models, and other ideas. In such circumstances, the variable of "trustworthiness" represents a trait as fictitious as those of alchemy.[10]

7. THE TALLY ANALOGY

The tally analogy is the second analogy underlying the arguments used to justify and explicate the amalgamation techniques. Recalling our test comparing the success rates F(M) and F(R), we can describe the tally analogy as the assumption that the test counts a series of instances (trials) of a class of phenomena -- "problems amenable to expert opinion pooling methods" -- from which it is meaningful to speak of a representative sample.

---

[10] At least the alchemical theories were grounded in a rich and surprisingly fruitful tradition of scientific theory and experimentation.(39,49)

In the test, we accumulate a relative frequency of success for methods M and R by applying them to problems as they confront us, or as we seek them out. But these fractions need not represent enduring characteristics of M, or by R; they need not approach a "real," stable value that could be thought of as the general effectiveness of M or R (or of M or R in a particular field).

The issue can be brought out in terms of some of the difficulties that undermine the recorder analogy. One can stack the test results by using M repeatedly for thousands of problems in areas where the recorder analogy holds -- i.e., where the three elements involved in the horse-race case are present. But such results are tainted because we cannot say whether these elements obtain in some typical proportion of potential applications of M. In addition, the test treats one major success of opinion synthesis on a par with a minor one, and treats ten successes in one area on a par with one success in ten areas. As before, regulatory politics, the funneling of research money, and other confounding factors may skew the test results. In other words, the test does not reflect an orderly browse among an independently characterizable set of situations (i.e., a reference class) potentially open to expert opinion pooling. Nor can the test be said to represent a random sample from such a set of situations. Recording F(M) is not like tallying (say) the frequency of success of a treatment for polio, where the standards for individuating one case of polio from another are much firmer. The suggestion is rather that F(M) and F(R) may not be measuring any characteristic -- "general effectiveness of M" or "M's effectiveness in toxicology" -- at all. There is a presumption, one amply supported by evidence, that cases of polio share some underlying common nature, so that, apart from sampling error, we would expect the success rate for the polio treatment in one series of traisl to come close to the success rate in another series of trials.[11] However, the existence of, let along justification for, such a presumption is highly questionable in opinion pooling -- what is the underlying common nature shared by situations in which the opinions of experts are sought? Without such a presumption, or without warrant for such a presumption, it becomes very unclear what it means to say that M outperforms R, or vice-versa.

8. BUT WHAT WOULD THE TEST RESULT BE?

Whatever their merits, as a practical matter, the general arguments concerning the recorder and tally analogies will be dismissed by the decision maker if it does turn out that F(M) significantly exceeds F(R) in an extended test. The decision maker may well reason, not unconvincingly, that so definite an experimental vindication of opinion pooling would provide *prima facie* grounds for thinking that something must be amiss with the theoretical arguments against opinion pooling (in areas where the recorder analogy fails).

So we must now consider what the actual outcome of the test would be. I believe that F(M) would not exceed F(R), and that the running graphs of both would appear chaotic. No doubt many others would be equally confident that F(M) would exceed F(R) if M is chosen with care. What accounts for the difference in outlook?

Two very different conceptions of science seem to be at stake. One camp sees it first and foremost as an activity that solves problems, eliminates error, and replaces uncertainty with knowledge. The rival camp sees science primarily as an enterprise that poses new problems, is as given to error as other human institutions, and increasingly reveals the scope of our ignorance (the more we learn, the more we realize there is to learn).[41] One side supports its case with appeals to ideal scientific practices such as critical discussion, open publication, and rigorous testing. The other side cites bandwagon effects, the vagaries of peer review, and the actual history of science, submitting that the ideals are honored more in word than deed, and that their observance is not always desirable and would not turn science into a primarily truth-producing activity in any case.[42] In short, the one group sees answers and a tendency toward consensus as the natural condition of science, while the other group finds questions, disagreement, and change to be typical.

This dispute colors the evaluation of opinion pooling. Consider first the "consensus" theory of science. An adherent might claim: because science proceeds by methods that on the whole resolve disagreements and eliminate errors, a scientific consensus, when and if reached, is more likely to stand up over time than is an individual scientist's opinion rendered when consensus is not present. Consensus in science is to be welcomed as increasing our confidence that any erroneous opinions have been winnowed out. Why not make a parallel claim about a group of experts? The crucial step comes in regarding the amalgamated opinion as an artificial surrogate for an actual consensus, as a means of downplaying (via low weights), if not winnowing out altogether, opinions unlikely to stand up. The very use of terms like "consensual," "pooled," and "group" judgments trades on this step. So there is a fundamental connection between the consensus picture of science and the expectation that an extended test would vindicate one amalgamation procedure or another. Opinion pooling as it were speeds up science (as portrayed by the consensus theory). This is very handy since many decisions can't wait for a real consensus to develop.

Nevertheless (granting the consensus theory for the moment), it is hard to see how the amalgamated opinion could be a legitimate substitute for actual consensus. According to the theory, actual consensus is, or should be, the outcome of a series of arguments and experiments concerning the topic under investigation. With opinion pooling, however, the only questions to be resolved concern which experts are best at forming true (or accurate) judgments. When those judgments depend crucially on the choice of a theoretical framework -- i.e., in cases in which the recorder analogy is invalid -- such questions cannot be resolved by argument or experiment because fictitious traits would have to be measured. One might as well ask which experts are best at precognition.

Be this as it may, the theory that disagreement and change drive science leads to a very different attitude towards opinion pooling. Applied to a group of experts, the theory implies that synthesized judgments do not mimic, much less hasten, the processes of science precisely because consensus is unwarranted, untrustworthy, and ephemeral in science, or science at its best (as the theory reads the history of science).[11] Since the synthetic opinion is supposed to be the next best thing to consensus, if there is no profit in the latter, there is none in the former. Accordingly, it becomes reasonable to expect that any opinion pooling method would do poorly in the extended test. In short, the disagreement theory of science implies F(M) would not exceed F(R), unless by luck.

9. PRACTICAL IMPORT: AN EXAMPLE

We have seen how different views on the potential practical success of opinion pooling can be traced to different ideas about the nature of science. These contending ideas can affect day-to-day science and technology policy decisions. For example, a recent report of the President's Office of Science and Technology Policy(43) claims to spell out the current state of knowledge in the field of carcinogenicity testing. 77 of 81 expert reviewers endorsed the report's conclusions.(44) 60 of the 77 rated them "very good" or "outstanding". Presumably, 60 experts would assign a fairly high probability to the truth of the conclusions, 4 would assign a fairly low probability, and the other 17 would assign one greater than 1/2. We may also suppose that by and large experts among the 77 would assign lower (or even equal) reliabilities to the 4 dissenters than to one another. Virtually all opinion pooling methods proposed so far would crank out a group estimate strongly at odds with the estimates of the dissenters. To all intents and purposes, these methods tell the decision maker not to take the dissenters' ideas seriously. And the consensus view of science makes it reasonable to accept the report's conclusions without further examining the substantive issues raised by the minority.

Not so with the disagreement view of science. It has us ask: <u>why</u> did four experts dissent? did they have good reasons? did the majority miss the forest for the trees, join a bandwagon of prevailing opinion, or fail to identify or pursue alternatives to the basic assumptions in their fields -- especially inasmuch as many frontier areas of science are

---

[11] The idea that consensus is undesirable, even suspect, presents both a methodological and a moral challenge to consensus approaches to science. The philosophers of science P. Feyerabend and K. Popper agree on this point (and little else). Feyerabend is an "epistemological anarchist," defying anyone to spell out the "scientific method". He says that unanimity among experts seldom rests on considerations internal to the subject matter, and instead is often the result of political decisions, shared prejudices, each expert's assumption that the other experts have checked out the details, and not having alternative theories to consider and do battle with. He says <u>science</u> will make more progress if many theories and methods are used simultaneously, even if some seem experimentally refuted or beyond repair at a particular time. He also argues that modern society and the individual scientist will be healthier and happier with such an opportunistic pluralism.

Popper is a "falsificationsit critical rationalist," insisting that a method of conjectures and refutations, testing one theory at a time and discarding it when the results so indicate, drives science and in fact makes it rational. However, he too takes unanimity or near-unanimity of opinion as a hallmark of second-rate science, of a failure to advance bold new ideas and to criticize old ones. See the discussion in the book cited in Ref. 30.

involved in carcinogenicity assessment? Different investigators may give different answers. But the point is that answering such questions required hard, detailed study of both (a) the substantive scientific and statistical debates over the nature, causes, and detection of carcinogenic and toxic effects, and (b) the human and institutional settings in which the science and statistics are formulated, paid for, and interpreted. There appears to be no way to subsume all of this in reliability ratings -- especially if they are set by the experts themselves.

The landslide of reviewer support surely makes it more likely that the OSTP report will influence national policy on toxic chemicals (in food, groundwater, consumer products, agriculture, industry, and waste management systems). The report's evaluations of the status of the relevant uncertainties -- the conditions under which structurally similar chemicals produce similar effects in similar species, dose-response relationships, molecular chemistry of carcinogenesis and mutagenesis, statistical indicators and laboratory tests -- may well become more entrenched, at least in the short run. Opinion pooling extends this imprimatur by seeming to certify that siding with the majority is the best bet in this case, and so may influence public policy.

How one regards such developments depends on how one thinks science works (or should work). My own view is that opinion pooling should not be used to evaluate, much less settle, the question of the viability of the minority position(s). The very lopsidedness of opinion could conceivably have a self-fulfilling, snowball effect, by making it easier to dismiss alternative points of view as surely unwarranted, to refuse to pay for research on them, to keep dissenters from review panels or prestigious jobs, etc. Too wide a use of opinion pooling could reinforce such tendencies by institutionalizing a kind of majority rule (with proportional representation embodied in weights) in science. The example of the OSTP report shows that the use and acceptance of opinion pooling matters because of the potential effect on our knowledge as well as on our public policies.

10. CONCLUSION

Distilling the arguments advanced so far, my advice to the decision maker amounts to this: "synthetic expert additives" are not good for you. Four indications support the diagnosis:

(1) Opinion pooling won't improve your decisions, because pooled judgments have no advantaged over individual ones (except in the comparatively insignificant cases in which the recorder analogy holds). This so, I have argued, mainly because the notion of an expert's reliability collapses when extended to judgments for which explicit or implicit theory assumptions are indispensible to each expert's verdict.

(2) You will open new cans of worms if you resort to opinion pooling to help analyze important issues: you will have to confront a controversy over the legitimacy of the method in each application. For suppose you propose to use some M in connection with some new issue, say in estimating a risk. Many parties -- government agencies, citizen and industry groups, ect. -- will be interested in the outcome. Some are sure to challenge M's validity even if M has already acquired a string of what hindsight reveals to have been successful applications. For even in that case the challengers can claim that M's earlier successes either were flukes (perhaps on the basis of the disagreement theory of science), or involved problems where the recorder analogy has real merit. And disputes over M's

validity will be still more contentious if M has rung up failures as well as successes, or is to be tried for the first time. The seemingly endless debates over the status and proper scope or probabilistic risk assessment[45] stands as a lesson to anyone who thinks opinion pooling will make the decision maker's road easier.

(3) Despite its explicit arbitrariness (that of M being better disguised), R is preferrable to M in some ways (if you have to use any formal mechanism operating upon a set of individual expert opinions). The more sophisticated pooling methods (intended to overcome the objections to the simplest methods) are cumbersome, time-consuming, and expensive, especially compared to R. For that matter, the more nearly unimodal or multimodal the distribution of expert estimates, the less likely it is that R will select an unpopular opinion. The more unimodal the distribution, the more likely it is that M and R will give similar results. When it is multimodal, R avoids the "smearing" effect of some amalgamation methods (e.g., those that produce a group estimate lying in between the two clusters of a bimodal distribution).[12]

(4) Opinion pooling should be avoided in democracies because it tends to institutionalize the status of and reliance upon experts. It encourages the decision maker to process rather than examine expert claims, and encourages the citizen to consume rather than judge expertise. And it pushes society in the direction of attaching far too much importance to questions of who is and is not an expert in the first place. Such tendencies are not healthy for free societies.

REFERENCES

1. Hallan, A. "Alfred Wegener and the hypothesis of continental drift," Scientific American 232, 88-98 (February 1975).
2. Kolata, G. "Dietary dogma disproved," Science 220, 487-8 (April 9, 1983).
3. Genest, C. and Zidek, J. Combining Probability Distributions: A Critique and an Annotated Bibliography, Technical Report No. 316, Dept. of Statistics, Carnegie-Mellon University: Pittsburgh, March 1984.
4. Embrey, D. Humphreys, P., Rosa, E., Kirwan, B., and Rea, K. SLIM-MAUD: An Approach to Assessing Human Error Probabilities Using Structured Expert Judgment, Vol. 1, DOE-BNL NUREG 51716 and NRC NUREG/CR-3518, Brookhaven National Laboratory: Upton, N.Y., March, 1984.
5. Lehrer, K. "Rationality as weighted averaging," Syntheses 57, 283-295 (1983).
6. Winkler, R. "The consensus of subjective probability distributions," Management Science 15, 2, b-61 - B-75 (October 1968).
7. Wagner, C. "Allocation, Lehrer models, and the consensus of probabilities," Theory and Decision 14, 207-220 (1982)
8. Wagner, C., "The formal foundations of Lehrer's theory of consensus,"

[12] The point being that the smearing can lead to a pooled opinion held by practically no-one and rejected by nearly everyone. Methods that avoid smearing often encounter other difficulties, such as a tendency towards producing bland amalgamated opinions that do not synthesize (see note 5). A discussion of the distributions of opinions and reliabilities with which these problems arise is beyond the scope of this essay.

165-180 in R. Bogdan, ed., Profile: Keith Lehrer, D. Reidel: Dordrechty Holland, 1980.
9. Brown, B. and Helmer, O. "Improving the reliability of estimates obtained from a consensus of experts," P-2986, the RAND Corporation, 1964.
10. Morris, P. "Decision analysis expert use," management Science 20, 9, 1233-1241 (May 1974).
11. Graham, J. and Henrion, M. "A probabilistic analysis of the passive-restraint question," Risk Analysis 4, 1, 25-40 (March 1984).
12. Segerberg, K. "A neglected family of aggregation problems in ethics, S Nous 10, 221-245 (1976).
13. Wagner, C., "Aggregating subjective probabilities: some limitative theorems," Notre Dame Journal of Formal Logic (in press).
14. Feyerabend, P. Against Method, Humanities Press: New York, 1975.
15. Feyerabend, P., Science in a Free Society, New Left Books: London, 1978.
16. Broad, W., "Paul Feyerabend: Science and the anarchist," Science 206, 534-537 (November 2, 1979).
17. Munevar, G., Radical Knowledge, Hackett: Champagne-Urbana, IL, 1981.
18. Quine, W., "Two dogmas of empiricism," 20-46 in Quine, From a Logical Point of View, Harvard University Press: Cambridge, 1961.
19. Black, M. "Probability," 464-479 in P. Edwards, ed., The Encyclopedia of Philosophy, vol. 6, Macmillan: New York, 1967.
20. Helmer, O., "The systematic use of expert judgment in operations research," P-2795, The RAND Corporation, 1963.
21. Sanders, F., "On subjective probability forecasting," J. Applied Meteorology 2, 191-201 (1963).
22. Eisenberg, E. and Gale, D. "Consensus of subjective probabilities: the pari-mutuel method," Annals of Mathematical Statistics 30, 165-168 (1959).
23. Ascher, W., Forecasting: An Appraisal for Policy-Makers and Planners, Johns Hopkins University Press: Baltimore, 1978.
24. Armstrong, J. "The seer-sucker theory: the value of experts in forecasting," Technology Review 83, 18-24 (June/July 1980).
25. Lehrer, K., "Social information," The Monist, 473-487 (1977).
26. Lindley, D., Tversky, A., and Brown, R., "On the reconciliation of probability assessments," J. Royal Statistical Society A, 142, 146-180 (1979); includes discussion.
27. Buckhout, R., "Eyewitness testimony, Scientific American 231, 6, 23-31 (December 1974).
28. Slovic, P., Fischhoff, B., and Lichtenstein, S., "Rating the risks," Environment 21, 3, 14-39 (April 1979).
29. Trommsdorf, G., "Group influences on judgment concerning the future," 145-166 in M. Irle, ed., Studies in Decision-Making: Social Psychological and Socio-Economic Analyses, de Gruyter: Berlin, 1982.
30. Kuhn, T., "Logic of discovery or psychology of research?" 3-39 in I. Lakatos and A. Musgrave, eds., Criticism and the Growth of Knowledge, Cambridge University Press: Cambridge, England, 1970.
31. Lakatos, I., "Falsification and the methodology of scientific research programmes," 91-195 in Lakatos and musgrave, op. cit.
32. Golding, S. and Rorer, L., "Illusory correlation and subjective judgment," J. Abnormal Psychology 80, 3, 249-260 (1972).
33. Whittemore, A., "Facts and values in risk analysis for environmental toxicants," Risk Analysis 3, 1, 23-34 (March 1983).
34. McCain, G. and Segal, E., The Game of Science, 4th ed., Brooks-Cole: Monterey, California, 1982.
35. Lewis, H., Budnitz, R., Kouts, H., Loewenstein, W., Rowe, W., von Hippel, F., and Zacharisasen, R., Risk Assessment Review Group Report to the U.S. Nuclear Regulatory Commission, NUREG/CR-0400, U.S. Nuclear Regulatory Commission: Washington, D.C., September, 1978.

36. U.S. Nuclear Regulatory Commission, Reactor Safety Study, WASH-1400, U.S. NRC: Washington, D.C., October, 1975.
37. Proceedings of the 1983 civilian Radioactive Waste Management Information Meeting, CONF-831217, U.S. Department of Energy: Washington, D.C., February, 1984.
38. Cole, S., Cole, J. and Simon, G., "Chance and consensus in peer review," Science 214, 881-885 (November 20, 1981).
39. Toulmin, S. and Goodfield, J., The Architecture of Matter, Harper: New York, 1963.
40. Dijksterhuis, E., The Mechanization of the World Picture, Oxford: Clarendon, 1961.
41. The Encyclopedia of Ignorance, Pergamon Press: Oxford, England, 1977.
42. Lakatos, I., and Musgrave, A., op. cit.
43. Chemical Carcinogens: Review of the Science and Its Associated Principles, President's Office of Science and Technology Policy: Washington, D.C., in Federal Register, May 22, 1984, 21593-21661.
44. Marshall, E., "Carcinogenesis without controversy," Science 224, 1078 (June 8, 1984).
45. Probabilistic Risk Assessment (PRA) Reference Document: Final Report, U.S. Nuclear Regulatory Commission: Washington, D.C., NUREG - 1050, September, 1984.
46. Lehrer, K. and Wagner, C., "Probability amalgamation and independence: a reply to Laddaga," Syntheses 55, 339-346 (1983).
47. Popper, K., The Logic of Scientific Discovery, Harper: New York, 1959.
48. Putnam, H., "Meaning and reference," in S. Schwartz, ed., Naming, Necessity, and Natural Kinds, Cornell University Press: Ithaca, N.Y., 1977.
49. Reaven, S., "Function, translation, and indexicality, Philosophical Studies 45, 31-50 (1984).

LEARNING FROM THE BLIND MEN AND THE ELEPHANT, OR SEEING THINGS WHOLE IN RISK MANAGEMENT

Steve Rayner

Oak Ridge National Laboratory
Oak Ridge, Tennessee 37831

ABSTRACT

Much of the literature on risk perception and management published over the last few years has asked how society should resolve the question, "How safe is safe enough?" There has been political and technical disagreement over the types of answers that may be given, as well as over the social values attached to perceived probabilities and magnitudes of various outcomes. Despite controversy, there seems to have been a large measure of consensus that, "How safe is safe enough?" is the right question to ask. This paper sets out to question that assumption. Various ingenious techniques of risk analysis have sought to discover the real risks inherent in various activities, but the picture each presents is like one of the proverbial blind men's perception of different parts of the elephant. From a socio-cultural viewpoint it can be seen that no single answer can be given to the problem of adequate safety in a complex society which contains a wide variety of perceptual biases about danger, expectations of the good life, and levels of demand for safety. The paper will argue that, from a societal risk-management perspective, we should be addressing a different range of questions that views societal risk as a whole rather than as the sum of individual hazards. The relevant questions would then be those of how consent is obtained from affected parties, what principles are being invoked for the distribution of liabilities that arise, and how trustworthy are the institutions that societies must depend on to manage technologies and the environment?

KEY WORDS: Risk: Epistemological Definition, Fiduciary Trust, Consent, Management of Liabilities.

Three notions have dominated discussion of societal risk management since Chauncey Starr's seminal article in 1969. The first of those is that there is a definable, i.e., measureable, phenomenon called risk. Second is that societal risk management is concerned with minimizing the probability and magnitude of undesired consequences without incurring excessive costs. Third is the assumption that the critical question facing risk managers charged with this delicate balancing act is "How safe is safe enough?" The third assumption can be interpreted in two ways, (a) safe enough to obviate any substantial danger to public safety (risk assessment), and (b) safe enough to satisfy public demands for safety (risk management). Today, I wish to focus exclusively on the application of this question in societal risk management.

It was clear fro Chauncey Starr's acceptance address at yesterday's SRA Awards ceremony, that jut as he was the first to approach risk management in this way, he is now among the first to shift his focus away from determination of acceptable risk and towards the social issues associated with institutional trust. Today I want to push on in the same direction and challenge all three of the dominant assumptions of risk management. I suggest:

- That we have reified the concept of risk to the extent that we have lost sight of many aspects of this multifacated phenomenon in pursuit of (a) bettering our own understanding of probabilities, and (b) correcting public misunderstanding about them.
- That while assessments of probabilities and magnitudes of undesired outcomes is essential to making engineering decisions about competing designs or alternative materials, they are largely irrelevant to societal technology choices.
- That the critical question facing risk managers is not "How safe is safe enough?," but "How fair is safe enough?"

Let us take each of these three dominant assumptions in turn:

1. That there is a single, if complex, phenomenon called "Risk."

Since Aristotle, Western thought has been dominated by a view of concept formation that I characterized some 5 or 6 years ago as the doctrine of essentialism (Rayner 1979). This states that a category can be defined by a set of necessary and sufficient conditions which much be satisfied for a particular case to be included in its extension. The principle of substitution (Jevons 1874) will, therefore, apply. That is to say, "Whatever we know of one object in a class, we know of the other objects, so far as identity has been detected between them."

I contend that this assumption is alive and well in the fledgeling discipline of risk analysis. In his paper presented to the General Motors conference on "Societal Risk Assessment," which, incidentally was subtitled "How safe is safe enough." William Lowrance (1980) defined risk as "a compound measure of the probability and magnitude of adverse effect." I recall a meeting of greybeards at the Russell Sage Foundation in 1980, which I was privileged to observe, in which we played with the formula:

$$\text{RISK} = \frac{\text{Probability x Magnitude}}{\text{Time}}$$

However and wherever it is discussed, it seems that there is a consensus that the essence of risk consists of the probability of an adverse event and the magnitude of its consequences. Other considerations, such as those of societal risk management, may be recognized as important, but are viewed as deriving from this Aristotlelian definition. I shall argue that this set of necessary or sufficient conditions may be adequate to define risk at the level of engineering-type calculations, but is quite misleading at the broader, more intractable, level of risk management. This requires a little more consideration of the epistemological problem raised by the type of definition that we can give to the concept of risk.

Wittgenstein offered the first serious critique of Aristotlelian essentialism in modern times. In rejecting the principle of substitution, Wittgenstein (1953) asked the question, "What is a game? The traditional response would be to enunciate the necessary and sufficient conditions for a game, but Wittgenstein demonstrated the impossibility of this approach:

"For if you look at them (games), you will not see something that is common to all, but similarities, relationships, and a whole series of them at that." (1953: 31^{e}.66)

For example, let us try to derive the essence of a game from the list of cricket, football, chess, and solitaire. Football and cricket are both played on a field by two teams of eleven players with a ball, but the constitutive rules, aims, and strategies are different. The only similarity that this pair has with chess is that they are all competitive, but solitaire is not. Solitaire does have in common with chess that it is played with pieces on a board. So there is a series of links which connect football, cricket, chess, and solitaire, but there is no single feature or set of necessary and sufficient conditions which is characteristic of, and common to all. It may be objected that solitaire is a puzzle, but since we have failed to detect an essence of games, any attempt to distinguish puzzles as anything but a set of close relatives in the wider family-resemblance category of games would seem to be fruitless.

Wittgenstein described this state of affairs by the celebrated analogy of a thread which does not take its strength from any single strand running the whole length, but from the overlapping of many strands (1953). The category of games is an open concept in Wittgenstein. We can recognize games as a form of characteristic human activity and it can be shown that individual games are similar sorts of things, but no single universal feature or essence can be stated. Wittgenstein claimed that there were many concepts similar to the category of games in this respect. Indeed he sometimes seems to imply that all categories are like this, and it is this hard version of the theory that I wish to uphold here today. For even if we do discover a single strand running the whole length of the thread, we would not be justified in regarding this to be the source of the thread's strength.

Similarly, the occurrence of a single feature in every member of a category is not sufficient to justify any claim that this is the essence of the category. For example, the fact that all games are goal-oriented activities (beating the opposition, solving the puzzle, or completing a number of rounds) cannot establish goal orientation as an essence since it is also a consistent feature of activities which are not games.

Wittgenstein called this method of concept formation family resemblances. Similar notions may be found in the works of the Soviet psychologist Vygotsky (1952), who employed an analogy of a chain rather than t thread; British philosopher Gilbert Ryle (1951), who coined the term polymorphous classification, and in modern microbiology where the terms polytypic and polythetic classification are used.

The main point I want to convey here is that in all of these systems of concept formation, items at one end of the chain which constitutes a category need not have any conditions in common with those at the other end. By the time I have finished speaking, I hope to have convinced you that this is the case with the category of risk. For example, common-sense definitions frequently do not depend on probabilities of loss. My own father used to puzzle me by insisting that he occasionally liked to bet, but he never gambled. His definition of risk was not dependent on probability of loss, but on the prospect of losing more than one can afford. On the other hand, economists sometimes like to talk of upside risk, where no losses are involved.

Like the "six men of Indostan," in J.G. Saxe's celebrated poem (1868), risk analysts of different scientific disciplines are grasping a different part of the beast and characterizing the whole by the parts.

Unlike the six, we are a tolerant community and usually allow that each is dealing with a different facet of the same problem. However, I really do begin to wonder if the scientific and the political communities are really tackling the same beast at all.

In order to make my point, let me move on to the second of the three assumptions I have been foolhardy enough to challenge today.

2. That the challenge of risk management is to minimize the probability and magnitude of adverse events without incurring excessive costs.

Economic rationalists argue that individuals decide to take a risk by first weighing its potential costs and benefits and then opting for the course of action that they think will maximize the advantages that will accrue. This assumption, popular among professional policymakers and sanctified by the U.S. Supreme Court, is well known as the utility principle.

The problems with the utility principle in public policymaking are well known. I shall not rehearse them today. Ken Arrow (1951 has shown that collective preferences cannot be derived by aggregating individual choices. Peter Self (1975) has illustrated the absurdity of trying to trade off incommensureables by reducing them to dollars or cents. Herbert Simon (1979) has shown us how decisions are made under conditions of bounded rationality and Mary Douglas (1978) has argued that the bounding of rationality is a cultural phenomenon based on differing experiences of social organization between the various institutions and constituencies that compose a complex industrial society.

Despite this overwhelming corpus by leading living social thinkers, including 2 Nobel prizewinners; risk managers, and the political scientists who study and advise them, are still asking, "How safe is safe enough?." Two factors immediately suggest themselves to account for the durability of this unanswerable (or in Weinberg's term, transcientific) question:

a. The absence of an alternative approach.
b. The predominance of the essentialist definition of risk as a question of probabilities and magnitudes.

I hope that you are beginning to see the link between my first 2 themes. It is this. The absence of an alternative to the cost-benefit, utility-principle, how-safe-is-safe-enough? approach to risk management is directly due to our persistent acceptance of the essentialist definition of risk.

Even the most enlightened of our colleagues in sociology, anthropology, and psychology have framed the problems of public acceptance of risks in terms of differential perceptions or probabilities. But, the choices between those probabilities are incomprehensible to most of the public and, I'll warrant, to a lot of policy makers as well. Fate has favored me with university education in three different disciplines and I am privileged to work in one of the world's premier scientific laboratories, but ask me the meaning of the difference between a probability of 10-5 and 10-8 and you might as well ask me to envisage how many more suitcases would be required to carry a billion dollars in $10 bills than in $1 bills. I can tell you, "ten times as many," but I can't begin to envisage how many would be needed in either case.

I would guess that the truth is that the public doesn't care about probabilities in choosing between two courses of action when the differences in probability are so small as they are in most of the risk-management decisions that policy makers currently face. These are not situations where the probabilities and magnitudes are indisputably high, but unclear cases involving very low probabilities, such as occupational radiation-exposure levels in medicine and industry or permissable levels of possible carcinogens in food or the environment.

This is the basis for my challenge to the third assumption:

3. That the critical question facing risk managers is "How safe is safe enough?"

Contrary to the accepted wisdom that I have often heard as SRA meetings, my experience has been that only a tiny minority of the public expect that life ought to be entirely free of involuntary danger. So, what do the vast majority of the policy makers' constituencies care about? Based on my own research, I would tentatively suggest at least 3 things:

a. Is the procedure by which collective consent is obtained for a course of action acceptable to those who must bear its consequences?

b. Is the principle that will be used to apportion liabilities for an undesired consequence acceptable to those affected?

c. Are the institutions that make the decisions that manage and regulate the technology worthy of fiduciary trust?

In a recently published article on institutional cultures of risk management (Rayner 1984), I have begun the process of identifying consistent packages of preferences for consent and liability principles with different organizational contexts. In collaboration with my colleague, Robin Cantor, I. am continuing to develop these, along with the different ways in which fiduciary trustworthiness is demonstrated in those same contexts. I will spare you the details of this work today in favor of a tentative, if adventurous, prediction that when all three of these factors are amenable to the risk-bearing population, they will hardly consider many of the low-probability, high-consequence events that currently obsess us to be worthy of the term risk at all. Chauncey Starr's example of the lack of public concern for the low-probability risk of the caged tiger devouring zoo visitors is one homely case. Public acceptance of nuclear power in Japan is another. The Americans trust their zookeepers while the Japanese trust their utility managements and regulatory institutions. Hence, these risks are not recognized as public policy issues. Yet a definition of risk as probability x magnitude cannot tolerate the conclusion that societal definitions of risk can be predicated entirely on issues of equity and trust to the exclusion of probabilistic considerations. Hence the need for a polymorphous definition that encompasses purely societal concerns about equity at the risk-management end of the elephant, and purely engineering-type concerns about probability and magnitude at the technical end. Such a definition would truly open new approaches to risk management for which the proper question would become not "How safe is safe enough?," but "How fair is safe enough?."

This is not merely a linguistic quibble, substituting fair for safe, for this would still leave us with a reified concept of risk as a thing "out there" in nature, rather like the vulgan-Marxist notion of dialectic in history or in nature (Rayner 1977). Risk is rather a way of classifying a whole series of complex interactions and relationships between people, as well as between man and nature. The point of changing the societal risk management question to "How fair is safe enough?" is to encourage risk managers to debate technology options on the explicit basis of social issues of trust and equity, rather than on rival estimates of the probability and magnitude of unlikely events.

REFERENCES

Arrow, K.J., 1951. Social Choice and Individual Values. New York, Wiley.

Douglas, M., 1978. Cultural Bias. Royal Anthropological Institute Occasional Paper No. 35. London.

Jevons, W.S., 1974. The Principles of Science: A Treatise on Logic and Scientific Method, Vol. II. London, Macmillan.

Lowrance, W.W., 1980. "The Nature of Risk" in R.C. Schwing & W.A. Albers (eds.) Societal Risk Assessment: How Safe is Safe Enough? New York, Plenum.

Rayner, S., 1977. "The Dialectics of Nature: A Sociology of Knowledge?" Marxism Today, Vol. 21, No. 5.

Rayner, S., 1979. The Classification and Dynamics of Sectarian Forms of Organization, Ph.D., University of London.

Rayner, S., 1984. "Disagreeing About Risk: The Institutional Cultures of Risk Management and Planning for Future Generations," in S. Hadden (ed.) Risk Analysis, Institutions, and Public Policy. Port Washington, New York. Associated Faculty Press.

Ryle, G., 1951. "Thinking and Language III," Proceedings of the Aristotelian Society (Supplementary Series) 25: 65-82.

Saxe, J.G., 1886. The Poems: Complete in one volume. Boston, Osgood.

Self, P., 1975. The Econocrats and the Policy Process: The Politics and Philosophy of Cost-Benefit Analysis. London, Macmillan.

Simon, H.A., 1979. "Rational Decision Making in Business Organizations." American Economic Review, 69, No. 4: 493-513.

Starr, C., 1969. "Social Benefits Versus Technological Risk," Science 165: 1232-38.

Vygotsky, L.S., 1962. Thought and Language. Kaufman & Vakar (trs). Cambridge, Massachusetts, MIT Press.

Wittgenstein, L., 1953. Philosophical Investigations, Oxford, Blackwell.

VALUE AND FUNCTION OF INFORMATION IN RISK MANAGEMENT

Joshua Menkes* and Lester B. Lave**

National Science Foundation*
Carnegie-Mellon University**
Washington, D.C. 20550*
Pittsburgh, PA 15213**

The motto "Better Living Through Chemistry" has its darker side as we discover the unwanted consequences of some very desirable applications of many pesticides, preservatives, and pharmaceuticals. The hazard that we encounter are in the form of toxicity, carcinogenicity, and mutagenicity. The hazards with their associated risks range from the negligible to substantial increases over standard morbidity and mortality.

It has become customary to think of risk in terms of frequency of occurrence or probability as well as a measure of the magnitude of the consequences should the hazard materialize. We shall refer to this bilinear construct (1) as theoretical risk to emphasize its scientific provenance as distinguished from the subjective perception of risk.

Theoretical risk levels in order of their severity can be organized in four categories: (a) zero risk; (b) de minimis risk; (c) acceptable risk; and (d) worst-case scenarios. In this scheme zero risk, a target actually specified in various statutes, occupies a privileged position: with zero risk the theoretical and the socially perceived risks are the same. However, perceptions of whether there can be zero risk vary. The statute instructs an agency, e.g., the FDA, to tolerate no risks in a specific area, such as cancer from food additives. For the no risk goal, the first two steps collapse into one--the identification of the offending hazard--the ban of the identified chemical follows automatically. Again, for the case of zero risk, data collection on hazard identification is simplified; there is no requirement to estimate the potency of the chemical or magnitude of the risk; there is no need for quantitative estimates of exposure. This approach affords a vast economy of information collection. Unfortunately, since the enforcement of the regulation is centralized, individuals have little knowledge or motivation to control their exposure and order their preferences.

The notion of zero risk has great intuitive appeal, and the fundamental theoretical concept of an empty class is not problematic. Problems do arise, however, when one tries to verify experimentally that the risk has in fact been eliminated. Any verification must be based on inductive inference rather than on logical necessity. For example, the statement "there is no arsenic in this soft drink" should more properly be

---

The views expressed in this paper do not necessarily reflect the views of the National Science Foundation, but are exclusively those of the authors.

phrased: Steps have been taken to eliminate all traces of arsenic, however, there are no reliable tests to measure extremely low levels of arsenic."

With this in mind one may define an almost-zero risk as an event that occurs extremely rarely, i.e. almost never, or as the negligible consequence of the exposure to a small but nonzero-hazard. Such a risk is called de minimis, from the dictum de minimis non curat lex, "the law is not concerned with trifles." The question is: what constitutes a trifling risk from both the theoretical point of view and the standpoint of public perception? Is there an absolute de minimis risk? EPA has proposed, a proposal since withdrawn, that in the case of Larvadex the risk of exposure to 0.4 ppm is a "trifle." (2) Does this constitute an absolute risk; and if so, from whence derives the moral and intellectual authority to designate this risk level as de minimis. And finally, what evidence has to be adduced from what is knowable to support the claim of the absolute proposition that a certain exposure level is harmless?

The emotional intensity of the arguments advanced for and against the legitimacy of the notion of acceptable risk, our third category, is indicative of the social conflicts that can be engendered by a policy that would accept a finite risk level as part of a cost/benefit calculus. In fact, one of the issues around which controversies cluster is the very notion of the existence of such a thing called acceptable risk. The issues are refreshingly fundamental: What is knowable about theoretical risks and the public perception of risk; who knows what and who needs to know what. The consistency with which these questions keep reappearing is remarkable. Not since the debates over Heisenberg's Uncertainty Principle have practical matters like the determination of level of risk engendered serious considerations of such arcane concepts as the ontology of knowledge.

The function of information in risk management comes into sharp focus when considering events that have a low probability of materializing but if it should materialize, the consequences would be devastating. Under these circumstances a worst-case analysis is called for in the NEPA legislation, in particular Section 1502.22. The courts have interpreted the requirement for a worst-case analysis to mean that there are situations where the bilinear form of risk representation cannot be applied. Consideration has to be given not to the probability of a catastrophic risk materializing but its possibility. This approach is tantamount to a simple linear--not bilinear--representation of risk.

WHAT IS KNOWABLE TO WHOM

Ordinarily, one does not think of knowledge in ontological categories, although the Greeks and notably Aristotle maintained that only the unchanging is knowable. This unchanging something is substance which is nothing less than God--the unmovable mover. Collingwood (3) notes that a theory of knowledge according to which only the unchanging can be known must be grounded in a metaphysics that recognizes as knowable only the residue which remains after the veil of human perception has been lifted. The quest for substance proceeded simultaneously in two directions. First, it was sought by removing from apprehension all that was incidental and obviously changeable, thereby uncovering the natural world embodied in knowledge now exempt from change. Secondly, it was sought by looking for unchanging relations between changeables. If the required changeless something can be found in one of these quests, the other becomes unnecessary. (4) Both approaches can be found in the risk analysis literature, where they are referred to as "actual" risk and "comparative" risk respectively.

The analytical objective determination of risk from laboratory experiments, actuarial data and models is identified with actual risk. This is in contradistinction to the application of human judgment which is referred to as perceived risk. Actual risk then is tied to the application of the scientific method which yields replicable results with predictable regularity. Deviations from regularity are attributed to imperfect measurement techniques but not to the intrinsic variability of the underlying phenomenon. Yes, Virginia, the world is stable it is only people who are fickle!! A weaker form of permanence is postulated by comparative risk analysis which avoids the necessity to attach a specific and permanent value to risks. It is the relationship between different risks that are assumed to be stable. For example: the risk from smoking a specific number of cigarettes is compared and equated to the consumption of a certain number of bottles of a soft drink. These comparisons are based on theoretical risks but only the equivalences are assumed to be stable.

Let us divert your attention for a moment from the discussion on risk levels to the schemes which serve to supply meaning to purported explanatory concepts such as risk categories. Toulmin argues convincingly that such categories, if universal, are necessarily embedded in a transcendent representation of nature which gives ultimate intellectual authority to "either an axiomatic system of propositions or a presuppositional system of concepts" (5). These presuppositions determine not only what kinds of conduct we consider right or wrong, but also what kinds of phenomena we regard as puzzling or self-explanatory.

But this intellectual authority is based on positive faith in the scientific method. Alas, the scientific method could not deliver the perennial truths it promised, arguably because of the real diversity in man's beliefs. Consequently there is no rational procedure that would allow us to assert that a risk level of one excess incidence of cancer per 10^6 exposed individuals is de minimis, while one in 10^5 is not. Things are getting pretty confusing by now when the scientific method is alleged to be grounded in faith whereas the perception of risk based on human judgment is non-falsifyable but real.

Toulmin suggests a wayout of those difficulties which are created by the commitment to a universal standpoint of rational judgment. He proposes that "Questions of rationality are concerned, precisely, not with the particular intellectual doctrines that a man -- or professional group -- adopts at any given time, but rather with the conditions on which, and the manner in which, he is prepared to criticize and change those doctrines as time goes on. The rationality of a science (for instance) is embodied not only in the theoretical systems current in it at particular times but also in its procedure for discovery and conceptual change through time."(6)

The link Toulmin observes between natural philosophy and empirical science--between the abstract analysis of possible explanatory forms and their application to actual classes of natural phenomena--is closely related to our central concern over the key relationship between the intellectual ideals of a scientific discipline and its explanatory procedures, concepts, and the theoretical problems. (7)

Concepts, schemes, and systems are implicit in and necessary for thought. A venture which may succeed or fail because we think forward, i.e., speculate, but understand backwards. Speculative thinking is an attempt to move from the concepts and conceptual schemes in one's own mind to other and novel schemes. Such speculative thinking is initially confusing as one seeks different interpretation; in fact, most all of these excursions go nowhere. Those that succeed, however, show up as changes in paradigms.

This brief excursion was meant to draw your attention to the danger of ignoring the fundamental limitations in our ability to acquire specific knowledge not only because such acquisition is beyond available techniques but also because this information does not exist in the form in which it is sought. To wit the formula the alchemists sought to turn base metals into gold could not be found because its existence was predicated on a view of nature that is, as we now know, incorrect.

THE MARKET: WHO NEEDS TO KNOW WHAT

Restricting himself to economics, Hayek observes that the peculiar character of the problem of a rational social order is determined "...precisely by the fact that the knowledge of the circumstances of which we must make use never exists in concentrated or integrated form but solely as the dispersed bits of incomplete and frequently contradictory knowledge which all the separate individuals possess." (8) It is the amazing ability of markets to operate with this incomplete knowledge that has made them the important institutions that they are. The market has the uncanny ability to integrate the two disparate types of knowledge, the specific knowledge in the hands of the entrepreneur and the decentralized knowledge of the consumer. The economic problem as Hayek points out and mutatis mutandis the rational risk assessment is not merely a problem of how to allocate given resources -- it is rather a problem of how to secure the best use of resources known to any of the members of society, for ends whose relative importance only these individuals know.(9) It is necessary to recognize and accept the fact that in the social problem solving that we have discussed one has to deal with the different categories of knowledge: centralized professional knowledge and diffuse social knowledge. This messy state of affairs is not a provisional imperfection--it is an essential feature of social problem solving.

The market is only one type of social problem solving institution and it is focused on another objective. However, many risks arise in the course of production, distribution, and consumption. Thus, any decisions concerning these activities have implications for risk. The market handles risk in just the same fashion as it handles other attributes of these activities. For example, the risk of an automobile is one attribute of the product along with its horsepower, size, fuel economy, price, etc. Consumers are presumed to prefer less risk and so prefer autos that have this attribute, other qualities held constant. However, it is difficult to get information about the risks of each model; it is also difficult to buy a car with precisely the mix of attributes that one desires. If risk is not a terribly important aspect of the automobile purchase decision, it might have little influence on either buyers or manufacturers. However, in theory, when consumers have information on risk, they can and will act on their preferences; their actions will lead manufacturers to change products to build in the "right" amount of safety (according to the preferences expressed by buyers).

Analagously, workers who face occupational risks are presumed to regard these risk as one attribute of a particular job, along with its wage rate, general working conditions, location, etc. Since workers care about safety, they can be expected to avoid jobs that are dangerous when they have an alternative or to demand higher wages to compensate for the risk. In theory, workers and employers would arrive at levels of occupational risk that would be correct in the sense that they took account of the preferences of workers and the costs of lowering risk.

One objective that is expressed to this line of reasoning is that workers and consumers really don't care about risk or: they may care about risks in ways that are different from those characterized by analysis. If this were so, they wouldn't care about riskin their voting behavior and regulatory agencies wouldn't have the political power to require risk reduction. Of course, there is a distribution of preferences about risk in the population; a political outcome represents some median safety preference. If a regulatory agency enforced this level, it would be less stringent than about half the population desired and more stringent than half the population wanted. However, if the safety conscious people felt more strongly, they might desire this level of safety. We suspect that this explanation is correct and that a major goal of federal agencies is to protect some people despite their preferences. This also provides a good explanation as to why some consumers or workers are hostile to federal agencies--and why these agencies find some regulations impossible to enforce in some cases.

A more salient objection concerns the cost of gathering an disseminating information in the market. It is difficult and expensive for individuals to acquire risk information about jobs and products. While institutions such as unions and consumer groups gather an disseminate this information, individuals are not as well informed as they could be. This sort of imperfection could lead to getting systematically too little safety or too much.

There are also risks that do not arise in normal economic activity. For example, earthquakes and storms impose risks and have nothing to do (at least directly) with economic activity. Are there market like institutions that can manage these risks? The answer is a partial yes, under some circumstances.

Insofaras actions can be taken to protect against adverse events, individuals are motivated to take the. For example, individuals would be motivated to build a sea wall against waves or a dam against floods. However, these actions require the agreement of many individuals and so there are problems with getting them all to agree, to pay their share, etc. In these circumstances, it is often simpler to use existing governmental institutions instead of having to create what is essentially a new institution for each risk.

Another aspect of protecting against risk is protecting against the financial loss of a storm or flood. This protection typically comes in the form of insurance. There are many companies offering policies to cover the financial losses of fire, flood, storms, etc. In the case of floods, a federal program was created to offer highly subsidized insurance. In general, there is little reason to believe that current insurance companies do not offer adequate protection against the financial loss associated with these sorts of adverse events.

One of the great accomplishments of microeconomic theory has been to explain the role and mechanisms of information collection and dissemination in a decentralized market economy. Since information is expensive to gather, process, and disseminate, major advantages accrue to institutions like the market that minimize the need for collection, processing and dissemination. Prices, which are easily observed, provide all of the information that individuals need to acquire other than that which each requires because of his special function.

An important characteristic of the market is that consumers' preferences can be manifested in a decentralized fashion and that producers can satisfy the individual preferences of each consumer. There is no need for consensus on a particular product in the sense that everyone has no alternative but to take on that is offered or nothing. Rather, aside from products where so few consumers desire them that production is unattractive, the market produces the types and quantities of each product that consumers desire.

Unfortunately, the more troublesome cases for production and allocation of goods and services are precisely the cases most like the situations involving risk management. Two difficult cases for economic theory are those of "public good" and "spillover effects." The first concern, public goods, occurs when a service cannot be provided to one individual without providing it to all. For such public goods, general concurrance is required, since the same level of national defense must be provided for the entire society. As Lave and Romer(10) show, many risk management decisions have a public good character and so management is inherently controversial.

The second difficulty concerns externalities. These are spillover effects that are unintended and generally undesired. Risk management is replete with spillover effects. It is not surprising that risk management is inherently more difficult than the simple economic models of goods and services.

Little attention has been given to the information requirements of risk management decisions. Since the cost of gathering, processing and disseminating this information is high, it is surprising that it has received so little attention.

The Risk Management Process

Shown in figure 1 in one characterization of the risk management process, beginning with hazard identification and ending with monitoring. The data requirements are characterized in the figure. These can be as informal as thinking about patterns in one's own experience or as formal as a public hearing. For every risk management decision, data are required for each step in this figure. For most public and virtually all private decisions, most of these steps are ignored or only the most rudimentary data are utilized. Major public risk management decisions may require an explicit, full scale analysis of available data or even the collection and analysis of new data. It should be noted that traditional markets perform reasonably well even if the available data are fragmentary--the market does not need perfect information to work--putting it differently: markets fail gracefully rather than catastrophically.

Under a market system, there is o centralized hazard identification, or indeed centralized action of any sort. The hazard would have to be identified by the individual producer, worker, or consumer--or by their agents, such as labor unions or Consumers Union. This is likely to be a slow, expensive process that never informs all relevant parties even about a major risk such as asbestos hazards. Once the hazard has been identified and the actors motivated to do something, each individual who feels the potential for harm must take action to determine how the risk can be lessened, whether the resulting cost is worthwhile, and to monitor

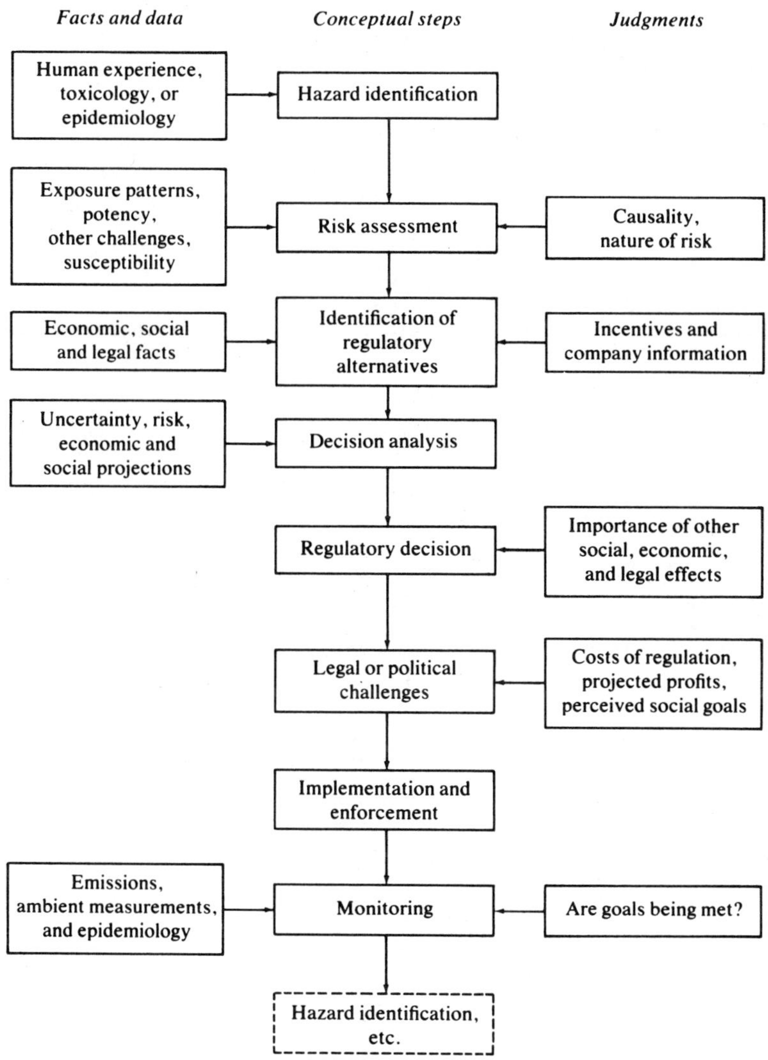

Figure 1

Hazard Management

to ensure that the correct actions continue to be taken. The decentralized nature of the process means that it will be uncoordinated and rough. However, once each actor requires the information, the process should work well since each will be motivated to ensure that the desirable changes are made. Unfortunately, there is no formal mechanism for hazard identification or solution worked out by one individual to be communicated to others.

To summarize, when the risks do not involve public goods or spillover effects, the market can make good risk management decisions. Since the market is a decentralized system, it is not economical in the gathering, analyzing, and dissemination of information. However, since these processes are done in a decentralized fashion, the individuals who will enforce the solution have the information and thus implementation of the outcome should be relatively easy.

REFERENCES

1. The representation of risk, R, is patterned after the von Neuman-Morgenstern "expected utility index." The index is the product of the probability of a certain outcome times the utility associated with this outcome. If p is the probability of a risk event occurring and C is the magnitude of the damage when the risky event takes place, we write R=pC. This index is linear in both p and C hence the term bilinear.
2. EPA Regulators Take on the Delaney Clause, SCIENCE, 25 May 1984, Volume 224, pp. 851-852.
3. R.G. Collingwood, The Idea of History, Oxford University Press, 1981 p. 42.
4. R.G. Collingwood, The Idea of Nature, Oxford University Press, 1981, p. 12.
5. Stephen Toulmin, Human Understanding, Princeton University Press, 1977 p. 81.
6. Ibid, p. 84.
7. Ibid, p. 6.
8. Friedrich A. Hayek, Individualism and Economic Order, University of Chicago, Midway Reprint, 1980, p. 77.
9. Ibid, p. 78.
10. Lester B. Lave and Thomas Romer, Specifying Risk Goals: Inherent Problems with Democratic Institutions, Risk Analysis, September 1983, p. 217.

COMMUNICATING SCIENTIFIC INFORMATION ABOUT HEALTH AND ENVIRONMENTAL RISKS: PROBLEMS AND OPPORTUNITIES FROM A SOCIAL AND BEHAVIORAL PERSPECTIVE*

Vincent T. Covello*, Detlof von Winterfeldt** and Paul Slovic***

National Science Foundation*
Washington, D.C. 20550
University of Southern California**
Los Angeles, CA 90089
Decision Research***
Eugene, Oregon 97401

INTRODUCTION

Risk communication takes place in a variety of forms, ranging from product warning labels on cigarette packages and saccharin bottles to interactions between officials and members of the public on such highly charged issues as Love Canal, AIDS, and the accident at Three Mile Island. Recent experience has shown that communicating scientific information about health and environmental risks can be exceedingly difficult and is often frustrating to those involved. Government officials, industry executives, and scientific experts often complain that laypeople do not understand technical risk information and that individual and media biases and limitations lead to distorted and inaccurate perceptions of many risk problems. Individual citizens and representatives of public groups are often equally frustrated, perceiving government and industry officials to be uninterested in their concerns, unwilling to take immediate and direct actions to solve seemingly simple and obvious health and environmental problems, and reluctant or unwilling to allow them to participate in decisions that intimately affect their lives. In this context, the media often play the role of transmitter and translator of information about health and environmental risks, but have been criticized for exaggerating risks and emphasizing drama over scientific facts.

These difficulties and frustrations point to the problems and complexities of communicating health and environmental information in a pluralistic, democratic society. A review of the literature (Covello, Slovic, and von Winterfeldt, 1986) on recent efforts to communicate information about health or environmental risks--such as the controversies over the risks of saccharin, the pesticide EDB, dioxin, AIDS, toxic wastes, smoking, driving without seat belts, and nuclear power plant accidents--suggests that communication problems arise from (1) message characteristics and problems (e.g., limitations of scientific risk assessments); (2) source characteristics and problems (e.g., limitations

*The views and conclusions expressed in this paper are solely those of the authors and do not necessarily reflect the views and conclusions of the National Science Foundation.

of risk communicators and risk assessment experts); (3) channel characteristics and problems (e.g., limitations in the means or media by which scientific information about health or environmental risks is transmitted); and (4) receiver characteristics and problems (e.g., characteristics of the intended recipients of the communication).

Message problems include:

- deficiencies in scientific understanding, data, models, and methods resulting in large uncertainties in risk estimates;
- difficulties in assessing accident sequences, human exposures, synergistic effects, and consequences for sensitive populations such as children, pregnant women, and the elderly;
- highly technical analyses that are often unintelligible to the average layperson.

Source problems include:

- disagreements among scientific experts;
- legal and resource constraints limiting the authority and actions of responsible officials;
- gaps between societal-level, aggregate data and individual fears and concerns;
- failures to disclose limitations of risk assessments and resulting uncertainties;
- limited understanding of the interests, concerns, fears, values, priorities, and preferences of individual citizens and public groups;
- use of bureaucratic, legalistic, and technical language;
- lack of trust and credibility.

Channel problems include:

- selective and biased media reporting that emphasizes drama, wrongdoing, disagreements, and conflict;
- premature disclosures of scientific information;
- oversimplifications, distortions, and inaccuracies in interpreting technical risk information;

Receiver problems include:

- inaccurate perceptions of levels of risk;
- lack of interest in risk problems;
- overconfidence in one's ability to avoid harm;
- strong beliefs and opinions that are resistant to change;
- exaggerated expectations about the effectiveness of regulatory actions;
- desire and demands for scientific certainty;
- a reluctance to make trade-offs between different types of risks or between risks, costs, and benefits;
- difficulties in understanding probabilistic information.

Given these problems and the widespread dissatisfaction with the current state of risk communication, this paper seeks to:

- review the findings and conclusions of the literature that bears on risk communication;
- assess the usefulness of this literature for improving risk communication;
- suggest ways that risk communicators can improve their performance;
- identify future research needs.

Risk communication, as defined in this paper, is any purposeful exchange of scientific information between interested parties regarding health or environmental risks. Interested parties include government agencies, corporations and industry groups, unions, the media, scientists, professional organizations, public interest groups, and individual citizens. In recent years, increased attention has been give to communications between government agencies and members of the public. Such communications are the subject of intense controversy and represent one of the most challenging and difficult aspects of risk management (Ruckelshaus, 1984; United States Environmental Protection Agency, 1984). As a result, many of the findings and conclusions reported in this paper specifically address this issue. However, many of the findings and conclusions are also relevant to communications between other interested parties.

Reflecting the broad scope of risk communication, the review of the literature encompasses such diverse fields as cognitive psychology, social psychology, consumer behavior, marketing, advertising, economics, mass communications, linguistics, anthropology, decision science, sociology, political science, health education, behavioral medicine, public health, environmental studies, law, and philosophy. A basic assumption of the review is that each of these disciplines can illuminate important aspects of the risk communication process.

RISK COMMUNICATION TASKS, PROBLEMS AND OPPORTUNITIES

As background for this study, and to help structure the literature review, interviews were conducted with officials and executives with risk communication responsibilities in industry and in five federal agencies: the Environmental Protection Agency, the Occupational Health and Safety Administration, the Nuclear Regulatory Commission, the Consumer Products Safety Commission, and the Food and Drug Administration. Officials interviewed unanimously agreed that risk communication is one of the most important tasks facing government and industry today. Officials disagreed, however, about the nature, importance, and priority of risk communication tasks. For example, some officials saw the main communication task as providing and explaining technical risk information to laypeople; others saw the task as effectively involving the public in risk management decisionmaking; still others argued that the primary task is to instill public trust and to restore or maintain the credibility of public and private sector institutions and organizations.

The data from the interviews were used to construct a typology of risk communication tasks (Table 1). As shown in the table, risk communication tasks can be organized into four general types according to the primary objective or intended effect of the communication:

- information and education;
- behavior change and protective action;
- disaster warnings and emergency information;
- joint problem solving and conflict resolution.

In the real world, these four types of risk communication tasks overlap substantially, but they still can be conceptually differentiated. The task of informing and educating the public can be considered primarily a non- directive, although purposeful, activity aimed at providing the lay public with useful and enlightening information. In contrast, both the task of encouraging behavior change and personal protective action and that of providing disaster warnings and emergency information can be considered primarily directive activities aimed at motivating people to take specific types of action. These three tasks, in turn, differ from the task of involving individuals and groups in joint problem solving and conflict resolution, in which officials and citizens

Table 1

A Typology of Risk Communication Tasks

TYPE 1: INFROMATION AND EDUCATION

- Informing and educating people about risks and risk assessment in general.
  Example: statistical comparisons of the risks of different energy production technologies.

TYPE 2: BEHAVIOR CHANGE AND PROTECTIVE ACTION

- Encouraging personal risk-reduction behavior.
  Example: advertisements encouraging people to wear seat belts.

TYPE 3: DISASTER WARNINGS AND EMERGENCY INFORMATION

- Providing direction and behavioral guidance in disasters and emergencies.
  Example: sirens indicating the accidental release of toxic gas from a chemical plant.

TYPE 4: JOINT PROBLEM SOLVING AND CONFLICT RESOLUTION

- Involving the public in risk management decisionmaking and in resolving health, safety, and environmental controversies.
  Example: public meetings on a possible hazardous waste site.

exchange information and work together to solve health and environmental problems. As shown by the descriptions of the four communication tasks provided below, each task is also associated with a different set of characteristic problems and a different set of recommendations for handling these problems.

Type 1: Information and Education

Examples. The following two examples are a composite of real and fictional cases and illustrate various aspects of this type of task.

- A government agency has just completed a large study comparing the risks from three methods of electricity production--coal, nuclear power, and solar energy. Government analysts have carefully assessed and compared all available information on the health and environmental risks associated with each technology. Efforts were made to examine a wide range of health and environmental risks--such as accidents, occupational diseases, and pollution--for every facet of the technology, from the mining of raw materials to transport and production. The study concluded that, at least for normal operations, the risks associated with nuclear power are lower than those for coal and solar energy. The findings of the study have been subjected to expert peer review and have been published in scientific journals. Following an internal assessment of the importance of the study, the agency decides to publish the findings in a pamphlet. In the preface to the pamphlet, the agency states that its primary objective is to provide the public with up-to-date information about the relative risks of various methods of electricity generation, and not to advocate a particular form of electricity production.

- A public health official is asked to investigate a report that the groundwater near a toxic waste site may be contaminated and that, as a result, the local drinking water may also be contaminated. The community newspaper learns of the investigation and reports that the local water may be dangerous to drink. The results of the agency's investigation indicate that the local drinking water may contain up to 50 parts per billion (ppb) of a toxic chemical that has been found to produce cancer in laboratory animals. Since the level of the chemical is above health standards, the agency proposes to shut down the water supply temporarily and to provide an alternative source of water. An agency spokesperson is asked to explain the agency's position to the press.

Problems. A variety of problems complicate the task of informing and educating people about risks and risk assessment.

- Risk information is often highly technical, complex, and uncertain. Because of uncertainties deriving from a lack of scientific data and from deficiencies in available methods and models, it is not uncommon to find substantial variations in risk estimates. For example, a committee of the National Academy of Sciences (1978) estimated that the expected number of bladder cancers resulting from the consumption of saccharin over the next 70 years ranges between .22 and 1,144,000 cases, depending on the assumptions that are made.
- Experts often disagree on the assumptions underlying a risk assessment and, as a result, often provide widely different risk estimates. One result of these disagreements is public confusion about the validity of risk estimates.
- Regulatory agencies sometimes lack public trust and credibility. Trust and credibility are intimately linked and can be undermined by numerous factors, including public perceptions that the agency lacks technical competence, that agency decisions are overly influenced by special interest groups, or that an agency is inappropriately biased in favor of a particular technology or political strategy.
- Experts and laypeople often define risk differently. Experts typically define risk strictly in terms of expected annual mortalities. Lay people almost always include other factors in their definition of risk, such as catastrophic potential, equity (that is, whether those receiving benefits from the technology bear their share of the risks), effects on future generations, controllability, and involuntariness. These differing conceptions often result in laypeople assigning relatively little weight to risk assessments conducted by technical experts and government agencies.
- Government officials often use technical, legalistic, or bureaucratic language. Besides being difficult to comprehend, such language gives the impression that officials are being unresponsive or evasive. For example, an official's statement that "groundwater contamination of 5 parts per billion is within the limits of acceptable safety standards set by the agency" may be technically correct but may also leave individuals suspicious and confused about the meaning and relevance to a particular situation. Government officials may argue that technical language is unavoidable, given the constraints placed on them by the nature of the data, by agency regulations, and by the law.
- People are frequently unwilling to believe government officials who claim that their decisions and actions are constrained by resource, statutory, and other limitations. Individuals who are directly affected by a government decision are especially reluctant to accept such claims and often demand that the risk agent be banned or that the hazardous activity be curtailed.

- People are often not as interested in risk problems as officials in government agencies. Given competing interests and priorities, issues that are high on the agendas of government agencies may be low on the agendas of average citizens. It may thus be difficult to get people to pay attention to risk information. Those that do pay attention may be highly selective and focus on unusual and dramatic aspects of the problem instead of on representative data and statistics.
- Risk information can be frightening. A statement by a government official meant to assure the public that its water is safe to drink, its air is safe to breathe, or its food is safe to eat may have exactly the opposite effect. Instead of alleviating concern, it may increase fear, anxiety, and avoidance of an activity that previously was assumed to be safe. The very fact that an official investigation is under way may be sufficient to create an atmosphere of fear and suspicion.
- People holding strong beliefs are exceedingly resistant to changing such beliefs, even when confronted with substantial and opposing scientific evidence. Motorcyclists, for example, often deny that they engage in a high risk activity, even when presented with statistics on the high incidence of motorcycle accidents. Such persons frequently question the accuracy of the statistics or the relevance of the statistics to their unique situation. For example, they may refer to their superior abilities and experience in handling motorcycles or to their own accident-free records.
- Although strong beliefs are difficult to change, weakly held beliefs can often be manipulated by subtle differences in the presentation of risk information.
- People have difficulty interpreting probabilistic information. Extremely small probabilities, such as a chance of one in a million or smaller, are especially difficult to comprehend.

Relevant Research. Of the various academic disciplines concerned with problems related to information and education, researchers in the fields of cognitive and social psychology have been among the most active.[1] Cognitive and social psychologists have explored questions such as: How do people define and perceive risks and what determines the risk's acceptability? Does newspaper coverage bias the perception of risks? What are the determinants of attitudes and attitude change? What factors influence the success or failure of educational efforts? How does the presentation of risk information influence people's perceptions and preferences?

Sociologists, economists, political scientists, anthropologists, health educators, and communications researchers have also been active in the study of problems related to information and education. While psychologists have usually focused on the mental processing and evaluation of risk information, sociologists, economists, political scientists, anthropologists, and health educators have focused on the influence of

---

[1]See, for example, Combs and Slovic, 1979; Slovic et al., 1978, 1979, 1980, 1981, 1982; Green, 1980; Johnson and Tversky, 1983; McNeil et al., 1982; Renn, 1981; Fischhoff et. al., 1978, 1979, 1984; Lichtenstein et. al., 1978; Otway, 1980; Otway and von Winterfeldt, 1982; Otway et al., 1978; Tversky and Kahneman, 1981; von Winterfeldt et. al., 1981; Vlek and Stallen, 1981; Squyres, 1980; Gardner et al., 1982; Lowrance, 1976; Kasperson and Kasperson, 1983; Covello, 1983; 1984.

social, economic, institutional, organizational, or cultural factors as well as the influence of the media on peoples' risk perceptions, preferences, and behavioral responses to risk- related information.[2]

Conclusions. Several general conclusions can be drawn from the literature on informing and educating people about risk. To perform this task effectively, communicators should:

- use simple, graphic, and concrete material, avoiding technical or specialized language wherever possible;
- compare risks within a carefully defined context that is relevant to the target audience;
- avoid comparisons of risks that may appear to the audience to be non-comparable because of different qualitative characteristics, for example, the risk of smoking compared to that of living near a nuclear power plant;
- understand and recognize people's qualitative concerns, such as concerns about catastrophic potential, dread, equity, and controllability;
- identify and explain the strengths and limitations of different risk measures; present, whenever possible, alternative measures and indices of risk, for example, measures of expected fatalities or incidences of disease for the entire population and for the most- and least-exposed individuals;
- identify, acknowledge, and explain uncertainties in risk estimates;
- provide opportunities for people to learn how to interpret risk information;
- relate on a personal level; when people ask personal questions--such as, "Can I drink the water?"--respond in a personal way without minimizing the risks and uncertainties;
- recognize the power of subtle changes in the way information is presented and use such knowledge responsibly;
- understand and recognize that health and environmental debates often involve much broader considerations, including political values and ideologies.

Type 2: Behavior Change and Protective Action.

Examples. The following two examples are a composite of real and fictional cases and illustrate various aspects of this type of task.

- A government agency has finished reviewing research that indicates a high-cholesterol diet increases arteriosclerosis and leads to heart attacks and strokes. Based on the evidence, the agency decides to launch a campaign encouraging people to adopt a low cholesterol diet. It contracts with a large advertising agency to develop a series of pamphlets aimed at several select population groups, such as the young and those who already have cardiovascular problems.
- A state highway safety agency is concerned about increasing numbers of highway deaths and plans a media advertising campaign aimed at

---

[2]See, for example, Conrad, 1980; Mazur, ; 1973; 1981; Nelkin, 1984; Nelkin and Brown, 1984; Perrow, 1984; Wildavsky and Douglas, 1982; Gross and Rayner, 1983; Twentieth Century Fund, 1984; Mitchell, 1980; Short, 1984; Douglas, 1966; Rothman, 1982; Sharlin, 1985; Levine, 1982; Shelanski et al., 1982; Winsten, 1985; Sandman, 1973; 1975; 1982; Sandman and Paden, 1979; Johnson and Covello, 1986; Morris et al., 1980; Lipset and Schneider, 1983; Barber, 1983; Short, 1984; Burger, 1984; Media Institute, 1985.

discouraging drunk driving during an upcoming holiday period. The agency enlists the support of local television stations in broadcasting a series of public service announcements. The announcements include messages that stress (1) the greater risks of being arrested for drunk driving during the holiday period because of increases in police surveillance and highway check points; (2) the responsibility of friends and relatives in preventing individuals from driving while intoxicated; and (3) the consequences of drunk driving, as shown through graphic photographs of highway automobile accidents.

Problems. In addition to problems already noted in relation to other communication tasks, a variety of problems complicate the task of encouraging behavior change and protective action.

- The losses incurred by changing behavior are tangible and immediate, such as the loss of the pleasure of smoking or the taste of a favorite food, while the gains are abstract, intangible, and remote in time.
- People frequently display an "optimism bias;" for example, a belief that fate or luck is on their side and that it "can't happen to me." This is especially true in activities that require skill and involve individual control, such as driving or skiing.
- People often resist government efforts aimed at changing behavior for political or ideological reasons. For many, such efforts represent unacceptable intrusions by government in their personal lives.
- The target audience of a behavior change campaign is often unmotivated and uninvolved, and, consequently, it may ignore the message. This has been shown in campaigns aimed at encouraging people to use seat belts and to improve their diet.
- People rationalize their resistance to behavior change. For example, smokers who need the psychological relief of smoking a cigarette often cite the pressures of modern life as their justification and minimize the risks.
- People seldom respond appropriately to high-threat or fear communications, such as photographs or films graphically depicting the physical symptoms of disease or the results of a disfiguring or fatal accident. Such communications may induce excessive fear and anxiety, which, in turn, may reduce people's attention, induce defensive responses, and evoke hostility toward the source of the communication.

Relevant Research. Of the various academic disciplines concerned with problems of behavior change and protective action, researchers in the fields of consumer behavior, marketing, advertising, social psychology, sociology, health promotion, and disease prevention have been among the most active.[3] Researchers in consumer behavior, marketing, and advertising have been concerned primarily with factors that underlie consumer choice. Social psychologists, sociologists, and health promotion specialists have been interested primarily in attitude change and

---

[3]See, for example, McGuire, 1985; Eagly and Chaiken, 1985; Earle and Cvetkovich, 1983; Earle, 1984; Vertinsky and Vertinsky, 1982; Weinstein, 1984; Maccoby et al., Maccoby and Solomon, 1977, 1981; Fishbein and Ajzen, 1975; Kiesler et. al., 1968, Gusfield, 1982; Robinson, 1976; Robinson et al., 1974; Sutton, 1982; Alcalay, 1983; Adler and Pittle, 1984; Evans and Clarke, 1983; Rice and Page, 1981.

persuasion, as well as in the social and institutional processes that precede behavior change.

Conclusions. Several general conclusions can be drawn from the literature on the task of encouraging behavior change and protective action. To perform the task effectively, communicators should:

- identify a specific target audience and tailor the communication to that audience;
- generate involvement by creating vivid, concrete images that the target audience can relate to on a personal level;
- use innovative ways to attract the attention of the target audience;
- avoid high-threat or fear campaigns; present a moderate position, and carefully develop the argument;
- use multiple channels and media, for example, newspapers, television, magazines, films, school presentations, and pamphlets in doctors' offices and hospital waiting rooms;
- present recommendations in the context of a balanced argument that accurately describes the strengths and weaknesses of both sides, especially if the target audience is likely to be exposed to strong counter-arguments and recommendations;
- build on expertise, trust, and credibility; people are more willing to accept a communication if the communicator is believed to be knowledgeable, respected, unbiased, and truthful;
- draw explicit conclusions and actively involve the audience in the argument;
- use peer and social relationships to support and reinforce the message, for example, enlist the aid of coworkers, fellow students, teachers, religious and community leaders, doctors, and employers.

Type 3: Disaster Warnings and Emergency Information.

Examples. The following two examples are a composite of real and fictional cases and illustrate various aspects of this type of task.

- A community is struck by a small earthquake and experts believe that a larger earthquake is likely to occur within six months. The exact date of the expected earthquake is, however, highly uncertain. Local police and fire departments are informed, and the media are provided information about emergency preparedness. When the earthquake finally occurs, it is much larger than anticipated, resulting in a massive breakdown in communications.
- A government agency is called in to deal with an accident that has occurred at a large industrial chemical facility. Emergency operations are under way, and a press conference has been called to inform the public about the state of the facility, reasons for the accident, actions being taken, extent of the damage, risks of further complications, and the need for possible evacuation. The situation is tense and compounded by rumors and conflicting information from several sources.

Problems. In addition to problems already noted in relation to other communication tasks, a variety of problems complicate the task of providing disaster warnings and emergency information.

- In most disasters and emergencies, the primary objectives of government officials are to minimize loss of life and to minimize property damage. These macro-objectives often come into conflict with the micro-objectives of local residents (and sometimes local disaster or emergency workers), who frequently assign highest priority to the protection of their own family members, friends, personal possessions, and property.
- Time pressures often compound an already difficult situation.

- Coordination between agencies and organizations frequently breaks down during disasters; confusion about responsibility and authority often results in multiple and competing sources of information
- Communication channels frequently break down during disasters and may result in confusion and the spread of rumors.
- Warning systems frequently produce false alarms, which confuse people, generate mistrust in the warning system, and may desensitize people to future warnings.
- People often deny the possibility of a disaster or that it may affect them personally. Thus, it is often difficult to capture public interest and attention before a disaster occurs. These observations appear, however, to apply more to natural disasters than to technological disasters.
- People are sometimes reluctant to evacuate, especially in natural disasters; concerns about the loss of personal belongings and of opportunities to save their homes and property often outweigh the motivation to evacuate.
- People usually require confirmation of the original emergency communication through several communication channels before taking action, for example, through telephone calls to the local police or to friends. However, these channels often break down during a disaster or emergency. Without the ability to confirm the original message, people may become confused and not act appropriately.

Relevant research. Of the various academic disciplines concerned with disaster warnings and emergency information, sociological and social psychological researchers have been among the most active. The literature includes general reviews,[4] case studies of risk communication prior to and during disasters and emergencies,[5] and studies of specific aspects of the disaster process, such as warning systems,[6] media coverage,[7] and evacuation behavior.[8]

Conclusions. Several general conclusions can be drawn from the literature on disaster warnings and emergency information. To perform this task effectively, communicators should:

- provide concrete information about specific actions people can take;
- transmit official communications through a single, reliable, respected, trustworthy, and highly credible spokesperson;
- be sensitive to the possibility of overloading people with too much information;
- use multiple channels (such as radio, television, pamphlets, sound

---

[4]See, for example, Mileti et al., 1975; Quarantelli and Dynes, 1977; Kreps, 1984; Saarnen, 1982.

[5]See, for example, Cutter and Barnes, 1982; Friedman, 1981; Greene et. al., 1980; Kunreuther et al., 1978; Lindell et al., 1983; Lagadec, 1982; Shelanski, 1982; Bowonder, 1985.

[6]See, for example, Mileti, 1975; Turner, et. al., 1981; Nilson and Nilson, 1981; Hodler, 1982; Pate-Cornell, 1986.

[7]See, for example, Mazur, 1981; et al., 1982; Rogers and Sood, 1986; Sandman, 1979; Peltu, 1985.

[8]See, for example, Perry et. al., 1980; Lindell et al., 1985; Quarantelli, 1980.

trucks, and door-to-door canvassing) for communicating;

- allow for the need of people to confirm disaster and emergency information;
- establish hot lines for rumor control;
- design communications that are sensitive to the fears and concerns of local residents and that provide information specifically addressing these concerns.
- design warning systems that are resistant to loss of credibility if the threat does not materialize;
- involve local citizens in the design of programs aimed at providing disaster warnings and emergency information.

Type 4: Joint Problem Solving and Conflict Resolution.

Examples. The following two examples are a composite of real and fictional cases and illustrate various aspects of this type of task.

- An agency official wishes to secure greater public involvement in the risk management decisions facing the agency. As preparation, the official reviews upcoming decisions facing the agency and identifies a particularly difficult one concerning a chemical plant in a small city that is highly dependent on the plant for jobs. According to agency experts, the chemical plant is not in compliance with agency regulations on toxic emissions and will not be able to meet prescribed levels even after installing advanced technological controls. The agency official sees two main policy options: tolerate the emissions or order the plant to close down. Given this dilemma, the official appeals to the citizens of the city to express their views, through workshops and public meetings, on the agency's policy options as well as on alternative approaches. The official hopes that these meetings will result in a broad and direct exchange of information between the agency and the community.
- A toxic waste site is declared hazardous to human health, and the department of public health sends in experts and public involvement specialists to work with local residents in finding a solution. In the first meeting, residents criticize the agency for foot dragging and demand that the site be immediately cleaned up, regardless of costs. The situation becomes especially tense when a young woman who lives near the site claims that drinking water contaminated by the site caused her recent miscarriage.

Problems. In addition to problems already noted in relation to other communication tasks, a variety of problems complicate the task of involving the public in joint problem solving and conflict resolution.

- Actions by a government agency aimed at involving the public in the regulatory decisionmaking process may be viewed as an attempt by the agency to abdicate its legal duties and responsibilities. For example, such attempts are sometimes viewed as an attempt by the agency to avoid enforcing health and environmental laws.
- It is difficult to hold public meetings in a highly charged, emotional atmosphere. A common characteristic of joint problem solving exercises is strong involvement by stakeholder groups, with each group bringing its own values and concerns to the arena. Frequently, the initial atmosphere is one of distrust and confrontation, rather than openness and cooperation.
- Agency officials often do not understand the nature of the conflict or sources of disagreement. Such disagreements can range from factual disputes about levels of risk to fundamental disagreements about values and ethical principles.
- Many communication strategies are inappropriate for specific types of conflict. For example, when conflicts are about facts and

statistics, information and education strategies are relevant. However, when conflicts are about the equity (fairness) of risk/benefit distributions or about basic values, education and information strategies are of little value. Such cases require careful diagnosis of the concerns of all interested parties and the design of options (such as safeguards or compensation schemes) that specifically address these concerns.

- The media may aggravate communication problems by highlighting personal fears and anxieties, by focusing only on the dramatic or sensational aspects of a story, and by emphasizing conflict rather than agreement.
- Individuals and groups involved in the process are often unwilling to compromise or accept tradeoffs. For example, local residents may exhibit a "not in my backyard" attitude when exposed to a new risk and frequently demand complete elimination of the risk regardless of costs, resources, or other constraints.
- Individuals and groups are often invited to participate in decisionmaking only after many of the most important decisions have been made.

Relevant Research. Of the various academic disciplines concerned with joint problem solving and conflict resolution, researchers in the fields of decision science, political science and sociology have been among the most active. Within these disciplines, specialists on public involvement and on social conflict have been especially productive. Public involvement studies have examined numerous examples of interactions between the public and government agencies.[9] Researchers have also drawn on the literature on game and decision theory[10] and on bargaining, negotiation, and mediation[11] to develop tools for diagnosing the nature of conflict among groups and mechanisms for resolving these conflicts. Three additional research traditions have also made substantial contributions: sociological research on the causes and the dynamics of technological controversies[12], research on environmental mediation and arbitration[13], and research on consent and decision making.[14]

Conclusions. Several general conclusions can be drawn from the literature on involving the public in joint problem solving and conflict resolution. To perform this task effectively, communicators should:

- involve the public early in decisionmaking, that is, before assumptions have been set, alternatives have been narrowed, key

---

[9]See, for example, Creighton, 1980; Delli Priscoli et al., 1983; Popper, 1985; Susskind, 1978; Susskind et al., 1978.

[10]See, for example, Luce and Raiffa, 1957; Keeney and Raiffa, 1976; von Winterfeldt and Edwards, 1985.

[11]See, for example , Raiffa, 1982.

[12]See, for example, Tribe et. al., 1976; Coser, 1956; Mazur, 1980; 1981; Nelkin, 1978; Perrow, 1984.

[13]See, for example, Busterud, 1980; Bingham, 1984; Cormick, 1980; Wellborn, 1979; O'Hare, 1977; 1984; Susskind and Wheeler, 1983; Bacow et al., 1983; Bacow and Wheeler, 1983; Mernitz, 1980.

[14]See, for example, Gibson, 1985; MacLean, 1986.

decisions have been made, and decisionmakers have become committed to a particular course of action;

- make the objectives for involving the public clear from the beginning;
- leave room for option invention by those directly involved or affected by the decision;
- understand and respect people's interests, emotions, values, priorities, preferences, and concerns;
- identify with the audience and avoid violations of community norms regarding dress, language, and demeanor;
- establish elaborate procedural safeguards to ensure that all voices with an interest or stake in the decision can be heard;
- carefully analyze the nature of the conflict and distinguish between different types of conflict, for example, distinguish factual disagreements from deeply rooted ideological conflicts;
- adopt different communication strategies for different types of conflict and disagreement.

Cross-Cutting Conclusions.

Several general conclusions can be drawn from the literature that cut across all four types of communication tasks.

- The roots of most risk communication problems lie in the complexities of the risk problem itself. What sometimes appears to be a simple and direct issue often turns out to be an issue of enormous scientific, economic, social, and political complexity. Limitations in the science of risk assessment, in public perceptions and understanding, in media coverage, in risk management institutions, and in economic and other resources pose serious obstacles to risk communication. Without awareness and recognition of these obstacles, effective communication is unlikely.
- Interactive and participatory approaches to risk communication appear to offer the greatest promise of better, less controversial, or less divisive decisions. Even the most directive types of risk communication tasks--such as encouraging personal behavior change and providing disaster warnings--can benefit substantially from public involvement, direct interaction, and exchanges of information. Such interactions are an important source of knowledge of public needs and concerns, without which communication efforts are likely to fail or be ineffective.
- There is no such entity as "the public;" instead, there are many publics, each with its own interests, needs, concerns, priorities, and preferences.
- Communication problems can often be viewed beneficially as decision problems. For example, the choice of one communication strategy over another can often be illuminated through a formal analysis of the multiple, competing objectives inherent in the decision, including the community's "right to know," the duty to protect public health, the costs of unnecessarily alarming people, and the possible repercussions of premature or delayed action.
- Government officials and individual citizens often hold different views of risk problems. Most federal regulatory agencies view risk problems from a societal or macroperspective. As a result, most analyses by government agencies provide only aggregate or population statistics for the community or nation as a whole. Aggregate or population statistics are, however, usually of little interest to individual citizens, who are more likely to view risks from a microperspective and to be concerned about the risks to themselves or to their loved ones than to society or the community as a whole. Given this divergence of viewpoints, government officials

are often at a disadvantage, since they often do not have immediate access to information that addresses the highly personalized questions asked by citizens.

- A large amount of research has been conducted that bears on problems of risk communication, but the literature specifically focused on risk communication is relatively small. Substantial progress has been made on some topics, such as psychological research on public perceptions of risk, but large gaps exist in our understanding of virtually every issue relevant to risk communication. Institutional arrangements and financial resources for research on risk communication issues are inadequate.
- Four main messages specifically directed at risk communicators emerge from the literature:
  (1) know your risk communication problem;
  (2) know your risk communication objectives;
  (3) use simple and non-technical language;
  (4) listen to your audience and know their concerns.

The first two messages point to the importance of understanding the unique characteristics of the four main risk communication tasks and underscore the importance of being clear about the nature of the problem and about the intent of the communication. The last two messages underscore the importance of adopting a communication approach that is appropriate to the specific problem and the specific audience.

- Many of the recommendations in this paper may seem obvious, but they are nonetheless continually and consistently violated. These violations make effective communication about health and environmental matters unnecessarily difficult.

BIBLIOGRAPHY

Adler, R., and R. D. Pittle, Cajolery or command: Are education campaigns an adequate substitute for regulation?, Yale Journal on Regulation, Vol. 1, 159-193, 1984.

Alcalay, R., The impace of mass communication in the health field. Social Science and Medicine., 17, 87-94, 1983.

Barber, B. The Logic and Limits of Trust. New Brunswick, New Jersey: Rutgers University Press, 1983.

Bingham, G. Resolving environmental disputes: A decade of experience. Washington D.C.: The Conservation Foundation, 1984.

Bowonder, B. Low probability event: a case study in risk assessment. Paper presented at the workshop "Risk analysis in developing countries", Hyderabad, India, October, 1985.

Burger, E., Health Risks: The Challenge of Informing the Public, Washington, D.C.: The Media Institute, 1984.

Busterud, J. Mediation: The state of the art. Environmental Professional, 1980, 2, 34-39.

Combs, B. and Slovic, P. Newspaper Coverage of Causes of Death. Journalism Quarterly, 1979, 56, 837-843.

Covello, V. T. The Perception of Technological Risks: A Literature Review. Technological Forecasting and Social Change, 1983, 23, 285-297.

Covello, V. T. Uses of social and behavioral research on risk. Environment International, June, 1984.

Covello, V., von Winterfeldt, D., and Slovic, P., Risk Communication: Background Report for the National Conference on Risk Communication. Washington, D.C.: Conservation Foundation, 1986.

Conrad, J. (Ed.) Society, technology, and risk assessment. New York; Academic Press, 1980.

Cormick, G.W. The theory and practice of environmental mediation.

Environmental Professional, 1980, 2, 24-33.
Coser, L.A. The functions of social conflict. New York: Free Press, 1956.
Creighton, J.L. Public involvement manual: involving the public in water and power resource discussions. Washington, D.C.: U.S. Government Printing Office, 1980.
Cutter, S. and Barnes, K. Evacuation behavior and Three Mile Island. Disasters, 1982, 6, 116-124.
Delli Priscoli, J., Creighton, J., Dunning, C.M. (ed.), Public involvement techniques: A reader of ten years experience of the Institute for Water Resources. U. S. Army Corps of Engineers, Institute for Water Resources, IWR Research Report 82-R1, May, 1983. Douglas, M. Purity and danger London: Routledge and Kegan Paul, 1966.
Douglas, M. and Wildavsky, A. Risk and Culture. Berkeley: University of California Press, 1982.
Eagly, A. H. and Chaiken, S. Psychological theories of persuasion. In L. Berkowitz (ed.) Advances in experimental social psychology. 1985. Earle, T.C. Risk communication: A marketing approach. Unpublished paper presented at the National Science Foundation/Environmental Protection Agency Workshop on Risk Perception and Risk Communication, Long Beach, CA., Dec. 1984.
Earle, T.C. and Cvetkovich, G. Risk judgement and the communication of hazard information: Toward a new look in the study of risk perception. BH ARC (400/83/017), Battelle Human Affairs Research Centers, Seattle, WA, 1983.
Evans, S.H. and Clarke, P., When cancer patients fail to get well: Flaws in health communication. In R.N. Bostrom (ed.), Communication Yearbook 7., Beverly Hills: Sage, 1983
Fischhoff, B., Slovic, P., Lichtenstein, S., Read, S. and Combs, B. How safe is safe enough? A psychometric study of attitudes towards technological risks and benefits. Policy Sciences, 1978, 8, 127-52.
Fischhoff, B., Slovic, P. and Lichtenstein, S. Weighing the risks. Environment, 1979, 21, 17-10, 32-38.
Fischhoff, B., Watson, S., and Hope, C. Defining risk. Policy Sciences, 1984, 17, 123-139.
Fishbein, M. and Ajzen, I. Belief, attitude, intention and behavior: An introuduction to theory and research. Reading, Mass.: Addison-Wesley, 1975.
Friedman, S. M. Blueprint for breakdown: Three Mile Island and the mass media before the accident. Journal of Communications, 1981, 31, 85-96.
Gibson, M. (ed.) To Breathe Freely: Risk, Consent, and Air. Totowa, New Jersey: Rowman and Allanheld, 1985
Green, C. H. Risk: Attitudes and beliefs. In D. V. Canter (ed.), Behavior in fires. Chichester: Wiley, 1980.
Greene, M., Perry, R. W., and Lindell, M. K. The March 1980 eruptions of Mount St. Helens: citizens perceptions of volcano hazard. Disasters, 1980, 49-66.
Gross, J. L. and Rayner, S. Measuring Culture: A Paradigm for the Analysis of Social Organization. New York: Columbia University Press, 1983.
Hodler, T. W. Residents' preparedness and response to the Kalamazoo tornado. Disasters, 1982, 2, 44-49.
Johnson, B. and Covello, V. (eds.) The Social Construction of Risk. Boston: Reidel, 1986 (forthcoming).
Johnson, E. J., and Tversky, A. Affect, generalization, and the perception of risk. Journal of Personality and Social Psychology, 1983, 45, 20-31.
Kasperson, R. and Kasperson, J., Determining the acceptability of risk: ethical and policy issues, in J. Rogers and D. Bates, eds., Risk: A Symposium, Ottawa: The Royal Society of Canada, 1983.
Keeney, R.L., and Raiffa, H. Decisions with multiple objectives:

Preferences and value tradeoffs. New York: Wiley, 1976. Kiesler, C.A., Collins, B.E., and Miller, N. Attitude change. New York: Wiley, 1968.
Kunreuther, H., Ginsberg, R. Miller, L., Sagi, P., Slovic, P., Borkan, B., and Katz, N. Disaster Insurance Protection: Public Policy Lessons. New York: Wiley, 1978.
Lagadec, P. Major Technological Disaster. Oxford: Pergamon, 1982.
Levine, A. G. Love Canal: Science, politics, and People. Lexington, Mass.: D.C. Heath, 1982.
Lichtenstein, S., Slovic, P., Fischhoff, B., Layman, M., & Coombs, B. Judged frequency of lethal events. Journal of Experimental Psychology: Human Learning and Memory, 1978, 4, 551-578.
Lindell, M., Bolton, P. A., Perry, R. W. Stoetzel, G. A., Martin, J. B., and Flynn, C. B. Planning concepts and decision criteria for sheltering and evacuationm in a nuclear power plant emergency. Technical report No. AIF/NESP-031, Batelle Human Affairs Research Centers, Seattle, WA, June, 1985.
Lindell, M., Perry, W. and Greene, M., Individual response to emergency preparedness planning near Mt. St. Helens, Disaster Management, January- March, 5-11, 1983
Lipset, S. and Schneider, W. The Confidence Gap: Business, Labor and Government in the Public Mind. New York: The Free Press, 1983.
Lowrance, W. Of Acceptable Risk: Science and the Determination of Safety. Los Altos, CA: Kaufman, 1976.
Luce, D. and Raiffa, H. Games and decisions. New York: Wiley, 1957.
Maccoby, N., Farquhar, J., Wood, P., & Alexander, J. Reducing the risk of cardiovascular disease: Effects of a community-based campaign on knowledge and behavior. Journal of Community Health, 1977.
Maccoby, N. and Solomon, D.S., Heart disease prevention: community studies. In R.E. Rice and W.J. Paisley (eds), Public Communication Campaigns. Beverly Hills, Ca. : Sage, 1981.
MacLean, D., (ed.) Values at Risk, Totowa, New Jersey: Rowman and Allanheld, 1986.
Mazur, A., Disputes between experts., Minerva.11, pp. 243-262, 1973
Mazur, A. The dynamics of technical controversy. Washington, D.C. : Communications Press, 1981.
Mazur, A. Media coverage and public opinion on scientific controversies. Journal of Communications Research, 1980, 31, 106- 115.
Media Institute, Chemical Risks: Fears, Facts, and the Media, Washington, D.C.: Media Institute, 1985.
Mileti, D. Natural hazard warning systems in the U.S.: a research assessment. Technical Report, Institute for Behavioral Science, University of Colorado, Boulder, CO, 1975.
Mileti, D., Drabek, T. and Haas, E. Human behavior in extreme environments. Boulder, Co.: University of Colorado, 1975.
McGuire, W. J. Attitudes and attitude change. In Lindzey and Aronson (eds.) Handbook of social psychology, 1985.
McNeil, B. J., Pauker, S. G., Sox, H. C. Jr. & Tversky, A. On the elicitation of preferences for alternative therapies. The New England Journal of Medicine, 1982, 306, 1259-1262.
Mitchell, R.C. Public Opinion on Environmental Issues: Results of a National Public Opinion Survey. Washington, D.C.: Council on Environmental Quality, 1980.
Morris, L., Mazis, M., and Barofsky, I., (eds.) Product Labeling and Health Risks. Banbury Report 6. Cold Spring Harbor, N.Y.: Cold Spring Harbor Laboratory, 1980.
National Academy of Sciences/National Research Council. Saccharin: Technical assessment of risks and benefits. Committee for a Study on saccharin and food safety policy. Washington, D.C. : National Academy of Sciences, 1978. (Second Edition.) Sage: Beverly Hills, 1984.
Nelkin, D. (ed.) Controversy : Politics of technical decisions, Beverly Hills: Sage, 1978.

Nelkin, D. and Brown, M. Workers at risk: Voices from the workplace, University of Chicago Press, 1984.
Nilson, L. B. and Nilson, D. C. Resolving the "sooner vs. later" controversy surrounding the public announcement of earthquake predictions. Disasters, 1981, 5, 391-397.
O'Hare, M. Not on my block you don't: Facility siting and the strategic importance of compensation. Public Policy, 197, 25, 1977, 407-458.
O'Hare, M. Bargaining and negotiation for a conflict resolution. In H. Kunreuther and P. Kleindorfer (eds.) Production, transportation, and storage of hazardous materials. New York: Springer, 1976.
Otway, H., Maurer, D. and Thomas, K. Nuclear power: The question of public acceptance. Futures, 1978, 10, 109-118.
Otway, H. J. Risk perception: A psychological perspective. In M. Dierkes, S. Edwards, and R, Coppock (eds.) Technological risk: its perspective and handling in Europe, Boston: Oelgeschlager, Gunn and Hain, 1980.
Otway, H.J. and v. Winterfeldt, D. Beyond acceptable risk : on the social acceptability of technologies. Policy Sciences, 8, 1982, 127-152.
Pate-Cornell, M.E. Warning systems in risk management. Risk Analysis, Vol. 6, No. 2, June 1986
Peltu, M. Risk communication: The role of the media. In H. Otway (Ed.), Risk and regulation. London: Butterworths, 1985.
Perrow, C. Normal accidents. New York : Basic Books, 1984.
Perry, R. W., Greene, M. R., and Lindell, M. K.. Enhancing evacuation warning compliance: suggestions for emergency planning. Disasters, 1980, 4, 433-449.
Popper, F. The Environmentalists and the LULU (Local Unwanted Land Use.) Environment, March 1985
Quarantelli, E. and Dynes, R. Response to social crisis and disaster. Annual Review of Sociology, 1977, 3 23-49.
Quarantelli, E. L. Evacuation behavior and problems: Findings and implications from the research literature. Department of Sociology, Disaster Research Center, Ohio State University, Columbus, Ohio, 1980.
Raiffa, H. The art and science of negotiation. Cambridge, MA: Harvard University Press, 1982.
Renn, O. Man, technology, and risk: A study on intuitive risk assessment and attitudes towards nuclear power, (Report Jul-Sept 115, Julich). Nuclear Research Center, 1981. Rice, R.E., and W.J. Paisley (eds.), Public Communication Campaigns., Beverly Hills: Sage, 1981.
Robertson, L. The great seat belt campaign flop. Journal of Communication., 26, pp. 41-45, 1976.
Robertson, L., Kelley, A., O'Neill, B., Wixom, C., Eiswirth, R., and Haddon, W., A controlled study of the effect of television messages on safety belt use. American Journal of Public Health, 64, 1071-1081, 1974
Rogers, E. M. and Sood, R. Mass media operations in a quick-onset natural disaster: Hurricane David in Dominica. Working paper, Annenberg School of Communications, University of Southern California, Los Angeles, 1981.
Rothman, S. Risk and nuclear power: Scientists, journalists, and the public. Public Opinion, 1982, 47-52.
Ruckelshaus, W., Risk in a Free Society, Risk Analysis, Vol. 4, No. 3, September, pp. 157-163, 1984.
Saarinen, T. (ed.) Perspectives on Increasing Hazard Awareness Boulder, Colorodo: Institute of Behavioral Science, 1982.
Sandman, P. M. and M. Paden, At Three Mile Island. Columbia Journalism Review, Vol 18, No. 2, July-August 1979, pp. 43-58.
Sandman, P. Environmental advertising and social responsibility, in D. Rubin and D. Sachs (eds.) Mass Media and the Environment, New York: Praeger, 1973.

Sharlin, H. I. EDB: A case study in the communication of health risk. Unpublished manuscript commissioned by the Office of Policy Agency, Environmental Protection Agency, Washington, D. C., January, 1985.
Shelanski, V., Sills, D. and Wolf, C. The Accident at Three Mile Island. Boulder, Colorado: Westview Press, 1982.
Short, J. The Social Fabric of Risk. American Sociological Review, December, 1984.
Slovic, P., Lichtenstein, S., and Fischhoff, B. Images of disaster: Perceptions and acceptance of risks from nuclear power. In G. Goodman and W. Rowe (eds.) Energy risk management London: Academic Press, 1979.
Slovic, P., Fischhoff, B. and Lichtenstein, S. Facts and fears: Understanding perceived risk. In R. Schwing and W.A. Albers (Eds.), Social risk assessment: How safe is safe enough? New York: Plenum, 1980. Revision in D. Kahneman, P. Slovic and A. Tversky (Eds.), Judgement under uncertainty: Heuristics and biases. Cambridge University Press, 1982, 464-489.
Slovic, P., Fischhoff, B. and Lichtenstein, S. Informing people about risk. In L. Morris, M. Mazis and I. Barofsky (Eds.), Product labeling and health risks, Banbury Report 6. The Banbury Center: Cold Spring Harbor, New York, 1980. Slovic, P., Fischhoff, B., & Lichtenstein, S. Perceived risk: Psychological factors and social implications. In F. Warner and D. H. Slater (Eds.), The assessment and perception of risk. London: The Royal Society, 1981.
Squyres, W.D. Patient Education. New York: Springer, 1980.
Susskind, L.E. The importance of citizen participation and consensus-building in the land use planning process. Cambridge: Massachusetts Institute of Technology, Laboratory of Architecture and Planning, 1978.
Susskind, L. Richardson, J.R., and Hildebrand, K. Resolving Environmental Disputes: Approaches to Intervention, Negotiation, and Conflict Resolution. Cambridge, MA: Environmental Impact Assessment Project, M.I.T., 1978
Sutton, S.R. Fear arousing communications: A critical examination of theory and research. In J.R. Eiser (ed.), Social Psychology and Behavioral Medicine. New York: Wiley, pp. 303-338, 1982.
Tribe, L.H., Corrine, S. Shelling T., and Voss (eds.) When values conflict: essays on environmental analysis, discourse and decision. Cambridge: Ballinger, 1976. Turner, R. H., Nigg, J. M., Paz, D. H., and Young, B. S. Community response to earthquake threat in Southern California. Institute for Social Science Research, University of California at Los Angeles, Los Angeles, Ca, 1981.
Tversky, A. & Kahneman, D. The framing of decisions and the psychology of choice. Science, 1981, 211, 235-271.
Twentieth Century Fund. Science in the streets. New York: Priority Press, 1984.
Vertinsky, I. and Vertinsky, P. Communicating environmental health assessment and other risk information : analysis of strategies. In Kunreuther, H. (ed.) Risk : a seminar series. IIASA-CP-82-S2, International Institute for Applied Systems Analysis, Laxenburg, Austria, 1982, 421-482. Vlek, C., and Stallen, D.J. Judging risks and benefits in the small and in the large, Organizational Behavior and Human Performance, 28, 1981, 235- 271.
United States Environmental Protection Agency, Risk Assessment and Risk Management: Framework for Decision Making., Washington, D.C.: Environmental Protection Agency, December, 1984.
von Winterfeldt, D., John R.S. and Borcherding, K. Cognitive components of risk ratings. Risk Analysis, 1, 1981, 277-287.
von Winterfeldt, D. and Edwards, W. Decision analysis and behavioral research. New York: Cambridge University Press, 1986.

Weinstein, N. D. Why it won't happen to me: Perceptions of risk factors and susceptibility. Health Psychology, 3, 1984, 431-457.
Weinstein, N.D. Seeking reassuring or threatening information about environmental cancer. Journal of Behavioral Medicine, 2: 125-139, 1979.
Wellborn, S. The potential of mediation for resolving environmental disputes related to energy facilities. American Management Systems, Inc., DOE/EV/10274-1, Dec.'79.

RISK ANALYSIS OR RISK ASSESSMENT; A PROPOSAL FOR CONSISTENT DEFINITIONS

Lawrence B. Gratt

IWG Corporation
San Diego, CA 92109

ABSTRACT

Definitions of risk analysis and risk assessment based on the dictionary definition of analysis and assessment are suggested. Proposed definitions of the terms used in risk analysis and risk assessment are presented to correct potential misuse and to encourage proper communications. The need for an overall glossary of terms for "The Society of Risk Analysis" is indicated.

KEY WORDS: Analysis; Assessment; Definition; Risk Analysis; Risk Assessment; Terminology.

INTRODUCTION

The purpose of this paper is to discuss some of the fundamental problems with the terminology used in risk analysis/assessment field. As an example, the National Research Council (1983) in its report on Risk Assessment in the Federal Government: Managing the Process addressed the problem of terminology. The report indicates "despite the fact that risk assessment has become a subject that has been extensively discussed in recent years, no standard definitions have evolved, and the same concepts are encountered under different names." The committee adopted its own terminology for the reported study.

As another recent example, Anderson (1984) presented a paper on modeling of risk at the Air Pollution Control Association in June, 1984 claiming that "some terms used in this report were found not to be well-rooted in tradition." He goes on to define various terms including dosage, exposure, hazard, and risk. The definitions he presents are non-standard and indicate a potential problem of establishing new definitions for every risk analysis.

A review of current terminology indicates the terms "risk assessment" and "risk analysis" have been used in various and interchangeable ways. The proliferation of new terms such as risk evaluation and risk estimation are not desirable and the time has arrived to establish a consistent set of definitions for use by practitioners in the field. This paper can serve as the initial basis for the preparation of a glossary of terms for The Society of Risk Analysis.

This paper will review the concepts of risk and proceed to a main concern, the definitions of risk analysis and risk assessment. The arguments for the suggested definitions are presented. Future actions by

The Society of Risk Analysis to prepare a glossary of terms are indicated. An initial glossary to aid in this effort is included in the Appendix.

RISK

Different discussions and definitions of RISK can be found in the risk literature. Several of these are considered below. The Webster's (1981) dictionary definition of RISK strongly suggests a probabilistic nature:

> The possibility of loss, injury, disadvantage or destruction; to expose to hazard or danger; to incur risk or danger.

The use by some authors of "possible risk" appears to be an improper use of the adjective "possible," since one would not want to refer to the "possible possibility of a loss." The use of the term RISK implies a possible, but not deterministic, outcome. Hammer (1972) defines RISK as follows:

> Risk is an expression or possible loss over a specific period of time or number of operational cycles. It may be indicated by the probability of an accident times the damage in dollars, lives, or operating units.

RISK has been used in ways similar to the word "probability." "Probability" can have several meanings distinguished by a common layman's use, the axiomatic mathematical definition, and the engineering approach using relative frequency. The U.S. Nuclear Regulatory Commission's (1975) report, WASH-1400, expressed a technical definition of RISK as follows:

> Risk (consequence/unit time) =
> Frequency(events/unit time) x Magnitude (consequence/event).

Lowrance (1976) defines risk in a similar vein:

> A measure of the probability and severity of adverse effects.

Starr, et al. (1976) adds the concept of cost in the definition of a risk measure:

> The probability per unit time of the occurrence of a unit cost burden. The cost burden may be measured in terms of injuries (fatalities or days of disability) or other damage penalties (expense incurred) or total social costs (including environmental intangibles). Risk thus involves the integrated combination of (a) the probability of occurrences, (b) the spectrum of event magnitudes, and (c) the spectrum of resultant personal injuries and related costs.

Rowe's (1977) definition of risk is succinct:

> The potential for unwanted negative consequences of an event or activity.

Others emphasize the probabilistic nature of risk such as in Wasserman and Wasserman's (1979) definition:

> The possible occurrence of an event that produces adverse effects on man and his environment. The degree of risk is related to both the probability of the event's occurrence and also to the estimated outcome in terms of the nature, intensity and duration of the adverse effects.

Ritter (1981) refers to risk as follows:

> The potential for realization of unwanted, negative consequences to human life, health, or the environment.

Morris (1982) uses probability in his definition:

> The probability of a specified adverse effect due to specific hazardous agent reaction.

Petak and Atkisson (1982) follow the WASH-1400 definition using probability for frequency:

A function of two major factors: (a) the probability that an event, or series of events of various magnitudes, will occur, and (b) the consequences of the event(s).

Mathematically, risk is represented by a random variable described by a probability distribution. A random variable is a mapping from a set of events to the real numbers. A probability density function can be described in terms of its parameters. Risk analysis is the process of estimating the parameters of the probability density function used to represent the random variable, risk. The risk can be represented by continuous, discrete, or combined functions, for example see McCormick (1981). For simplicity, only continuous relationships are indicated as generalizations to discrete variables are analogous to the relationship of tions to discrete variables are analogous to the relationship of continuous and discrete probabilities. A number of continuous relationships have been used depending on the type of relationship desired and the number of parameters used to describe the probability distribution.

Let X be the random variable representing the events describing the adverse consequence or risk and P(x) the probability density function that represents the risk. The risk (adverse consequence) is a function of time, t. The risk random variable is a function of time, X(t) and probability density function representing the risk is p(x(t)). A random variable that is a function of time is a stochastic process. The treatment of risk as a deterministic quantity instead of a stochastic process has and will lead to misconceptions regarding the risk estimation process.

A probability density must have certain mathematical properties for validity. The conditions for a one-dimensional continuous random variable are Lass and Gottlieb (1971):

1) X is a random variable which takes on numerical values c, $-\infty < x < \infty$ and is a nonnegative function p(x) such that

$$dP = p(x)dx \qquad (1)$$

is the probability that X will assume a value in the interval (x, x+dx).

$$2)\ dP \to 0 \text{ as } dx \to 0 \qquad (2)$$

$$3)\ \int_{-\infty} p(x)dx = 1 \qquad (3)$$

Measures of the risk can be derived in many ways from the basic probability density function.

Letting y be the event or effect and t represent time, some of the risk measures used are as follows:

<u>Expected Risk</u>

$$\langle x \rangle = \int_0^\infty \int_{-\infty}^\infty x(t)p(x(t))dxdt \qquad (4)$$

or the consequence for all time. This is the expected value of the random variable representing the risk (adverse consequence).

Expected Risk over Period T

$$\langle x\rangle_T = \int_0^T \int_{-\infty}^{\infty} x(t)r(x(t))dxdt \qquad (5)$$

or the consequence for time period 0 to T. This is the expected value of the random variable represents the risk over time period T.

Expected Risk Incidence

$$x_I = \frac{\langle x\rangle_T}{\int_0^T tdt} \qquad (6)$$

or the time-averaged expected risk over time period T.

Frequency for Consequence Greater Than R

$$F(x>R) = \int_R^{\infty} p(x(t))dx \qquad (7)$$

or the risk is the probability that the adverse event will occur at time t having a consequence of magnitude equal to or greater than R. This is a form of cumulative probability function for X.

The risk measures listed above are not mean to be all inclusive, as one can observe that different measures can be derived for the particular application under analysis.

Although there are various measures of risk, what one would like to know is the complete probability distribution. The factors of the functional form of the risk ramdon variable in an actual analysis are not necessarily independent; that is, the exposure and consequence are not really separable variables. The simple definitions emerging from the WASH 1400 study have tried to treat risk as a product of a frequency and magnitude (consequence). The problem arising from this has been the separation of the frequency (which is really a conditional probability distribution), magnitude (consequence), and their ensuing individual treatments as risk measures. Avoiding mathematical terms, the following definition of RISK is proposed.

> RISK: The potential for realization of unwanted, adverse consequences to human life, health, property, or the environment; estimation of risk is usually based on the expected result of the conditional probability of the event occurring times the consequence of the event given that it has occurred.

The proposed definition encompasses both a common, understandable form and a scientific form.

RISK ANALYSIS AND RISK ASSESSMENT

The Webster (1980) dictionary definitions of ANALYSIS and ASSESSMENT are:

> ANALYSIS: Separation of a whole into its component parts; an examination of a complex, its elements, and their relationship.
> ASSESSMENT: The act or an instance of determining the importance, size, or value of something.

These definitions have become confused when used in conjunction with the adjective "risk." The risk literature uses RISK ANALYSIS and RISK ASSESSMENT as individual terms with a variety of definitions. The time has come to resolve the proper use of these terms and to establish a consistent set of risk analysis/assessment terminology for use by practitioners in the field. The proliferation of new terms such as "risk estimation," "risk evaluation," and "risk characterization" tends to further confuse the existing situation.

I suggest that The Society for Risk Analysis initiate an effort to prepare a glossary of terms that members can use for a common basis of definition. The effort needs to start with RISK ANALYSIS and RISK ASSESSMENT.

Surveying various definitions in the risk literature, it appears that RISK ASSESSMENT was first in use and referred to the estimation process coupled with a judgment regarding the significance of the risk. Later, ASSESSMENT and ANALYSIS we used interchangeably. The question arises as to whether analysis is included in the process of assessment or assessment is included in the process of analysis.

Common usage, as illustrated by the name of our society, indicates RISK ANALYSIS may become the more encompassing term. I believe now is the time to agree to a set of definitions to correct potential misuse and to encourage proper communications. To simply illustrate the possibilities, consider the Venn diagrams below.

I propose the relationship in Figure 1-A for the following reasons:

1) Analysis is apparently more restrictive than assessment.
2) Assessment includes analysis and judgment.
3) Analysis implies a heavy reliance on quantitative aspects to achieve understanding.
4) Assessment implies an overall evaluation including unquantifiable facts.

Difficulties arise in trying to establish simple distinctions for the relationship because risk analysis and risk assessment are really

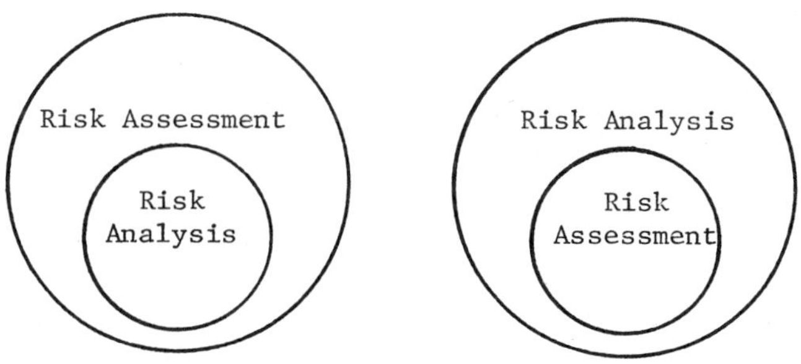

Figure 1

Venn Diagram for Risk Analysis/Assessment Relationships

processes that, depending on the particular circumstances, incorporate various degrees of detailed examination, judgment, and evaluation. The difference I propose are based on the following:

1) Risk Analysis will be concerned with the examination of a complex process, facility, or concept in order to understand and quantify the nature of the risk it presents.
2) Risk Assessment includes risk analysis in a critical manner with a judgment of the risk presented; an evaluation using both quantified and unquantified information.

Many authors may disagree with the proposed differentiation between ASSESSMENT and ANALYSIS. The proposed relationship between ANALYSIS and ASSESSMENT is opposite to some authors (e.g., Ritter (1981) and Douglas (1982)) and consistent with others (e.g., Kates (1981) and Rowe (1977)). Some authors tend to use ANALYSIS and ASSESSMENT as interchangeable words. The proposed relationship between ANALYSIS and ASSESSMENT herein is based on analysis leading to quantification and assessment including factors beyond the quantification. Simply put, analysis is the scientific objective process of risk estimation; assessment is the combining of the analysis with subjective considerations.

Based on the previous discussions, the following definitions of RISK ANALYSIS and RISK ASSESSMENT are proposed:

RISK ANALYSIS: A detailed examination performed to understand the nature of unwanted, negative consequences to human life, health, property, or the environment; an analytical process to provide information regarding undesirable events; the process of quantification of the probabilities and expected consequences for identified risks.

RISK ASSESSMENT: The process, including both risk analysis and risk management alternatives, of establishing information regarding and acceptable levels of that risk for an individual, group, society, or the environment.

The proposed definitions do not address other concepts that are also adding to the proliferation of terminology, such as risk management. Additional terms, such as risk evaluation, risk estimation, risk characterization can be avoided by first agreeing to the relationship between risk analysis and assessment. Risk management becomes the "control loop" that uses risk assessment in a decision process.

RECOMMENDATIONS

Comments are invited regarding the proposed definitions and the Society's sponsorship of the development of a glossary.

ACKNOWLEDGMENTS

This paper is a partial report on the development of a risk analysis/assessment glossary of Gratt and Perry (1983) performed under sponsorship of the U.S. Environmental Protection Agency. Dr. A.A. Moghissi was the project officer. The author is thankful to B. Perry, C. Whipple, P. Cho, H. Inhaber, R. Farnham, W. Suttcliffe, L. McCray, V. Covello, R. Blond, D. Layton, and S. Morris for aid and comments in an attempt to develop a glossary.

REFERENCES

Anderson, G.E., 1984, "Modeling of Risk to Population from Multi-Species Exposure to Atmospheric Toxic," Annual Meeting of the Air Pollution Control Assoication, Paper 84-102.2, San Francisco, California, June 24-29.

Douglas, J., R. Wyzga, and P. Ricci, 1982, "Toward Better Methods of Risk Assessment," EPRI Journal, March.

Gratt, L.G. and B.W. Perry, 1983, Development of a Risk Analysis/Assessment Glossary, IWG-FR-003-02, February 28.

Hammer, W., 1972, Handbook of System and Product Safety. Prentice-Hall, Inc., Englewood Cliffs, New Jersey.

Kates, R.W., 1981, Risk Assessment of Environmental Hazard. Scope 8. John Wiley and Sons, New York, New York.

Lass, H. and P. Gottlieb, 1971, Probability and Statistics. Addison-Wesley Publishing Company, Reading, MA.

Lowrance, W.W., 1976, Of Acceptable Risk. William Kaufmann, Inc., Los Altos, CA.

McCormick, N.J., 1981, Reliability and Risk Analysis. Academic Press, New York.

Morris, S., 1982, Letter to A. Moghissi, dated June 3.

National Research Council, 1983, Risk Assessment in the Federal Government: Managing the Process. National Academy Press, Washington, D.C..

Petak, W.J. and A.A. Atkisson, 1082, Natural Hazard Risk Assessment and Public Policy. Springer-Verlag, New York, NY.

Ritter, D., 1981, Risk Analysis Research and Demonstration Act of 1981. 97th Congress H.R. 3441.

Rowe, W.D., 1977, An Anatomy of Risk. John Wiley and Sons, New York, NY.

Starr, C., R. Rudman and C. Whipple, 1976, "Philosophical Basis for Risk Analysis." In: Annual Review of Energy 1: 629-662.

U.S. Nuclear Regulatory Commission, 1975, Reactor Safety Study: Main Report. WASH-1400 (NUREG 75/014), October.

Wasserman, M. and D. Wasserman, 1979, "Risk Assessment in Geographical Occupational Health," Geographical Medica 9: 8-27.

Webster's, New Collegiate Dictionary. G. and C. Merriam Company, Springfield, MA 1980.

Webster's, 3rd International Dictionary. G. and C. Merriam Company, Springfield, MA 1981.

APPENDIX: PROPOSED GLOSSARY OF TERMS FOR RISK ANALYSIS

accident: occurrence in a sequence of events which usually produces unintended injury, death or property damage.

acute toxicity: Any poisonous effect produced within a short period of time following exposure, usually up to 24-96 hours, resulting in severe biological harm and often death.

attributable disease risk: The rate of the disease in exposed individuals that can be attributed to the exposure. This measure is derived by subtracting the rate (usually occurrence or mortality) of the disease among nonexposed persons from the corresponding rate among exposed individuals.

cancer: An abnormal, potentially unlimited, disorderly new tissue growth.

case-fatality rate: A ratio of the number of deaths due to a disease to the number of cases of that disease in a specified period of time.

comparative risk: The ratio between risks associated with two (or more) actions leading to the same goal.

damage: Damage is the severity of injury or the physical, functional, or monetary loss that could result if control of a hazard is lost.

danger: Expresses a relative exposure to a hazard. A hazard may be present, but there may be little danger because of the precautions taken.

De minimis risk: From the legal maxim 'de minimis non curat lex' or the law is not concerned with trifles'. The risk below which the law is not concerned.

disease: A general term describing a morbid condition which can be defined by objective, physical signs (e.g., hypertension), subjective symptoms or mental phobias, disorder of function (e.g., biochemical abnormality), or disorders of structure (anatomic or pathological change).

dose: The amount of energy deposited from radiation or the amount of a given chemical at the site of effect.

effect: A biological change caused by an exposure.

failure modes and effects analysis: A tool to systematically analyze all contributing component failure modes and identify the resulting effects on the system.

fault tree analysis: A technique by which events that interact to produce other events can be related using simple logical relationships permitting a methodical building of a structure that represents the system.

hazard: A condition with the potential of causing an undesirable consequence.

health effect: An adverse deviation in the normal function of the human body.

in vivo: Within living organisms.

in vitro: Outside of living organisms (in glass).

LC50: Lethal concentration fifty.

LD50: Lethal dose fifty.

mortality rate: The number of deaths that occur in a given population during a given time interval.

mutagen: Any substance that causes changes in the genetic structure in subsequent generations.

pollutant: Any material entering the environment that has undesired effects.

population at risk: A specific population subjected to the hazard of concern. A limited population that may be unique (with respect to susceptibility to the effect or to the dose) for a specific dose-effect relationship.

premature death: A death that occurs before statistical expectation, usually attributable to a specific cause.

prevalence: The number of existing cases in a population who have the disease at a given point (or during a given period) of time.

proportionate mortality ratio (PMR): The fraction of all deaths from a given cause in the study population divided by the same fraction from a standard population.

relative potency: A comparison of the potency of two or more reference chemicals.

relative risk: The ratio of the rate of the disease (usually occurrence or mortality) among those exposed to the rate among those not exposed.

risk: The potential for realization of unwanted, adverse consequences to human life, health, property, or the environment; estimation of risk is usually based on the expected result of the conditional probability of the event occurring times the consequences of the event given that it has occurred.

risk analysis: A detailed examination performed to understand the nature of unwanted, negative consequences to human life, health, property, or the environment; an analytical process to provide information regarding undesirable events; the process of quantification of the probabilities and expected consequences for identified risks.

risk assessment: The process, including both risk analysis and risk management alternatives, of establishing information regarding and acceptable levels of that risk for an individual, group, society, or the environment.

safety: Relative protection from adverse consequences.

source term: The hazardous agent released from a facility or activity.

standardized mortality ratio (SMR): The ratio of observed deaths in a population to the expected number of deaths as derived from a standard population rates with adjustment of age and possibly other factors such as sex or race.

surrogate: Something that serves as a substitute.

synergism: A cooperative action of two substances that results in a greater effect than both of the substances could have had acting independently.

teratogenic: Substances that are suspected of causing malformations or serious deviations from the normal type, which can not be inherited in or on animal embryos or fetuses.

EXPOSURE TO ELECTROMAGNETIC FIELDS AND BRAIN TUMORS

Ruey S. Lin, Patricia C. Dischinger and Katherine Farrell

Maryland State Department of Health and Mental Hygiene
Baltimore, Maryland 21201

ABSTRACT

An epidemiologic study was conducted on occupation and industry entries listed on the death certificates of 951 adult white male Maryland residents who died from brain tumor between 1969 and 1982. Compared to controls who were matched with the cases on age and date of deaths and who died of causes other than malignancy, a significantly higher proportion of primary brain tumor persons (gliomas and astrocytomas) was found to have been employed in electrical occupations such as electrician, electric or electronic engineer, and utility company serviceman. To further explore the relationship, the occupations of cases and controls were grouped according to a level of potential job exposure to electromagnetic (E-M) fields i.e., definite, probable, possible and no exposure. An increase in the odds ratio for brain tumor was found to be positively related to potential E-M exposure levels. Furthermore, the mean age at death was found significantly younger among the cases in the presumed definite E-M exposure group than those with no apparent E-M exposure group. Similar findings of an increase of brain tumors in electrical workers have been reported elsewhere. These findings suggest that E-M exposure may be associated with the pathogenesis of brain tumors, particularly in the promoting stage.

KEY WORDS: Brain Tumors; Death Certificates; Electrician; Electromagnetic Fields; Non-ionizing Radiation; Occupation; Oncogenesis

In recent years, several epidemiological studies have suggested that environmental factors may be related to the occurrence of brain tumors. Specifically, concern has been raised over the possibility of increased risk of brain tumor among employees in the petrochemical industry (1,2). A few studies on workers of some refineries suggest that petrochemical compounds such as vinyl chloride may be responsible for the increase (3-5). However, other studies showed no significant increase in brain tumor deaths among petrochemical industry employees (6-8).

The current study, utilizing vital records in the State of Maryland for 1969-1982, was conducted to further explore the association between occupation and brain tumor mortality.

MATERIALS AND METHODS

Death certificates of white male Maryland residents, aged 20 and over, were collected from the following ICD Codes for review:

ICD8 (1969 - 1978)	ICD9 (1979 - 1982)	TYPE OF TUMOR
191	191.0 - 191.9	Malignant neoplasm of brain
192.9		Malignant neoplasm of other part of nervous system, site unspecified
225.0	225.0	Benign neoplasm of brain
238.1	239.6	Neoplasm of unspecified nature, brain

All females and black males were not included because of the relative lack of occupational information on their death certificates.

From the total of 1,043 death certificates collected, ninety-two certificates were excluded. Sixty-eight were metastatic tumors from other sites and twenty-four were rare brain tumors such as melanoma, lymphoma, medulloblastoma and ventricular tumor. Rare brain tumors were excluded because of their different histopathology and natural history, which could introduce a bias in a study of gliomas (9).

As a result, 951 brain tumor deaths were included in this study: 370 glioma or glioblastoma multiforme, 149 astrocytomas, and 432 brain tumors (histologic type not specified). It is suspected that some of the non-specified brain tumors may be secondary tumors or other rare types of brain tumors. Table 1 shows the distribution of cases by the type and ICD code on the certificates. Because of the overlap between the ICD codes and different histologic types of brain tumors, subsequent analyses were based on the histologic type recorded on the death certificate and not on ICD code. Gliomas and astrocytomas were grouped together as a single category of primary brain tumor. Brain tumors with histologic type not specified were treated as a separate category in the analysis.

Occupations recorded on death certificates among brain tumor deaths were compared to those of a control group of death certificates of white adult male Maryland residents with causes of death other than malignant neoplasms. Controls were matched with the cases on age (±1 year) and date of death (as close to the case as possible) and were selected from a ten per cent computer-randomized roster of Maryland deaths.

Statistical significance was assessed using a chi-square test on the distribution of various occupations, the t-test on age distribution as well as 95 percent confidence intervals around the odds ratios using the logit method (10).

RESULTS

Table 2 shows the age distribution of the 951 white male brain tumor deaths by histologic type. The peak age is in the 60-69 year age group for all three types of brain tumors. The mean age at death for the glioma and astrocytoma groups is similar. However, it is higher for the non-specified brain tumor group which may be a result of the inclusion of secondary brain tumor cases in that category.

Preliminary analysis of the data on cases disclosed a disproportionate representation of workers employed in occupations associated with electricity or electromagnetic fields. In Table 3, the number of deaths in each of these occupational groups is compared to the

Table 1

Distribution of 951 White Male Brain Tumor Deaths in Maryland, 1969-1982, by Type and ICD Code, as Recorded on Death Certificates

ICD Code	Glioma/Astrocytoma		Brain Tumor (Non-Specified)		Total	
	No.	%	No.	%	No.	%
191	388	74.7	129	29.8	517	54.3
192.9	127	24.5	0	0.0	127	13.4
225.0	0	0.0	2	0.5	2	0.2
238.1 or 239.6	4	0.8	301	69.7	305	32.1
Total	519	100.0	432	100.0	951	100.0

Table 2

Age Distribution of 951 White Male Brain Tumor Deaths in Maryland, 1969-1982, by Type

Age	Gliomas		Astrocytomas		Brain Tumors (Non-Specified)	
	No.	%	No.	%	No.	%
20-29	11	3.0	5	3.4	10	2.3
30-39	25	6.8	14	9.4	28	6.5
40-49	63	17.0	24	16.1	57	13.2
50-59	98	26.5	37	24.8	102	23.6
60-69	116	31.3	55	36.9	129	29.9
70-79	44	11.9	12	8.1	84	19.4
80 & over	13	3.5	2	1.3	22	5.1
TOTAL	370	100.0	149	100.0	432	100.0
Mean Age (yrs.)	57.6		56.2		60.0	
S.D.	13.2		12.7		13.5	

Table 3

Observed Number of Electricity-Related Occupations Among White Male Brain Tumor Deaths in Maryland, 1969-1982, as Compared with the Expected Number from 1970 Maryland Census Data, by Type

Occupation	1970 Maryland Census white male employees aged 20 and over		Glioma & Astrocytoma		Brain Tumor (Non-Specified)	
	No. of Employees	%	Obs. No.	Exp.* No.	Obs. No.	Exp.* No.
Electrician	9,300	1.15	13‡	5.7	8	4.7
Electric or electronic Engineer technician	11,600	1.43	18‡	7.2	12‡	5.8
Utility (electric & telephone) lineman, serviceman	8,400	1.04	19‡	5.1	8	4.2
3 occupations together	29,300	3.62	50‡	18.0	28‡	14.7
Total	809,600	100.0	499†	499.0	407†	407.0

* Expected number calculated according to 1970 census white male employee's occupation proportional distribution

† Excluded the number with unknown occupation

‡ $P < 0.01$

number expected based on the proportion of these occupations surveyed and reported in the 1970 Maryland population census. There is a greater-than-expected number of brain tumor deaths in all of these occupational groups, especially for the glioma/astrocytoma group. These data, however, may be biased because they represent the prevalence of these occupations at one point in time (1970 census), where as the mortality data span a 14-year period. Also age is not adjusted. In addition, some differences may be attributable to occupation recording practices for the census as compared to death certificates.

To overcome this possible bias, a case-control study was conducted. Death certificates were obtained from a control group of white adult Maryland residents who died from causes other than malignancy, matched with the cases on age and date of death. Table 4 presents the case/control comparison of occupational distribution for several occupational groups. Occupations related to electricity or apparent exposure to electromagnetic fields showed an excess representation among the cases of the glioma/astrocytoma group as compared with the controls; but no significant differences were noted between the cases and controls in the non-specified brain tumor group. No significant excess of brain

Table 4

Occupation Distribution Between Brain Tumor Cases and Controls

Occupation	Glioma & Astrocytoma		Brain Tumor (Non-Specified)	
	No. of Cases	No. of Controls	No. of Cases	No. of Controls
Electrician	13	10	8	8
Electric or Electronic Engineer & Technician	18*	6	12	9
Utility Co. Employees	19	11	8	8
Carpenter	15	13	14	17
Mechanic, Machinist or Repairman	33	27	21	22
Petrochemical Ind.	5	4	7	6
High Professionals (physician, lawyer, etc.)	24	17	18	18
Farmer	14	12	15	9
Navy Personnel	11	4	8	3
Other Occupations	347	381	296	314
Unknown	20	34	25	18
Total	519	519	432	432

* $P<0.05$

tumor deaths was found in the petrochemical industries, although the numbers were small. There were only twelve cases in this job category, as compared with ten controls.

Based on these preliminary findings, occupations were then grouped according to level of possible exposure to electromagnetic fields -- i.e., definite, probable, possible or no exposure. Exposure category for each occupation was determined in consultation with an industrial hygienist, an occupational physician and a radiation physicist. Since there are no available data on duration of occupational exposure on death certificates, the categories of electromagnetic fields exposure are based simply on job title and industry entry. Table 5 shows the rankings. Cases and controls were assigned to categories independently by two persons blinded as to their status. Any discrepancies in ranking were solved by a third opinion.

Table 5

Category of Occupation According to
Assumed Electromagnetic Field (E-M) Exposure:

Category A (Definite exposure to E-M field):

(1) Electric and Telephone Company: Serviceman, lineman, foreman, and engineer;

(2) Railroad and telecommunication: engineer;

(3) Electrician, electric and electronic engineer in industry

Category B (Probable exposure to E-M field):

(1) Electrician, non-specified, retail;

(2) Dispatcher, highway patrolman;

(3) Engineer (non-specified) in electric, electronic, aerospace and telecommunication industries;

(4) Repairman in appliance (TV, radio, electronic device, etc.) and telecommunication industries;

(5) Welder

Category C (Possible exposure to E-M field):

(1) Policeman, security guard, airline pilot;

(2) Employee (non-specified) in utility company (telephone and electric) railroad, and telecommunication industries;

(3) Mechanic, machinist, repairman, maintenance man, stationary engineer, boiler, etc.;

(4) Steelworker, tool maker, plumber, pipefitter, carpenter;

(5) Engineer (non-specified) in other industries;

(6) Officer (non-specified) in Navy, Air Force and Coast Guard

Category D (no exposure to E-M field):

All occupations other than the above stated 3 categories

Table 6 shows an analysis of case/control differences based on these exposure categories for glioma and astrocytoma. There are 27 cases as compared with 14 controls in category A (definite E-M exposure), and 21 cases versus 12 controls in category B (probable E-M exposure) a significant difference. The odds ratios for these two exposure groups are 2.15 and 1.95, respectively. It is of interest to note that there is an increase in the odds ratio positively related to E-M exposure levels. However, only the categories A and C have 95 percent confidence intervals

Table 6

Occupation Distribution Between Primary Brain Tumor Cases (Glioma and Astrocytoma) and Controls, by Category of Electromagnetic Field (E-M) Exposure

Category of Occupation According to E-M Exposure	Cases No.	Controls No.	Odds Ratio	95% (Confidence) Intervals
A (Definite E-M Exposure)	27	14	2.15	(1.10-4.06)
B (Probable E-M Exposure)	21	12	1.95	(0.94-3.91)
C (Possible E-M Exposure)	128	99	1.44	(1.06-1.95)
D (No E-M Exposure)	323	360	1.00	
Unknown Occupation	20	34		
Total	519	519		

above 1.0. Table 7 presents the same analysis for non-specified brain tumors. Although there are a greater number of cases than controls in both categories A and B, but none of them has an odds ratio significantly different from 1.0.

Table 8 shows the mean age at death of the cases by E-M exposure category. Cases of glimoa/astrocytoma group in the "definitely E-M exposure" category died at significantly younger ages than those with no E-M exposure (mean age 52.8 versus 57.9, respectively, $t = 2.23$; $P < 0.05$). However, there are no significant differences in mean age at death, by E-M exposure levels, in those with non-specified brain tumors.

DISCUSSION

Weakness of death certificate data as material for epidemiologic studies are widely acknowledged (11). However, as a source for generating hypotheses regarding occupational or industrial hazards associated with specific types of cancer, such data can be quite valuable. The difficulty with the small numbers of cancer cases often met in retrospective and prospective epidemiologic studies can be overcome in mortality statistics studies by combining related occupational categories and by analyzing data from a larger time period.

In the present study, there were two possible sources of error: 1) misdiagnosis or misclassification of cause of death, and 2) misclassification of occupation. The first type of error is probably less common for brain tumors than for other cancers since both malignant and benign brain tumors usually result in death (12). In a study of the accuracy of death certificates, mortality rates for brain tumors were judged to be among the most accurate based on both detection and confirmation (11). Two measures were taken in order to reduce error in

Table 7

Occupation Distribution between Brain Tumor Cases (non-specified) and Controls, by Category of Electromagnetic Field (E-M) Exposure

Category of Occupation According to E-M Exposure	Cases No.	Controls No.	Odds Ratio	95% (Confidence) Intervals
A (Definite E-M Exposure)	15	10	1.54	(0.68-3.38)
B (Probable E-M Exposure)	19	15	1.30	(0.60-2.78)
C (Possible E-M Exposure)	87	95	0.94	(0.68-1.31)
D (No E-M Exposure)	286	294	1.00	
Unknown Occupation	25	18		
Total	432	432		

Table 8

Mean Age at Death and Standard Deviation by Category of E-M Exposure and Type

Category of E-M Exposure	Glioma & Astrocytoma		Brain Tumor (Non-Specified)	
	No. of Cases	Mean age at death±SD	No. of Cases	Mean age at death ±SD
A (Definite E-M Exposure)	27	52.8±11.1*	15	59.1±17.2
B (Probable E-M Exposure)	21	61.2±13.5	19	60.8±11.1
C (Possible E-M Exposure)	128	56.2±12.5	87	60.9±13.1
D (No E-M Exposure)	323	57.9±13.1	286	59.6±13.6
Unknown Occupation	20	54.4±16.9	25	61.5±13.7
Total	519	57.2±13.0	432	60.0±13.5

* $P < 0.05$, as compared with no E-M exposure group age distribution

secondary brain tumor cases. First, all death certificates collected were reviewed by the authors (RL and PD). Any death certificates that indicated that the brain tumor was due to metastasis were removed from the analysis. Second, cases were classified according to histologic type, i.e., glioma or glioblastoma multiforme, astrocytoma, and non-specified "brain cancer" or "brain tumor." Gliomas ans astrocytomas, diagnosed either by biopsy or autopsy, were considered as primary brain tumors. Analyses were performed separately for the group of non-specified brain tumors.

Errors stemming from occupational misclassification may come from several sources: inaccuracies as a result of lack of knowledge of next-to-kin, or incomplete information. The underlying assumption of our study was that the recorded occupation was the principal job and that the duration of exposure was sufficient to cause an effect in the decedent's health. However, since the information on occupation of controls was also obtained from death certificates, the potential for error should be shared equally by cases and controls (13).

Mortality data from the State of Maryland show an association between primary brain tumor and employment in occupations with exposures to potentially high electromagnetic fields. Several studies in the literature have also mentioned a higher risk of brain tumor among electrical and maintenance workers, but the majority of such findings have been reported as part of a particular industry study, such as the petrochemical industry. Generally, researchers concentrated on hypotheses linking brain tumors to chemical exposures. In a study in Los Angeles, Preston-Martin and coworkers found a significant increase in brain tumors among electricians and engineers, particularly aircraft employees (14). The authors, however, concluded that the data "fail to implicate any specific exposures within the (aircraft) industry that might be important etiologically, since the majority of workers were classified at diagnosis as engineers or some other type of white collar worker."

In another epidemiologic study of CNS tumors conducted among industrial workers in Ohio, Mancuso also observed that individuals working with electrical machinery had the highest CNS tumor mortality rate (15). Nicholson et al. (16) reported an excess of brain tumors among operating engineers in chemical and petrochemical industries in Texas and Louisiana. The category of operating engineers included stationary engineers, stationary firemen, and construction engineers who were engaged in the operation and maintenance of heavy equipment.

Alexander et al. (2) reported on an investigation of an apparent increased prevalence of brain tumors in a Texas petrochemical plant. Although examination of work histories of the 20 affected employees revealed no common chemical exposures or work assignments, seven were maintenance workers, one boiler maker, one machinist and one operator in energy systems. All of these workers would be classified, according to our current ranking category, as being Risk Category C (possible electromagnetic field exposure).

In addition, Seidman et al. (17), in a study of brain tumor mortality among asbestos insulation workers in the United States and Canada, observed that the excess of brain tumors among insulators was concentrated in the younger ages (those under 50) and during the early period after onset of work (15-24 years) in contrast with the age distribution and latency in other asbestos-associated neoplasms. The authors concluded that "this may have relevance to theoretical concerns about questions of initiation and promotion in the etiology of cancer, particularly with regard to brain tumors." Since insulators or pipefitters frequently work in close proximity to electrical wiring, these occupations would also be ranked as being Risk Category C in our study.

Bond et al., (18) conducted a case-control study of brain tumor deaths among employees at a large chemical manufacturing plant and found an association between risk of brain tumor and employment in a magnesium production department, particularly noted for the workers assigned to this department for a short period (1-5 years). Although the authors stated that some exposure to electromagnetic fields could be expected in both chlorine and magnesium production, they discounted this possibility because of a lack of other epidemiologic support and because the odds ratios tended to decrease with an increasing duration of employment in this department.

Another industry which utilizes high levels of electricity is aluminum production; Milham (19) found a significant increase of mortality from benign brain tumors (five observed versus 0.7 expected) among aluminum reduction plant workers, particularly for workers categorized as nonexposed to chemicals, such as mechanics, maintenance electrical, metal products, etc.

In a follow-up study of U.S. enlisted Naval personnel exposed to microwave radiation during the Korean War period (20), veterans were categorized as either high exposure (electronic technician, fire control technician or aviation electronics technician) or low exposure (radioman, radarman, and aviation electrician mate). No significant differences were noted in the rate of brain tumors between the two exposure groups. However, there were sixteen cases that died from malignancy of eye, brain and other nervous systems (ICD Codes 190-192), representing 7.9 percent of all malignant deaths (16 out of 202) as compared with a 3.8 percent proportional mortality ratio of brain tumors to all malignancies among U.S. males, aged 20-64, in 1970 (21) ($P < 0.01$). Both groups of veterans were exposed to electromagnetic fields and would be classified in Risk Category A or B in the current study. It is of interest to find that navy personnel were more represented among cases than controls in the present study (11 versus 4 in the glioma/astrocytoma group, and 8 versus 3 in the brain tumor non-specified group, see Table 4).

Wertheimer and Leeper (22) have reported that adults living in Colorado close to wires carrying high-flow alternating electric current have an increase in the incidence of all cancers. Cancer types most significantly associated with current wire configuration included cancer of the nervous system, uterus, breast and lymphoma. In addition, they observed that the strongest high current wire configuration-cancer association appears at periods after six to nine years exposure suggesting a rather short latency between alternating magnetic field exposure and cancer manifestation. In a similar study, they also found an association between electrical wire configuration and childhood cancer incidence (23). On the other hand, a study from Rhode Island did not show an association between childhood leukemia and electric wire configurations (24). A recent report from Sweden, however, is compatible with the study by Wertheimer and Leeper, showing an association between high tension wires and the occurrence of childhood tumors (25).

Several reports have already shown an increased risk of leukemia among electricians and other workers regularly exposed to electromagnetic fields (26-29).

Electrical and magnetic fields surround us, as they are emitted from high tension wires, radio transmitters, and household appliances. This form of non-ionizing radiation extends from infrared to microwaves and radiowaves, and down to emissions from alternating electrical currents which have extremely low frequencies and long wave lengths. All except infrared rays readily penetrate the body; however, the amount of absorption varies depending on frequency of waves and type of tissue. Systems of the body which are particularly vulnerable to electromagnetic waves include the central nervous system, hematopietic-immunologic system, cardiovascular and endocrine system (30).

The knowledge that pulsing electromagnetic fields can induce weak electric current in bone and stimulate osteogenesis has been well applied in the treatment of ununited fractures (31). Experimental studies on animals have also shown a proliferation reaction of the microglia in the brain (gliosis) following exposure to electromagnetic fields in various ranges of frequencies (32). Several investigators have reported the effects of electromagnetic fields on the efflux of calcium ion in biological system, principally brain tissue (33-35). Changes in calcium binding in chicken cerebral tissue have been demonstrated even with exposure to weak environmental electric fields oscillating at low frequency (33). Calcium ions play a prominent role in maintaining cellular membrane integrity and function. A large number of experimental observations suggest that a significant correlation may exist between the level of the electrical transmembrane potential difference in somatic cells and the intensity of their mitotic activity. Based on the experimental findings, Cone (36) has postulated a unified theory of mitosis control and oncogenesis associated with membrane potential difference. In fact, recent laboratory studies have demonstrated that weak pulsing electromagnetic fields can modify biological processes and induce RNA transcription (37).

Easterly (38) has hypothesized a mechanism for a relationship between exposure to magnetic fields and the induction of cancer: alternations in the mitotic processes caused by exposure to magnetic fields can provide a proliferative stimulus to latent tumor cells, thereby leading to the expression of malignant neoplasia. Wertheimer and Leeper (22) considered that their findings of a distinct pattern of latency between exposure to high current wires and cancer diagnosis in Colorado cancer cases is consistent with the hypothesis of cancer promotion produced by electromagnetic fields exposure. Our findings that glioma cases whose occupations were more likely to have exposed them to electromagnetic fields developed and died from brain tumor at significantly younger ages than did those who were not occupationally exposed to electromagnetic fields, also strongly support the theory of non-ionizing radiation as a promoting agent. The observation of Seidman and coworkers (17) that the excess of brain tumor among insulation workers was concentrated in the younger ages and during the early period after exposure can also be explained by this hypothesis. Furthermore, recent studies on experimental animals indicate that non-ionizing radiation may promote the development of several types of cancer, spontaneous or induced, in mice (39).

If oncogenic effects of non-ionizing radiation are limited only to the facilitation of the proliferation of tumor cells but not to the initiation of the malignant transformation of normal cells, then one may expect the promoting effects of electromagnetic field exposure to be more apparent with respect to the mortality from certain types of tumor such as brain tumors and leukemia. These conditions usually result in death even without metastasis because of their vital location or incessant proliferative capacity. Indeed, thus far epidemiologic evidence shows only the brain tumors an leukemia mortality to be associated with electromagnetic field exposure.

In recent decades there has been a rapid increase in the use of radiofrequency in industry, military and civilian settings. During the same period of time, the mortality rate for brain tumors in the United States also rose from 3.80 to 5.80 in whites and 2.15 to 3.85 in non-whites per 100,000 population for years from 1940 to 1977 (40). Although it is not known whether this reflects an actual increase in incidence of brain tumors or a difference in reporting and coding practices over time, if our data reflect a causal relationship between brain tumors and some characteristics of exposure to non-ionizing radiation, such a trend would be expected.

Despite this accumulation of epidemiologic evidence, the interpretation remains uncertain. It is not known whether the increased risk of brain tumors and leukemia among electrical workers is due to magnetic or electric fields themselves or possibly to a common chemical exposure such as polychlorinated biphenyls, other organic solvents, or metal fumes in these occupations.

If a causal association can be proved, the implications would be quite significant from the economic, social and political standpoint, as non-ionizing radiation is an important technology which permeates almost every aspect of life in modern, industrial society.

REFERENCES

1. Lewin R. Government/Industry Dispute Brain Tumor Risk. Science 1980; 210:996-7.
2. Alexander V., Leffingwell S.S., Lloyd, J.W., Waxweiler, R.J., and Miller, R.L. Investigation of an Apparent Increased Prevalence of Brain Tumors in a U.S. Petrochemical Plant. Ann, NY Acad. Sci. 1982:381:97-107.
3. Theriault, G., and Goulet, L. A Mortality Study of Oil Refinery Workers, J. Occup. Med. 1979; 21:367-70.
4. Thomas T.L, Docoufle P, and Moure-Eraso R. Mortality Among Workers Employed in Petrochemical Refining and Petrochemical Plants. J. Occup. Med. 1980; 22:97-103.
5. Thomas, T.L., Waxweiler, R.J., and Moure-Eraso, R. Mortality Patterns Among Workers in Three Texas Oil Refineries. J. Occup. Med. 1982;24:135-41.
6. Wen, C.P., Tsai, S.P., McClellan, W.A. and Gibson, R.L. Long-term Mortality Study of Oil Refinery Workers. I. Mortality of Hourly and Salaried Workers. Am. J. Epidemiol. 1983;118:526-41.
7. Hanis, M., Holmes, V.M., Shallenberger, L.D., Smith, L.G., and Jones K.E. Epidemiologic Study of Refinery and Chemical Plant Workers. J. Occup. Med 1982;24:203-12.
8. Alderson, M. and Rushton, L. Mortality patterns in Eight U.K. Oil Refineries. Ann. NY Acad. Sci. 1982;381:139-45.
9. Schoenberg, B.S., Christine, B.W., and Whisnant, J.P., The Descriptive Epidemiology of Primary Intracranial Neoplasms: The Connecticut Experience. Am. J. Epidemiol. 1976;104:499-510.
10. Grant, J.J., and Thomas D.G., The Performance of Three Approximate Confidence Limit Methods for the Odds Ratio. Am. J. Epidemiol. 1982;115:453-70.
11. Percy, C., Stanek, E., and Gloeckler, J., Accuracy of Cancer Death Certificates and Its Effect on Cancer Mortality Statistics. Am. J. Public Health 1981;71:242-50.
12. National Cancer Institute. Cancer Patient Survival, Report No. 5.

Bethesda, MD: National Cancer Institute, 1976. (DHEW Publication No. (NIH) 77-992).
13. Gordis, L. Should Dead cases be Matched to Dead Controls? Am. J. Epidemiol. 1982;115:1-5.
14. Preston-Martin, S. Henderson, B.E., and Peters J. Descriptive Epidemiology of Central Nervous System Neoplasms in Los Angeles County. Ann. NY Acad. Sci. 1982;381:202-8.
15. Mancuso, T.F., Epidemiological Study of Tumors of the Central Nervous System in Ohio. Ann. NY Acad. Sci., 1982;381:17-39.
16. Nicholson, W.J., Seidman, H., Selikoff, I.J., Tarr, D., and Clark, E. Brian Tumors Among Operating Engineers in the Chemical and Petrochemical Industry in Texas and Louisiana. Ann NY Acad. Sci. 1982; 381: 172-80.
17. Seidman, H., Selikoff, I.J., and Hammond, E.C. Mortality of Brain Tumors Among Asbestos Insulation Workers in the United States and Canada. Ann. NY Acad. Sci. 1982:381:160-71.
18. Bond, G.G., and Cook, R.R. Case-control Study of Brain Tumor Deaths Among Employees at a Chemical Manufacturing Plant. Ann. NY Acad. Sci. 1982;381:73-82.
19. Milham, S., Mortality in Aluminum Reduction Plant Workers. J. Occup. Med. 1979;21:475-80.
20. Robinette, C.D., Silverman, C., and Jablon, S. Effects Upon Health of Occupational Exposure to Microwave Radiation (Radar). Am. J. Epidemiol. 1980;112:39-53.
21. National Center for Health Statistics. Vital Statistics of the United States, 1970, Vol. 2, Mortality. Rockville, MD: National Center for Health Statistics, 1974. (DHEW Publication No. (HRA) 74-1102).
22. Wertheimer, N. and Leeper E. Adult Cancer Related to Electrical Wires Near the Home. Int. J. Epidemiol. 1982;11:345-55.
23. Wertheimer, N. and Leeper, E. Electrical Wiring configurations and Childhood Cancer. Am. J. Epidemiol 1979;109:273-84.
24. Fulton, J.F., Cobb, S., Preble, L., Leone, L., and Foreman, E. Electrical Wiring Configurations and Childhood Leukemia in Rhode Island. Am. J. Epidemiol. 1980;111:292-6.
25. Tomenius, L., Hellstrom, L. and Enander, B. Electrical Constructions and 50HZ Magnetic Field at the Dwellings of Tumor Cases (0-18 years of age) in the County of Stockholm. Presented in International Symposium on Occupational Health and Safety in Mining and Tunnelling. Prague, June 21-25, 1982.
26. Milham S., Mortality from Leukemia in Workers exposed to Electrical and Magnetic Fields (letter). N. Engl. J. Med. 1982;307:249.
27. Wright, W.E., Peter, J.M., and Mack, T.M. Leukemia in Workers Exposed to Electrical and Magnetic Fields (letter). Lancet 1982;2:1160-1.
28. McDowall, M.E., Leukemia Mortality in Electrical Workers in England and Wales (letter). Lancet 1983;1:246.
29. Coleman, M., Bell, J., and Skeet, R. Leukemia Incidence in Electrical Workers (letter). Lancet 1983;1:982-3.
30. Food and Drug Administration. Biological Effects of Electromagnetic Waves. Rockville, MD: Food and Drug Administration, 1977. (DHEW Publication No. (FDA) 77-8010).
31. Bassett, C.A.L., Mitchell, S.N., and Graston, S.W. Pulsing Electromagnetic Field Treatment in Ununited Fractuers and Failed Arthrodeses. JAMA 1982;247:623-8.
32. Kholodov, Y.A., The Effects of Electromagnetic and Magnetic Fields on the Central Nervous System. Moscos: Academy of Science, USSR, 1966 (NASA Technical Translation, 1967).
33. Bawin, S.M., and Adey, W.R. Sensitivity of Calcium Binding in Cerebral Tissue to Weak Environmental Electric Fields Oscillation at Low Frequency. Proc. Nat'l. Acad. Sci. 1976;73:1999-2003.
34. Blackman, C.F., Elder, J.A., Weil, C.M., Benane, S.G., Eichinger, D.C.

and House, D.E. Induction of Calcium-ion Efflux from Brain Tissue by Radiofrequency Radiation: Effects of Modulation Frequency and Field Strength. Radio Sci 1979;14(6S):93-8.
35. Adey, W.R., and Bawin, S.M. Binding and Release of Brain Calcium by Low-Level Electromagnetic Fields: A Review. Radio Sci 1982;17(5S):149-57.
36. Cone, C.D. Unified Theory on the Basis Mechanism of Normal Mitotic Control and Oncogenesis. J. Theor. Biol. 1971;30:151-81.
37. Goodman, R. Bassett, C.A.L., and Henderson, A.S. Pulsing Electromagnetic Fields Induce Cellular Transcription. Science 1983;220:1283-5.
38. Easterly, C.E. Cancer Link of Magnetic Field Exposure: A Hypothesis. Am. J. Epidemiol. 1981;114:169-74.
39. Szmigielski, S., Szudzinski, A., Pietraszek, A., Bielec, M., Janiak, M., and Wrembel, J.K. Accelerated Development of Spontaneous and Benzopyrene-induce Skin Cancer in Mice Exposed to 2450-MHZ Microwave Radiation. Bioelectromagnetics 1982;3:179-91.
40. Garfinkel, L. and Sarokhan, B. Trends in Brain Cancer Tumor Mortality and Morbidity in the United States. Ann. NY Acad. Sci 1982;381:1-5.

UNCERTAINTY IN POPULATION RISK ESTIMATES FOR ENVIRONMENTAL CONTAMINANTS*

C. Richard Cothern, William A. Coniglio and
William L. Marcus

US Environmental Protection Agency
Washington, D.C. 20460

ABSTRACT

Risk assessments of environmental contaminants involve input information that in many cases contain large uncertainties. The contribution of these uncertainties to the risk assessment of several volatile organic compounds (VOCs) in drinking water are listed and tabulated. The contribution of each of these uncertainties is determined or estimated. The largest contribution to the uncertainty is due to the model used to extrapolate bioassay data in the hazard assessment part of the risk assessment. For the VOCs in drinking water the uncertainty in the hazard assessment is four to six orders or magnitude.

KEY WORDS: Risk Assessment, Population Risk, Volatile Organic Compounds, VOCs, Uncertainty, Drinking Water

INTRODUCTION

The risk of individuals experiencing an adverse health effect after being exposed to an contaminant may be calculated by multiplying the level of personal exposure times a measure of the hazard resulting from the toxic potency of the contaminant. A more complete discussion of the calculations discussed here are contained in a report from the US Environmental Protection Agency's Office of Drinking Water (Cothern, 1984). The total estimated number of cases of disease or death expected among an entire society for one lifetime is called the population risk. This estimate is used as background information in the development of regulatory standards in the USEPA's Office of Drinking Water.

For this paper the present exposure level for drinking water contaminants among the US population is calculated and expressed as a population-concentration estimate. Population concentration is the sum of the products of the number of people times their estimated exposure level. Typical units of population-concentration found in drinking water contaminants are persons x micrograms/liter. The carcinogenic potency for individual contaminants has been expressed as the number of cases of

---

*The ideas and thoughts expressed in this paper are those of the authors and do not necessarily reflect the thoughts and policies of the US Environmental Protection Agency.

cancer per person per microgram per liter. This paper considers only the carcinogenic risk for specific contaminants and does not attempt to deal with other manifestations of toxicity nor does it consider the question of what is a carcinogen. The volatile organic compounds (VOCs) analyzed here were selected for study because they offered a range of data of potency and population exposure via drinking water. The compounds include; benzene, carbon tetrachloride, 1,1-dichlorethane, 1,1-dichloroethylene, perchloreothylene, trichlorethylene, and vinyl chloride. The analysis provides a basis for determining the direct effectiveness of national drinking water standards to reduce cancer illness and death from present levels of these chemicals in water supplies. This portion of the analysis could be called risk reduction analysis.

The occurrence estimates were evolved from monitoring data developed in several national drinking water surveys. The surveys included are; the National Organic Reconnaissance Survey, the National Organic Monitoring Survey, the National Screening Program for Organics in Drinking Water, the Community Water Supply Survey, the Rural Water Survey and the Groundwater Supply Survey. Taken together these surveys provided data on over 1,000 ground water supplied and 1,000 surface water supplies. This data base was drawn from 48,458 groundwater supplies and 11,202 surface water supplied and are shown in Table 1.

Drinking water contributes to the three routes of exposure; viz, ingestion, inhalation and dermal. An estimate of the relative importance of these routes for different age groups for VOC's in drinking water is shown in Table 2. Due to the chemical characteristics of the VOCs considered here the dermal exposure appears to be very small compared to the ingestion and inhalation exposure potentials which may be close in magnitude. Respiration exposure is of increased importance due to the enrichment of indoor air from containment degassing during water use.

The carcinogenic potency of these contaminants was estimated from bioassay data developed from animal tests conducted at high dose levels. The biological response at lower levels of exposure was estimated using mathematical models. Figure 1 shows a typical extrapolation. The starred points are the experimental animal data and subsequent lines represent the projections from four different models. The error bars on each line represent the upper 95% confidence calculations from the model.

Table 3 is a typical example of the calculations of population risk and the risk reduction analysis. As seen in Table 3, the population risk is in the range of less than one to 400 per lifetime. (Fractional population risks resulting from calculations have been rounded off to the designation of less than one). In most cases the largest contribution to population risk is from the population exposed below the detection limit. The contribution to the uncertainty as a result of toxicological design and results are discussed in a later section where it is shown that the choice of model predominates.

The ranges of uncertainty in population risk estimates and risk reduction analysis are quite large. This paper will describe in detail the contributions to this uncertainty in the areas of both exposure and hazard estimates. Although the discussion here is specific for VOCs in drinking water, it can in general be applied to other environmental contaminants. If the actual shape of the dose-response curve were known, the uncertainty in these estimates could be reduced to about two to three orders of magnitude. Data for human exposure especially at the lowest exposure category could improve the uncertainty and data at lower dose levels could also reduce the uncertainty. Neither of these sources of

Table 1

Total Estimated Population (In Thousands) Exposed to the Contaminant in Drinking Water at the Indicated Concentration Range Micrograms/Liter

Contaminant	Population Concentration (see footnote *) Mean best total	Mean worst total	Drinking Water Concentration (microgram/liter) <0.5	>0.5-5	>5-10	>10-20	>20-30	>30-40	>40-50	>50-60	>60-70	>70-80	Other
Benzene	$1.6x10^7$	$1.2x10^8$	209,166	4,535	123	62	16	6	3	2	1	1	<100
Carbon Tetrachloride	$9.5x10^7$	$1.9x10^8$	187,609	24,460	1,389	43	655	0	0	0	0	0	0
1,2-Dichloro-ethane	$3.8x10^7$	$1.4x10^8$	202,187	12,890	0	143	0	0	0	0	0	0	0
1,1-Dichloro-ethylene	$6.7x10^6$	$2.2x10^8$	209,630	2,295	52	0	0	0	0	0	0	0	0
Tetrachloro-ethylene	$4.5x10^7$	$1.5x10^8$	202,989	10,556	434	254	82	0	0	0	105	0	0 >100
Trichloro-ethylene	$1.3x10^8$	$2.2x10^8$	189,288	23,287	426	212	0	738	257	42	0	128	42
Vinyl Chloride	$1.5x10^7$	$1.2x10^8$	212,497	1,331	472	0	0	0	0	0	118	0	0

* The population concentration is the sum of the products of the population and the average drinking water concentration to which that population was exposed. The mean best total assumes that all in the lowest concentration were exposed to 0.0 micrograms/liter and the mean worst assumes that all were exposed to the maximum or 0.5 micrograms/liter.

Table 2

Comparative Model of Absorbed Dose From A
Volatile Pollutant (100 micrograms/liter) In Drinking Water

Absorbed Dose(micrograms/day)

	Formula fed Infant (4 kg)	Pre-teen (32 kg)	Adult Female (60 kg)	Adult Male[a] (70 kg)	Male[b] (70 kg)
Fluid Ingestion	80	150	200	200	200
Inhalation of enriched indoor air	10	10	70	50	50
Inhalation of enclosed shower air	--	100	300	200	--
Dermal absorption bathing	.02-.06	--	--	--	2.5
Swimming	--	10-300	--	--	--
Total Absorbed Dose micrograms/kg/day	20 micrograms/	10-20 micrograms/	10 micrograms/	7 micrograms/	4 micrograms/

q showering adult male
b adult male that does not take showers

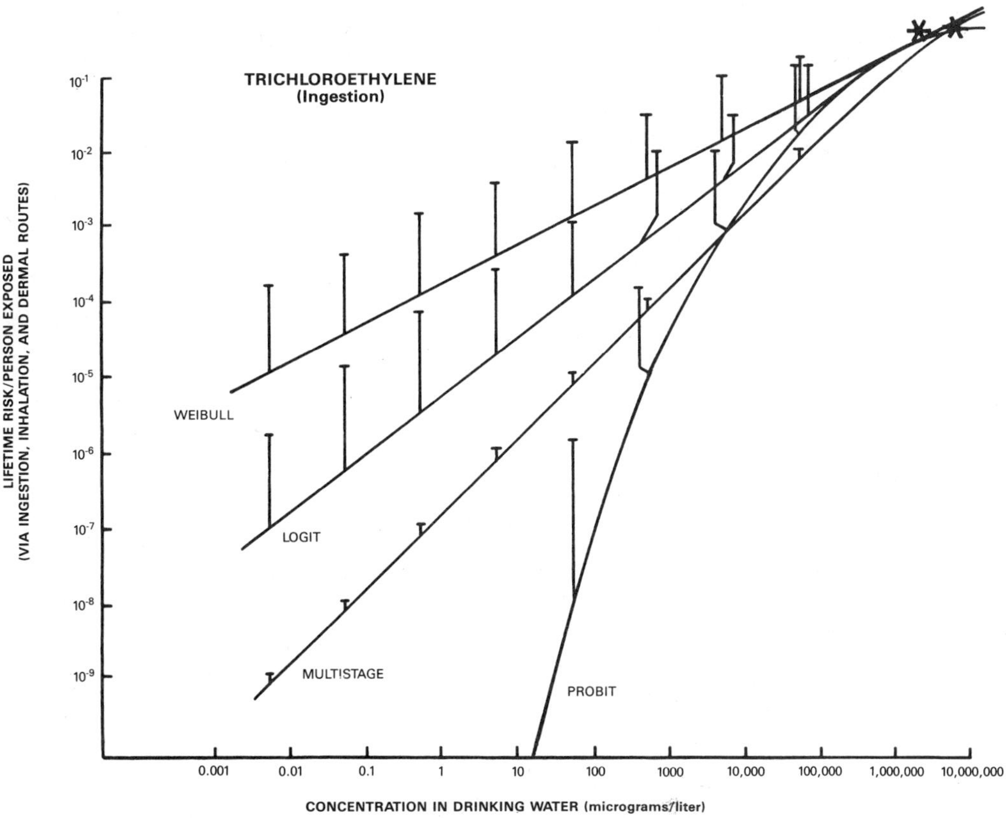

Figure 1

Bioassay data (starred points in upper right hand corner) and model extrapplations for the logit, multistage, probit and Weibull models from those bioassay data for trichloroethylene. The bioassay data was obtained by ingestion exposure. The lifetime risk/person exposed was calculated assuming that the risk due to inhalation and dermal exposure is equal to that due to ingestion exposure. The error introduced by this simplifying assumption is less than the widths of the lines shown. The lines are the point estimates and the error bars indicate the upper 95% confidence limit. The lower 95% confidence limit is the horizontal axis.

uncertainty can be influenced easily due to the extreme cost involved and because of the unethical nature of human experimentation.

UNCERTAINTY

A scientific or technical endeavor such as a risk estimation produces numbers and the most important part of the number is the uncertainty involved. The actual number has little meaning without an indication of the error or uncertainty. For example, to give an estimate as 100 cases/lifetime, it has a different meaning if it is 100 + 1 or 100 + 80. In the first case, the tens and units places have significance while in the second case they do not.

There are many ways to determine the uncertainty in the number, however, we will consider three here: viz, a statistical, practical, and the so-called Delphi approach. Too often these and other approaches are not used. It has been recommended that the level of concern about uncertainty should be increased (MOrgan et al., 1983).

The stistical approach requires knowing the frequency distribution involved. To determine this, many measurements must be made. When the data are sufficient and the frequency distribution becomes defined the

Table 3

Benzene-Population Risk Estimates
For Current Levels of Drinking Water Exposure

Mean Drinking Water Concentration (Micrograms/Liter)	Number of People Being Served	Total Lifetime Individual Risk For the Mean Concentration*		Lifetime Population Risk
		Low (Probit)	High (Weibull)	
0.25	2.1×10^8	$<10^{-10}$	1.7×10^{-6}	<1 - 360
2.75	4.5×10^6	$<10^{-10}$	1.3×10^{-5}	<1 - 60
7.5	1.2×10^5	$<10^{-10}$	3.1×10^{-5}	<1 - 4
15	6.2×10^4	2×10^{-9}	5.1×10^{-5}	<1 - 3
25	1.6×10^4	1×10^{-8}	8.9×10^{-5}	<1 - 1
35	6×10^3	3×10^{-8}	1.3×10^{-4}	<1 - 1
45	3×10^3	4×10^{-8}	1.6×10^{-4}	<1 - 1
55	2×10^3	5×10^{-8}	1.7×10^{-4}	<1 - 1
65	1×10^3	7×10^{-8}	2.2×10^{-4}	<1 - 1
75	1×10^3	1×10^{-7}	2.4×10^{-4}	<1 - 1
100	1×10^3	5×10^{-7}	3.2×10^{-4}	<1 - 1
			Total†	<1 - 400

Benzene-Risk Reduction Analysis
For Limiting Drinking Water Concentration

Maximum Allowable Drinking Water Concentration (Micrograms/Liter)	Approximate Individual Risk Rate for Maximum Concentration	Cummulative Cases Averted°
75	$1 \times 10^{-8} - 2 \times 10^{-4}$	<1 - 2
35	$7 \times 10^{-9} - 1 \times 10^{-4}$	<1 - 6
7.5	$<1 \times 10^{-10} - 8 \times 10^{-5}$	<1 - 10
2.75	$<1\ 1\ 10^{-10} - 7 \times 10^{-5}$	<1 - 70

† Rounded to one significant figure

* The total individual risk was determined by assuming that the risk due to inhalation exposure is equal to that due to ingestion

°Number of cases averted for concentrations shown in the first column

standard deviation can be calculated and used as a measure of the error or uncertainty. This approach is the ideal one as it would allow the development of a risk profile or actual distribution of probable risks as a function of the magnitude of the risk. The luxury of such a plethora of information is unlikely for risk estimation such as those considered in this paper. It is more likely that the data is so sparse that the frequency distribution is unknown and the statistical approach cannot be used.

The practical approach to determine the uncertainty is well known to those who have taken the laboratory in Physics I. By a practical approach is meant the determination of the contribution to the uncertainty from observation and experience with the measurement tools. If we are measuring the length of a table, the uncertainty can be determined by noting the smallest measure on the measuring device.

A third way of estimating the uncertainty is to ask the experts--sometimes called the Delphi approach. This can be used to develop an estimate of the nature and magnitude of uncertainty. As an example of this approach a recent study sought the judgments of experts concerning the environmental effects of sulfur dioxide (Morgan et al. 1983). In that study the experts were asked to estimate the plume flight time of sulfur dioxide from coal fired power plants and an estimate of the health impacts. There was a small range of disagreement (or uncertainty) among the experts concerning the flight time but considerable uncertainty with regard to the health aspects.

Potential difficulties with the Delphi approach are due to the different way people make judgments in the face of uncertainty (Whittmore, 1983). It has been found that anxiety about uncertainty leads to overconfidence, that the seriousness of the risk can lead to overestimates and the method of expressing the risk can influence judgement (for example people react differently if the risk is expressed as 1 in 1000 of dying or a chance of 99.8% of living).

This discussion will use a combination of the practical and Delphi approaches to determine the uncertainty. Further, the contributions to uncertainty that can not be quantified will be listed and discussed.

HEALTH EFFECTS UNCERTAINTY

This section is an attempt to list, define, and quantitate the uncertainty that may be an integral part of toxicological experimental design. Biology by definition is not an exact science. When an experiment is designed, a carefully reasoned question is asked. If the answer is known, then one has an 80% chance of getting the same experimental results. Rarely, and with good reason, are experiments repeated that take years to perform. Therefore, the presentation below attempts to catalogue the potential and actual uncertainty that is inherent in the design, extrapolation and interpretation of such experiments. The results of this analysis are shown in Table 4.

The probability of choosing the appropriate end-point is less than 10% using a chemical from a well characterized class such as the VOC's. The chance of being wrong is larger in an unknown or poorly characterized chemical class. This is illustrated by the unexpected occurrence of brain cancer that was exposed to acrylonitrile (Quast, 1980 a and b).

The probability of choosing appropriate dose levels to produce the

Table 4

Contributions to Uncertainty From Toxilogical Studies

Source	Contribution
Many Orders of Magnitude(three or more)	
Extrapolation of animal data from high doses(producing lifetime risks of 0.1 to 0.01) to low doses(producing lifetime risks in the range 10^{-4} to 10^{-6})	Indifferent to choice of model based on biological grounds-difference between models can be 5 to 6 orders of magnitude at the low dose extrapolated levels.(Krewski and Van Ryzin, 1981, Smith, 1983 and Bolton, 1983)
Distribution of animals among dose levels(E.G. more at lower dose levels) and total number	Money and space limit number to about 1500 animals. If choose doses poorly, the extrapolated data could be off by as much as three orders of magnitude(Land,1980 and OTA, 1981)
Basing risk estimate on a curable cancer-e.g. skin cancer	If the cancer is 99.99% curable, the risk estimate is overestimated by four orders of magnitude(USEPA,1983)
Surroundings(noise, air flow, cage cleanliness, purity of drinking water)	This could lead to an overestimate of orders of magnitude
Animal variability	Since the analysis centers only on the average animal at each dose group, then errors of several orders of magnitude could be introduced
Interspecies comparison	Can introduce errors of a few orders of magnitude(Crouch, 1979 and 1981)
One to Two Orders of Magnitude	
Purity of the test compound-the vehicle used could be the cause of the deliterious effect or via contamination or decay	If so this could lead to an overestimate of as much as two orders of magnitude(Henschler, 1976 and Nutrition Foundation)

Table 4 continued

Source	Contribution
Good laboratory procedures (GLP's)	If not followed can lead to an underestimate of one to two orders of magnitude
Synergism/antagonism	Although little information exists concerning the effect of multiple chemicals, it could be a significant contribution to underestimation
Little Change Up to One Order of Magnitude	
Selection of dose levels	Once a rough idea of the toxic response is known, these could be poorly chosen and introduce errors of as much as 20% if chosen as 1/2 of 1/4 of the MTD arbitrarily
Examination of animals for cancer pre-coursers	If do not look for onset of hyperplasia to help characterize history, can get errors of the order of a factor of 3 leading to an underestimate
Dietary considerations	If the animal chow is not formulated the same way errors of the order of a factor of 2 can can occur plus can have as much error if the food is unpalatable(Revis, 1983)
Time-to-tumor	Could miss the effect of could have an under-estimate by as much as a foctor of 2
Adding benign tumors to malignant ones to improve the statistics	Can introduce an uncertainty of a factor of 2(NCI, 1977a and c and USHEW, 1970)
Interference of diseases in animals	Can introduce an uncertainty pf a factor of 2
Statistical noise	If one had only a few additional cases for a common cancer it may be difficult to detect them above the backround. This could introduce an uncertainty of a factor of 2

Table 4 continued

Source	Contribution
No corresponding human tissue	Assumption that will have effect on some other organ in humans can introduce an error as large large as an order of magnitude(Maltoni, 1983 and Tomatis, 1979)
Body weight vs surface vs ppm	Up to one order of magnitude difference(OTA, 1981, and Anderson, 1983)
Use of upper 95% confidence limit rather than a point estimate	By the usual stastical confidence limit proce- edures the uncertainty can be up to an order of magnitude
Housekeeping and mechanical procedures	Cage location and randomization if not considered can introduce errors that lead to an underestimate of as much as an order of magnitude(NCI, 1976a and NCI, 1976)
All or Nothing	
Choice of endpoint	May miss toxicity if the wrong endpoint is chosen
Dose levels	With little or no information to go on the dose levels could be too high or low to demonstrate an demonstrate an effect or produce dose response information
Capabilities of personel	Choose incompetent or dishonest personel(Moore, 1983 and US v Kleppinger)
Choice of species, strain, age and sex of animals	Choose one that did not show the effect(Weisburger, 1983 and Casarett, 1980)
Statistical test must be chosen first-otherwise bias will enter the analysis	Forcing the desired conclusion by statistical manipulation(Salsburg, 1977)
Arbitrary choice of p level	Whether or not the toxic effect is detected based on arbitrary signifance level(Fisher, 1957)

desired effect is dependent upon several factors. These include the number of dose levels, the regimen of dosing the the route of exposure. One can give the animals one or more doses daily. The material can be given via ingestion (in food or other vehicle) or inhalation. The inhalation exposure could be for up to 24 hours a day for seven days a week. The most common method used in inhalation, however, is 6 hours a day for 5 days a week. This latter commonly used regimen introduces an uncertainty because all of the extrapolations are for lifetimes of continuous exposure. There are many examples of choosing the wrong dose. The NCI bioassay program adjusts their dose requirement if the animals begin to die or exhibit morbidity at an unacceptable rate (NCI, 1978b). Choice of dose level can contribute one to two orders of magnitude uncertainty to the risk estimate.

The facilities in which acceptable experiments are performed have any components. Most important are the quality of personnel and the adequacy of the physical plant. An example of unacceptable data due to poor plant facilities is from Gulf South Research Institute (Moore, 1983). In IBT (U.S. vs Kepplinger, 1983) the personnel though qualified were dishonest. In the case of GSRI (Moore, 1983) the personnel were honest but manifested performance deficiencies.

The experimental model (choice of species, strain, age and sex is an integral part of the design of the experiment. NCI recommends the use of specific species and strains of rodents (Weisberger, 1983). The NCI has determined that the use of the Fisher 344 rat and the B6C3F1 mouse produce the best combination of sensitivity and low background for use in carcinogenicity studies. This is not true in every case. It is entirely likely that for a particular chemical these animal models may be insensitive. An example of this kind of phenomena is thalidomide. Only certain strins of rabbits and some monkeys were capable of demonstrating phocomelia (Casarett, 1980). Improper attention to these details can cause the analysis to miss the effect totally.

One should select the statistical procedure before performing the study. This is to eliminate the possibility of introducing the personal bias and desire of the experimenter. Without this objectivity, the experimenter may search for the statistics that confirm the significance of the already subjectively determined result.

The experimental design may play a major role in reducing uncertainty. One can choose an unusual design such as a truncated or pyramidal model in which the lower dose levels have significantly more animals. The increases the sensitivity approximately equal to the square root of the increase number (Land, 1980 and OTA, 1981). One must know ahead the shape of the dose-response curve for this design to be successful. Therefore as the numbers of animals are increased (by squaring) one can reduce the dose level being investigated by one order of magnitude (Land, 1980). This procedure is limited to a total of 20,000 animals because only a small number of facilities can handle such large numbers of animals. The cost is approximately 7 million dollars for an experiment using 20,000 animals. Using this large number of animals the lowest measured dose level is reduced by only one order of magnitude compared to that achieved with the normal 1500 animal experiment. Errors in choosing the experimental design can lead to uncertainties of one to two orders of magnitude in the risk estimates.

The maximum tolerated dose (MTD) is derived from an experiment. It is the dose that produces a 10% depression in weight gain over 90 days. The MTD is a surrogate that attempts to predict the dose level at which no more than 20% of the animals exposed will die over the duration of a lifetime experiment. The next standard level chosen is usually some fraction of the MTD-say 1/2 or 1/4. This ignores the dose-response curve. It may be wiser to choose the 5% and 1% projected response for 90 days as the second and third dose levels. If the MTD is not a sufficiently high dose level, then the true carcinogenic response will be missed. If it is too high, we will lose the entire dose group, and not have enough dose level information to use for extrapolation. The NCI, therefore, has made interim changes in dose levels to compensate for the incorrect selection. The NCI then reports the dose level the animals were exposed to as a time weighted average. It is difficult to determine the actual dose level that will reproduce the effect using the time weighted average. Using this approach can contribute up to two orders of magnitude to the uncertainty in the risk estimates.

Prior knowledge of the historical background incidence of the species and strain of animals under test is critical. Initially one would like to design an experiment in which the target organ, if known, normally has a low tumor incidence. This is because a high background incidence of tumors in the target organ system would make a low-level carcinogenic response impossible to detect (Watanabe, 1983).

Many of the early NCI reported experiments were called into question because animals who were exposed to different volatile organic carcinogens were housed in the same room as animals exposed to other volatile organic chemicals (NCI, 1976a). This phenomena can contribute orders of magnitude to the uncertainty estimate as an overestimate.

The NCI has kept track of the historical control background cancer levels. They have segregated the population on the basis of different laboratories who routinely performed their experiments. Over time and many experiments later, it became obvious that certain of these laboratories had a significantly higher incidence of background cancers (Haseman, 1984). One must assume, since the NCI supplied randomly selected animals of identical species and strain, that differences in the way the animals were handled or housed were the genesis of the laboratory difference. These differences could be the result, inter alia, of; noise stress, laminar flow vs. HEPA filtered air, purity of drinking water and cage cleanliness. These effects can contribute one to two orders of magnitude to the uncertainty in the risk estimate.

There are certain housekeeping aspects to experimental procedures that may have significant effects. The first and most ignored is the randomization of cage placement of animals in their racks. There is some evidence that the location in the racks had greater significance than the experimental treatment.

A source of problems addressed by Good Laboratory Practices (GLPs) was the handling of recently dead and moribund animals. Significant amounts of needed tissue were lost to gross and microscopic examination due to inadequate procedures. GLPs now require frequent enough examination to prevent cannibalism and autolysis. Lack of these procedures produced major effects on the experimental outcomes.

GLPs also specify that experienced and qualified personnel must supervise and perform autopsies. It is likely that subtle effects would be missed if only a technical alone performed necropsies. (A small footnote to necropsy procedure is included. The `standard' is to preserve tissue in a formalin solution. If one is looking for immune or autoimmune effects, they will be destroyed by formalin preservation). Poor laboratory practices can contribute one to two orders to magnitude t the uncertainty in the risk estimate.

Microscopic characterization of grossly observed lumps and bumps is used to determine the carcinogenic character of the lesion. Today, panels of `experts' are assembled, duplicate, double blind, coded slide sets distributed, and a consensus is used as the final result. This does not take into account what the `true' answer really is (NCI, 1977 a and b). It is not that these talented experts differ in the characterization of the structural architecture that they are examining. It is that a clear definition of what a carcinogenic lesion in that tissue is lacking. Benign tumors are added to any malignant tumors and used in the calculation of risk (Surgeon General, 1970 and Plant, 1975). Another example is to consider curable skin cancer due to arsenic exposure. (USEPA 1983). This characteristic can contribute a factor of one or two orders of magnitude in the risk estimate.

Another source of uncertainty is the lack of well defined decision criteria. For instance 1-5 Napthlenediamine is called a carcinogen while Clonitracid and 2-(chloromethyl)pyradine were not, even though all three had significant increases in uterine tumors (Salsburg, 1983).

There is a major error in the evaluation of carcinogenicity data, that is all tumors that are cancer are considered equal. This is shown in its most unreasonable form when one considers the cancers produced by arsenic (cancer of the skin) and those produced by benzene (leukemia). Skin cancer is curable 99.9% of the time. Benzene induced leukemias are incurably fatal. The risk model as applied predicts as much predicts a much higher incidence of skin cancer vs. leukemia. The CAG, therefore, ranks the risk from arsenic as higher than that for benzene (USEPA, 1983). This phenomena can contribute as much as two orders of magnitude to the uncertainty in the risk estimate.

Two tissues for which there is no human equivalent are the Harderian gland and Zymbal's gland. However experience has shown that these are predictive of cancer although in some other tissue (Maltoni, 1983). Animals and humans may not show the same effect. For example, benzene causes cancer of the hematopietic system in humans and not in animals, benzidene causes bladder cancer in humans and liver cancer in animals and cadmium causes lung and prostate cancer in humans and cancer of the tests in animals (Tomatis, 1979). Cigarettes have been shown to cause lung cancer in humans and not in animals and dioxin (2,3,7,8-tetrachlorodibenzo-p-dioxin) is one of the most lethal chemicals to animals but the effect observed to date in humans is only chloroacne and some possible soft tissue cancers. It would appear that to use animal bioassay data to predict the effects of humans will be wrong for some environmental contaminants. This phenomena can mash the risk estimate.

There is a tendency to use the most sensitive animal model to determine the potential carcinogenicity of chemicals. This may overestimate the potential of actual carcinogenicity in humans by several orders of magnitude.

One of the major considerations in the extrapolation of data from animal to man are the differences in metabolic patterns. The animal model may not produce a proximal carcinogen and one would miss characterizing a true carcinogen. An example is arsenic (USEPA, 1983).

Animals may produce more of the active proximal carcinogen and less of the material to deactivate or bind the carcinogen than does man (USHEW,

1978). The distribution of the chemical may be different in man than in animals. An example would be the binding of heavy metals in plasma and red blood cells (USEPA, 1983).

By far the largest contribution to the uncertainty due to contributions from aspects of the hazard assessment is due to that related to the choice of dose-response model. As shown in table 4 and figure 1 this uncertainty is in the range of four to six orders of magnitude.

UNCERTAINTIES IN OCCURRENCE ESTIMATES AND POPULATION CONCENTRATION ESTIMATES

The assessment of the frequency of finding VOC's in drinking water supplies across the United States and the concentration being served to various portions of the American population is a major input to the risk assessment process. This is called an occurrence estimate. The assessment process involves the assembly of monitoring data from national studies in which the criteria of system selection did not later the probability of the system being representative of the pollution in other systems within its class: thereby allowing the formation of a distribution curve which can be used to project national occurrence estimates and populations being served. The uncertainties resulting from this process begin with conditions surrounding sampling design which ultimately determines sensitivity to low frequencies of occurrence and moves through all of the technical procedures for sampling, analysis, data verification and interpretation. Each of these procedures contribute some uncertainty to the final estimates. The overall list of uncertainties in estimating exposure and an estimate of their impact on the risk analysis is shown in Table 5.

Uncertainty in the Occurrence Assessment

Two sources of uncertainties must be dealt with during the procedure for selecting sampling locations. These uncertainties are:

-- Was the selection of systems to be sampled conducted using criteria which provided an unbiased representation of the national universe of community water supplies?

-- Was the study designed so as to detect levels of pollution which may occur infrequently?

The information used in this risk assessment to project national occurrence rates was formed as an amalgam from national survey data. In some studies the sampling locations were selected randomly. In others sampling locations were selected independently by individuals within different states and regions throughout the country with the only guidance being not to select systems known to be contaminated. Prior to forming an amalgam of these data sets data were compared with data from "random" sites, and it was found there was not statistical difference between the two data sets. Thereafter, the data sets were merged to increase the number of samples available from which to assess national occurrence.

It would appear that there is less uncertainty surrounding the body of the national frequency distribution formed from these data than exists when one examines the high and low ends of that curve. The exact number of systems experiencing an extremely high level of contamination is unknown. Information exists from state agencies demonstrated that such places exist yet were undetected in the national monitoring survey. One possible source of this "false negative" could be caused by the lower limit of sensitivity created by randomly selecting 1,000 sampling locations from a population of 48,000 candidates.

The number of systems experiencing these high levels and the number of people they serve is probably quite small and is expected to have a minor impact on population concentration estimates. However if this distribution of systems having elevated concentration was affected by some

Table 5

Sources of Uncertainty for Occurrence, Population Concentration and Exposure Estimates Used in the Assessment of Volatile Organic Chemicals in Drinking Water

Factors	Impact on estimate of†: Occurrence	Population Concentration	Risk
Generation of monitoring data			
Proportion of population sampled	5% (U)		
Representativeness of systems selected			
Geographic distribution, system size and source of water	10% (E)		
Sampling methods		50%	Factor of two
Site of sample collection	20% (E)		
Time of sample collection	20% (E)		
Method of sample collection	10% (U)		
Container type	10% (U)		
Stability during storage	100% (U)		
Sample analysis			
% recovery from sample	10% (U)		
Compound identification	10% (E)		
Accuracy of quantitative determination	40% (E)		
Assumptions during data analysis			
Lower limits of quantification		factor of 3-4	factor of 2 (E)
Oral exposure rates			
Intake rate of water			10% (E)
Pollutant level in consumed water(hot vs cold)			50% (O)
% absorption for oral intake			10% (O)
Respiratory exposure rates			Factor of 3
Dermal exposure			Negligible

†U = leads to an underestimation of the risk; O = leads to an overestimation of the risk; E = could lead to an overestimation or an underestimation of the risk.

factor causing a "clustering" of systems within a category not sampled during the survey the underestimate could be larger.

Uncertainty is also created because of our inability to detect true concentration, especially at the low end of the frequency distribution. For this analysis it has been assumed that places where the pollutant was not detected at a "minimum quantifiable level" (<0.5 micrograms/liter) contained 0.25 micrograms/liter [the median concentration between 0 and 0.5 micrograms/liter].

We know that certain solvents are ubiquitous in societal use and appear at the micrograms/liter level in rain water. For these substances there are probably few, if any, water supplies which are absolutely free of contamination. However, for other less ubiquitous chemicals, a number of water supplies are probably absolutely clean. By far the largest number of people fall into the "less than" category.

The data available from national monitoring studies were generated from a single sample taken at one point in time at one location within the distribution system of each water supply. How representative is data from a grab sample of the drinking water quality being provided to users throughout the system?

The problem of partial or complete disappearance of the chemical pollutant contained in drinking water prior to analysis due to physical or biological processes may produce a large uncertainty in the analytical results. When partial disappearance is suspected, the concentrations represented in data may be taken as "minimum values" from an unknown but larger level present at the time of sampling. A problem arises in interpreting data reported as "zero or minimum detectable" found in such a data set for the possibility exist that higher levels were originally present which diminished in sample handling produce false negative readings at the time of analysis. Under these conditions, it would be impossible to identify low occurrences from false negatives. Preliminary data reported by Glick in 1979 demonstrated a rapid loss of benzene, ethybenzene and toluene from spiked tap water to background in about 10 days. Chlorobenzene, m-xylene and p-dichlorobenzene levels were found to diminish in 2-4 weeks of storage. The loss of benzene was especially rapid with 50% loss being found within 2 days (see table 5). The situation with respect to the degradation of unsaturated aliphatic hydrocarbons is not well understood.

What level of uncertainty has been created for chemicals discussed in this report by loss during the storage of samples prior to analysis? To address this question (see table 5) one must realize that the only data used to project national occurrence were those developed from water samples drawn after treatment in the distribution system. Most of these samples, especially those from systems using surface water, would have contained chlorine disinfectant. Data on the level of VOCs in surface water supplies was developed in studies predating the recognition of possible degradation problems in which holding times were long. Therefore, there exists a level of uncertainty surrounding surface water contamination with each of the VOCs, but especially benzene. Data from which groundwater predictions have been made are also samples drawn from the distribution system. However, since a significant number of smaller groundwater systems do not disinfect their water it is indeed fortunate that data were derived from a more recent study (GWSS) which was designed to ensure minimal pollutant loss during sample storage.

The process of chemical analysis is filled with technical

uncertainties. Procedures have been established to control the quality of the resultant data. Confirmation via alternate analytical techniques of a percent of the qualitative findings has been practiced on the data used on these surveys. Therefore, this source of uncertainty is relatively low. In addition, quantitative determinations may become a large surface of uncertainty. Once again, laboratory procedures were instituted for these studies to assure that the error was minimized.

Uncertainties Exposure Assessment

The drinking water contribution to respiratory and oral exposure has been estimated using assumptions about the rate of water used during fluid intakes--bathing, the transfer rate to air of VOCs transported in the drinking water, the turnover rate for air within the closed confined of a bathroom and house and the length of time with a corresponding volume of air inhaled within each environment (see table 5).

Definitive data are not available to define the variance or average of any of these factors as they affect the United States population. The assumptions made are based on "reasonable" extensions data confirmatory of a qualitative process but insufficiently precise to validate quantitative conclusions. It was assumed that the average intake of drinking water over a lifetime approximates 2 liters/70 kg rate, established for "average adults" resulting in 0.028 micrograms/liter/kg absorbed dose per day. This consumption rate (0.03 liters/kg/day) has been selected as the population mean for this report. Comparison with theoretical average rates reveals that fluid intakes per unit body weight generally decrease with age, beginning at 7-9 times as high in infants and reaching a plateau by early teens and adulthood. A weighted average derived from a theoretical water consumption curve indicates that the lifetime average may be 20% higher than the current factor being used. Population studies on the variability of water consumption have not been conducted in the United States. Canadian data indicates that 13-16% of the adults may consume more than 2 liters of water per day.

The vast majority of the population does not remain in one city and is not served by a single water supply over their lifetime. However, the rate of movement from locations with higher pollutant concentrations and vise versa is unknown. Thesame lack of knowledge exists for the stability of drinking water concentrations over time, therefore the uncertainty associated with these factors can not be quantified.

It has been assumed that what is present in cold tap water is present in consumed fluids made with drinking water. We know from the Canadian study that the preponderance of drinking water consumed by children is in the form of cold drinks and the amount of hot drinks increases with increasing age. We also know that while the VOCs are transported in drinking water they are rapidly transferred into air with aeration, heating, or upon standing. These factors would diminish oral exposure of adults significantly but tend to create point source liberations into indoor air which might increase respiratory intake. Since the percent of absorption from respiratory intake is assumed to be 50% of the oral absorption rate the overall exposure of adults might be reduced by 30-50%.

This risk analysis recognized the transferance of drinking water borne pollutants into indoor atmosphere and the resultant exposure of household members to an enriched pollutant atmosphere during time spent in the dwelling. It assumes that respiratory uptake is equivalent to oral uptake with the major contribution to exposure taking place during showering or bathing and the time spent within the enclosed bathroom immediately after these activities prior to dispersal of the pollutant concentration. This conclusion is based on several assumptions and their associated uncertainty.

Laboratory and field testing data is available to support this assumption. Engineering documents produced for the USEPA by Environmental Sciences and Engineering of Gainesville, Florida, document a high efficiency of aeration processes for removing the VOCs from contaminated raw water. In addition, laboratory tests show that hot water has lost virtually 100% of its VOC contamination upon standing for 3 hours or being poured between containers 20 times. Cold water exhibits 20-80% loss under the same conditions. Since bathing and showers involve a large percent of heated water which isaerated during impact on body or tub the volatility is enhanced. An assumption of 100% transference to air may be in excess of 5-10%.

Physical factors used to define the volume of air in a bathroom into which the pollutant would be dispersed were chosen instinctively. The volume of air assumed to be present in the bathroom was 9.4 m or 6'x 7' x 8'--the 30 gallons of water assumed to be used in a shower. The result of 100% transfer of the water pollutant was 113 micrograms of pollutant present at 1 microgram/liter in water for 9.4 m of air or a resulting concentration of 1.2 micrograms/m. It was further assumed that the individual spending 30 minutes in th bathroom inhaled .42m of air containing .5 micrograms of pollutant of which 50% was absorbed. The net absorbed dose was estimated to be .0036 micrograms/kg/day.

There is some data from simulation studies at the University of Pittsburgh (Couch, 1984) as well as field studies in New Jersey to indicate that pollutant concentration in a shower stall is greatest at the breathing zone. However, there is no data showing that the pollutant is removed from the air by binding to fabric, wood or plastics in the room. Since the scenario assumes 100% availability it is likely to tend a worse case estimate. However, the assumed rate of absorption (50%) was an average from animal tests (30-80%) conducted at higher air levels.

An average of 250 gallons of water pass through an American household each day. At a pollutant concentration of 1 microgram/liter, 946 micrograms of a pollutant become available for dispersal within the household. Knowing that the VOCs are easily transferred from the water column into air, we also know that the bulk of VOCs in drinking water get dispersed throughout the house. There are no measurements of the indoor air concentration resulting from this process. By analogy, to estimates made for radon it was assumed that level in household air (in liters) would be equivalent to 10 of the level found in water (liters). At an initial concentration of 1 microgram/liter in water concentration in general household air would be .0001 microgram/liter. (.001 microgram/liter equals .1 microgram/m). Individuals spending 12 hours per day within the home environment inhaling 10 m of air would be exposed to 1.0 micrograms of pollutant of which 50% would be absorbed. The net absorbed dose is equivalent to .07 micrograms/kg day.

CONCLUSION

The risk assessment for volatile organic compounds found in drinking water involves large uncertainties. The contribution to uncertainty in the exposure analysis is no greater than an order of magnitude. The contribution of uncertainty from the hazard assessment has been shown to be four to six orders of magnitude. The predictions of the models used to extrapolate the bioassay data from the realm of experimental reality to the unknown region of 10/lifetime contribute the largest part of the uncertainty.

It is important to note that there is no sound scientific reason for choosing one model over another. No one model is consistently more conservative than another, nor is it possible to chose a number in the range that is most likely to be correct. The choice of the model to be used therefore is arbitrary though necessary. Until it is known which model is the correct one it is not likely that risk assessments such as those described here will have uncertainty ranges less than the four to six orders of magnitude indicates. The most likely information that will lead to the correct choice of model is to determine the mechanism of cancer.

Until the contribution to the uncertainty due to the choice of model can be reduced, it is not likely that efforts to reduce other uncertainties will be very useful. At that time efforts should be directed to reducing the uncertainties from other contributions to uncertainty in the toxicology area as listed in Table 4.

REFERENCES

Andeson, E.L., 1983, Quantitative Approaches in Use Assess Cancer Risk, Risk Analysis, 3:277.

Bolton, J.G., Morrison, P.F. and Soloman, K.A., 1983, Risk-Cost Assessment Methdology for Toxic Pollutants from Fossil Fuel Power Plants, RAND Corporation Report R-2993-EPRI, Santa Monica, CA 90406.

Casarett, L.J., and Doull, J., 1980, Toxicology, Macmillan Publishing Co., NY.

Cothern, C.R., W.A. Coniglio and W.L. Marcus, 1984, Techniques for the Assessment of the Carcinogenic Risk to the U.S. Population Due to Exposure From Selected Volatile Organic Compounds From Drinking Water Via the Ingestion, Inhalation and Dermal Routes. This report is available from the National Technical Information Service (NTIS), Springfield, under the number PB 84 213941.

Couch, A.F., 1984, Assessing Human Inhalation Exposure to Trichloroethylene Volatilization from Contaminated Residential Water Supplies, MS Hygiene Thesis, University of Pittsburgh.

Crouch, E. and Wilson, R., 1979, Interspecies Comparison of Carcinogenic Potency, J. Toxicol. Environ. Hlth., 5, 1095-1118.

Crouch, E. and Wilson, R., 1981, The Regulation of Carcinogens, Risk Analysis, 1, 47-57.

Fisher, R.A. and Yates, F., 1957, Statistical Tables, 5th Edition, Tables VI, VI and VI, Oliver and Boyd, Edinburgh.

Haseman, J., Huff, J.A., and Borman, G.A., 1984, Historical tumor incidence in the B6C3F1 mouse and F 344 Rat, CFN Press, See G.A. Borman, National Cancer Institute, Box 27709, Research Trinagle Park, NC 27709.

Henschler, D., Eder, E. Neudecker, T. and Metzler, M., 1976, Carcinogenicity of Trichloroethylene: Fact or Artifact?, Arch. Toxicol., 37: 233-236.

Krewski, D. and Van Ryzin, J., 1981, Dose Response Models for Quantal Response Toxicity Data, In Statistics and Related Topics, M. Csorgo, D.A. Dawson, J.N.K. Rao and A.K. Md. E. Saleh, eds. North Holland Publishing Co., New York.

Land, Charles, 1980, Estimating Cancer Risks From Low Doses of Ionizing Radiation, Science, 209:1197.

Morgan, M.G., Morris, S.C., Henrion, H., Amoral, D.A.L., and Rish, W.R.,

1983, Technical Uncertainty in Quantitative Policy Analysis--A Sulfur Air Pollution Example, Manuscript submitted to Science from the Department of Engineering and Public Policy, Carnegie-Mellon University, Pittsburgh, PA 15213.

Maltoni, C., Conti, B. and Cotti, G., 1983, Benzene: A Multipotential Carcinogen, Results of Long-Term Bioassays Performed at the Bologna Institute of Oncology, Am. J. of Industrial Medicien, 4:589-6301.

Moore, J., 1983, Memorandum, Withdrawal of Gulf South Research Performed NCI Studies, USEPA, Office of Pesticides and Toxic Substances, TS-788, Washington, DC, 20460.

NCI (National Cancer Institute), 1976a, Carcinogeneses Bioassay of Trichloroethylene, Technical Report Series No. 2, Technical Information Section, Carcinogeneses Testing Program, National Toxicology Program, Landow Bldg., Rm. A306, Bethesa, MD 20205.

NCI (National Cancer Institute) 1977a, Bioassay of 1,1,1-Trichloroethane for Possible Carcinogenicity, National Cancer Institute Carcinogenesis Technical Report Series No. 3. NCI-CG-TR-3, January 1977, Technical Information Section, Carcinogenesis Testing Program, National Toxicology Program, Landow Building, Room A306, Bethesda, MD 20205.

NCI (National Cancer Institute) 1977b, Bioassay of Tetrachloroethylene for Possible Carcinogenicity, DHEW Publication No. 77-813, Technical Information Section, Carcinogenesis Testing Program, National Toxicology Program, Landow Building, Rm A306, Bethesda, MD 20205.

NCI (National Cancer Institute), 1977c, Bioassays of Nitriloacetic Acid, Nitriloacetic Acid, Trisodium Salt and Monohydrate for Possible Carcinogenicity, Technical Report Series No. 6, US Dept. of Health, Education and Welfare, Bethesda, MD 20014.

NCI (National Cancer Institute) 1978, Bioassay of Picloram, Technical Report Series No. 23, U.S. Dept. of Health, Education and Welfre, National Institutes of Health, Bethesda, MD, 20014. See also reports 2,3 & 5.

Nutrition Foundation, 1983, Report of the Ad Hoc Working Group on Oil/Gavage in Toxicology, 888 Seventeenth St., NW, Wahington, DC 20006.

Office of Technology Assessment (OTA), 1981, Assessment of Technologies for Determining Cancer Risks From the Environment, US Government Printing Office, Washington, DC 20402.

Plant, P.A., 1975, A Dangerous Definition, Chem., Engineering News, Oct. 20, 1975.

Quast, J.F., C.E. Wade, C.G. Humiston, R.M. Carreon, E.A. Herman, C.N. Park and B.A. Schwetz, 1980a, A Two-Year Toxicity and Oncology Study with Acrylonitrile Incorporated in the Drinking Water of Rats, Toxicology Research Laboratory, Dow Chemical USA, Midland, MI 48640, Jan. 22, 1980.

Quast, J.F., D.J. Schuetz, M.F., Balmer, T.S. Gushow, C.N. Park and M.J. McKenna, 1980b, A Two-Year Toxicity and Oncology Study with Acrylonitrile Following Inhalation Exposure of Rats, Toxicology Research Laboratory, Dow Chemical USA, Midland, MI 48640, Dec. 9, 1980.

Revis, N.W., Douglas, B.H., McCauley, P.T. Witshi, H.P. and Bull, R., 1983, The Relationship of Drinking Water Chlorine to Coronary Atheroschlerosis, Pharmacologist, 25 Abstract 732.

Salsburg, D., 1983, The Lifetime Feeding Study in Mice and Rats-An Examination of Its Validity as a Bioassay for Human Carcinogen, Fundamental and Applied Toxicology, 3:63-67.

Schneiderman, M.A., 1980, The Uncertain Risks We Run: Hazardous Materials, ED. by R.C. Schwing and W.A. Albers, Jr., Societal Risk Assessment--How Safe is Safe Enough?, Plenum Press.

Smith, C.S., 1983, Risk Assessment of DBPC, Unpublished interoffice memorandum, Shell Oil Company, One Shell Plaza, P.O. Box 4320, Houston, TX 77210.

Surgeon General, 1970, Report on the Ad Hoc Committee on the Evaluation of Low Levels of Environmental Chemical Carcinogens, Evaluation of Environmental Carcinogens.
Tomatis, L., 1979, The Predictive Value of Rodent Carcinogenicity Test in the Evaluation of Human Risks, Annual Reviews of Pharmacology and Toxicology, 19:511-530.
USEPA, 1983, Health Assessment Document for Inorganic Arsenic, EPA-600/8-83-021A, Washington, DC.
US vs Kepplinger et al, 1983, Northern District of Illinois, #81CR235.
U.S. Department of Health, Education and Welfare, 1978, Special Occupational Hazard Review of Trichloroethylene, Center for Disease Control, NIOSH, Rockville, MD 20857.
Wantanabe, P.G., 1983, Personal Communication, Concerning Sprague-Dawley Rats, Spartan Substrain, Dow Chemical Co., 1803 building Midland, MI 48640.
Weisburger, J.H., and G.M. Williams, 1983, Chemical Carcinogens, Chapter 2, Bioassay of Carcinogens: In Vitro and In Vivo Test, Washington, DC.
Whittmore, A.S., 1983, Facts and Values in Risk Analysis for Environmental Toxicants, Risk Analysis, 3:23-33.
Westrick, J.J., Mello, J.W., And Thomas, R.F., The Gound Water Supply Summary of Volatile Organic Contaminant Occurrence Data, EPA Technical Support Division, Office of Drinking Water, Cincinnati, OH 1983.

ANALYSIS OF UNCERTAINTIES IN CRAC2 CALCULATIONS:

THE INHALATION PATHWAY

G.G. Killough,* and D.C. Dunning, Jr.**

Oak Ridge National Laboratory*
Oak Ridge, Tennessee 37831
The MAXIMA Corporation**
Oak Ridge, Tennessee 37830

ABSTRACT

CRAC2 is a computer code for estimating the health effects and economic costs that might result from a release of radioactivity from a nuclear reactor to the environment. This paper described tests of sensitivity of the predicted health effects to uncertainties in parameters associated with inhalation of the released radionuclides. These parameters are the particle size of the carrier aerosol and, for each element in the release, the clearance parameters for the lung model on which the code's dose conversion factors for inhalation are based. CRAC2 uses hourly meteorological data and a straight-line Gaussian plume model to predict the transport of airborne radioactivity; it includes models for plume depletion and population evacuation, and data for the distributions of population and land use. The code can compute results for single weather sequences, or it can perform random sampling of weather sequences from the meteorological data file and compute results for each weather sequence in the sample. For the work described in this paper, we concentrated on three fixed weather sequences that represent a range of conditions. For each fixed weather sequence, we applied random sampling to joint distributions of the inhalation parameters in order to estimate the sensitivity of the predicted health effects. All sampling runs produced coefficients of variation that were less than 50%, but some differences of means between weather sequences were substantial, as were some differences between means and the corresponding CRAC2 results without random sampling. Early injuries showed differences of as much as 1-2 orders of magnitude, while the differences in early fatalities were less than a factor of 2. Latent cancer fatalities varied by less than 10%.

KEY WORDS: Consequence Analysis; Nuclear Reactor; Dose Conversion Factor; Particle Size; Respiratory Clearance; Latin Hypercube Sampling; ICRP Lung Model

1. INTRODUCTION

In this paper we discuss some sensitivity tests of calculations performed with the CRAC2 computer code to uncertainties in parameters for inhalation dose. CRAC2 estimates early and latent health effects and economic costs that might result from a hypothetical release of radioactivity from a nuclear reactor to the environment.

The CRAC2 code (Ritchie et al., 1983; Ritchie et al., 1984) is a revision of CRAC (calculation of Reactor Accident Consequences), which was developed for use in the Reactor Safety Study (U.S. Nuclear Regulatory Commission, 1975). Consequence analysis codes such as CRAC2 are normally used in conjunction with the last stage of a Probabilistic Risk Assessment (PRA) of the engineered safety design of a nuclear reactor. CRAC2 uses hourly meteorological data (wind speed and direction, atmospheric stability class, and precipitation rate) and a straight-line Gaussian plume model to predict the transport of airborne radioactivity; models for plume depletion by wet and dry deposition; a model of population evacuation; and data for the distributions of population and land use. The data base also includes dose conversion factors for external and internal exposure to radionuclides, risk data for converting estimates of collective dose to numbers of radiation-induced health effects, and economic data for land use. The code can compute results for single weather sequences, or it can perform random sampling from a data file containing one year's weather data and simulate the release repeatedly for each weather sequence in the sample. In this paper, where meteorology is concerned, we are interested only in the former nonrandom mode; we describe simulations in which the meteorology is fixed, but random sampling is introduced in connection with uncertainties in the dose conversion factors for inhalation of radionuclides.

Sensitivity studies reported in this paper were carried out using meteorological data from a file containing one year's observations, a nonuniform distribution of population, a set of representative land use data, and a single evacuation model. Both a "large" and a "small" release of radionuclides were used as source terms. With the exception of the noble gases and organic iodine, the release fractions for the different radionuclide leakage groups were a factor of 3-5000 less for the small release than for the large one. Most of our calculations were carried out with the large release. A zero release height and a sensible heat release of 10^4 cal/s were used in all of our calculations; the zero release height was sometimes required to obtain nonzero early fatalities.

In our analyses, we consider uncertainties in the particle size of the carrier aerosol and in the rate of clearance of the various radionuclides from the respiratory passages. In the standard CRAC2 data base, the particle size is fixed at 1 μm. This parameter determines the deposition of inhaled particulate material in each of the major regions of the respiratory passages. The rates of clearance from the lungs are based on estimates of the solubility of the expected chemical forms of released radonuclides; these estimates represent a consensus of groups of experts (Morrow et al., 1966; U.S. Nuclear Regulatory Commission, 1975). The relevant portion of the CRAC2 data base consists of dose conversion factors (rems to target organs per curie of inhaled radionuclide) in which these assumptions are implicit and fixed. In effect, the fixed particle size and clearance rates have been replaced by probability distributions, and for three fixed weather sequences, repeated runs of CRAC2 have been performed with joint random sampling from these distributions. Thus, we are able to compile corresponding probability distributions of output quantities (numbers of early and latent health effects) and assess their sensitivity to uncertainties in the inputs. Details and results of these calculations are discussed in the following sections.

2. DOSE CONVERSION FACTORS FOR INHALATION

Factors for converting unit amounts of inhaled radionuclides to dose to thirteen target organs are part of the data base of CRAC2. These dose conversion factors depend on two parameters that characterize the physical and chemical properties of the radionuclide and the carrier aerosol. The first of these parameters is the aerosol particle size, expressed as activity median aerodynamic diameter, or AMAD (μm), which determines the deposition of the inhaled particles in the regions of the respiratory tract. The second parameter has three discrete values that characterize the rate of clearance of inhaled particulate material from the respiratory passages as rapid, intermediate, or slow. These clearance rates have been estimated from considerations of solubility of particular chemical compounds of each radionuclide. The appropriate values for these parameters in a CRAC2 analysis are subject to some uncertainty, and we have carried out sensitivity studies to determine how these uncertainties propagate into the estimates of health effects.

In order to be more precise about the nature of these parameters and the sensitivity tests, it is necessary to give some details of the dose conversion factors and the respiratory deposition and clearance model on which they are based. For each particulate radionuclide in the source term, the CRAC2 data base contains factors for dose to thirteen target organs due to inhalation of the radionuclide. These factors have been calculated for brief periods of acute exposure (1 year for lungs; 7 days for marrow, bone, endosteal cells, stomach wall, small intestine, upper large intestine, and lower large intestine; and 2 days for the thyroid, whole body, testes, ovaries, and other tissues); and for the time intervals 0-1, 1-10, 10-20, 20-30, 30-40, and 40-50 years for chronic exposure. Corresponding factors are included for noble gases, but these factors are not dependent upon the parameters that concern us and therefore were not varied in the sensitivity analyses to be discussed.

Each dose conversion factor for inhalation of particulate matter depends on the particle size of the carrier aerosol, AMAD, (μm). The dependence is expressed in terms of deposition fractions for the major respiratory regions (nasal pharynx or NP, tracheobronchial tree or TB, and pulmonary region or P). For any value of AMAD in the applicable range, the deposition fractions may be read from the curves in the graph of Fig. 1. In general, we note that larger particles tend to be trapped in the nasal pharynx, while smaller particles are breathed into the pulmonary region and tend to deposit there. The tracheobronchial tree receives the same fraction throughout the size range. But a point of particular importance is this: the total respiratory deposition fraction (i.e., the sum of the fractions for the three regions) is larger for large particles than for small ones (98% for 10-μm particles compared with 63% for 0.2-μm). Figure 1 shows the total deposition fraction for several values of AMAD. Thus, while larger particles tend not to penetrate deeply into the respiratory passages, a larger fraction of those inhaled are retained than is the case for small particles. Some dose conversion factors, such as those routinely used in CRAC2, do not include the dose to the nasopharyngeal tissues in computing the dose to the lungs; but the material deposited in the nasal pharynx is available for absorption into the blood and thus contributes to the dose to other systemic organs. Hence the effect of the greater total deposition in the respiratory passages resulting from larger particle sizes is increased dose to some organs and therefore potentially higher estimates of numbers of health effects. In Section 4, we shall return to the question of how the lung dose is computed and how an alternate definition changes the CRAC2 estimates of health effects.

The first parameter that concerns us, then, is the measure of aerosol particle size of AMAD. All dose conversion factors for particulate inhalation that are contained in the CRAC2 data base were calculated for

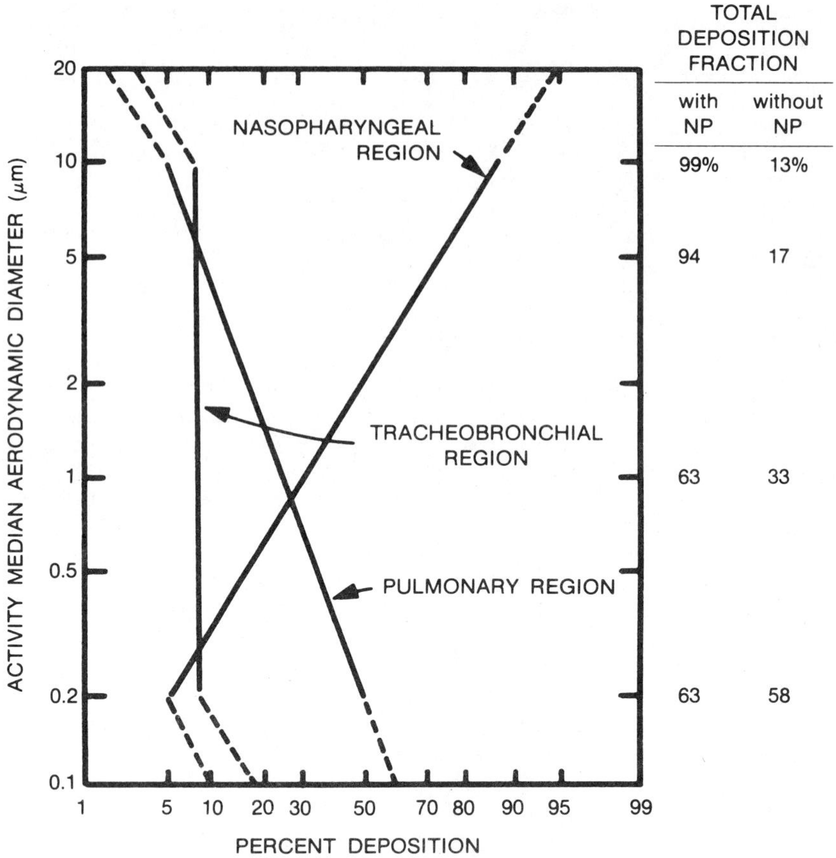

Figure 1

Analysis of Uncertainties in CRAC2 Calculations

Respiratory deposition model for particular material. The radioactive fraction of an aerosol that is deposited in the nasopharyngeal (NP), tracheobronchial (TB), and pulmonary (P) regions is shown as a function of the activity median aerodynamic diameter (AMAD) of the particle size distribution. The model is intended for use with aerosol distributions that have an AMAD between 0.2 and 10 μm with geometric standard deviations less than 4.5. Provisional deposition estimates further extending the size range are given by the broken lines. For the unusual distribution having AMAD greater than 20 μm, complete deposition in the nasal pharynx can be assumed. The model does not apply to aerosols with AMAD below 0.1 μm. Several total deposition fractions are shown, some including and others excluding the nasal pharynx. While the dose to the nasal pharynx is not included in the lung does in CRAC2 calculations, radioactive material deposited there is available for uptake to the blood. Therefore, the large total-fractional deposition associated with the larger particle sizes is potentially important if the code should be extended to include more realistic ranges (at present, AMAD is fixed at 1 μm). If the definition of lung dose were revised to include the nasal pharynx, the contribution of larger particle sizes would assume an even greater importance.

AMAD = 1 μm. Our sensitivity analysis deals, in part, with uncertainty about the choice of this parameter's value.

Once the radioactive material is deposited in the respiratory passages, its rates of clearance from the respective regions are strongly dependent upon the physical and chemical composition of the material. The lung model on which the calculation of the dose conversion factors was based was developed by the International Commission on Radiological Protection (ICRP) (Morrow et al., 1966; ICRP, 1972; ICRP, 1979). Figure 2 summarizes the compartmentalization and parameterization of the model. Rather than treating a multidimensional continum of dynamic behaviors, the ICRP, in defining its lung model, has simplified the complex topic of respiratory clearance by assuming three discrete classes of materials. These classes are labeled as class D (in which clearance is accomplished in a matter of days), W (weeks), and Y (years). The clearance-rate coefficients of the model depend on the class that is assumed for a particular radionuclide. These clearance-rate coefficients are related in Figure 2 to the clearance pathways labeled a, b, c, ..., i by way of which the deposited radioactivity is moved from the sites of deposition into the blood, the lymph nodes, and the GI tract. The table in Figure 2 quantifies the rates at which these migrations of radioactivity are

COMPARTMENT		CLASS D		CLASS W		CLASS Y	
		T	F	T	F	T	F
N-P	a	0.01	0.5	0.01	0.1	0.01	0.01
(D_3 = 0.30)	b	0.01	0.5	0.4	0.9	0.4	0.99
T-B	c	0.01	0.95	0.01	0.5	0.01	0.01
(D_4 = 0.08)	d	0.2	0.05	0.2	0.5	0.2	0.99
	e	0.5	0.8	50	0.15	500	0.05
P	f	n.a.	n.a.	1.0	0.4	1.0	0.4
(D_5 = 0.25)	g	n.a.	n.a.	50	0.4	500	0.4
	h	0.5	0.2	50	0.05	500	0.15
L	i	0.5	1.0	50	1.0	1000	0.9

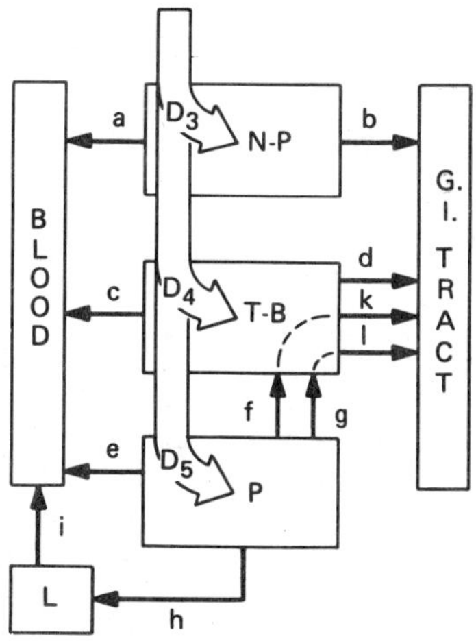

Figure 2

Analysis of Uncertainties in CRAC2 Calculations

ICRP model for clearance of particulate material from the respiratory tract. The columns labeled "D," "W," and "Y" correspond, respectively, to rapid, intermediate, and slow clearance of the inspired material. The symbols T and F denote biological half-time (days) and fractional removal coefficient for the indicated clearance pathway. The regional deposition fractions are denoted by D_3, D_4, and D_5 for nasal pharynx, tracheobronchial tree, and pulmonary region, respectively. The values shown are for AMAD = 1 μm; values appropriate for other AMAD may be read from Fig. 1. The respiratory lymph nodes are denoted by L. Retention functions for radioactivity in the major regions depend on the foregoing parameters and are derived from systems of ordinary differential equations (e.g. Killough et al., 1978a; ICRP, 1979).

assumed to take place. Each pathway out of a region is assigned a fraction F and a half-time T (days). For each region, the values of F sum to one. The radioactivity remaining after t days is

$$A = DF \exp[-((\ln 2)/T+\lambda_R)t]$$

where λ_R is the radioactive decay-rate coefficient (day^{-1}) for the radionuclide, and D is the deposition fraction D_3, D_4, or D_5 that corresponds to the particular region from which the clearance pathway emerges. The parameters T and F vary with the choice of clearance class (D, W, or Y); the deposition fraction D is determined by the value of AMAD. In practice, the lung model is translated into a system of ordinary differential equations which can be solved by numerical methods.

To illustrate the notations of Figure 2, we note that in the case of Class D material deposited in the nasal pharynx, 50% is available for direct absorption into the blood with half-time 0.01 days, while the other 50% is swallowed into the gastrointestinal (GI) tract with the same half-time. For a Class Y material in the nasal pharynx, only 1% is available to be absorbed into the blood with half-time 0.01 days, while 99% is cleared to the GI tract but at a much slower rate, with half-time 0.4 days. The contrasts in behavior among the D, W, and Y materials are substantial in all three major respiratory regions of the model. Because of these differences, the dose to the lungs is generally less for Class D than for Class Y materials; a factor-of-five difference is typical of short integration periods (comparable to those associated with the estimation of acute effects), but the variation from one radionuclide to another is substantial. The rapidly cleared and absorbed Class D materials are introduced more promptly into systemic organs and therefore tend to give a higher dose to those organs than do Class Y materials, which are detained for longer periods in the respiratory tract.

Each dose conversion factor for inhalation of particulate material that is stored in the CRAC2 data base has implicit in it the assumption of a particular respiratory clearance class: 23 of the 54 nuclides are assigned to Class Y, 11 to Class W, and 14 to Class D [the remainder are noble gases, for which the ICRP lung model is inappropriate and for which the factors for internal dose are based on an alternate model (Bernard and Snyder, 1975); these factors will not be included in our sensitivity analysis]. The clearance class assignments are based on assumptions about the chemical form of the material in a release and have generally been made from guidance offered by ICRP task groups and other experts (Morrow et al., 1966; ICRP, 1972) and from assumptions made about releases in the Reactor Safety Study (U.S. Nuclear Regulatory Commission, 1975). We should emphasize that all isotopes of the same element are considered to belong to the same respiratory clearance class. Even though the determination of clearance class depends on the particular chemical compound of the material, assumptions about the predominant chemical form of each element must be assumed to have been made when the CRAC2 data base was prepared. We note that this assumption might possibly be an implicit restriction on application of the code to releases of a kind not contemplated in the Reactor Safety Study.

Our sensitivity analysis is concerned with uncertainty about (1) the value of the particle diameter AMAD for the carrier aerosol and (2) the information that determined the D, W, or Y class for each element in the release.

Uncertainty in AMAD

All dose conversion factors for inhaled particulates in the CRAC2 data base were calculated for AMAD = 1 μm. But there is considerable uncertainty in this value, and in consultation with NRC staff, we chose to replace it by a probability distribution that is uniform over the range 0.2-10 μm (Fig. 3), which is the normal range spanned by the deposition model shown in Fig. 1. It should be noted that this distribution has a mean value (5.1 μm) that is substantially greater than the CRAC2 default value, so that, on the average, the total fraction of inhaled radioactivity that is deposited in the respiratory passages will be greater than that associated with the default value of 1 μm.

Uncertainty in D, W, Y.

The assignment of a radionuclide to class D, class W, or class Y depends on properties of the expected chemical form. Groups of experts (Morrow et al., 1966) have made such assignments for numerous chemical compounds, and these assignments have been applied to the radionuclides in the CRAC2 data base on the basis of assumptions about the probable chemical form of each element in the release (U.S. Nuclear Regulatory Commission, 1975). Hence, in the CRAC2 data base, all isotopes of a given element are assigned the same clearance class. Table 1 shows the assignment of class D, class W, or class Y to each radionuclide in the CRAC2 data base (no assignment is shown for noble gases, because the model is not applicable to them).

In place of the three discrete choices D, W, and Y, we have defined a continuum consisting of the interval [0,1], with correspondence as follows: 0→D, 0.5→W, and 1→Y. For each element, we define a triangular probability density function over [0,1] with its mode (Sometimes called the "most probable value") at D, W, and Y according to the class assigned to the element by expert consensus (Fig. 4). Thus, for a class-D nuclide, we sample from the triangular density function with its mode at class D (i.e., zero). This arrangement expresses uncertainty in the assignment but biases the sampling toward the assigned clearance class. We note, however, that the mean of the class-D distribution is 1/3 rather than

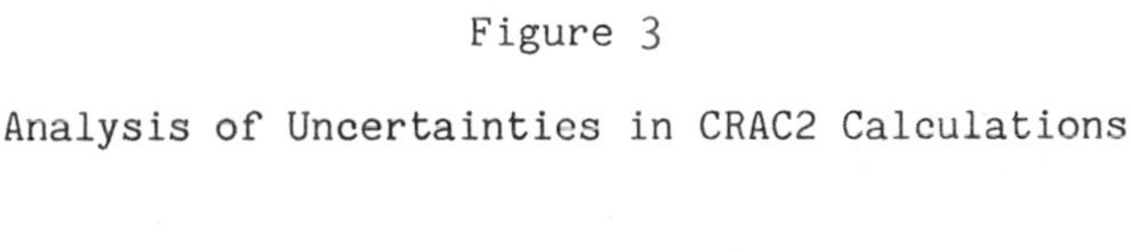

Figure 3

Analysis of Uncertainties in CRAC2 Calculations

Probability density (or frequency) function for the uncertainty in AMAD. The default value of AMAD for CRAC2 calculations is 1 μm, but this value is subject to uncertainty. The probability density function shown above was chosen to express maximum uncertainty within the applicable range (0.2-10 μm). Notice that its mean value is 5.1 μm, a substantially larger value than the default assumption.

Table 1

Respiratory Clearance Classes for the Radionuclides in the CRAC/CRAC2 Data Base

Index	Nuclide	Clearance	Index	Nuclide	Clearance
1	Co-58	Y	28	Te-131m	W
2	Co-60	Y	29	Te-132	W
3	Kr-85	-	30	I-131	D
4	Kr-85m	-	31	I-132	D
5	Kr-87	-	32	I-133	D
6	Kr-88	-	33	I-134	D
7	Rb-86	D	34	I-135	D
8	Sr-89	D	35	Xe-133	-
9	Sr-90	D	36	Xe-135	-
10	Sr-91	D	37	Cs-134	D
11	Y-90	W	38	Cs-136	D
12	Y-91	W	39	Cs-137	D
13	Zr-95	Y	40	Ba-140	D
14	Zr-97	Y	41	La-140	W
15	Nb-95	Y	42	Ce-141	Y
16	Mo-99	Y	43	Ce-143	Y
17	Tc-99m	D	44	Ce-144	Y
18	Ru-103	Y	45	Pr-143	Y
19	Ru-105	Y	46	Nd-147	Y
20	Ru-106	Y	47	Np-239	Y
21	Rh-105	Y	48	Pu-238	Y
22	Sb-127	W	49	Pu-239	Y
23	Sb-129	W	50	Pu-240	Y
24	Te-127	W	51	Pu-241	Y
25	Te-127m	W	52	Am-241	Y
26	Te-129	W	53	Cm-242	Y
27	Te-129m	W	54	Cm-244	Y

zero; similarly, the mean of the class-Y distribution is 2/3 and not 1. Thus, on the average, both extremes are shifted somewhat toward class-W.

The Sampling Scheme and Resultant Dose Conversion Factors

We have devised a scheme to sample a joint probability distribution of AMAD and a clearance value in [0,1] for each of the 23 particulate elements in the CRAC2 data base. Each trial in the sampling process results in a random vector of 24 stochastically independent components from the joint distribution A number N (=25 in our investigations) of these random vectors is selected in the sampling. For each choice of a random vector, CRAC2 is executed with data from that vector; each execution produces an output value for each of the quantities of interest (numbers of early and latent health effects), and means and measures of dispersion (e.g., standard deviations) can be estimated from the aggregate of outputs resulting from the N random input vectors. Figure 5 shows the general arrangement of the random input vectors in relation to the rectangular distribution for AMAD and the triangular distributions for the clearance classes of the 23 elements.

From the variety of available sampling strategies, we have chosen Latin hypercube sampling. A detailed discussion of the theory and implementation of this technique has been given by Iman et al., 1980.

The data base for our sampling scheme includes dose conversion

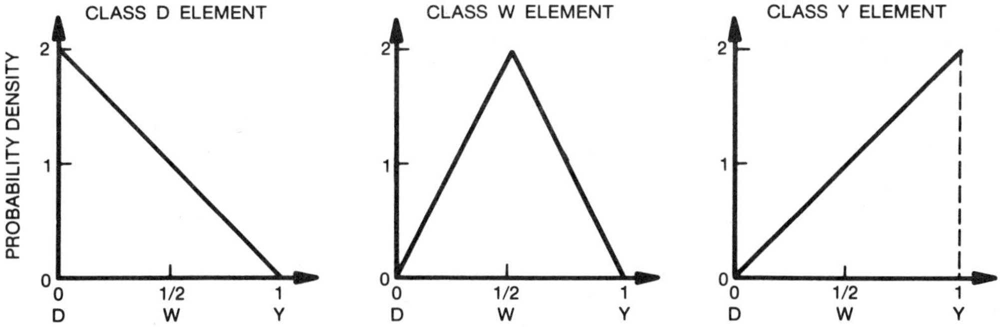

Figure 4

Analysis of Uncertainties in CRAC2 Calculations

Triangular probability density functions of the sampling distributions for Class D, Class W, and Class Y inhaled particulates. Each element in the release is assigned to the distribution for which the mode (sometimes called the "most probable value") corresponds to the clearance class assigned to that element by expert consensus. This consensus is expressed in tables of D, W, and Y assignments to a variety of chemical forms (e.g., U.S. Nuclear Regulatory Commission, 1975). The assignments that are implicit in the dose conversion factors in the CRAC2 data base involve further assumptions about the expected chemical forms of the nuclides in the release.

factors for all three classes (not just for the class assigned to the element, as in the ordinary CRAC2 data base). Let us denote such factors generically by d_D, d_W, and d_Y. When our sampling scheme chooses the number x from the interval [0,1] for one of the elements, with probability density assigned by the triangular distribution that is appropriate for the particular element, the dose conversion factor d_x will be computed as a suitable combination of d_D, d_W, and d_Y as follows:

$$d_x = 2[d_D(1/2 - x) + d_W x] \text{ if } 0 \le x < 1/2,$$

$$= 2[d_W(1 - x) + d_Y(x - 1/2)] \text{ if } 1/2 \le x \le 1.$$

CRAC2 uses the dose conversion factors d_x in its computation of numbers of early and latent health effects resulting from the release. Each trial in the sampling scheme gives different numbers of effects because of the different value of AMAD and the perturbation of each element from its D, W, or Y lung clearance category.

It was not possible to complement the ordinary CRAC2 data base with dose conversion factors that would be entirely compatible with the methodology that produced the CRAC2 factors; consequently, we used an alternate data base derived from the INREM II and RADRISK computer codes (Killough et al., 1978; Denning et al., 1980). Some background for this data base is given in Sect. 5.

RANDOM SAMPLING SCHEME

We generate a sequence of *N* random vectors; each vector has 24 stochastically independent components:

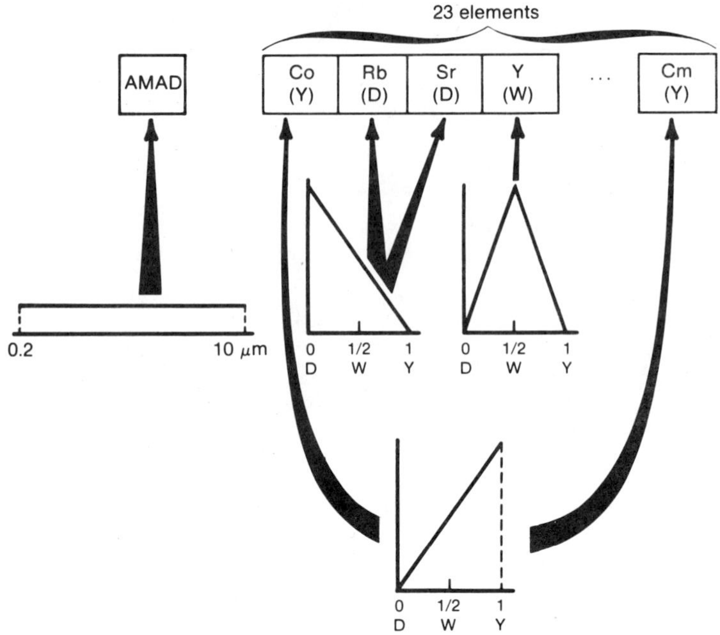

Each component is sampled from the appropriate probability distribution, with Latin hypercube stratification.

Figure 5

Analysis of Uncertainties in CRAC2 Calculations

Schematic portrayal of the random sampling procedure. At each "trial," a random vector of 24 stochastically independent components is constructed by sampling the distribution of AMAD and, for each element in the release, the appropriate triangular distribution for the assumed clearance class of that element. At each trial, CRAC2 is executed with the data from the current random vector, and at the end of N trials, the aggregates of estimated health effects are compiled into probability distributions that represent their sensitivity to the uncertainties in the inputs.

3. RESULTS OF THE SIMULATIONS

Table 2 collects the results of the CRAC2 consequence estimates for the Latin hypercube sensitivity runs and the corresponding "default" estimates that represent standard runs of CRAC2. The results that are based on Latin hypercube sampling are presented as follows in the tables:

$$\text{mean} \times (1 \pm \text{C.V.}),$$

where C.V. is the fraction (standard deviation)/mean. In one set of runs, AMAD was kept fixed at 1 μm, and the results are given in brackets in Table 2.

The measures of effect that we have chosen for display are the

Table 2

Sensitivity of Predicted Health Effects to Uncertainty in AMAD and D, W, Y Classification of Nuclides

		Weather seaquence #1		Weather sequence #2		Weather Sequence #3	
		Default	L. hypercube	Default	L. hypercube	Default	L. hypercube
A	Early fatalities	53.7	68.5×(1±.20) [52.1×(1±.09)]	0.413	0.748×(1±0.33)	2.31	2.82×(1±.11)
A	Early injuries	2450	4439×(1±.08) [3877×(1±.03)]	1.54	228×(1±.42)	150	403×(1±.10)
A	Latent/initial	1090	1202×(1±.12) [1336×(1±.11)]	1320	1420×(1±.10)	93.9	100×(1±.11)
A	Latent/total	2670	2787×(1±.05) [2929×(1±.05)]	2140	2227×(1±.07)	161	167×(1±.07)
B	Early fatalities	0	0				
B	Early injuries	4.35	69.4×(1±.20)				
B	Latent/initial	13.4	10.3×(1±.06)				
B	Latent/total	90.6	87.4×(1±.01)				

A = large release; B = small release.

Latin hypercube results are given as mean×(1±C.V.), where C.V. = standard deviation/mean.

Numbers in brackets [] are results of Latin hypercube simulations in which AMAD was held constant at 1 μm.

following:

EARLY FATALITIES -- Number of fatalities occurring within one year due to initial exposure to the radioactive cloud

EARLY INJURIES -- Number of injuries or illnesses occurring within one year due to initial exposure to the radioactive cloud

TOT. LATENT/INIT. -- Total latent cancer fatalities occurring due to initial exposure to the radioactive cloud

TOT. LATENT/TOT. -- Total latent cancer fatalities occurring due to both initial and chronic exposure.

The Latin hypercube simulations estimate substantially greater numbers of early injuries than the CRAC2 default runs, and in all categories of health effect (with one exception, where the effect is zero), the Latin hypercube calculation estimates a larger number. Some of the increase is due to the greater mean value of AMAD in the random sampling runs (a mean of 5.1 μm vs. a default value of 1 μm). The parenthesized quantities in Table 2 are the results of Latin hypercube sampling calculations in which the clearance classes were varied with each trial, as previously described, but AMAD was kept fixed at 1 μm. In this comparison, most of the difference disappears for early fatalities, but the mean value for early injuries still exceeds the default value by 58%. Fixing AMAD increases the mean latent cancer fatalities.

In the column of Table 2 under weather sequence #2, the Latin hypercube mean for early injuries exceeds the default value by two orders of magnitude. We attribute this large discrepancy to the effects of thresholds in the dose-response models used in CRAC2 (Ritchie et al., 1983). For the three weather sequences represented in Table 2, the early injuries show the most substantial increase over their default values of all health effects considered.

For each weather sequence and health effect studies, estimates of the variability arising from uncertainties in AMAD and respiratory clearance class are rather modest: the maximum coefficient of variation (standard deviation relative to the mean) is 42%. This variability is considerably exceeded by the differences between means and the corresponding default values, and by the variation of means among the weather sequences that were used.

4. RESPIRATORY REGIONS INCLUDED IN LUNG DOSE DEFINITION

Another aspect of the dose conversion factors that we considered is the extent of respiratory tissue to which the definition of lung dose is applied. Assumptions vary among different tabulations of dose conversion factors. The dose to the lungs is computed as some weighted average of doses to the major respiratory regions -- NP (nasal pharynx), TB (tracheobronchial region), P (pulmonary region), and L (lymph nodes). The original INREM II tabulations (Killough et al., 1978; Dunning et al., 1978; Dunning et al., 1981) used the mass-weighted average of all four regions. The EPA-sponsored RADRISK work (Dunning et al., 1980) used the dose to the P region as its basis for health effects. The dose conversion factors prepared for the Reactor Safety Study (U.S. Nuclear Regulatory Commission, 1975) by W.S. Snyder and his co-workers used the mass-weighted average of TB, P, and L. The CRAC/CRAC2 data base is generally derived from the Reactor Safety Study numbers, although there are numerous discrepancies. Thus the lung dose numbers in this data base also use the mass-weighted average of TB, P, L. ICRP Publication 30 (ICRP, 1979) has adopted the practice of computing the lung dose as the mass-weighted average of TB, P, and L. It should be understood that the question of the most appropriate interpretation is a matter of some controversy.

For this study, the INREM II and RADRISK computer codes were used to prepare special data bases of dose conversion factors for the sensitivity studies of CRAC2 involving the inhalation pathway. We prepared two data bases, one for each of the following assumptions:

- The lung consists of NP + TB + P + L
- The lung consists of TB + P + L (consistent with the CRAC2 data base).

The latter data base (without NP) is considered the standard one, but some results using the former (with NP) are included because of the significantly larger numbers of predicted health effects that are calculated when it is used. The number of early fatalities is the most affected measure of impact, differing by more than a factor of three when the two assumptions are applied. Table 3 gives the details. The INREM II/RADRISK numbers were generated with Latin hypercube sampling. The parenthesized numbers are results from runs in which AMAD was held constant at 1 μm.

Figure 6 shows a quantitative comparison of lung dose estimates based on the alternate assumptions displayed above. The ratio

$$\tilde{A}_{NP+TB+P+L} / \tilde{A}_{TB+P+L}$$

is plotted vs. the radioactive decay-rate coefficient, λ_R (day^{-1}), for various combinations of AMAD and clearance class. The symbols $\tilde{A}$ denote time integrals of the radioactivity burdens of the lung tissue indicated by the respective subscripts. This ratio is equal to the ratio of doses to these tissues. The numerator uses all four regions of the respiratory tract, while the denominator omits the nasopharyngeal (NP) region. All values are larger than 1; the excess above 1 is a measure of the effect

Table 3

CRAC2 Calculations with and without Counting Dose to Nasal Pharynx (NP) as Part of Lung Dose

Large release, Weather sequence # 1			
	INREM II/RADRISK		
	With NP	Without NP	Default
Acute fat.	222 (121)	68.5 (52.1)	53.7
Acute inj.	4498 (3916)	4439 (3877)	2450
Latent/Init.	1451 (1754)	1202 (1336)	1090
Latent/Tot.	3069 (3377)	2787 (2929)	2670

that the inclusion of the NP region has on the lung dose estimate. But we remind the reader that material deposited in the NP region, whether it is considered in the calculation of lung dose or not, contributes to the dose to other organs as a result of its absorption into the blood, either directly from the NP or indirectly by way of the gastrointestinal (GI) tract.

We conclude this section with several observations based on Fig. 6:

- The importance of including NP in the dose-receiving lung tissue increases with increasing particle size.
- The effect of including NP is most sensitive to Class W and least sensitive to Class D for a given half-life.
- This NP effect, for a given clearance class and particle size, diminishes most rapidly for half-lives between about 100 minutes and seven days. For longer half-lives, the effect is negligible.
- For very short half-lives, the lung dose per unit intake of 5-μm particles exceeds that for 1-μm particles by a factor of about five.

5. SOME BACKGROUND FOR THE INREM II/RADRISK DATA BASE

The dose conversion factors for inhalation (dose to a reference adult at specified times following an acute unit intake) used in the sensitivity analyses of CRAC2 were computed using a computer code that was adapted from the computer codes INREM II (Killough et al., 1978) and RADRISK (Dunning et al., 1980). The model reference adult is based on the ICRP Reference Man Report (ICRP, 1975). The ICRP lung model (Morrow et al., 1966; ICRP, 1972; ICRP, 1979) was used to simulate the deposition and retention of inhaled aerosols in the respiratory tract. The GI tract is represented by a catenary model of four segments (ICRP, 1979). Retention of activity in systemic organs is modeled by linear combinations of exponential terms; all metabolic data conform to current ICRP recommendations (ICRP, 1980; ICRP, 1981a; ICRP, 1981b).

Estimates of dose were computed separately for activity deposited initially in the nasopharyngeal, tracheobronchial, and pulmonary regions

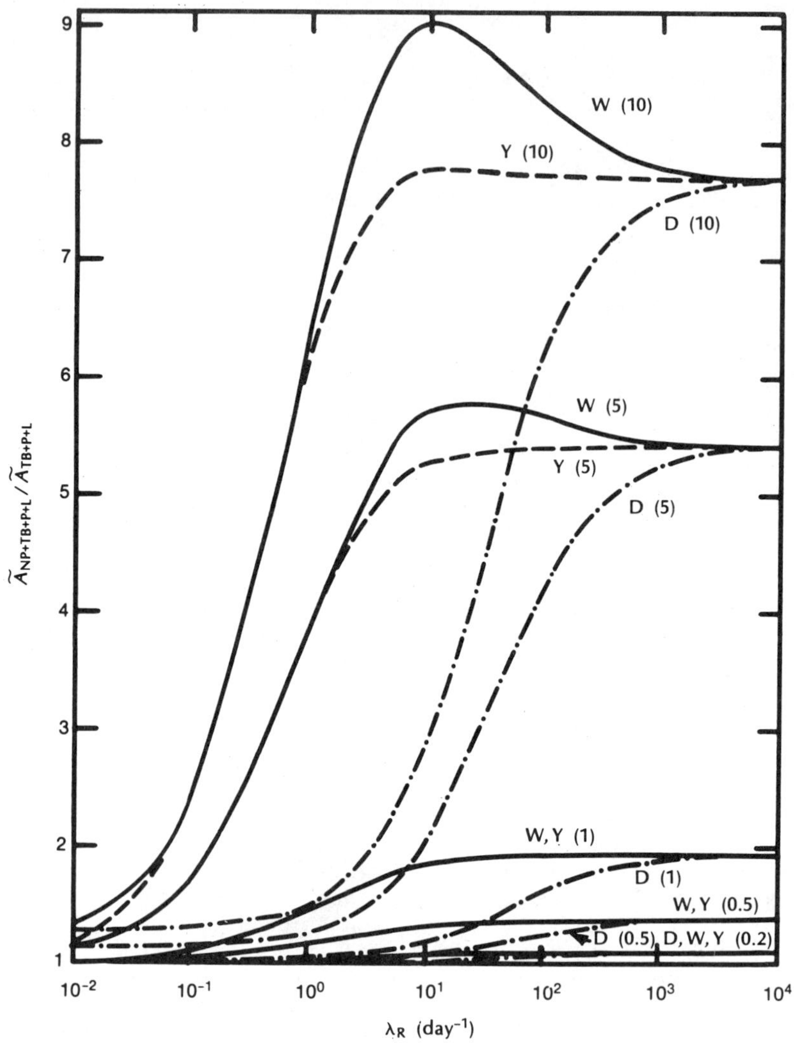

Figure 6

Analysis of Uncertainties in CRAC2 Calculations

Comparison of two definitions of dose to the lungs. The dose conversion factors in the ordinary CRAC2 data base estimate dose to the lungs as a weighted average of the doses to the tracheogronchial, pulmonary, and lymphatic regions (TB+P+L). If the dose to the nasal pharynx is included (NP+TB+P+L), the estimates of health effects by CRAC2 are increased significantly. The curves in this figure represent the ratio of time-integrated levels of a radionuclide in all four over TM+P+L; these ratios, which are equal to the corresponding dose ratios, are plotted as functions of the radioactive decay-rate coefficient for the radionuclide. Curves are shown for D, W, and Y materials and for a range of AMAD values (the numbers in parentheses, in μm). For radionuclides with long half-lives, the differences become unimportant. For extremely short-lived nuclides, the differences are pronounced and depend essentially on Amad. In the intermediate range, the situation is more complex; for fixed AMAD, the values tend to be monotonic, increasing in the order D, W, Y.

of the lung; estimates of dose from inhalation of aerosols with activity median aerodynamic diameter (AMAD) of 0.2 to 10 μm can be computed from a function that relates deposition in each of the three regions to AMAD (Fig. 1). Doses were computed for each of the three clearance classes (D, W, and Y) of the ICRP lung model (Fig. 2). A complete tabulation of these dose estimates, as well as a more complete description of the methods, has been prepared (Dunning and Eckerman, 1984).

These dose estimates differ from those used previously in the CRAC and CRAC2 computer codes in several ways. For many of the radionuclides considered, the metabolic models and parameters have changed significantly as additional data have become available. Also, the quality factor recommended by the ICRP for alpha particles has been revised from $Q_\alpha = 10$ to $Q_\alpha = 20$, and the use of the additional modifying factor for alpha particles in bone ($N_\alpha = 5$) has been discontinued. Several discrepancies also exist in the definition of some of the thirteen organs used in the CRAC code: (1) estimates of dose to the active red bone marrow have been substituted for total marrow; (2) the effective dose equivalent as defined in ICRP Publication 26 (ICRP, 1977) has been used to represent the dose to whole body; (3) doses to skeleton have been approximated by multiplying the endosteal cell dose by the ratio of the bone dose to endosteal dose in earlier tabulations, because the total skeleton is not included in the current dosimetric data base (use of the endosteum rather than the skeleton is recommended for future revisions of the CRAC/CRAC2 data base); and (4) dose to the "organ" OTHER is computed as a mass-weighted mean of the doses to the remaining organs in our dosimetric data base not explicitly included in the CRAC2 organ list (adrenals, bladder wall, kidneys, liver, breast, pancreas, brain, skin, spleen, thymus, and uterus). The cumulative effect of these discrepancies between the current dose estimates and those used previously in CRAC2 is difficult to estimate, but differences exceeding a factor of two occur between some of the dose conversion factors.

6. SUMMARY AND CONCLUSIONS

We have described calculations with the CRAC2 computer code to examine the sensitivity of estimated health effects to uncertainties in particle size and rate of clearance of radionuclides from the respiratory passages. For two of the three fixed weather sequences that we examined, our results indicated increases of one to two orders of magnitude in early injuries, in comparison with calculations using the standard CRAC2 data base. We also obtained increases of nearly a factor of two in early fatalities in some of the runs. But changes in latent cancer fatalities were minor. Economic costs were not affected by changes in the dose conversion factors for the inhalation pathway.

It is possible that a more coherent picture of these sensitivities in predicted health effects might emerge from simulations that combine random sampling of the AMAD and clearance class distributions with random sampling of meteorological sequences. Such a sampling plan was beyond the reach of our resources for the present study, but the results we have obtained suggest that it merits future consideration.

7. ACKNOWLEDGMENTS

We are very grateful to D.C. Kocher and R.W. Leggett, Health and Safety Research Division, Oak Ridge National laboratory, for helpful discussions and reviews of a draft of this paper. This work was sponsored by the Division of Systems Integration, Office of Nuclear Reactor Regulation, U.S. Nuclear Regulatory Commission, under Interagency Agreement DOE 40-550-75 with the U.S. Department of Energy under contract DE-AC05-84OR21400 with Martin Marietta Energy Systems, Inc.

REFERENCES

Bernard, S.R., and Snyder, W.S. 1975. "Metabolic Models for Estimation of Internal Radiation Exposure Received by Human Subjects from the Inhalation of Noble Gases," p. 197 in Health Physics Division Annual Progress Report for Period Ending June 30, 1975, ORNL-5046, Oak Ridge National Laboratory.

Dunning, D.E., Jr., Bernard, S.R., Walsh, P.J., Killough, G.G., and Pleasant, J.C., 1978. Estimates of Internal Dose Equivalent to 22 Target Organs for Radionuclides Occurring in Routine Releases from Nuclear Fuel-Cycle Facilities, Vol. II, NUREG/CR-0150 Vol. 2, ORNL/NUREG/TM-190/V2, Oak Ridge National Laboratory.

Dunning, D.E., Leggett, R.W., and Yalcintas, M.G., 1980. A Combined Methodology for Estimating Dose Rates and Health Effects from Exposure to Radioactive Pollutants, ORNL/TM-7105, Oak Ridge National Laboratory.

Dunning, D.E., Jr., Killough, G.G., Bernard, S.R., Pleasant, J.C., and Walsh, P.J., 1981. Estimates of Internal Dose Equivalent to 22 Target Organs for Radionuclides Occurring in Routine Releases from Nuclear Fuel-Cycle Facilities, Vol. III, NUREG/CR-0150 Vol. 3, ORNL/NUREG/TM-190/V3, Oak Ridge National Laboratory.

Dunning, D.E., Jr., and Eckerman, K.F., 1984. Estimates of Radiation Dose at Selected Times Following Acute Inhalation and Ingestion of Radionuclides, NUREG/CR-3799, ORNL/TM-9192, Oak Ridge National Laboratory (in press).

Iman, R.L., Conover, W.J., and Campbell, J.E., 1980. Risk Methodology for Geologic Disposal of Radioactive Waste: Small Sample Sensitivity Analysis Techniques for Computer Models, with an Application to Risk Assessment, NUREG/CR-1397, SAND80-0020, Sandia National Laboratories.

International Commission on Radiological Protection, 1972. The Metabolism of Compounds of Plutonium and Other Actinides, ICRP Publication 19, Pergamon Press, Oxford.

International Commission on Radiological Protection, 1975. Report of the Task Group on Reference Man, ICRP Publication 23, Oxford, Pergamon Press.

International Commission on Radiological Protection, 1977. ICRP Publication 26, Ann. ICRP 1(3), 1-47.

International Commission on Radiological Protection, 1979. Limits for Intake of Radionuclides by Workers, ICRP Publication 30 Part 1, Ann. ICRP 2(3/4), 1-116.

International Commission on Radiological Protection, 1980. Statement and Recommendations of the 1980 Brighton Meeting of the ICRP -- Limits for Intakes of Radionuclides by Workers, ICRP Publication 30 Part 2, Ann. ICRP 4(3/4), 1-71.

International Commission on Radiological Protection, 1981a. Limits for Intakes of Radionuclides by Workers, ICRP Publication 30 Supplement to Part 2, Ann. ICRP 5(1-6), 1-751.

International Commission on Radiological Protection, 1981b. Limits for Intakes of Radionuclides by Workers, ICRP Publication 30 Part 3 Including Addendum to Parts 1 and 2, Ann. ICRP 6(2/3), 1-124.

Killough, G.G., Dunning, D.E., Jr., and Pleasant, J.C., 1978a. INREM II: A Computer Implementation of Recent Models for Estimating the Dose Equivalent to Organs of Man from an Inhaled or Ingested Radionuclide, NUREG/CR-0114, ORNL/NUREG/TM-84, Oak Ridge National Laboratory.

Killough, G.G., Dunning, D.E., Jr., Bernard, S.R., and Pleasant, J.C., 1978b. Estimates of Internal Dose Equivalent to 22 Target Organs for Radionuclides Occurring in Routine Releases from Nuclear Fuel-Cycle Facilities, Vol. 1, NUREG/CR-0150, ORNL/NUREG/TM-190, Oak Ridge National Laboratory.

Morrow, P.E., Bates, D.V., Fish, B.R., Hatch, T.F., and Mercer, T.T., 1966. "Deposition and Retention Models for Internal Dosimetry of the human Respiratory Tract," Health Phys. 12(2), 173.

Ritchie, L.T., Johnson, J.D., and Blond, R.M., 1983. Calculations of Reactor Accident Consequences, Version 2, CRAC2: Computer Code User's Guide, NUREG/CR-2326, SAND81-1994, Sandia National Laboratories.

Ritchie, L.T., Alpert, D.J., Burke, R.P., Johnson, J.D., Ostmeyer, R.M., Aldrich, D.C., and Blond, R.M., 1984. CRAC2 Model Description, NUREG/CR-2552, SAND82-0342, Sandia National Laboratories.

U.S. Nuclear Regulatory Commission, 1975. Reactor Safety Study: An Assessment of Accident Risks in U.S. Commercial Nuclear Power Plants, WASH-1400, NUREG 75/014.

AUDITING WASTE DISPOSAL FACILITIES

David I. Brandwein, Brian L. Murphy, Bruce D. Cowen,
Mark S. Halvorsen, William W. Beck, Jr.,
and Harold C. Elston Jr.

TRC Environmental Consultants, Inc.
800 Connecticut Blvd.
East Hartford, Connecticut 06108

ABSTRACT

In order to minimize their potential liability for the disposal of hazardous wastes, many companies are evaluating how they select their waste disposal vendors. Companies fear being named in the future as a potentially responsible party at a facility which might become a Superfund site. This has prompted the development of a number of programs to audit a waste disposal facility before placing it on the list of "acceptable" vendors. Most audits are based primarily on technical evaluations carried out on a more or less subjective or "engineering judgment" basis. This paper describes a procedure which TRC has developed to audit waste facilities based on financial viability, management attitudes, and technology/site-related factors. Financial viability criteria include solvency, insurance, and RCRA financial mechanisms. Management attitudes include experience, planning abilities, competence, quality control, and reputation. Technical concerns addressed within a risk assessment framework include waste handled, site location, and appropriateness and adequacy of technology used. The procedure is designed to achieve a numerical profile of disposal or recycling vendors and to compare facilities based on a more quantitative basis. This audit system has been tested, and the audit criteria selected based on actual experience will be discussed.

KEY WORDS: Facility Audits; Environmental Audits; Waste Disposal Facilities; TSDS's; Hazardous Waste Disposal Contractors

INTRODUCTION

Legal entanglements for those involved at any point in the waste management chain - from generators to disposers - who may have, intentionally or unintentionally, mishandled their wastes are growing by leaps and bounds. One of the most complex problems so far is the scope of liability. What is clear is that the burden to pay the bills for resolving problems caused by hazardous waste disposal is usually falling on the waste generator. Generators are usually more visible and more likely to be financially able to pay the high price placed on past practices - unlike waste haulers or facility or property owners who are often out of business, bankrupt or otherwise unable to pay. The holding of individual generators responsible for entire site cleanups, regardless of their degree of involvement, has found favor with the courts. For example, since October of 1983, six federal district courts have issued opinions ruling that "joint and several" liability may be applied under Superfund in appropriate circumstances.[1]

In response to the concern that their company may be held responsible in the future for problems at a waste disposal or recycling facility or at a waste broker now being used or considered for use, many waste generators are initiating programs to evaluate their waste vendors. This concern should be present regardless of whether the facility has all relevant permits and regardless of what type of wastes the generator ships there or the facility is authorized to receive because permits are not necessarily reflective of adequate management controls or financial ability to prevent problems. In fact, the generator may be just as likely held responsible for paying for damage caused by someone else's waste as his own.

As a result, concerned generators have instituted a variety of audit programs to screen and approve their vendors. This paper describes an audit procedure in the form of a scoring system developed by TRC Environmental Consultants, Inc. that assesses the financial viability and management attitude of a facility, in addition to the technical or site audit more typically being conducted today.

OVERVIEW OF THE AUDIT APPROACH

A waste generator's interest in evaluating waste disposal vendors is generally twofold: first, to prevent injury to people or the environment caused by improper waste handling; and second, to eliminate or limit the generator's financial liability because of actions or omissions on the part of a vendor.

Although most audits encompass only a technical evaluation of a vendor's operation, we believe it is essential that an audit also assess the financial and management strength of the facility (and its parent organization, if any) being reviewed. The technical and management capabilities of a disposer are important for determining the likelihood of a problem or risk occurring, but the financial ability of the company to devote resources to capital improvements, good personnel, adequate insurance, and to taking care of problems is at least equally important. The key questions are: does the company have the financial strength to do things right now; will it be there in the future if things go wrong; and will it have the resources then to solve the problem without substantial involvement on the generator's part?

Figure 1 provides a flow diagram for how the audit procedure is intended to be used. The procedure entails two general stages. First, a prospective facility must "pass" certain minimum financial and insurance criteria. If it fails these criteria (or, as an option, refuses to provide the information), the facility is eliminated from consideration, and a site audit becomes unnecessary. If a facility meets these criteria, the second stage is undertaken: the site visit, data gathering, and evaluation and scoring of management and technology/site-related criteria.

One of our primary goals in developing this audit program has been to develop an <u>objective</u> procedure to evaluate contractors. We believe that any audit procedure that relies mainly on individuals receiving information and inspecting a facility, without detailed evaluation criteria, will be more subjective than objective. Furthermore, any audit report that relies on the auditor's narrative assessment alone rather than on efforts to score or rate detailed criteria will be more subjective than objective.

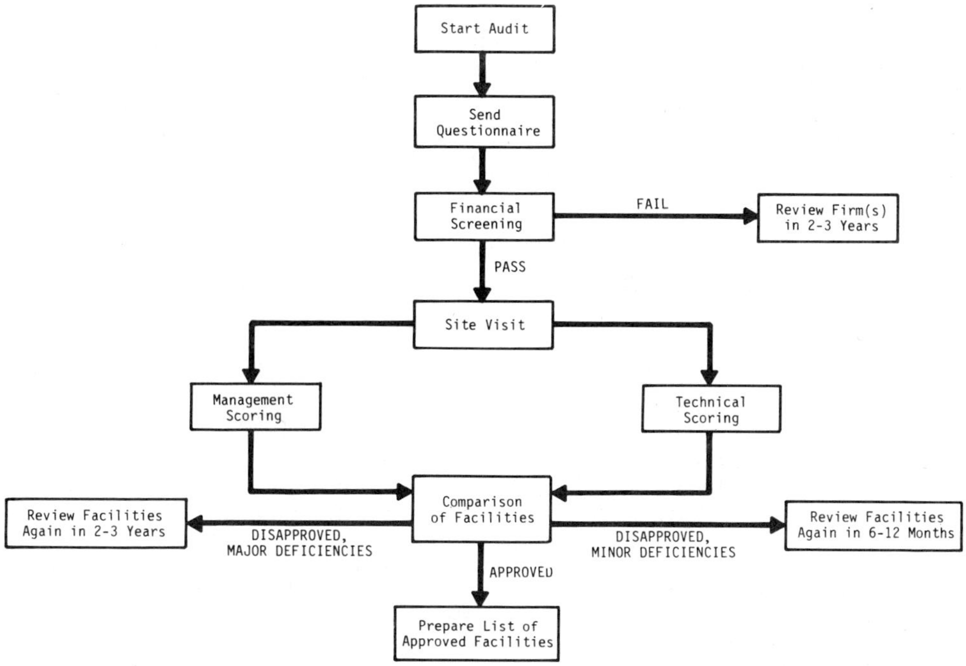

Figure 1

Schematic of Overall Audit Procedure

For these reasons, TRC decided to develop quantitative scoring procedures. An evaluation of a facility's financial viability or physical risk factors is, by its nature, already quantitative. Quantitative evaluation of a facility's management can also be done, although this requires rating intangibles. For example, liability can be influenced by perceptions as well as by reality. All other factors being equal, there is undoubtedly a greater likelihood of liability being incurred at a facility with a poor public or regulatory agency reputation. Part of the management audit procedure, therefore, necessarily deals with issues of perception.

The system consists of questionnaires to gather the necessary site and company information and a scoring methodology to process the information. Some of the information collected by questionnaire, for example, quantity of waste handled, enters directly into the scoring system. Other information collected serves as a background for forming judgments in "intangible" areas such as reputation or management attitudes.

In any case, the ratings are, by definition, relative. The scores do not mean anything in the absolute sense but rather provide a basis for more objectively assessing the relative merits and risks of one facility against another.

FINANCIAL EVALUATION

The procedures and specific tests we developed are designed to determine whether or not a potential hazardous waste disposal firm meets certain minimum financial standards, rather than to measure the overall financial strength of an organization. The financial evaluation is conducted before inspecting the facility; if the facility "fails" the audit (i.e., does not meet minimum criteria), the inspection is not necessary because the facility has been eliminated from consideration.

In the development of these procedures we considered the use of a number of financial ratios and other standard tests generally used when analyzing a commercial concern. However, because of the inherent problems of interpreting the results of such tests, we concluded that it would be more appropriate to establish procedures that are easily understandable to those without a strong background in financial analysis. In addition, we believe that it would be difficult and in many cases impossible to obtain certain financial data prepared on a comparable basis because of complex corporate structures.

The procedures developed cover three basic areas. The first part of the procedures relate to general organizational information and financial strength; the second area relates to financial information the firm provides to satisfy the financial tests under the Resource Conservation and Recovery Act (RCRA); and the third area provides procedures for determining the adequacy of the firm's insurance coverage.

General Organizational and Financial Information

We being with a series of questions which have been answered in a questionnaire distributed to the facility and its parent organization upon its being considered as a vendor candidate. Typical questions will include:

1. What type of business organization does the firm operate under?
2. When was the firm incorporated and in what state?
3. Has the corporate charter ever been repealed in any state?
4. How long has the firm been in the hazardous waste business?
5. Is the firm a subsidiary of another?
6. If a corporation, obtain most recent annual report and SEC Form 10-K (if publicly traded).
7. Obtain any other audited financial statements.
8. Obtain the Dun & Bradstreet (D&B) report on the firm and its "capital and credit" rating. Determine overall acceptable rating.

9. What is the firm's policy on indemnification of its clients?
10. What is the financial capability of other clients?
11. Determine approximate percentage of firm's annual revenue the generator will provide and assess against recommended maximum levels.

RCRA Financial Mechanism

1. If applicable, obtain copy of documentation submitted to EPA or state as evidence of satisfying RCRA financial mechanism(s). What mechanism was used for each?
2. Obtain chief financial officer's letter including required data from the independently audited financial statement.
3. Obtain a copy of the independent CPA's report on examination of the financial statements for the last completed fiscal year.
4. If firm is unwilling to provide the above, at a minimum request that they have their CPA(s) provide a letter opinion stating that the firm currently meets the minimum RCRA requirements.

Insurance Coverage

The purpose of this section is to review the firm's liability insurance coverage to determine that coverage and limits of liability meet certain minimum requirements. In addition to reviewing the firm's insurance coverage, the carrier's insurance rating should also be reviewed.

If the contractor's liability coverage does not equal or exceed minimum requirements, the firm should not be considered a viable hazardous waste disposal or recycling vendor. Insurance coverage should include environmental impairment liability as well as other insurance such as:

- Comprehensive automobile liability
- Worker's compensation
- Comprehensive general liability
- Umbrella policy (optional)

This information and data is compared against minimum standards. If a potential disposal facility passes this stage of the audit, we go on to the management and technology/site evaluations. If the facility fails the financial evaluation, it is removed from consideration but may be reconsidered in two to three years or as circumstances otherwise direct.

MANAGEMENT EVALUATION

This evaluation is of management's attitudes, philosophy, reputation, competency, and commitment to proper waste disposal. Intangible areas such as these may be evaluated in part through generally objective criteria such as the adequacy of training programs, contingency plans, and security procedures and through criteria which are more attuned to perceptions such as management's experience, philosophy, and commitment to the business.

In order to minimize the subjectivity of a straight "narrative assessment" procedure, TRC developed a list of detailed criteria for each of the three sub-areas we believe comprise a management evaluation:

1. Attitude - What is the company's management philosophy, experience, character, responsiveness and commitment to running an environmentally sound facility?
2. Planning - How well does the company do the things necessary to

run a well-operated facility (e.g., security, training, emergency response, quality control)?

3. Reputation - How well do others -- regulatory agencies, other waste generators, the community, the media -- think the facility is run?

Within each of these three sub-areas, TRC identified additional criteria - sixteen in all. These are listed in Table I. Within each of these sixteen criteria, TRC prepared a list of up to eight questions. For example, Table II provides the series of questions that will be answered during the site inspection about "Housekeeping and Appearance." Based on information gathered from a questionnaire and a site inspection, the auditor obtains answers to these questions and rates the answers on a 1-5 scale (1 = poor, 5 = excellent). Averages and subtotals are then calculated, the total normalized, and a final score assigned.

A copy of the form is taken by the auditor to the site inspection and notes made thereon. Based on these notes and a Facility Questionnaire which preferably should be completed by the facility and received by the auditor in advance of the site visit, the auditor will score each "lettered" sub-criterion. The scoring may be done during the site visit or, more likely, as immediately thereafter as is possible. Some of the criteria, for example, those dealing with reputation (no. 12-16), may require post-site visit telephone calls and will be scored then.

Through experience with a variety of waste storage, treatment and disposal operations, with the judgment that comes from such experience, and with use of this system, the auditor will develop a sense of how facilities compare and how scores should be assigned. In order to provide guidance on scoring, we have developed, for each of the sixteen management criteria, examples of what constitute a poor (score = 1), average or good (score = 3), or excellent (score = 5) rating. Table III provides an example of how each question should be scored. Obviously, different people may assign different scores for the same facility. This is why it is important that consistency be sought in who does the audits or that a new auditor familiarize himself or herself with the "baseline" of past scoring history in order to minimize these differences.

Table IV shows how the results for five actual facilities may be presented. The normalization factor as 6.7 is chosen to produce a maximum possible score of 100.

TECHNOLOGY/SITE EVALUATION

Facility evaluation on the basis of technology and site/related factors, such as the particular chemicals disposed of or the particular technologies employed, differs in character from the management and financial evaluations in that the evaluation process developed is based on simple mathematical models. The models describe the risk at a facility; in fact, the technical factors' score is taken as $\log_{10}$ (risk). Risk is defined here as the expected number of people in the community who will be injured by either a chemical release or by a fire/explosion. For example, if the probability of a storage tank rupturing and releasing trichlorethylene into ground water is 10^{-6}/year and the number of people who would then drink the water at injurious concentrations is 10, then the risk on an annual basis is 10^{-5} and the score as we have defined it is -5. In practice, rather than trying to evaluate facility scores in absolute terms, we only compare relative scores.

Table I

Management Evaluation Criteria

A. Attitude

1. Management Philosophy
2. Experience and Knowledge
3. Responsiveness to the Audit
4. Liability Insurance Coverage
5. Housekeeping and Appearances

B. Planning

6. Security
7. Personal Training
8. Emergency Response Capabilities
9. Medical Surveillance
10. Backlog Management
11. Quality Control

C. Reputation

12. Regulatory Agencies
13. Clients
14. Community
15. Media
16. Legal

Two mathematical models have been developed, one for chemical releases and one for fires and explosions. In each case the equation describing the risk is the basis for our scoring system. That is, the physical quantities that are scored, the range of scores that is assigned, and the way in which scores are combined are based on the risk model equation.

Table II

Sample of Management Evaluation Questionnarie

	Score
5. <u>Housekeeping/Appearance</u>	
a. On approach, is facility screened from view, landscape, or otherwise made aesthetically acceptable?	________
b. Are roads paved and kept clear? Are there signs clearly informing waste haulers or visitors where to go?	________
c. How clear and well maintained is the facility: roads, building interiors/exteriors, locker rooms, etc.? Does the facility show a concern for housekeeping details?	________
d. Are storage/treatment/disposal areas in apparent good condition under visual inspection? Are drums and pallets in good condition? No leaking pipes or containers? Any apparent spills that have not been completely cleaned up?	________
e. Any significant noise, odors, or visible emissions?	________
f. Is safety equipment/personnel protection equipment visible during site tours?	________

Avg.	________

The chemical release score, S_R, is given by combining scores for:

- Population density near the site (S_P).
- Toxicity of the particular chemical (S_β).
- Quantity of wastes handled (S_Q).
- Lifetime in the environment (S_τ) up to some maximum value.
- Appropriateness of the treatment technology (S_τ).

Certain scores, particularly, S_T, depend on pathway. In general, all scores are defined as log base ten of the parameter in question. The score S_t was determined by applying engineering judgment for different generic waste streams (e.g., heavy metals, acids/bases, chlorinated materials, etc.) versus different treatment, disposal or storage technologies (e.g., chemical precipitation, incineration with varying destruction efficiencies, landfills, with different degrees of liners, etc.).

Table III

Sample of Management Evaluation Scoring Criteria

5. Housekeeping/Appearance

a. 1 = Facility is a real eyesore and is not compatible with surrounding land use

3 = Industrialized area, facility does not do anything special to make itself presentable

5 = Facility is completely screened via landscaping

b. 1 = Roads not maintained; no direction signs

3 = Roads maintained adequately but not spotless; some signs but could be better

5 = Roads paved, swept and clearly marked

c. 1 = Poor

3 = Good

5 = Outstanding

d. 1 = Drums not well stacked; pallets broken; discolored puddles

3 = No visible leaks but area, pallets, shelves could be in better condition

5 = Excellent condition

e. 1 = Poor

3 = Some problems, relatively minor

5 = No problems

f. 1 = No equipment visible or accessible

3 = First aid equipment, eyewash, etc. available but could be better located or marked

5 = Equipment readily available, inspected recently, well-marked, etc.

The fire and explosion score $S_{F/E}$ is given by combining scores for:

- Population density near the site (S_P).
- Flammability of waste (S_F).
- Quantity of waste (S_Q).

For each facility, S_R can be computed by chemical, treatment process and pathway; as a result we develop a prioritized list of risk areas, by

Table IV

Management Scoring For Five (5) Facilities

MANAGEMENT CRITERIA	No. 1	No. 2	No. 3	No. 4	No. 5
A. ATTITUDE					
1. Philosophy	3.6	1.5	1.7	3.8	4.0
2. Experience	4.3	2.1	3.4	4.3	3.8
3. Responsiveness	4.5	1.0	1.0	1.5	4.0
4. Insurance	3.0	1.0	3.0	4.0	5.0
5. Housekeeping	2.8	1.1	1.9	3.8	4.8
A. Subtotal	18.2	6.7	11.0	17.4	21.6
A. Average	3.6	1.3	2.2	3.5	4.3
B. PLANNING					
6. Security	4.0	2.4	1.3	3.9	4.4
7. Training	4.4	3.5	2.5	3.0	2.2
8. Emergency Response	3.9	3.5	2.5	4.0	2.6
9. Quality Control	4.0	1.5	3.5	4.2	4.1
10. Medical Surveillance	nsf	nsf	nsf	5.0	3.5
11. Backlog Management	2.7	1.0	4.0	4.0	2.0
B. Subtotal	19.0	11.9	13.8	24.1	18.8
B. Average	3.8	2.4	2.8	4.0	3.1
B. REPUTATION					
12. Regulatory Agency	3.5	2.5	2.0	4.3	4.5
13. Clients	4.5	2.5	3.3	4.0	4.0
14. Community	4.7	2.5	2.5	3.0	4.5
15. Media	3.5	1.0	nsf	nsf	nsf
16. Legal	nsf	nsf	nsf	nsf	nsf
C. Subtotal	16.2	8.5	7.8	11.3	13.0
C. Average	4.0	2.1	2.6	3.8	4.3
Total of A+B+C Averages	11.4	5.8	7.6	11.3	11.7
Normalized Total = 6.7 x Average Total	76.4	38.9	50.9	75.7	78.4
Adjusted Normalized Score	76	39	51	76	78

"nsf" = insufficient information

Table V

Chemical Release Scores

Facility	S_{μ}	S_Q	Air	S Surface Water	Ground Water	S_T	S_P	Air	S Surface Water	Ground Water
No. 1										
TCE	4	2.9	2	1	7	-3	3.5	11.4	8.4	14.4
Pb (Lead Solution)	5	3.0	2	3	4	-3	3.5	12.5	11.5	12.5
Cd (Cadmium Solution)	5	3.0	2	3	6	-3	3.5	12.5	11.5	14.5
Cr^{+6} (Hexavalent Chromium Solution)	5	3.3	2	3	7	-3	3.5	12.8	11.8	15.8
CN (Cyanide Solution)	4	3.5	2	1	7	-3	3.5	12.0	9.0	15.0
No. 2										
Solvents, Alcohols, Polymers	2	5.1	2	1	7	-2	2.9	12.0	9.0	15.0
No. 3										
TCE	4	3.7	2	1	7	-3	3.1	11.8	8.8	14.8
PCB's (Liquid)	5	3.4	2	3	7	-3	3.1	12.5	11.5	15.5
Pb	5	3.1	2	3	4	-3	3.1	12.2	11.2	12.2
Cd	5	3.1	2	3	7	-3	3.1	12.2	11.2	15.2
Cr^{+6}	5	3.1	2	3	7	-3	3.1	12.2	11.2	15.2
No. 4										
Pb	5	?	2	3	4	-3	2.1	NSF	NSF	NSF
Cd	5	?	2	3	7	-3	2.1	NSF	NSF	NSF
Cr^{+6}	5	3.9	2	3	7	-3	2.1	12.0	11.0	15.0
CN	4	3.5	2	1	7	-3	2.1	10.6	11.0	15.15
No. 5										
PCB's (1% of Solids)	5	2.9	2	3	7	-3	3.2	12.1	13.1	17.1
Other Materials		?						NSF	NSF	NSF

Table VI

Most Important Risks At Each Site

Facility	Chemical	Pathway	Score	Comment
No. 1	TCE	A	11.4	
No. 2	Solvents, Alcohols, Polymers	A	12.0	
No. 3	PCB's	A	12.5	Possibly from fire
	TCE	A	11.8	
	PCB's	SW	11.5	Score may be increased
	Total		12.6	by food chain uptake
No. 4	-	-	-	Risk much less than other facilities
No. 5	PCB's	A	12.1*	Possibly from fire

*Some other materials, e.g., asbestos, unscored due to insufficient information.

Table VII

Fire and Explosion Scores

Facility	S_P	S_F	S_Q	$S_{F/E}$
No. 1	3.7	2	1.3	6.6
No. 2	2	3	2.3	6.5
No. 3	3	2	2.4	6.6
No. 4	2	-	-	N/A
No. 5	3	2	4.6	8.07

Table VIII

Scoring Rank by Facility

	Chemical Release Risk	Fire/ Explosion Risk	Management Risk
No. 1	4	2-4	1
No. 2	2-3	2-4	2
No. 3	1	2-4	3-5
No. 4	5	5	3-5
No. 5	2-3	1	3-5

score, for each site. For example, a list might be:

1. Trichlorethylene from a distillation process entering ground water (highest score).
2. PCB's from a storage tank entering the atmosphere (second highest score).
3. And so on for other pollutants, processes and pathways.

Pathway scores are combined to obtain an overall chemical release site score. See Tables V and VI. Fire and explosion score can be broken down by chemical and process in a similar way. See Table VII. No attempt is made to combine fire and explosion with chemical release scores.

It is usually convenient to do the scoring in two steps. In the initial step, crude values or assumptions may be used to develop the prioritized list or risk areas. In the final step more refined estimates are made for the top scoring areas on the list. It is generally necessary only to identify the few key areas to assess total risk at a site.

In the final scoring four general areas are examined more closely:

- The actual management and operation of the technology used at the site; i.e., whether it is well run or poorly run.
- Accessibility of surface and ground water to a chemical release, based on site geology, terrain, etc.
- Surface and ground water use for drinking or irrigation.
- Ability of materials to be mobilized into specific pathways; for example, a wetted storage site will not be subject to fugitive particulate emissions whereas a dry one will be.

Based on considerations in these areas a particular pathway or chemical may be disregarded or the score in an area may be adjusted.

REFERENCES

U.S. v. Chem-Dyne, 572 F. Supp. 802 (S.D. Ohio 1983);
U.S. v. Wade, 20 ERC 1277 (E.D. Pa., December 20, 1983);
U.S. v. A & F Materials Co., 20 ERC 1353 (S.D. Ill.), January 20, 1984;

U.S. v. North Eastern Pharmaceutical & Chemical Co., 20 ERC 1401 (W.D. Mo. January 31, 1984);
U.S. v. Conservation Chemical Co., 20 ERC 1427 (W.D. Mo. February 3, 1984); and
U.S. v. South Carolina Recycling and Disposal, No. 80-1274-6, D.S.C., February 23, 1984.

ANALYSIS OF UNCERTAINTIES IN CRAC2 CALCULATIONS: WET DEPOSITION AND PLUME RISE*

R.C. Ward†, and D.C. Kocher**

Oak Ridge National Laboratory
Oak Ridge, Tennessee 37831

B.B. Hicks, R.P. Hosker, Jr., J-Y. Ku,*** and K.S. Rao

Atmospheric Turbulence and Diffusion Laboratory
National Oceanic and Atmospheric Administration
Oak Ridge, Tennessee 37831

ABSTRACT

We have studied the sensitivity of results from the CRAC2 computer code, which predicts health impacts from a reactor-accident scenario, to uncertainties in selected meteorological models and parameters. The sources of uncertainty examined include the models for plume rise and wet deposition and the meteorological bin-sampling procedure. An alternative plume-rise model usually had little effect on predicted health impacts. In an alternative wet-deposition model, the scavenging rate depends only on storm type, rather than on rainfall rate and atmospheric stability class as in the CRAC2 model. Use of the alternative wet-deposition model in meteorological bin-sampling runs decreased predicted mean early injuries by as much as a factor of 2-3 and, for large release heights and sensible heat rates, decreased mean early fatalities by nearly an order of magnitude. The bin-sampling procedure in CRAC2 was expanded by dividing each rain bin into four bins that depend on rainfall rate. Use of the modified bin structure in conjunction with the CRAC2 wet-deposition model changed all predicted health impacts by less than a factor of 2.

KEY WORDS: CRAC2, Wet Deposition, Plume Rise, Uncertainty, Sensitivity, Bin-Sampling, Meteorological Models, Health Effects.

---

*Research sponsored by the Division of Systems Integration, Office of Nuclear Reactor Regulation, U.S. Nuclear Regulatory Commission under Interagency Agreement DOE 40-550-75 with the U.S. Department of Energy under contract DE-AC05-94OR21400 with Martin Marietta Energy Systems, Inc.

†Computing and Telecommunications Division

**Health and Safety Research Division

***Permanent address: Department of Atmospheric Sciences, State University of New York at Albany, Albany, NY 12222.

INTRODUCTION

The CRAC2 computer code[1,2] estimates the health impacts and economic costs that could result from a hypothetical nuclear reactor accident. CRAC2 is a revised version of the CRAC code that was used in the Reactor Safety Study.[3] The code utilizes hourly meteorological data (i.e., wind speed and direction, atmospheric stability class, and precipitation rate) in a straight-line, Gaussian-plume model to describe radionuclide transport in the atmosphere, and includes models for plume depletion via wet and dry deposition. Results can be obtained either for single weather sequences, which we refer to as single trails, or for importance sampling[1] from one year's weather data, which we refer to as meteorological bin-sampling. Bin-sampling runs produce statistical distributions of output, from which mean values averaged over the results for each sampled weather sequence can be obtained.

This paper presents a study of the sensitivity of health impacts predicted by the CRAC2 code to uncertainties in selected meteorological models and parameters. The sources of uncertainty studies include the use of alternative models for plume rise and wet deposition and an alternative procedure for meteorological bin-sampling. These models can affect predicted early fatalities, early injuries, and latent cancer fatalities.[1]

The sensitivity studies were performed with a prescribed release of radionuclides, a non-uniform population distribution, a single evacuation model, and one year's meteorological data from a reactor site. Three release heights (RH = 0, 25 and 50 m) and two sensible heat rates ($F = 10^4$ and 10^6 cal/s) were used in the calculations.

ANALYSIS OF PLUME-RISE MODELS

The plume-rise model contained in CRAC2[4] and the alternative model investigated in this study[5] were both developed by Briggs. The two models are identical for stable atmospheric conditions. For unstable and neutral conditions, the maximum plume rise, ΔH, in the CRAC2 model[4] is given by:

$$\Delta H = 6.11(FP)^{3/5} (RH)^{2/5}/U,$$

where RH is the release height (m), U is the wind speed (m/s) at the release and FP is the sensible heat rate (m^4/sec^3)[$FP = 3.7 \times 10^{-5}$ F where F is in cal/sec]. To avoid zero plume rise when RH is zero, the program resets RH to 1 meter. The maximum plume rise in the alternative model[5] is given by the minimum of ΔH_1 or ΔH_2:

$$\Delta H_1 = 30(FP/U)^{3/5}$$

$$\Delta H_2 = 24(FP/U^3)^{3/5}(RH + 200FP/U^3)^{2/5},$$

where the first formula will generally apply in convective or unstable conditions and the second in neutral conditions. The plume heights are obtained by adding the release heights to the plume rise.

For unstable and neutral conditions, the alternative model generally gives greater plume heights than the model in CRAC2, with the largest increases occurring for zero release height. However, in both single-trial and bin-sampling runs, predicted health impacts usually showed little sensitivity to the plume-rise model; i.e., the results changed by 40% or less. Exceptions occurred for RH = 0 and $F = 10^6$ cal/s, where the alternative model decreased early fatalities from 48 to 0 in a single-trial run with no rain (the plume height increased from 16 to 120 m in this case) and by a factor of 3 in a bin-sampling run.

ANALYSIS OF WET-DEPOSITION MODELS

The wet-deposition model in CRAC2 uses a scavenging rate, Λ, of the form $\Lambda = CR$, where R is the precipitation rate in mm/hr and the washout coefficient, C, is 10^{3} (mm-s/hr)$^{-1}$ for unstable and neutral conditions and 10^{-4} (mm-s/hr)$^{-1}$ for stable conditions.[1] This model does not properly account for certain cases such as a strong convective storm at night, where unstable conditions might prevail at ground level, or a stratiform system in daytime, e.g., a winter rainstorm. Thus, we felt that a model with constant scavenging rates that depend only on storm type rather than ground-level stability and rainfall rate would be a reasonable alternative. On the basis of measurements of precipitation scavenging for sub-cloud "washout" conditions,[6,7] which are expected in most continuous or frontal rain storms, and for in-cloud "rainout" conditions,[6,8,9] which are likely with convective storms, we chose scavenging rates of $\Lambda = 5 \times 10^{-5}$ s^{-1} for continuous rain and $\Lambda = 10^{-3}$ s^{-1} for convective rain. The scavenging rate in the alternative model is usually less than in the CRAC2 model, particularly for continuous rain during unstable or neutral conditions. In our analysis, the assignment of storm type to precipitation events was based on actual meteorological observations.

We studied the sensitivity of CRAC2 results to the wet-deposition model in single-trial runs with rain and in bin-sampling runs. The weather sequences for the single-trial runs were chosen to give different combinations of storm types, atmospheric stability classes, wind speeds, and rainfall rates, durations, and locations. In the single-trial runs, the predicted health impacts often were very sensitive to the wet-deposition model, and the effects of the alternative model on the results often were quite different for the different weather sequences. These results are discussed below.

Three weather sequences involving continuous rain were investigated in single-trial runs. The first was studied only for RH = 50 m and F = 10^{6} cal/s. This sequence involved unstable and neutral conditions and continuous rain which began at the distance from the release point to which the evacuated population was assumed to be relocated. Use of the alternative model decreased predicted early fatalities from 3200 to nearly 0, increased early injuries from 0 to more than 10,000, and increased latent cancer fatalities by a factor of 40. In this and other cases where large changes in early fatalities or injuries were obtained, the changes are also due in part to the thresholds in the dose-response models.[1] For the second weather sequence, which involved unstable and neutral conditions and continuous rain throughout the sequence, early fatalities were nearly zero for all values of RH and F, but early injuries and latent cancer fatalities increased with the alternative model by a factor of 10-20. The third weather sequence involved mostly stable conditions and continuous rain throughout, with heavy rain beginning at a distance of about 30 km from the release point. In this case, early fatalities changed by less than 15% with the alternative model, early injuries by 10-50%, and latent cancer fatalities by 25-50%.

Two weather sequences involving convective storms and unstable or neutral conditions were studied in single-trial runs. The sequences differed mainly in the distance from the release point at which the storm began. For a sequence where the storm began at about 30 km, use of the alternative model increased predicted early fatalities by more than an order of magnitude for all values of RH and F. Early injuries and latent cancer fatalities also increased, but only by about 40%. For a weather sequence where the storm began at about 5 km, the changes in all predicted health impacts were insignificant.

In bin-sampling runs, use of the alternative wet-deposition model increased predicted mean early fatalities for most values of RH and F, but only by 40% or less. However, for RH = 25 or 50 m and F = 10^6 cal/s, mean early fatalities decreased by a factor of 5-7 when the same set of weather sequences was sampled in comparing the two wet-deposition models, but increased by a factor of 2 when a different set of weather sequences was sampled with the alternative model. The important feature of these runs was that mean early fatalities using either wet-deposition model were small, i.e., less than 20, and most of the weather sequences sampled gave a null result. Thus, we are observing the sensitivity of mean early fatalities in bin-sampling run to whether or not the sampling scheme selects those rare single trials which give unusual large early fatalities. For all values of RH and F, mean early injuries decreased between 10% and a factor of 3 with the alternative model, and mean latent cancer fatalities increased by less than 30%.

ANALYSIS OF MODIFICATIONS IN METEOROLOGICAL BIN-SAMPLING

In the meteorological bin-sampling procedure in CRAC2, each of the 8760 hours of weather data in one year is sorted into one of 29 bins, 7 of which represent weather sequences with rain occurring at specified distances from the release site.[1] In our alternative procedure, we divided each of the seven rain bins into four bins that depend on rainfall rate, thus increasing the total number of bins to 50. Sorting of weather sequences on the basis of the amount of rainfall is a potentially important consideration if the CRAC2 wet-deposition model is used, since the scavenging rate depends on rainfall intensity.

We investigated the sensitivity of predicted health impacts to the choice of rain-bin structure using only the CRAC2 wet-deposition model. Four weather sequences were sampled from each of the 29 original bins and two weather sequences from each of the 50 modified bins, so that approximately the same number of trials were selected in each case. For the different values of RH and F, the alternative procedure changed mean early fatalities by 0-90%, mean early injuries by 3-60%, and the mean latent cancer fatalities by less than 15%. However, the changes in early fatalities and injuries may not be significant, because they are similar to variations that result only from selecting different weather sequences with the same bin structure. Nonetheless, we feel that the modified bin structure should be incorporated into the code if the CRAC2 wet-deposition model is used.

CONCLUSIONS

The most important results of this study involved the sensitivity of health impacts calculated by the CRAC2 code to changes in the wet-deposition model. In the CRAC2 model, the scavenging rate depends on rainfall rate and atmospheric stability class, whereas the alternative model used a scavenging rate that depends only on storm type, i.e., continuous (frontal) or convective. In single-trial runs involving rain, the alternative model often produced large changes in all health impacts and the effects depended greatly on the characteristics of the weather sequence. The large changes in early fatalities and injuries were also due in part to thresholds in the dose-response models. In bin-sampling runs, changes of nearly an order of magnitude in predicted mean early fatalities were observed and remove changes in mean early injuries were often a factor of 2-3, but changes in latent cancer fatalities were small. The large changes in early fatalities occurred when most of the weather sequences sampled gave a null result, and the effects of the wet-deposition model on mean early fatalities also depended greatly on whether or not the same weather sequences were sampled in the model comparison. Use of an alternative plume-rise model usually had little effect on health impacts. Modification of the meteorological bin structure in CRAC2 to include rain bins that depend on rainfall rate changed mean health impacts by less than a factor of 2.

REFERENCES

1. L.T. Ritchie, D.J. Alpert, R.P. Burke, J.D. Johnson, R.M. Ostmeyer, D.C. Aldrich, and R.M. Blond, CRAC2 Model Description, NUREG/CR-2552, SAND82-0342, Sandia National Laboratories (1984).
2. L.T. Ritchie, J.D. Johnson, and R.M. Blond, Calculations of Reactor Accident Consequences, Version 2, CRAC2: Computer Code User's Guide, NUREG/CR-2326, SAND81-1994, Sandia National Laboratories (1983).
3. Reactor Safety Study, Appendix VI: Calculation of Reactor Accident Consequences, WASH-1400 (NUREG 75/014), U.S Nuclear Regulatory Commission (1975).
4. G.A. Briggs, Plume Rise, TID-25075, U.S. Atomic Energy Commission (1969).
5. G.A. Briggs, "Plume Rise Predictions," p. 59 in Lectures on Air Pollution and Environmental Impact Analyses, American Meteorological Society, Boston, MA (1975).
6. K.P. Makhon'ko, S.G. Malakov, and M.P. Nekhorosheuva, "Washout of Fission Products from the Atmosphere," p. 167 in Radioactive Isotopes in the Atmosphere and Their Use in Meteorology, AEC-TR-6711, U.S. Atomic Energy Commission (1967).
7. A.S. Avramenko and K.P. Makhon'ko, "Atmospheric Washout of Radioactive Contaminant in the Mountains of Central Caucuses," p. 154 in Nuclear Meteorology, National Technical Information Service, Springfield, VA (1969).
8. N. Bhandari and Rama, "Study of Atmospheric Washout Processes by Means of Radon Decay Products," J. Geophys. Res. 68, 3823 (1963).
9. B.I. Styra, E.J. Verbra, and K.K. Benninger, "Geochemistry of Atmospheric Radon and Radon Products," Ann. Rev. Earth Planet, Sci. 5, 227 (1977).

USE OF HEALTH EFFECT RISK ESTIMATES AND UNCERTAINTY IN FORMAL REGULATORY PROCEEDINGS: A CASE STUDY INVOLVING ATMOSPHERIC PARTICULATES

L.J. Habegger* and A.H. Özkaynak**

Argonne National Laboratory*
Argonne, Illinois 60439
Harvard University**
Cambridge, Massachusetts 02138

ABSTRACT

Coal combustion particulates are released to the atmosphere by power plants supplying electrical energy to the nuclear fuel cycle. This paper presents estimates of the public health risks associated with the release of these particulates at a rate associated with the annual nuclear fuel production requirements for a nuclear power plant. Utilization of these risk assessments as a new component in the formal evaluation of total risks from nuclear power plants is discussed.

KEY WORDS: Particulate Health Effects; Nuclear Fuel Cycle; Risk Assessment; Risk Management; Regulatory Proceedings.

1. INTRODUCTION

It has been estimated that 323 thousand megawatt-hours (Mwhr) of electricity are required to manufacture and process the average annual fuel requirements of a 1000 megawatt nuclear power plant.[1,2] Because of the magnitude of this requirement, the environmental impacts of the air emissions from facilities generating this electricity have been included as a consideration in the overall impact of the operation of nuclear power plants. The results of these considerations have, however, previously in general been that these these air emission impacts are not significant on the basis of their small contribution (less than 0.02%) relative to the total of other national atmospheric emissions.[3] The impact of the atmospheric emissions on the ambient air quality would not be expected to cause violations of Clean Air Act requirements.

In the operating license hearings for the Harris nuclear power plant, the Atomic Safety and Licensing Board (ASLB) ruled in favor of the argument that further analysis is necessary to determine the environmental impacts of affluents released in the uranium fuel cycle. Specifically, further testimony was considered necessary to quantitatively address the health effects of the coal combustion particulates released during the generation of electricity utilized in the uranium fuel cycle.[4] The purpose of this paper is (1) to describe the scientific basis and results of a health effects risk assessment performed in response to the ASLB directives,[5] and (2) to discuss some of the constraints or factors that must be considered in cases such as this in which a risk assessment that involves uncertainty provides direct input to the formal risk management (regulatory) process.

The foremost constraint imposed on the analysis was that the particulate emissions to the atmosphere are 1154 metric tons (MT) per year as listed in Table S-3, "Table of Uranium Fuel Cycle Environmental Data," given in Title 10, Part 51 of the Code of Federal Regulations (CFR). This listed emission rate is based on out-dated information, and, as is discussed later, actual emissions are currently considerably lower. Also, the analysis was constrained by the ASLB hearing process to consider only primary particulate emissions, even though Table S-3 also listed emissions of sulfur oxides and other pollutants.

2. AIR QUALITY AND EXPOSURE ANALYSIS

Of the 323 thousand megawatt-hours (Mwhr) of electricity given in Table S-3 as the total annual requirement for the fuel cycle for the model LWR plant, approximately 96%, or 310,00 Mwhr, is required to perform the uranium isotope enrichment step of the fuel cycle.[2] At present, there are three facilities that could provide uranium enrichment for the fuel for the Harris plant. These are the gaseous diffusion plants (GDPs) located near Portsmouth, Ohio; Paducah, Kentucky; and Oak Ridge, Tennessee. Power for the GDPs is drawn from utility grids, and because of the interconnections it is not consistent with actual conditions to completely associate increased electrical generation, and thus increased particulate emissions, with any specific individual power plant. Allocating the generation to the mix of plants in the utility grid, including hydro, oil, gas, and nuclear would, however, be inconsistent with the constraints that the particulate emissions are fixed at a level of 1154 MT/yr and associated with coal-fired generation.

The approach used in this analysis was to compare the results of three separate analyses that each assume the total electrical generation, and thus the total 1154 MT/yr particulate emissions, occurs alternatively at one of three coal-fired power plants that were constructed primarily to provide power to the GDPs. One of these is the Joppa plant in Massac County, IL, which has 735 Mw of its 1100 Mw capacity "dedicated" to the Paducah GDP. The additional two plants considered were the Kyger Creek and Clifty Creek plants with capacities of 1086 and 1304 Mw, respectively.[6] The Kyger plant in Gallia County, OH, and the Clifty Creek plant in Jefferson County, IN, are dedicated to the Portsmouth GDP, which had a 1980 total power requirement of 2260 Mw. In the remainder of this paper, only the analysis and results related to the Kyger Creek plant are presented; similar results were obtained for the other two plants.

An initial aspect of the risk assessment for this plant was the further characterization of the emission source. The value of stack parameters used in the estimation of plume rise for the Kyger Creek plant[6] were: stack height, 305 m; inside stack diameter, 8.94 m; exit velocity, 33.5 m/s; exit temperature, 149° C.

The emission rate associated with the electrical generation for the uranium fuel cycle was assumed to be 1154 MT/yr as indicated in Table S-3. Table S-3, however, does not provide data on particulate size distribution, which is critical in the estimation o related health risks. Using available data for pulverized-coal fired boilers,[7] an estimated 7900 MT of fly ash is emitted without controls, and approximately 10%, or 790 MT of the particles have diameters 2.5 microns (μm) or less; 35%, or 2760 MT of the particles have diameter in the range of 2.5 to 15 μm; and the remaining 4350 MT of particles have diameters greater than 15 μm.

To obtain the Table S-3 particulate emission level of 1154 MT/yr, it must be assumed that only 85.4% of the 7900 MT/yr of ash carried in the flue gas is removed. This modest level of emission control can be achieved by various types of control devices. However, the relative rate of removal of different particle sizes is dependent on details of control device design, which are not specified in Table S-3 or cited references. Because of the unknowns in the control device·design used as a basis for Table S-3, a conservative approach was used in the analysis for this testimony in which it was assumed that all of the least harmful particulates greater than 15 μm are collected, none of the most harmful particulates less than 2.5 μm are collected, and the level of control on the midsize 2.5 to 15 μm particles is adjusted to give total emission rate of 1154 MT/yr. The result of this approach gives 790 MT/yr emissions of less than 2.5 μm and 364 MT/yr of emissions in the 2.5 to 15 μm range.

The above approach to characterizing the particulate emissions was designed to given an overall conservative results that overestimates health effects of the coal particulates. The actual Joppa, Clifty Creek, and Kyger Creek power plans emit particulates at a considerably lower rate than that given in Table S-3. For example, the Kyger Creek plant has an ESP with greater than 99.8% test efficiency and 99.4% design efficiency. This range of efficiencies gives estimated total particulate emissions of 16 to 114 MT associated with 323 thousand Mwhr generation. With the high removal efficiency of ESPs, a significant fraction of particles with diameters less than 2.5 μm is also removed. For example, using the calculation procedures given in Reference 7, it is estimated that approximately 70% of the 0-2.5 μm particle size range is removed with an ESP that has an overall removal efficiency of 96.5% for all particulates.

The air pollutant dispersion modeling conducted in the analysis was based on hourly data for wind speed and direction, temperature, and height of the surface mixing layer. For the Kyger Creek plant, this input data was obtained from surface and upper air data collected at Huntington, WV. The U.S. Environmental Protection Agency (EPA) recommends that five consecutive years of data be used to obtain representative conditions and 1971-1975 data was used for the Kyger Creek site. A small difference in estimated concentration among the five years of analysis, as described below, indicates that selection of year of meteorological record does not change the overall conclusions of this testimony.

Topography can affect ground-level concentrations and was an input to the air pollutant dispersion model. At each location where the atmospheric particulate concentration was estimated, the elevation relative to the power plant was obtained from area maps compiled by the U.S. Geological Survey.

Ambient particulate concentration increments based on this data was estimated using the Industrial Source Complex (ISC) computer model.[8] This is a standard model recommended by the U.S. EPA for use in air dispersion analysis for regulatory purposes. The ISC model has the capability to simulate particle removal by deposition, which results in lower concentrations, especially at distant receptor points. However, this feature was not utilized, which adds further to the conservatism of the analysis.

The particulate concentration and population exposure analysis covered a circular area of 50 miles radius with the power plant emission source at the center. The circular areas were divided into 360 grid cells with boundaries formed by radials at 10 increments and 10 concentric rings.

The accuracy of the ISC model applied to a single source is dependent on the application. The annual average concentration is used to quantify health risks and it is for such long-term averages that the better model accuracy usually results. As an example, a recent comparison between ISC predictions and monitored annual averages for particulates was performed in the vicinity of an industrial source in Minnesota[9]. After the computed background from other sources was subtracted from the observed monitoring data, the ISC-predicted annual averages at 6 locations ranged from an underprediction of 17% to an overprediction of 6%. For shorter term averages, such as the 24-hour maximums, a lesser predictive accuracy of within a factor of approximately 3 to 5 is usual.[10]

The averaging times that correspond to the National Ambient Air Quality Standards (NAAQS) for particulates are annual averages and maximum 24-hour average concentrations. For Kyger Creek, the maximum annual averages using the available 1971-1975 meteorological data were 0.010, 0.009, 0.013, 0.011 and 01013 μg/m^3, respectively, and 0.60, 0.53, 0.51, 0.47, and 0.71 μg/m^3 for the 24-hour maximum. Assuming the particulates are conserved in the atmosphere (no deposition or transformation) the particle size distribution in the atmosphere is the same as for the emission source, i.e., 68% less than 2.5 μm in diameter and 32% in the 2.5 to 15 μm diameter range.

For comparison the NAAQS primary standards (intended to protect public health) for particulates is 75 μg/m^3 for annual averages and 260 μg/m^3 for 24-hour maximums. Based on the estimates obtained in this analysis, the concentrations associated with the Table S-3 level of emissions would contribute only a very insignificant increment relative to the NAAQS.

An additional base for evaluating the significance of these concentrations is in comparison to increments allowable in "pristine" areas designated as Class I areas under the regulations for prevention of significant air quality deterioration.[11] In Class I areas the maximum increment in particulate concentrations allowable from new development is 5 μg/m^3 for annual average and 10 μg/m^3 for 24-hour maximum. Although there are no designated Class I areas within 100 miles of any of these power plants considered, the increment in particulate concentration from the Table S-3 level of emissions would not be cause for concern relative to these stringent standards even if they did apply.

No estimates were made of incremental particulate concentration beyond the 50-mile radius surrounding the power plants. This was not done for reasons similar to those expressed in EPA statements related to implementation of regulations for prevention of significant deterioration (PSD)[11]: "the (EPA) Administrator intends to limit generally the application of air quality models to a downwind distance of no more than 50 kilometers. This is because dispersion parameters commonly in use are based on experiments relatively close to sources, and extending these parameters to long downwind distances results in great uncertainty as to the accuracy of the model estimates at such distances. Also, since the air quality impact of many sources falls of rapidly to insignificant levels, EPA does not intend to analyze the impact of a source beyond the point where the concentrations from the source fall below certain levels 1 μg/m^3 for annual average particulates)." From the computer calculations, for the 1154 MT/yr emissions, nowhere within the vicinity of the three power plants does the estimated annual average incremental concentration approach the 1 $\mu g/m^3$ annual average recommended cut-off for PSD modeling and in the outer 40-50 mile (65-80 kilometer) band the mean of the annaul average incremental concentration is only 0.003 $\mu g/m^3$ for the Kyger Creek plant. Thus, based on EPA guidelines for PSD, estimation beyond 50 miles (80 kilometers) is not recommended.

The basic approach for computing human exposure resulting from the incremental ambient particulate concentrations was to estimate the population in each of the 360 receptor grid cells surrounding each of the three power plants. The estimated population in each cell is assumed to be exposed to the particulate concentration increment estimated for the cell midpoint.

Estimates of the population in each grid cell are derived from data collected by the U.S. Bureau of Census for block groups (in urban areas) and enumeration districts (in rural areas). The population for each of the geographic units whose centroid is located within a grid cell is assumed to be totally located within that cell. The various other census data, e.g., population by age, sex, occupation, education, is also available. For this analysis the 1980 census data was used.

Within a 50-mile radius of the Kyger Creek plant the total estimated 1980 populations was 870,000. The total computed population exposure (sum over all grid cells of population times concentration) using the annual average concentrations was 2174 persons $\mu g/m^3$ for Kyger Creek (mean of 1971-1975). The total computed population exposure using the maximum 24-hour concentration in each grid cell was 47,200 persons $\mu g/m^3$ for Kyger Creek (mean of 1971-1975). The population-weighted average (sum of exposures divided by population) over all grid cells was 0.0025 $\mu g/m^3$ for the incremental annual average particulate concentration and 0.054 $\mu g/m^3$ for the maximum incremental 24-hour concentration (mean of 1971-1975). The population-weighted averages can be interpreted as the concentration to which the average individual in the area of analysis is exposed.

3. ASSESSMENT OF POPULATION HEALTH EFFECTS

In this section the likely extent of population health risks within the 50-mile radius of the power plants resulting from the primary particle emissions of 1154 MT/yr is quantified. Various morbidity and mortality health outcomes were considered, including both, chronic and acute* effects. Acute respiratory disease incidents and hospital respiratory disease emergency admissions were among the morbidity health outcome variables studied. For chronic morbidity, chronic respiratory disease prevalence was considered.

For acute mortality effects of particle pollution, daily long-term (time-series) mortality risk coefficients were used. For chronic mortality effects, cross-sectional mortality coefficients relating annual mortality and annual average pollution in a large number of Standard Metropolitan Statistical Areas (SMSAs) in the U.S. were employed. Finally, toxicologic evidence regarding exposures to particles at or near the projected pollution levels were reviewed.

---

*Acute (respiratory) morbidity indicates short-term illness such as pneumonia, influenza and common coughs, while chronic (respiratory) morbidity indicates persistent, long-term illness such as chronic bronchitis, bronchial asthma or other obstructive lung disease.

Basically there were two available approaches to characterize the health impacts associated with the emitted particles. One was to rely heavily on EPA's recent criteria document on particulate matter and sulfur oxides,[12] and the second was to conduct or rely on a risk analytic approach oriented toward providing numerical exposure-response information.

The second approach was selected primarily because neither the EPA's criteria document nor the EPA Office of Air Quality Planning and Standards (OAQPS) staff review of the criteria document[13] provide quantitative exposure/response information for concentrations as low as those projected for this study. The EPA documents were, however, used to guide the selection of toxicologic and epidemiologic data bases that could be used to derive exposure-response coefficients or relationships.

If the EPA criteria document had been used solely we could only have concluded, and only qualitatively, that the projected health effects associated with the particulate emissions would most likely be negligible and undetectable. To illustrate, Table 1 summarizes the levels of baseline and incremental particle pollution impacts in the geographic area studied. Since most of the projected concentration impacts are extremely small, the changes in the ambient particulate levels surrounding the three plants are negligible, and the baseline or background plus incremental concentration is on the average, are less than 70 $\mu g/m^3$ (annual average) for Total Suspended Particulates (TSP) and less than 25 $\mu g/m^3$ for Fine Particulates (< 2.5 μm). A recent Federal Register notice by the EPA[15] proposes an annual PM_{10} particle (those particles with an aerodynamic diameter smaller than or equal to a nominal 10 μm) standard in the range of 50 to 60 $\mu g/m^3$ based on its staff assessment of long-term epidemiologic studies which support "no significant effects" in the concentration range of 80 to 110 $\mu g/m^3$ TSP or in equivalent PM_{10} levels, 40 to 55 $\mu g/m^3$.

Most of the up-to-date information on airborne particle toxicity and epidemiologic information regarding the population-based exposure-response relationships used in the risk and analytic approach in this study are contained in the recent Harvard Health and Environmental Effects Document (HEEDs) on airborne particles.[15,16,17] The airborne particle HEEDs are the product of a multi-disciplinary DOE study and are intended to summarize results to-date of continuing analyses of the nature, magnitude and uncertainty of potential health impacts of airborne particles. One of the main objectives of Harvard's assessments, reported in these HEEDS, is the identification of: limitations, gaps in knowledge, and research needs in the area of health risks resulting from population exposures to ambient particles. Also used were results from a recent EPA study on benefit analysis of alternative ambient standards for particulates.[18]

3.1 Morbidity Effects Analysis

In an attempt to quantify the morbidity effects of 1154 MT/yr coal particle emissions, several well regarded studies that provided quantitative estimates for air pollution concentration and morbidity outcomes were selected. A study of hospital visits conducted by Samet and his co-workers[19] addresses a concentration range for particles that is relevant to air pollution levels in the study area. The other studies (Ferris et. al.,[20,21] and Saric et. al.,[22]) address relatively high particulate levels, and one must extrapolate using linear exposure-response models beyond the range of observed concentrations in these studies in order to derive a risk coefficient. However, this approach must be taken cautiously as very little evidence exists to indicate whether or not the relationships observed at higher particle

Table 1

Comparison of Area-Specific Baseline Particle Pollution Concentrations and Population Exposures to Impacts Associated with the Kyger Creek Power Plant*

Baseline Values**				Projected Increments			
Annual Average (ug/m^3)		Population Exposures (ug/m^3 x persons)		Annual Average*** (ug/m^3)		Population Exposures (ug/m^3 x persons)	
TSP	FP	TSP	FP	IP	FP	IP	FP
68	24	5.9×10^7	2.1×10^7	0.003 - 0.004 (mean = 0.0033)	0.002 - 0.003 (mean = 0.002)	1,990 - 2,436 (mean = 2,174)	1,353 - 1,656 (mean = 1,478)

* 50-mile radius around the power plant defines the study area. Projected range of impacts is based upon 5 years of meteorology (1971-1975).

IP refers to the mass of Inhalable Particles (aerodynamic diameter < 15 um) and typically, IP is 54% of TSP in the ambient atmosphere.[23] FP refers to the mass of Fine Particles (aerodynamic diameter < 2.5 um.)

** Mean TSP concentrations are based on: 1) NASN data for counties surrounding and including the power plants (data for years 1978-1981), 2) EPA Inhalable Particle Network data for (urban) sites near the power plants (data for years 1978-1982), and 3) EPRI/SURE measurements at nearby (rural) class lsites (data for 1977-1979).

*** Annual averages are calculated by averaging over all grid cells.

concentrations hold at lower concentrations. Although the EPA has excluded this study from its criteria document, other researchers have included it to allow the projection of conservative estimates.[16,18]

With these qualifications, the coefficients used for predicting acute morbidity included a coefficient for excess emergency room visits for respiratory disease, derived from the Samet <u>et. al.</u>, study,[19] with a value of 13 ± 6.5 cases/year/100,00 persons per $\mu g/m^3$ TSP. The conservative (i.e., near upper-bound) age-specific coefficients used to project acute respiratory disease incidents obtained from the Saric <u>et. al.</u>, study[22] were:

Age	Risk Coefficient*
0 - 24 years	540
25 - 54 years	100
55+ years	121

* Acute respiratory disease incidents/year/100,000 per $\mu g/m^3$ TSP.

Due to uncertainties in the application of the Saric results to this study, zero was used as the lower-bound limit and 1.5 times the estimated coefficient was used to compute the upper 95% confidence limit.

In addition to the extrapolation problem, the range of coefficients presented above logically include zero because there are other important contributors to the uncertainties of the morbidity risk estimates. These can be categorized into sampling and non-sampling errors associated with the epidemiologic studies of both the morbidity and the mortality effects of air pollution. Sampling errors refer to the lack of precision of a sample result. If a sample were to be collected without sampling error, then one could reproduce from the sample the results which would have been obtained if the entire population had been included. To the extent that this is not true, sampling error exists. Non-sampling errors include a variety of factors that influence the uncertainty of the estimated airborne particle concentration-response relationships. They include: confounding factors (e.g., cigarette smoking, socio-economic status, occupational exposures, race, prior exposures and residence); collinearities with other pollutants (e.g., particles and sulfur dioxide); changing measures of particle pollution that are not entirely comparable [e.g., British Smoke (BS)* or Coefficient of Haze (COH)** versus TSP]; oversimplifications in estimating personal exposures from data collected at fixed-site monitors; and biases due to historical and cross-community differences in particle and source composition.

Due to potential contributions by the factors noted, the standard errors of the estimates have typically been defined half as large as the estimates themselves.[15] Using these standard errors, the 95% confidence limits are estimated by adding (or subtracting) twice the standard error to the mean risk estimate. Negative lower-bound estimates, if obtained, are set at zero, since beneficial effects of air pollution are not readily plausible.

Table 2 summarizes the most likely acute morbidity effects associated with airborne particles emitted from the Kyger Creek power plants. Table 2 provides the corresponding uncertainty limits for these projected health effects. For the area surrounding the Kyger Creek facility, the incremental excess emergency room visits for respiratory disease would be about one case every two years (0.5 per year). Whereas, the expected number of incremental annual acute respiratory disease incidents for the same areas are about 11 per year.

Estimates for chronic morbidity effects were not included since the available coefficients are derived from the studies of Ferris and his colleagues[20,21] based on ambient levels typically greater than 130 μg/m^3, and thus significantly larger than the baseline plus incremental in this analysis.

3.2 Results of Mortality Effects Analysis

In order to quantify the acute and the chronic exposure mortality effects of airborne particles, mortality coefficients derived from time-series and cross-sectional analysis reported in the Harvard[15,16] and the Mathtech[18] reports were chosen. In addition, an upgraded cross-sectional mortality analysis was performed using the 1980 census and vital statistics data in conjunction with FP (fine particles, i.e., aerodynamic diameter smaller than 2.5 μm) data for the same period.

Only the effects of particle pollution on total mortality i.e.,

---

*British Smoke. An optical aerosol measurement used more commonly in europe and expressed in μg/m^3. It is obtained by measurement of the decreased reflectance of a filter paper after the passage of a measured volume of air.

**Coefficient of Haze. An optical aerosol measurement used to quantify the mass of particle deposit on a filter tape sampler by

Table 2

Summary of Most Likely Projected Population Health Effects Associated with Airborne Particles Emitted from the Kyger Creek Power Plant Supplying Electricity to the Uranium Fuel Cycle*

Acute Morbidity Effects		Mortality Effects	
Annual No. of Emergency Room Visits for Respiratory Disease	**Annual Acute Respiratory Disease Incidents****	**Daily Mortality**	**Annual Mortality**
0.5	**11**	**$0.8 \times 10^{-5} - 1.9 \times 10^{-5}$**	**0 - 0.03**

*** Uncertainties due to sampling and non-sampling errors are not included here. Therefore, zero risks are also possible.**

**** These are tentative results and should be considered to be conservative due to inherent problems associated with the interpretation of the Saric _et al._ study[22] which was used in developing the health risk coefficients.**

changes in the mortality from all causes, were analyzed. From time-series analyses providing daily mortality risk coefficients, the following coefficients derived from the analysis of 14 years of daily New York City mortality and air pollution records were considered:

0.02 ± 0.005 deaths/day/100,000 persons per unit COH[15]

0.046 ± 0.007 deaths/day/100,000 persons per unit COH[17]

Both of these time-series coefficients were used to present a range. Before application, the coefficients were converted to representative TSP levels by scaling with a factor of 0.01. (This was based upon site-specific analysis using 500 same-day measurements, conducted at the City College of New York, of COH and TSP.) The coefficient was further adjusted to account for the difference in the size distribution of the projected power plant particle emissions. The conversion used was a scaling upward by a factor of 1/0.54, since IP (i.e., smaller than 15 μm) mass is, on the average, 54% of the TSP mass.[23]

From cross-sectional mortality analysis, the following sets of coefficients were considered:

0 - 0.47 deaths/year/μg/m^3 TSP per 100,000 persons[18]
0.4 ± 0.24 deaths/year/μg/m^3 TSP per 100,000 persons[16]
2.31 ± 0.81 deaths/year/μg/m^3 FP per 100,000 persons[17]

The final range of cross-sectional mortality coefficients selected were 0 to 2.31 ± 0.81 deaths/year/100,000 persons per μg/m^3 FP (since all the other coefficients, if used, would project mortality impacts within the range generated by these risk coefficients).

Although the latest fine particle coefficient[17] should provide an improvement over the previously reported cross-sectional mortality coefficients of particle pollution, there are still large uncertainties surrounding any such damage coefficient. Indeed, despite the fact that the coefficient is statistically greater than zero, uncertainties which are not considered by such analyses [e.g., non-sampling errors such as in the measurement of the exposure variable] make it possible that the particle mortality risk coefficient might be zero.

Tables 2 and 3 summarize the predicted mortality impacts and their associated uncertainties, respectively. The most likely annual mortality risks associated with emissions from the Kyger Creek plant are less than 0.03 per year within the 50-mile radius of the plant. Predicted average daily mortality risks from the Kyger Creek plant would likely be 0.8×10^{-5} to 1.9×10^{-5} deaths per day associated with the 1154 (MT/yr) particle emissions. These daily projections, if multiplied by 365, are still less than one-third of the annual mortality projections based upon using risk coefficients derived from cross-sectional mortality studies.

Based upon analysis of the daily population exposure projections within the 50-mile radius surrounding the Kyger Creek plant, the maximum 24-hour IP exposure is 49,000 person-$\mu g/m^3$ to give a maximum daily mortality of 4.2×10^{-4} excess deaths.

3.3 Estimated Incremental Health Risk Compared to Baseline Values

Scientifically, the significance of the health risk projections can usually be demonstrated by comparison with the baseline morbidity and mortality rates in the areas studied. However, even in terms of absolute effects, both the concentration and the mortality effects are significantly below the detectable or observable levels in the region studied. Projected morbidity effects are also very small and well below the sampling error in most epidemiologic surveys.

To compare the projected mortality and the morbidity risks to expected baseline values, the socio-demographic profile of the population near these three plants was thoroughly examined. It was determined that: (a) the population characteristics are typical of the national average; and (b) national disease incidence rates and the cross-sectional mortality model developed using a national data base can be applied reliably in projecting the baseline risks in the region studied.

The 1980 cross-sectional model was first used to predict the area-specific total mortality rates (TMR) using the socio-demographic information obtained from the 1980 census for the areas surrounding the three plants. These predictions were then checked for consistency with the observed mortality rates in 14 cities in Indiana and Ohio. It was found that the results were in excellent agreement.[5]

The model that was used to predict the annual TMR was obtained from a multiple-regression analysis of the 1980 U.S. census and air pollution data. The regression coefficients of the model chosen accounted separately for the contribution of the following independent variables to the predicted TMR: percentage of the population over 65 years old; median age of population; percentage of population which is non-white; decimal logarithm of the population density; percentage of population with more than 1 year of college; percentage of families with an annual income less than $5,000; and estimated fine particle mass. The predicted total mortality rates are noted to vary between 814 and 1,084 deaths per 100,000 persons per year and are consistent with the observed Midwest average, 895 $\pm$ 108.

Table 3

Uncertainty Limits (95% Confidence) for the Projected Health Effects (Kyger Creek)

Acute Morbidity Effects		Mortality Effects	
Annual No. of Emergency Room Vists for Respiratory Disease	Annual Acute Respiratory Disease Incidents	Daily Mortality*	Annual Mortality
0 - 1.0	0- 17	0.4×10^{-5} - 2.5×10^{-5}	0 - 0.05

* Uncertainties due to non-sampling errors not included, i.e., the possibility of zero as a lower limit is not excluded.

A rate for the daily hospital emergency room visits from all causes was derived from the Samet, et. al., study[19] to be 24.5/20,000, or 1.22×10^{-3} visits per day per person.* For annual acute respiratory disease incidence rates, the age-specific rates used were:[18] 1.52 (0 - 24 years), 1 (25 - 55 years), and 0.7 (55+ years) annual acute respiratory disease incidents per person in the U.S.

Expected excess emergency room visits due to respiratory conditions were 389,000. The expected annual acute respiratory incidents are 979,000 for this area. Total annual mortality are estimated to be 7.500 deaths in the Kyger Creek area. These absolute values in the Kyger Creek area are quite large due to the size of the population and the magnitude of the expected incidence or prevalence rates.

The percentage of change in the expected incidence of total morbidity resulting from the incremental 1154 MT/yr particle emission in the Kyger Creek area is very small (0.0001 to 0.005 percent). The likely percentage of change in the expected annual total mortality resulting from the incremental particle emissions is less than 0.0004 percent. Both in terms of absolute numbers and in terms of relative proportions compared to baseline health impacts in the areas analyzed, the projected impacts are very small. Furthermore, concentration as well as health impacts are so small that they could not even be detected with the state-of-the-art monitoring, survey design, and analysis techniques. Finally, all of the projected health impacts are much smaller than the uncertainties associated with the available risk assessment models and the predictions resulting from their use.

4. DISCUSSION AND CONCLUSIONS

The following additional general conclusions can be made regarding the coal particulates health risk analysis presented in this paper, and in particular the use of similar analysis in formal regulatory proceedings:

1. Risk management through the formal regulatory hearing, or similar legal process, is the mechanism by which scientific risk assessment has a direct impact on societal decisions. Scientists involved in risk assessment should thus have an awareness of the demands and constraints of the process.
2. The regulatory process can force resolution of scientific issues. A foremost example is the extent to which the linear dose-response model should be utilized. Exposures of large populations to low concentrations of hazardous pollutants can lead to exaggerated estimates of health effects with the linear model if caution is not used. Also, during cross-examination in this hearing the need for dose-response data based specifically on coal particulates was emphasized, although the uncertainty limits used for the coefficients in the analysis were intended to take into account variability in particulate characteristics.
3. Legally defined constraints of the regulatory process may limit development of more comprehensive perspectives. For example, in the risk assessment presented in this paper, the analysts were constrained to consider only previously defined obsolete emission estimates; effects of other pollutants, such as secondary particulates from sulfur emissions, were not considered; and the health effect benefits/costs from coal plants being displaced by the nuclear plant were excluded from consideration at this hearing.
4. Uncertainty analysis is a necessary component of risk assessments that are to be used in the risk management (regulatory) process, however, there is a natural tendency to misuse extreme values in the ranges to support particular views. Central or most-likely estimates should thus be derived when possible and clear explanations of the basis for the uncertainty limits should be provided.

At the date of the writing of this paper no decision on the significance of the health effects of the coal particulates in the uranium fuel cycle has been made by the ASLB.

ACKNOWLEDGMENT

The support for this study by the United States Nuclear Regulatory Commission is gratefully acknowledged.

REFERENCES

1. Code of Federal Regulations, Title 10, Part 561, Licensing and Regulatory Policy and Procedures for Environmental Protection.
2. "Environmental Survey of the Uranium Fuel Cycle," WASH-1248, Directorate of Licensing, Atomic Energy Commission, Washington, D.C. (1974).
3. "Draft Environmental Statement, Shearon Harris Nuclear Power Plant, Units 1 and 2," NUREG-0972, U.S. Nuclear Regulatory Commission (1983).
4. Atomic Safety and Licensing Board Order of January 27, 1984, in the Operating License Proceedings for Harris Units 1 and 2.
5. NRC Staff Testimony of Dr. Loren J. Habegger, Dr. A. Hâluk Ozkaynak

and Mr. Ronald L. Ballard Regarding Eddleman Contention 8F(1) (Health Effects of Coal Particulates at the Table S-3 Level), before Atomic Safety and Licensing Board in the Operating License Proceedings for the Harris Units 1 and 2 (June 15, 1984).
6. U.S. Federal Power Commission, Form 67, "Steam-Electric Plant Air and Water Quality Control Data," Kyger Creek Electric Station (1980).
7. "Assessment of Energy and Economic Impacts of Particulate-Control Technologies in Coal-Fired Power Generation," ANL/ECT-9, Midwest Research Institute and Argonne National Laboratory, Argonne, IL (1980).
8. "Industrial Source Complex (ISC) Dispersion Model User's Guide," EPA-450/4-79-030, U.S. Environmental Protection Agency, Research Triangle Park, NC (1979).
9. Heron, T.M., J.F. Kelly, P.G. Haateja, "Validation of the Industrial Source Complex Dispersion Model in a Rural Setting," Jour. of the Air Poll. Control Assoc., Vol. 31, No. 4 (April 1984).
10. Londergan, R., D. Minott, D. Wackter, T. Kincaid, and D. Bonitata, "Evaluation of rural Air Quality Simulation Models," EPA-450/4-83-003, U.S. Environmental Protection Agency, Research Triangle Park, NC (1982).
11. "1977 Clear Air Act Amendments to Prevent Significant Deterioration," 43 FR 26380 (June 19, 1978).
12. "The Revised Criteria Document: Air Quality Criteria for Particulate Matter and Sulfur Oxides" (three volumes), U.S. Environmental Protection Agency, EPA-600/8-82-029a-c (December 1982).
13. "Review of the National Air Quality Standards for Particulate Matter: Assessment of Scientific and Technical Information", OAQPS staff paper, U.S. Environmental Protection Agency, EPA-450/5-82-001 (January 1982).
14. Federal Register, "Proposed Revisions to the National Ambient Air Quality Standards for Particulate Matter", V49, N55 (March 20, 1984).
15. Ozkaynak, H., G.D. Thurston, T.D. Tosteson, C.M. Smith, P.L. Kinney, B. Beck, W. Skornick, S.D. Colume and A. Schatz, "Analysis of Health effects Resulting from Population Exposures to Ambient Particulate Matter," prepared by Harvard University for the U.S. Department of energy, Contract No. DE-AC02-81EV10731 (October 1983).
16. Ozkaynak, H., J.A. Warren, T.D. Tosteson, J.S. Evans, G.D. Thurston and A.D. Colome, "Analysis of Health Effects Resulting from Population Exposures to Ambient Particulate Matter," prepared by Harvard University for the U.S. Department of energy, Contract No. DE-AC02-81EV10731 (October 1982).
17. Ozkaynak, H., "Analysis of Health Effects Resulting from Population Exposures to Ambient Particulate matter," Harvard University, 1984 Revisions (in preparation).
18. Manuel, E.H., R.H. Horst, Jr., K.M. Brennan, J.M. Hobart, C.D. Harvey, J.T. Bentley, M.C. Duff, D.E. Klinger and J.K. Tapiero, "Benefit and New Benefit Analysis of Alternative National Ambient Air Quality Standards for Particulate Matter," prepared by Mathtech, Inc. for Benefits Analysis Program, Economic Analysis Branch, Office of Air Quality Planning and Standards, U.S. Environmental Protection Agency, Contract No. 68-02-3826 (March 1983).
19. Samet, J.M., F.E. Speizer, Y. Bishop, J.D. Spengler and B.G. Ferris, Jr., "The Relationships Between Air Pollution and Emergency Room Visits in an Industrial Community," J. Air Pollut. Control; Assoc., 31: 236-240 (1981).
20. Ferris, B.G., Jr., H . Chen, S. Puleo and R.L.H. Murphy, Jr., "Chronic Non-Specific Respiratory Disease in Berlin, New Hampshire, 1967-1973: A Further Follow-up Study," Amer. Rev. Resp. Dis., 113: 475-485 (1976).

21. Ferris, B.G., Jr., I. Higgins, M.W. Higgins and J.M. Peters, "Chronic Non-Specific Respiratory Disease in Berlin, New Hampshire, 1961-1967: A Follow-up Study," Amer. Rev. Resp. Dis., 107: 110-122 (1973).
22. Saric, M., M. Fugas and O. Huristic, "Effects of Urban Air Pollution on School-Age Children," Arch. Environ. Health, 36: 101-108 (1981).
23. Ozkaynak, H., G.D. Thurston, A.D. Schatz, R.G. Isaacs and R.B. Husar, "Relationships between Aerosol Extinction Coefficients Derived from Airport Visual Range Observations and Alternative Measures of airborne Particle Mass, paper no. 84-59.2, 77th Annual Meeting of the Air Pollution Control Association, San Francisco, CA (June 24-29, 1984).

APPLICATION OF ESTIMATED EXCESS LIFETIME CANCER RISK IN FIELD SITUATIONS

Stan C. Freni

U.S. Department of Public Health Service
Atlanta, Georgia 30333

ABSTRACT

The term "excess lifetime cancer risk" connotes the risk of death of cancer in excess of the "natural" background risk, resulting from a lifetime exposure to carcinogens. In practice, exposure is not for a lifetime, competing causes are prominent, the exposure usually comprises more than one carcinogen, and the risk of cancer death may be substantially reduced by therapy. Therefore, the following points and questions and questions may be considered. (1) The estimated excess unconditional risk based on animal bioassays overestimates the risk for humans. A conditional risk, however, would require life-table techniques, for which age-specific mortality rates are essential but difficult to obtain. (2) For a shorter-than-lifetime exposure, the Weibull time model appears to be inadequate because it does not accommodate for the probable significant reduction of additionally accrued incremental risk after cessation of exposure. (3) Should the level of risk used for initiating public health activities be the same as that for initiating remedial actions? (4) When populations are exposed to multiple chemicals, should decisions concerning public health be based on the sum of the risk estimates for all of the chemicals or on the highest risk estimate for one chemical? (5) When small populations are exposed, should the predicted actual excess number of cancer deaths rather than a predicted risk level be the sole basis for decisions on public health and remedial action?

KEY WORDS: Risk Analysis; Risk Assessment; Cancer Risk; Risk in Epidemiology; Cancer Risk Estimate.

1. INTRODUCTION

Cancer risk estimates (CREs) obtained through mathematical modeling, in which results of high-dose animal tests are extrapolated to low-dose situations in humans, are based on controversial assumptions and theoretical concepts. Discussions on this subject usually concern the theoretical problems associated with extrapolation between species and outside the applied dose range. In addition to these problems, extrapolated CREs may present difficulties in another area. This regards the application of CREs as a measure of the excess lifetime risk of cancer deaths in field situations, where a population of limited size is exposed to a mixture of chemicals present in soil, water, or air. Some of these difficulties are related to the theoretical concepts underlying the derivation of CREs, but most are associated with the practice of real life, which often has little connection with the research world. The following sections deal with groups of related problems, some of which are well known, other sections may appear new or unusual, since they concern dilemmas that health officials may have when assessing the health hazard at a site contaminated by chemical wastes.

2. CANCER RISK ESTIMATES AS UNCONDITIONAL RISKS

CREs are essentially unconditional risk estimates extrapolated from animal tests. They are applied on the assumption that the decrement over time in the animal test population, due to causes other than cancer, is comparable to that in humans. However, in a well-performed bioassay, spontaneous morbidity and mortality are minimal, and most animals will survive to the end of the experiment. In contrast, an abundance of factors cause premature death in humans, the age-specific mortality is entirely different from that in animals. For instance, a substantial part of premature mortality in humans is caused by neonatal and infant diseases and by accidents. Rodents entering a cancer bioassay usually are no younger than 5 to 8 weeks of age, which is comparable to a lower grade school age in humans, and accidental death among laboratory rodents is rare.

In each age group, the actual population at risk of exposure to a carcinogen is much smaller in humans than in animals, and the decrement over time follows different curves. Therefore, CREs will overestimate the number of cancers attributable to exposure. One possible way of dealing with species-specific mortality patterns might be to apply a decremental life-table technique when calculating a CRE for humans.

3. SHORTER-THAN-LIFETIME RISK

Risk estimates are based on theoretical concepts that generally have little similarity with real-life situations. Rarely are workers in the field faced with the question of what the excess lifetime risk (ELR) would be for the general population if it is exposed for a lifetime to a single carcinogen. Instead, one is confronted with questions such as may arise from the following real situation:

A city well field, serving a population of 35,000 and private wells in adjacent neighborhoods are contaminated by solvents, including several carcinogens, originating from a known source nearby. Since the contamination level is in the tens to hundreds of parts per billion, the following emergency remedial actions have been taken: the most heavily contaminated city wells have been closed, the production of water from the remaining few clean wells has been increased, a bottled water program has been started for the less than 250 residents with contaminated private wells, and residences with contaminated wells have been connected to the city water system. Two options are available for the interim period before a definitive remedial action can be implemented: (1) drill new city wells adjacent to the endangered field and pump to waste the water from barrier wells after treatment with activated carbon or (2) maintain the status quo, since the time of the interim period would not be longer than about 1 year. The field health official is placed for a difficult problem: to determine whether the prevailing hazard is serious enough to warrant the first option, a multimiillion-dollar interim solution, or whether the ELR posed by an exposure for only 1 year is within the limit of acceptability, which would lead to the no-action option. This would be a situation par excellence for applying a CRE to avoid or minimize emotional and subjective considerations.

However, an effort to calculate a CRE for 1 year would be impeded because usually a CRE is available only as an ELR for a lifetime exposure. To deal with shorter-than-lifetime exposures, the Carcinogen Assessment Group (CAG) of the U.S. Environmental Protection Agency (EPA) has developed the multistage-Weibull model, which can be expressed by the following equation:

$$R = 1 - \exp - (B \cdot D \cdot W^{1/3} \cdot ((t-s)^{k-1} - (t-f)^{k-1}))$$

in which R = risk for the period of concern, usually the average life span

B = chemical-specific potency (the slope of the model)
D = dose in mg/kg/day

W = average weight (in kg) during the period of exposure
t = age at end of observation period, usually 70 year (life span)
s = age at start of exposure
f = age at which exposure stopped
k = number of stages involved in the transformation to a cancer cell

From the above equation, we can see that for fixed values of B, D, t, an k, the risk will be greater for lower values of s --that is, the earlier in life the exposure starts. This quantitative implication of the risk model has been worked out in Table 1 for various values of s and exposure periods of 1 and 10 years, with k = 5. This model assumes that if the exposure is discontinued, the past exposure continues, for a lifetime, to accrue incremental risk. However, there is evidence from human and animal studies that post-exposure accrual of risk may be arrested or substantially slowed down, and that the probability of a beneficial effect of discontinuation of exposure may increase if the cessation starts earlier in life (1,2).

Figure 1 shows that the risk of bladder cancer in mice fed 2-acetylamino-fluorine is reduced dramatically if the exposure is stopped at various points in time, even as late as 15 months after exposure, which is comparable to the fifth decade in humans. The reduction is close to complete, in other words, little additional incremental risk is accrued once the exposure is stopped (1). The reduced risk of lung cancer when a person stops smoking is well established (2), and is one of the arguments in stop-smoking campaigns. Reduction of additional incremental risk following discontinuation of exposure has also been observed for nasal cancer caused by occupational exposure to nickel (2), DDT-induced liver cancer in mice (3), and skin cancer in mice caused by local application of benzopyrene (2).

In the light of these facts, the use of a model that does not recognize the possibility of a reduction of additional incremental risk accrued after cessation of exposure, may constitute an application of a safety factor of unknown magnitude. Interestingly, risk reduction has not been observed after radiation given to humans for the treatment of pulmonary tuberculosis, ankylosing spondylitis, and uterine bleeding has been discontinued (2). This suggests that the mechanism leading to the development of cancer by radiation may fundamentally differ from the pathway of chemical carcinogenesis. At least one example is known in which the ELR is already determined early in life. Mice fed 2-AAF showed only little reduction of the ELR of liver cancer after the exposure was discontinued after a period of 9 months (1). It would considerably increase our understanding of the process of carcinogenesis if efforts

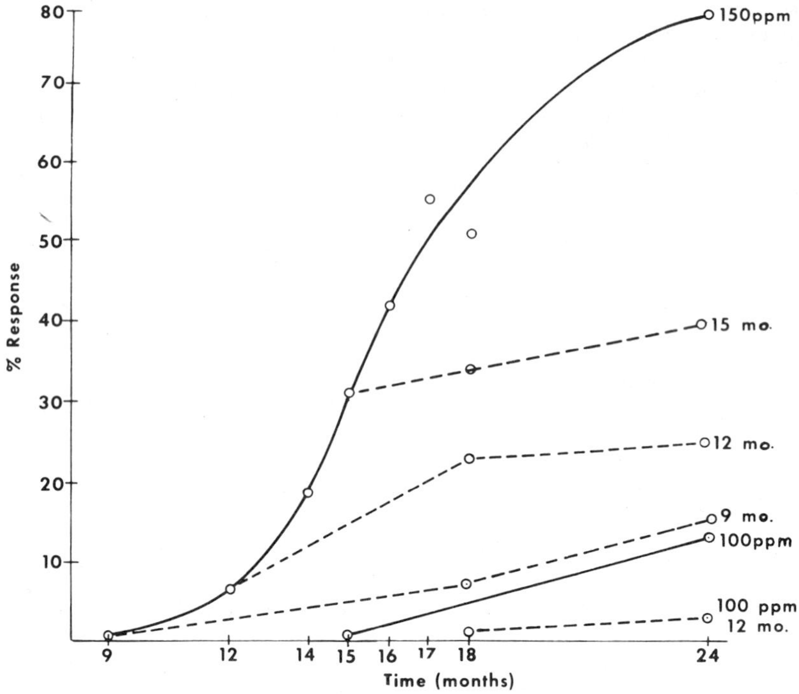

Figure 1

Cumulative incidence of bladder cancer in female Balb/c mice fed 2-acetylamino-fluorene, starting at 4-5 weeks of age and continuing for periods of 9, 12, and 15 months, with observation periods of 18 and 24 months. The solid lines represent the period of exposure, the broken lines the period since cessation of exposure.

NOTE: the authors acknowledges the authors and the publisher (Pathotox Publishers) of reference 1 for the use of their figure 5.

could be made to investigate what type of animal cancers and associated carcinogens would fit or not fit the multistage-Weibull model. The difference in cancer mortality in animals and humans due to therapy will not be discussed in this paper.

4. CREs AS A CRITERION FOR REMEDIAL ACTIVITIES AT HAZARDOUS WASTE SITES

Each time environmental pollution causes a hazardous situation for residents or workers, local public health and environmental authorities must decide whether or not to begin remedial actions. Health authorities can do little more than restrict access to the contaminated area, an offer bottled water or a connection to clean public water mains in case of contaminated private wells. Often, health studies and environmental surveys are contemplated when authorities receive urgent requests from people in the endangered areas. Such activities and the potential cleanup may be extremely expensive in relation to the magnitude of the predicted or expected health effects. CREs again seem to be the preferred tool to

Table 1

Excess lifetime cancer risk caused by exposure to carcinogens for periods of 1 and 10 years relative to risk for lifetime exposure, calculated with the multistage-Weibull model (k=5) at various starting ages of exposure.

Age at exposure	Risk for 1 year relative to lifetime exposure	Age at exposure	Risk for 10 years relative to lifetime exposure
0 - 1	2.733×10^{-2}	0 - 9	2.667×10^{-1}
10 - 11	2.786×10^{-2}	10 - 19	2.254×10^{-1}
20 - 21	2.021×10^{-2}	20 - 29	1.426×10^{-1}
30 - 31	1.081×10^{-2}	30 - 39	7.175×10^{-2}
40 - 41	4.504×10^{-3}	40 - 49	2.699×10^{-2}
50 - 51	1.301×10^{-3}	50 - 59	6.373×10^{-3}
60 - 61	1.508×10^{-4}	60 - 69	4.384×10^{-4}

use in deciding if a study or a cleanup is warranted. However, it appears that in such situations public health and environmental authorities often apply CREs differently.

Public health authorities may start actions when citizens begin voicing their concerns, which are not necessarily related to the concentration of the contaminants or to the magnitude of the CRE. Environmental authorities tend to let their actions be determined by several factors, among which the CRE and the Acceptable Daily Intake (ADI) are probably the most important criteria, because they are considered objective and chemical-specific measures of the seriousness of the hazard. In other instances, authorities and the public recognize that many uncertainties are associated with CREs and interpret these uncertainties as evidence that CREs are not adequate quantitative measures of a carcinogenic hazard. These differences in opinion not only open the gate to emotional and never-ending discussions, but also may result in problems caused by the inconsistent use of CREs in health assessments.

It is insufficiently understood that CREs are not intended as absolute estimates of risk, but rather as rough indicators of the carcinogenic potency of a chemical. It is EPA's view that CREs "should not be regarded as accurate representations of the true cancer risks even when the exposures are accurately defined" (4). Derivatives of CREs, such as the Relative Potency Index, may be used in ranking chemicals according to their estimated capacity to induce cancers, in setting priorities for

Cleanup, and in making regulatory decisions. That CREs are nevertheless used in practice as accurate data that can be used as a criterion of acceptability of a contamination may have been promoted by CREs presented as simply estimates, thus suggesting that they would be point estimates. Presentation of point estimates of the risk rather than the upper confidence limit is necessary to prevent improper use of CREs.

Public health authorities tend to incorporate other issues in addition to CREs in their assessment of a health hazard. First, people are also concerned about health outcomes other than cancer. If carcinogens are present many officials ignore or deemphasize the hazard of noncancerous effects. It is usually argued that concentrations leading to remedial actions based on CREs are much lower than concentrations causing other toxicity. However, very little is known of noncancerous toxicity at low doses. The ADI, developed by the National Academy of Sciences (5), is based on the Lowest Observed Effect Level (LOEL) or the No Observed Effect Level (NOEL), to which an arbitrary safety factor is applied. Other than in cancer, no attempt has been made to predict toxic effects at levels below the NOEL, although this cannot be considered the true NOEL, based as it is on an observation of no effects in an animal or human population in the order of a few tens to a few hundreds.

It may well be possible that contamination levels considered to be acceptable with regard to cancer would yield an unacceptable risk of, say, liver or kidney toxicity. Further, it is likely that the compound effect of multiple chemicals, closely related in chemical structure and/or biochemical activity, on the same target organ is at least additive. This might render the hazard of, for instance, liver toxicity quantitatively more significant than that of cancer. This would simply follow from the fact that, in the typical field situation, there are more liver toxicants than carcinogens.

A second issue of consideration in a health evaluation is the likelihood of actual exposure. For instance, a CRE for volatile organic chemicals erroneously assumes an average daily intake of 2 liters of water, whereas the true average is more likely to be 1 liter (5). Further, by far the largest part of the daily water intake, certainly in this country, is in the form of soft drinks, hot beverages, or as a component of prepared solid food. In all these cases the water has been subjected to heat or to intensive contact with air, resulting in rapid evaporation. On the other hand, showers may pose a poorly quantified risk of exposure by inhalation.

Third, when the exposed population is of a limited size, should the predicted actual excess number of cancer deaths be preferred as a criterion for decisions on public health and remedial actions over the CRE? A risk expressed as a probability is poorly understood by the lay public, which is more responsive to absolute numbers. This is even more true for the concept of the cumulative ELR. To calculate, instead, a predicted absolute number of cancers on the basis of a CRE, one would need for each individual chemical the specific information on the slope of the model and the exponent of the time in the risk equation. Such information is very difficult to obtain, and we have already discussed the questionable value of the time model.

However, let us assume that these problems have been solved, an optimistic assumption indeed. The question then arises whether numbers of excess cancers should be estimated for each age and sex category in the population. This would be scientifically the only sound option, because the time model predicts different risk for each age group. The fact that males and females usually have different sensitivities to carcinogens reduces the value of an estimate based on a CRE for the most sensitive sex. The maximum error that could be made by the lack of sex-specific CREs is an overestimation with a factor of 2. The overestimation caused by ignoring the age distribution is larger, probably one or two orders to magnitude, dependent on the k-value in the risk model and on the age pattern of the actual exposed population.

Setting aside the issues related to the validity of the CRE and of the CRE-based excess number of cancer deaths, should public health authorities prefer either one of these measures? What would be the difference in impact on the public health and on the approach to the citizen's concerns of, say, an upper confidence limit of the CRE of 10^{-5} ELR and zero excess cases in 70 years in a population of 1,000? These figures are compatible and may therefore be compared. More precisely, would implementing a 10-million-dollar interim remedial action be justified in a situation where the ELR exceeds an arbitrary threshold of 10^{-6}, even though not one excess case in the actual population in a period of 70 years is predicted? Would the response to this question be different if it concerned the implementation of a health study at 10% of the above cost? What if the exposure is limited to 1 year, as in the example of a field situation given earlier? The irrelevance of an ELR of 10^{-7} would then become even more obvious.

Alternatively, many authorities counter by saying that the individual risk is the same regardless of the size of the population exposed. The lack of an unequivocal opinion from the scientific community on these questions explains the inconsistency of approaches of public health officials to similar situations, whereas environmental authorities prefer to let the value of the CRE be decisive.

Finally, most health authorities are familiar with working with occupational standards. Such measures often have the advantage of being based on a LOEL or a NOEL and take into account a safety factor and feasibility or economic consideration. Such a measure is, therefore, appealing to the intuitive preference of the public for a measure based on the practice of daily life rather than on scientifically sophisticated but unprovable theoretical concepts. Remarkably, CREs have not been used in occupational situations, whereas occupational standards have not been a topic of serious discussions for residential populations. It seems extremely worthwhile to examine the applicability and justification of occupational criteria for residents, as such criteria can easily be adjusted for age, the difference between a 40-hour workweek and a 7 x 24-hour week, the between a working life span of 40 years and the average human life span.

5. THE CANCER RISK POSED BY MULTIPLE AND SIMULTANEOUS EXPOSURES

In most actual situations of nonaccidental contamination people are exposed to a mixture of chemicals, which may include several carcinogens. It seems odd to evaluate the health hazard on the basis of the chemical with the highest CRE at the prevailing concentration. Yet, this approach is common and is reflected in EPA's current effort to promote discussions on how else to assess the risk at a site posed by a mixture of carcinogens (4).

Possibilities for the compound effect of multiple exposures include additivity, multiplicativity, synergism, and antagonism. However, discussions on this topic can contribute little to the solution of the problem, since only very few experiments have been done with a combination of carcinogens. Further, the issue of the risk of a combination of one (or more) carcinogen with one (or more) other chemical has rarely been raised. Yet, such a combination may be even more hazardous than the carcinogen itself. For example, alcohol itself does not cause cancer of the upper digestive tract (mouth, pharynx and esophagus), but the abuse of alcohol among smokers substantially increases the risk of cancer at these sites.

From a public health standpoint, it is reasonable to assume at least an additive effect if people are exposed to more than one carcinogen acting on the same target organ. This approach, however, is full of practical an theoretical problems. Consider the real situation of exposure to drinking water from an aquifer contaminated with trichloroethane (TCA), tetrachloroethylene (PCE), dioxane, benzene, and fluoranthene. Such a number of carcinogens is certainly not unusual. The following problems arise. Although a CRE has been issued (4), TCA has not been classified by EPA as a carcinogen because its Scientific Advisory Board considered the outcome of the cancer bioassay inconclusive. The validity of the results of the bioassay of PCE, and thus of the CRE, has been questioned because of observed irregularities in the test performance (4). Dioxane is an animal carcinogen (6), but it is not included in the carcinogens mentioned in EPA's Water Quality Criteria Document (7), and a CRE is not available. Fluoranthene is neither included in the former reference, nor is it listed as a carcinogen by EPA. However, the chemical is a polynuclear aromatic hydrocarbon (PAH), one of a group of chemicals that include many strong animal and human carcinogens, and EPA has calculated a CRE for PHAs as a group (7). Further, the group of six carcinogenic PAHs for which a World Health Organization European Standard for Drinking Water has been developed includes fluoranthene. These contradictory evaluations leave it to one's personal judgement whether fluoranthene is a carcinogen and whether the CRE for the group of PAHs can be used. In summary, of the five chemicals above, only benzene appears to have a CRE free of practical complications.

A more basic problem associated with the use of CREs in any form of arithmetic, and this includes the summation of CREs for the purpose of estimating an aggregate risk, is that the EPA's CREs are upper confidence limits of the point estimate of the risk. Since CREs are directly derived from test outcomes, it is unavoidable that the confidence limits of the point estimate differ for each individual chemical. Further, there appears to be no consistency in the determination of the basic entries into the risk model (mortality rate and dose) if premature noncancer mortality and changes in dosing occur. Adding up nonuniformly derived CREs renders the credibility of the aggregate risk dubious. The summation of CREs that are upper limits of different confidence ranges would be an unacceptable statistical flaw.

The evaluation of the aggregate effect of exposure to multiple carcinogens, therefore, may initially seem a simple task, but in reality it abounds with problems for which, as yet, there are no generally accepted solutions. Nevertheless, the summation of CREs might be a possible solution to the problem, warranting investigation if:

(a) CREs are presented as point estimates rather than as confidence limits;
(b) information is released with regard to the basic entries used for the derivation of the CRE, to ensure uniformity of methodology;
(c) details of the performance and the results of the animal test or the human study used for deriving the CRE are published, which would enable investigators to evaluate the quality of the evidence for carcinogenicity.

6. CONCLUSION

Although the concept of the CRE as an objective quantitative measure of the hazard posed by carcinogens is attractive, it appears that the CRE involves many inherent uncertainties. Some of these may lead to an underestimation of the true risk and others would result in overestimation. The total result is likely to be an overestimation, the extent of which is unknown. The essential deficiencies in the CRE appear to be the inadequacy in dealing with the effect of time on the risk, and the format of the CRE that does not allow for age- and sex-specific sensitivities. These deficiencies render the CRE in its present form very difficult to apply to practical situations of a contaminated area with a limited residential population, an shorter-than-lifetime exposure, and a mixture of chemicals. The White House Office of Science and Technology Policy concluded that (8) "It is apparent that human cancer risk assessment is still in an evolutionary state and that a number of emerging areas of science will impact on this process in the future." It is obvious, that more interaction is needed between those who develop and those who are called upon to apply CREs so that these issues can be fully explored.

REFERENCES

1. Littlefield, N.A., Greenman, D.L., Farmer, J.H., and Sheldon, W.G. "Effects of Continuous and Discontinued Exposure to 2-AAF on Urinary Bladder Hyperplasia and Neoplasia. J. Environ. Pathol. Toxiocl. 3:17-34, 1979.
2. Day, N.E., and Brown, C.C., Multistage Models and Primary Prevention of Cancer. J. Natl. Cancer Inst. 64:977-989, 1980.
3. Tomatis, L., Tutosov, F., Charles, R.T., Boiocchi, M., and Gatt, E. "Liver Tumors in CF-1 Mice Exposed for Limited Periods to Technical DDT." Z. Krebsforsch 82:-25-35, 1974.
4. Environmental Protection Agency. "National Primary Drinking Water Regulations; Volatile Synthetic Organic Chemicals; Proposed Rule Making." Federal Register, June 12, 1984, pp. 24330-24355.
5. National Academy of Sciences. "Drinking Water and Health." National Academy Press, Vol. III, 25-61, 1980.
6. National Cancer Institute, "Bioassay of 1,4-dioxane for Possible Carcinogenicity." NCI Technical Report Series 80-1978, DHEW (NIH) 78-1330.
7. Environmental Protection Agency. "Water Quality Criteria Documents; Availability." Federal Register, November 28, 1980, pp. 79318-79379.
8. Executive Office of the President, Office of Science and Technology Policy. "Chemical Carcinogens; Review of the Science and its Associated Principles." Federal Register, May 22, 1984, pp. 21594-21661.

MODEL UNCERTAINTY:

IMPLICATIONS FOR ANIMAL LOW-DOSE CANCER RISK ASSESSMENT EXPERIMENTS

S.C.Y. Wong

c/o E.Y. Chai
Shell Development Company
Houston, Texas 77001

ABSTRACT

Animal experiments often provide part of the data base for assessing the carcinogenic potential of low levels of environmental carcinogens. The actual low-dose risk cannot be measured directly from quantal data experiments. Yet model uncertainty poses difficulties to risk assessors for low-dose extrapolation. Three implications of model uncertainty for risk assessors are: (1) there is a need to use a rich model in risk analysis for valid inferences on the extrapolation, (2) it is necessary to include both the model and the sampling uncertainties in the indicator of uncertainty for extrapolations from risk analysis, and (3) experimental design needs to be investigated under rich model and realistic sample size assumptions. Based on the first two implications, an inference procedure is proposed. The procedure has a wider scope of applicability, includes a goodness of fit test for the model, and provides a better description and confidence interval for the extrapolation than the traditional approach. Based on the third implication, experimental design is examined with a design criterion which allows for both the model and the sampling uncertainties. Issues on how best to reduce the total uncertainty and the trade-offs between the model and the sampling uncertainties are examined.

KEY WORDS: Quantal Animal Carcinogenicity Bioassays; Model Uncertainty; Sampling Uncertainty; Inference Design

1. INTRODUCTION

Animal experiments often provide part of the data base for assessing the carcinogenic potentials of low levels of environmental carcinogens. Typically, quantal data based on some carcinogenic endpoint are observed during a suitable time period for animals exposed to high doses of the carcinogen. The number of responding animals and the total number of animals are recorded for each group. Extrapolations to the low-dose region within the same animal species are then made based on some model of the dose response curve. Of interest to the experimenter is the cancer risk or added risk at a given low dose, or the dose corresponding to a certain low risk or added risk level.

This paper examines some issues of model uncertainty for the low-dose extrapolation problem. A number of difficulties for risk assessors due to model uncertainty are described in Sec. 2. Implications of model uncertainty on low-dose inferences are discussed in Sec. 3. An inference procedure is proposed in Sec. 4 and its advantages to risk assessors for the low-dose extrapolation problem are discussed in Sec. 5. Implications of model uncertainty on the design of animal low-dose cancer risk assessment experiments are described in Sec. 6. Some design results based on a `rich' model are given in Sec. 7.

2. MODEL UNCERTAINTY: DIFFICULTIES FOR RISK ASSESSORS

The actual low-dose risk cannot be measured experimentally. Some form of mathematical model, be it simple or elaborate, is needed for extrapolation to low doses.

Many parametric models are currently used for cancer risk assessment. Each parametric model is described by one functional form. Examples are the probit, logistic, Weibull (extreme value), multistage, 1-hit (linear), gamma multi-hit, and the threshold models. Different models give low-dose extrapolations often differing by many orders of magnitude.

The choice of a parametric model is highly uncertain. Opinions on these parametric models vary (Crump, et al., 1976; Hartley and Sielken, and discussion, 1977; Food Safety Council, 1980; Carlborg, 1981; Armitage, and discussion, 1982; Environmental Protection Agency, 1982; Hoel, et al., 1983; Park and Snee, 1983). The diversity in the parametric models and in the opinions about them suggest a lack of solid information about the functional form of the true dose response curves in general. It is generally agreed that current parametric models can have little biological bases, with the pharmacokinetics and carcinogenic mechanisms not very well understood (Park and Snee, 1983). The empirical data generally fit the parametric models equally well and cannot identify the specific parametric model to be used.

The model uncertainty has therefore posed great difficulties to risk assessors in the choice of a specific parametric model for low-dose extrapolation.

3. IMPLICATIONS OF MODEL UNCERTAINTY ON LOW-DOSE INFERENCES

The first implication of model uncertainty for risk assessors is the need to use a rich model in risk analysis.

The need is apparent if we examine the role of a model in risk anlysis. There are several ways in which a model can be viewed and used. A model can be used (1) for description of the dose response curve, (2) as a representative of theories to be tested, or (3) as a summary of the information and uncertainties prior to the experiment, so prior information summarized in the model is utilized in the low-dose inferences and uncertainties can be removed with information from the experimental data. Since inferences on the extrapolation are always made within the framework of a model, prior information summarized in the model affects the low-dose inferences. Valid low-dose inferences can only result if the model includes the true dose response curve. The little knowledge about the carcinogenic processes means that a rich model provides a more accurate description of the knowledge and ignorance about the true dose response curve prior to the experiment. The use of a rich model maintains a higher chance for valid low-dose inferences. The variety of dose response curves observed and the diversity of mechanisms of action of carcinogens also rquire the model to include a wide variety of descriptions and hyptheses for investigation as to what might be consistent with the data.

One example of a rich model is the pool of all the proposed parametric models. Taken as a diverse collection of descriptions, hypotheses, or approximations to the possible true dose response curve, the rich model circumvents the problem of having to choose with little basis a specific parametric model for extrapolation.

The second implication of model uncertainty for risk assessors is the necessity to include both the model and the sampling uncertainties in the indicator of the uncertainty for extrapolations from a risk analysis. The issue is first discussed in Deisler (1982).

Model uncertainty is the uncertainty in the true dose response curve when the experimental response rates are exactly known. It arises from not knowing which is the correct parametric form for the prior information and reflects the inadequacy in the number of experimental doses.

Sampling uncertainty is the uncertainty in the response rates at the experimental doses. It arises from the fluctuation of the data and reflects the inadequacy in the number of samples at the experimental doses.

Accurate assessment of the uncertainty in the extrapolation is essential for decision making and directing efforts. If a specific parametric model is chosen, the derived indicator of uncertainty reflects only the sampling uncertainty based on the validity of the assumed parametric model. The model uncertinaty is therefore not accounted for in the traditional uncertianty indicator. Whereas the sampling uncertainty from the quantal data may already be large (Park and Snee, 1983), a large part of the uncertainty in the extrapolation derives from the uncertainty in the specific parameteric model (Armitage, and discussion, 1982). A combined uncertainty index therefore offers a more realistic description of the uncertainty associated with an extrapolation.

4. AN INFERENCE PROCEDURE

Based on the two implications of model uncertainty mentioned, an inference procedure extending Korn's procedure (1982) is proposed.

The procedure provides useful tools for risk assessors to extract various types of information needed for cancer risk assessment and subsequent risk management analysis. The procedure includes a goodness of fit test for the model and provides a best description and a confidence interval for the extrapolation. The goodness of fit test checks the plausibility of the model against the data. Once the model is accepted, the best description and the confidence interval estimate the extrapolation. The latter accounts for the sampling uncertainty in the data. The upper limit of the risk estimate, if low enough, serves as an indication of safety. The lower limit, if high enough, is an indication of risk. If the model is appropriately formulated, the width of the interval in a logarithmic scale indicates the uncertainty in the low-dose inference. More weight can therefore be given to assessments with lower uncertainties.

We will illustrate the inference procedure with the two dose case in Figure 1. The mathematical descriptions and details for an arbitrary number of experimental doses are given in Wong and Chai (manuscript, 1984).

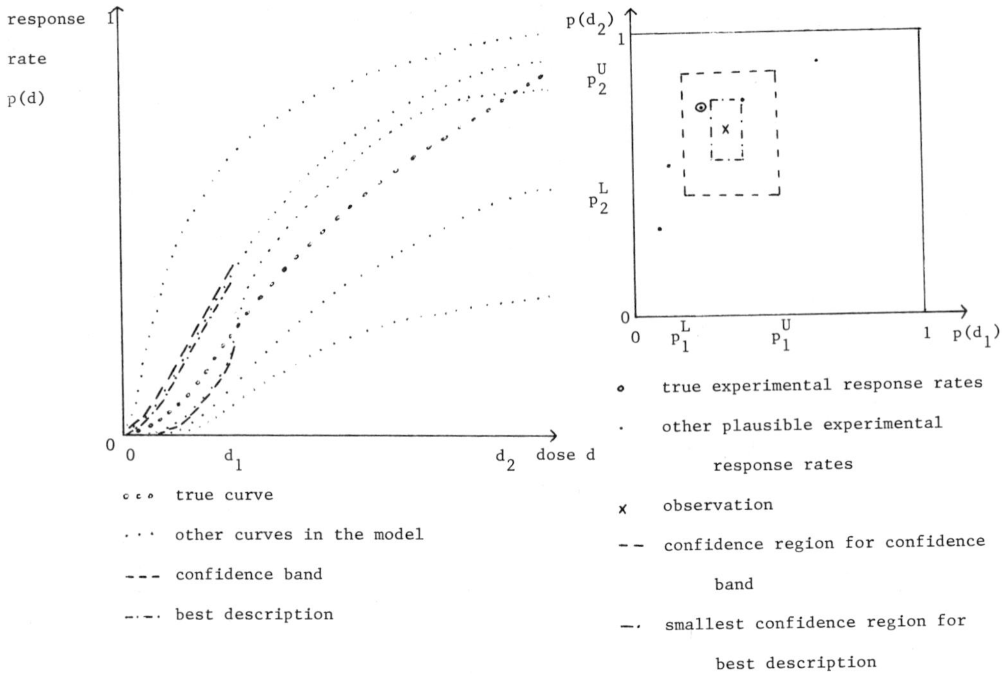

Figure 1

The Proposed Inference Procedure

Basically, a model is a collection of dose response curves considered plausible prior to the experiment. The model profile set R_M contains all the profiles of the dose response curves in the model at the experimental doses. The arbitrary model in the illustration has six dose response curves, one of which is the true curve. There are two experimental doses d_1 and d_2. The profile (i.e., the response rates at the experimental doses) for each curve is plotted in the unit square on the right hand side with the response rate at the lower dose $p(d_1)$ on the horizontal axis, and that at the higher dose $p(d_2)$ on the vertical axis. The model profile set R_M has all the combinations of experimental response rates considered possible under the model prior to the experiment.

The experiment leads to an observation marked X in the unit square. A rectangular confidence region R_D can be constructed to include all experimental response rate combinations consistent with the data. Random fluctuation of the data is allowed for in the confidence region, but no consideration is given to the model. The method of construction of the confidence region R_D is to form confidence intervals (p_1^L, p_1^U), (p_2^L, p_2^U) for the experimental response rates independently and at a fixed joint confidence level to allow for the sampling uncertainty. Such confidence intervals for the binomial parameters can be constructed based on the F-distribution (Johnson and Kotz, 1969), so they hold even for small samples.

The model profiles are therefore further qualified for consistency with the data allowing for sampling uncertainty in the data. Dose response curves in the model with profiles consistent with the data are used to span the confidence band for the extrapolations. Prior and data information are therefore combined in the assessment. In the illustration, the two model profiles within the confidence region are compatible with both the prior and the data information. Corresponding to these two profiles are three dose response curves. The highest and the lowest response rates at each inference dose can then form a confidence band for the extrapolations. Extrapolations within the band are possibilities under the prior information and are not rejected based on the information and the sampling uncertainty in the data.

The goodness of fit test for the model checks for coherence between the model profile set R_M and the confidence region R_D. There may not be any model profile within the confidence region for some experimental outcomes. The data is then incompatible with the model.

The best description, not necessarily a point estimate, estimates the extrapolation based on the curve(s) in the model in closest agreement with the data. The size of the confidence region R_D depends on the confidence level. Given that the model is accepted, the dose response curve(s) with profile(s) in the smallest R_D which overlaps R_M provide(s) the best descritpion for the extrapolation.

5. ADVANTAGES OF THE INFERENCE PROCEDURE TO RISK ASSESSORS

5.1 On the Model Issues

The procedure widens the scope for risk analysis. It applies to a wide variety of models representing different levels of prior information that may be appropriate for different spescific cases. The model can be parametric or nonparametric, with one or more functional forms, or with or without constraints on the model parameters. The prior information for a specific case can therefore be more appropriately represented. As discussed above, taking the pool of all proposed parametric models can enhance the validity of the low-dose inferences and avoid the problem on the choice of a specific parametric model.

5.2 On the Types of Inferences

Unlike some inference procedures which provide only an upper limit for the response rate or added risk and a lower limit for the dose with a low risk or added risk, our procedure provides a more complete assessment. It tests for goodness of fit of the model and provides an upper limit, a lower limit, and a best description for the extrapolation.

5.3 On Design and Data Restrictions

The procedure can be applied to any set of quantal data, with no restriction on the experiment with regard to the sample size or the number of experimental doses.

Many traditional inference procedures require certain design and data restrictions. These limit the scope in which inferences can apply. Examples are: data with all responses in some groups and all non-responses in the other groups; occasional non-monotonicity in the observed response rates with increasing dose; large sample assumptions which may not hold for practical sample sizes; the requirement that the number of experimental doses have to exceed the number of model parameters for a goodness of fit test to apply; the requirement that the number of experimental doses have to be at least that of the model parameters for a feasible point and resulting interval estimate.

In a rich model situation with many parameters, the requirement on a large number of doses may not be feasible or optimal given finite resources. The less than required number of experimental doses often make traditional inferences impossible, or force the experimenter to limit the choice of the model to fairly restrictive types of models. The choice of the model and the choice of the number of doses are two separate issues. The model should be formulated according to its use. The number of experimental doses may depend on the model because of efficiency considerations. Neither choice should be dictated by the limitation of the inference procedure.

5.4 On the Best Description and the Interval Estimate

The best description and the interval estimate from the procedure have features more desirable than the traditional estimates, especially for rich model situations.

In a rich model situation, there is often some arbitrariness in the best description from the traditional approach. The traditional best description is often a maximum likelihood point estimate. In the multistage model when the number of stages is large, numerous estimates are equally qualified as best descriptions (Carlborg, 1981). The same applies to a rich model situation. There may be many dose response curves in the model equally best in describing the data. The choice of a point estimate as the best description is hence arbitrary. The best description from our procedure is not necessarily a point estimate. The arbitrariness in the point estimate is therefore made explicit.

The best descriptions from the traditional approach are sometimes difficult to interpret. The risk estimate from the linearized multistage procedure partially accounts for the sampling uncertainty. It is hence a partial upper estimate, but is often treated as a best description.

Descriptions from unbiased assessment are needed for proper risk management analysis. Cornfield is against built-in conservatism for cancer risk assessment (Armitage, and discussion, 1982). There is also the need to separate scientific from policy-related activities (Park and Snee, 1983). The aribtrariness in the point estimate or mis-interpretation in the role of the descriptor should hence be avoided.

The bias in the point estimate and the poor large sample approximation for its sampling distribution often affect the interval estimate from the traditional approach. The interval estimate from the traditional approach is generally based on the maximum likelihood point estimate and the large sample approximation to its sampling distribution. The properties of the point estimate and the large sample approximation under the multistage model have been examined (Portier and Hoel, 1983). The point estimate for the dose with a given low risk or added risk is shown to be biased even for large samples for certain true curves. The sampling distribution of the point estimate is multi-modal in general. The large sample approximation to the sampling distribution is poor even for large samples. The interval estimate and the coverage probability of the interval are hence affected.

Contrary to the traditional approach, our interval estimate does not depend on the feasibility to obtain a point estimate and applies to any sample size without using any large sample approximation. The arbitrariness of the point estimate in a rich model situation, the bias in the point estimate, and the performance of the large sample approximation hence do not affect the interval estimate. The interval estimate has an exact coverage probability for any sample size. This exactness provides some guarantee on the performance of the inference procedure.

5.5 On the Uncertainty in the Extrapolation

The width of the confidence interval, in a logarithmic scale, accounts for both the model and the sampling uncertainties. The indicator therefore gives a more realistic assessment of the uncertainty in an extrapolation from risk analysis.

The inclusion of the model and the sampling uncertainties in our uncertainty indicator can be shown easily if we take the pool of the proposed parametric models as the model. The dose response curves within a vicinity of the data are used to span the confidence band irrespective of the specific parametric model of origin. The uncertainty from not knowing which is the correct parametric model is hence included. The vicinit of the data within the confidence region accounts for the sampling uncertainty. Increasing the sample size to infinity at each experimental dose would reduce all the sampling uncertainty. But even if we have an infinite sample at each experimental dose, in a rich model situation, we may still have many functions corresponding to those true experimental response rates. The uncertainty about the true function when we know exactly our experimental response rates is the model uncertainty.

6. IMPLICATIONS OF MODEL UNCERTAINTY OF EXPERIMENTAL DESIGNS

The third implication of model uncertainty for risk assessors is on experimental designs.

The uncertainty in the parametric models and the practicality of experiments with a reasonable number of doses and samples necessitate the study of practical experimental designs under rich model assumptions. Previous designs are investigated under rather restrictive types of models and are often based on large sample assumptions. Our inference procedure does not depend on any large sample assumption, has no minimum requirement on the number of experimental doses, and applies to rich models. Experimental designs can therefore be investigated under the rich model and practical sample size situation without being limited by the inference procedure.

The expected width of the confidence interval, in a logarithmic scale and conditional on the acceptance of the model, can be used as an uncertainty index to evalute experimental designs. The uncertainty index, by incld
uing both the model and the sampling uncertainties, is therefore a more realistic measure for evaluating experimental designs under rich model situations.

Given finite resources, the distribution of the samples over a number of experimental doses necessarily imply a trade-off between the model and the sampling uncertainties post-data. Recall that sampling uncertainty can be reduced by increasing the sample sizes at the experimental doses and model uncertainty can be reduced by increasing the number of experimental doses. With a few exceptions, the current practice is to experiment with one or two doses besides the control. The use of a design criterion which includes the model and the sampling uncertainties would enable us to investigate how best to reduce the total uncertainty in the extrapolation and the trade-offs between the model and the sampling uncertainties under a rich model situation.

7. SOME DESIGN RESULTS UNDER A 'RICH' MODEL

Investigation can be made on the experimental design problem under a rich model situation. The rich model may be the pool of all or some of the proposed parametric models as deemed appropriate for the situation. Work done so far is based on the k-stage model with an arbitrarily fixed large k. The model is being used rather extensively for cancer risk assessment and has been recognized as a fairly rich model. We will therefore take this model as the first approximation to a rich mode.

Even though computer programs for the analyses and the experimental design investigations have been written for any arbitrarily specified k and any number of experimental doses, only a limited number of cases have been studied in relation to the experimental design problem. The study assumes a 6-stge model with or without background. Four true functions of the 6-stage model have been examined: a 1-hit function without background,

$$p(d)=1-\exp(-2.3026d)$$

a 1-hit function with background,

$$p(d)=1-\exp(-.10536-2.1972d)$$

a 6-hit function without background,

$$p(d)=1-\exp(-2.3026d^6)$$

and a 6-hit function with background,

$$p(d)=1-\exp(-.10536-2.1972d^6)$$

Because of the practicality and the current usage, one or two dose designs with or without the control are investigated. Doses with cancer risks of .1, .3, .5, .7, amd .9, and sample sizes of 12, 60, and 240 are used in the investigation. The expected width of the confidence band of the extrapolation, in logarithmic scale, is used as the uncertainty index to evaluate experimental designs. Sometimes the reciprocal of the uncertainty index, an information index, is used instead.

The following results illustrate some of the considerations for reducing the total uncertainty in the extrapolation when designing for low-dose cancer risk assessment experiments. Details of the study are given in Wong and Chai (manuscript in preparatioh, 1984).

It is well known that uncertainty can be reduced by increasing the sample size. Given any design with fixed number of doses and dose levels, an increase in the sample size at any of the experimental doses would reduce the average size of the confidence region. On the average, fewer dose response curves in the model will qualify for the extrapolation leading to a narrower average band width for the extrapolation.

The degree to which a larger sample helps to reduce the uncertainty should be carefully examined for efficient use of resources.

One factor for consideration is the extrapolation distance. The extrapolation distance is defined as the inference dose relative to the highest experimental dose. Generally speaking, uncertainty increases with extrapolation distance. Depending on the true function, uncertainty can increase very rapidly with the extrapolation. Much of the information could be lost early on in the extrapolation. Short extrapolations are generally susceptible to improvements from larger samples. Distant extrapolations, on the other hand, may be very insensitive to the increase in samples at the experimental doses.

Another situation in which the increase in samples may not

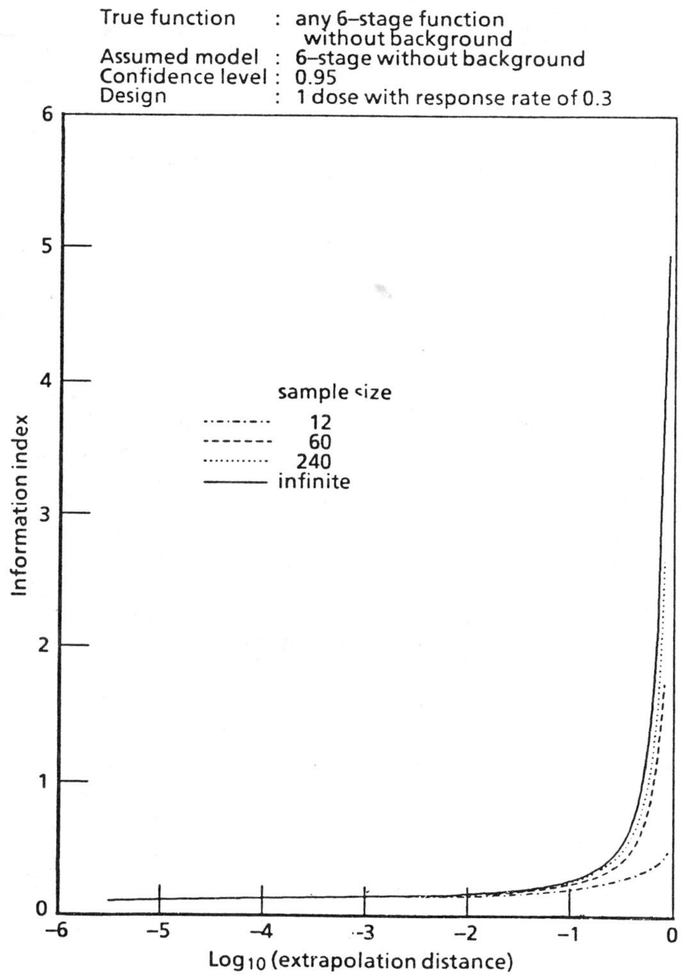

Figure 2

Information Index for Various Extrapolation Distances and Sample Sizes

effectively reduce the uncertainty in the extrapolation is in cases with model uncertainty. Given a number of doses and the dose levels, the model uncertainty represents the uncertainty in the extrapolation at infinite samples. The total uncertainty therefore approaches the model uncertainty as the sample size increases. The rate of such convergence may depend on the design. Fast convergence implies that a relatively small sample may have achieved much of what a large sample offers.

An illustration of the two situations above regarding extrapolation distance and model uncertainty is given in Figure 2. Under the conditions specified in the graph, the information index drops below 0.2 after extrapolating one order of magnitude below the experimental dose. An information index of 0.2 implies that the upper and lower limits of the confidence band are apart by 5 orders of magnitude. Hence, for long extrapolations, the increase in sample size does not improve the information. For short extrapolations, the improvement after a moderate sample size is rather limited due to model uncertainty.

Another factor for consideration is the true underlying dose response curve. When the underlying dose response curve is a 1-hit function, with or without background, the true function is fairly close to the average upper limit. A larger sample helps to reduce the uncertainty for both short and distant extrapolations essentially by raising the average lower limit rapidly when a proper design is used. An over-stated k in the k-stage model can therefore be compensated by a larger sample and a proper design. On the contrary, when the underlying dose response curve is a 6-hit function, with or without background, the true function is closer to the average lower limit. There can be little improvement from a larger sample due to slow lowering of the average upper limit, and the necessity for the average lower limit to remain below the true function. Hence for highly non-linear true dose response curves, the increase in samples may not be efficient in reducing the uncertainty in the extrapolation due to conservatism of the multistage model. These two situations are illustrated in Figure 3 and Figure 4.

Another way to reduce the total uncertainty is to increase the number of experimental doses since it reduces the model uncertainty. The additional experimental dose however means that a sample of a given size has to be distributed over more doses. As a result, the individual experimental response rates cannot be determined as precisely. Some of the two dose designs are not as efficient as some of the one dose designs. However, the second dose generally helps. There are always two dose designs more efficient than the one dose designs given appropriate setting of the dose levels.

Good designs allow more information to be acquired. The optimal location of the doses may depend on the sample size, the true function, and the inference dose. Prior information is therefore needed for optimal dosing. For one dose designs, with or without the control, the non-zero dose can be located at a medium risk level for small samples, and at a lower level as the sample size increases. Generally wide spacing between the doses in a two dose design is preferred. Even allocation of the samples, though not always optimal in every situation, is generally reasonable.

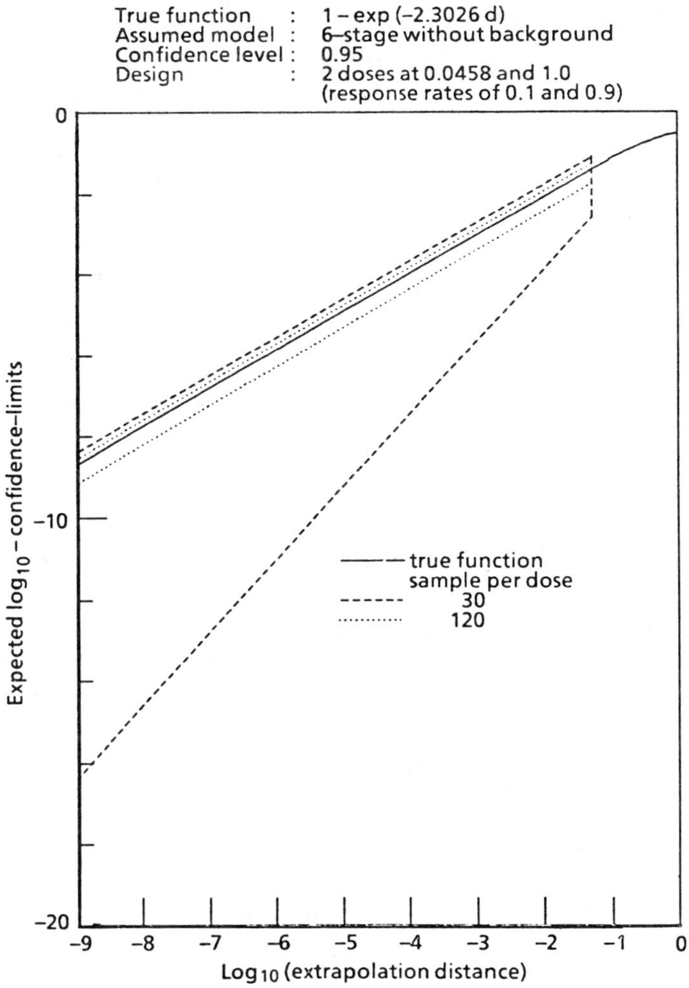

Figure 3

Expected Confidence Band Under The True 1-Hit Function

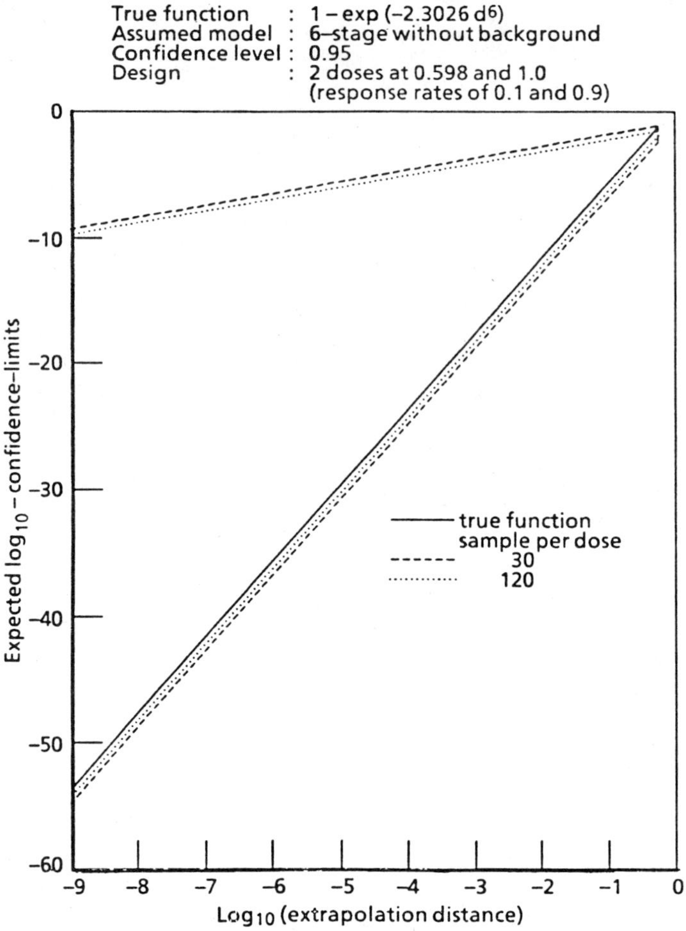

Figure 4

Expected Confidence Band Under The True 6-Hit Function

REFERENCES

P. Armitage, "The Assessment of Low-dose Carcinogenicity," Biometrics Supplement: Current Topics in Biostatistics and Epidemiology, 119-129 (1982).

F.W. Carlborg, "Multi-stage Dose-response Models in Carcinogenesis," Food and Cosmetics Toxicology 19, 361-365 (1981).

K.S. Crump, D.G. Hoel, C.H. Langley, and R. Peto, "Fundamental Carcinogenic Processes and Their Implications for Low Dose Risk Assessment," Cancer Research 36, 2973-2979 (1976).

P.F. Deisler, Jr., "A Goal-oriented Approach to Reducing Industrially-related Carcinogenic Risks," Drug Metabolism Reviews 13, 875-911 (1982).

Environmental Protection Agency, "Report on Workshop on Estimating Ambient Water Quality Criteria for Epigenetic Carcinogens," (1982).

Food Safety Council, "Quantitative Risk Assessment," In Proposed System for Food Safety Assessment, 137-160. Washington, D.C.: Food Safety Council (1980).

H.O. Hartley and R.L. Sielken, Jr., "Estimation of 'Safe Dose' in Carcinogenic Experiments," Biometrics 33, 1-30 (1977).

D.G. Hoel, N.L. Kaplan, and M.W. Anderson, "Implication of Nonlinear Kinetics on Risk Estimation in Carcinogenesis," Science 219, 1032-1037 (1983).

N.L. Johnson and S. Kotz, "Discrete Distributions," New York: Wiley (1969).

E.L. Korn, "Confidence Bands for Isotonic Dose-Response Curves," Journal of the Royal Statistical Soceity, Series C 31, 59-63 (1982).

C.N. Park and R.D. Snee, "Quantitative Risk Assessment: State-of-the-Art for Carcinogenesis," The American Statistician 37, 427-441 (1983).

C. Portier and D. Hoel, "Low-dose Rate Extrapolation Using the Multistage Model," Biometrics 39, 897-906 (1983).

S.C.Y. Wong and E.Y. Chai, "A Quantal Bioassay Inference Procedure and the Low Dose Extrapolation Problem," Manuscript (1984).

S.C.Y. Wong and E.Y. Chai, "Low Dose Extrapolation: Inference and Design Under the Multistage Model," Manuscript in Preparation (1984).

MULTIPLE TIME MEASURES ARE NECESSARY TO REDUCE UNCERTAINTY IN DOSE-RESPONSE MODELING: TIME- AND DOSE-MECHANISMS OF THE ED_{01} STUDY*

T.D. Jones, C.S. Dudney and P.J. Walsh

Oak Ridge National Laboratory
Oak Ridge, Tennessee 37831

ABSTRACT

Prompt cellular proliferation, resulting from toxic hyperplasia and/or normal tissue homeostasis, is required to complete carcinogenic initiation processes. Also, cell proliferation is required for carcinogenic potentiation. We have developed a mathematical model for tumorigenesis incorporating four dimensions: (1) dose, (2) exposure time, (3) sacrifice time, and (4) response. The model is based on the mechanism of compensatory cell proliferation. The fit was tested against data for tumors observed in 24,192 BALB/C mice fed 2-AAF at concentrations of 0, 35, 45, 60, 75, 100, and 150 ppm to time-of-sacrifice at 9, 12, 14, 15, 16, 17, 18, 24, and 33 months at the National Center for Toxicological Research (NCTR). Because this study was designed to measure a dose of 2-AAF needed to induce a 1% tumor rate, it has become commonly known as the ED_{01} study. The coefficients of the mathematical model were determined from the ED_{01} data and the model was then used to predict the time- and dose-development of tumors in a second NCTR study of mice fed 2-AAF at concentrations of 60, 75, 100, and 150 ppm for 9, 12, and 15 months, and then sacrificed at 18 and 24 months. Previous dose-response models have not described the ED_{01} study adequately. In contrast, this four-dimensional analysis fits the dose-response quite adequately (when time-on-feed is the same as time-to-sacrifice) and is also totally predictive (without any redetermination of numerical coefficients) for the feed-wait-sacrifice protocol.

KEY WORDS: 2-Acetylaminofluorene; ED_{01} Study; Dose-Response; Time-to-Tumor; Bladder Tumors; Liver Tumors; Unifying Concept; Carcinogenesis; Mechanisms; Promotion.

MODEL DEVELOPMENT

The general dose-response function (Jones et al., 1983) is assumed to be

$$R[D(\Delta)] = [I\{D(\Delta)\} \cdot A \cdot P\{D(\Delta)\} + R_o(\tau)] \cdot C\{D(\Delta),T,\tau\}, \quad (1)$$

*Research sponsored by the Office of Health and Environmental Research, U.S. Department of Energy, under contract DE-AC05-84 OR21400 with Martin Marietta Energy Systems, Inc.

where Δ is time on treatment, T is time from first treatment, τ is age, $I\{D(\Delta)\}$ is the neoplastic-initiation of cancer trending cells, A is a function comprising other biological parameters such as endocrine balance,, immumologic alterness, etc., $D(\Delta)$ is dose as a function of time, $R_o(\tau)$ is the age-profile of natural tumor incidence in a cohort, $P\{D(\Delta)\}$ describes the tumor promotion gained from the proliferating cells at a specific pathological site, and $C\{D(\Delta),T,\tau\}$ is a competing-risk factor which describes the survival of the population to the time where neoplasia is counted.

$I\{D(\Delta)\} = 0$ implies that initiated cells other than those expressed in the spontaneous tumor rate $R_o(\tau)$ are absent in the tissue-at-risk, while $I\{D(\Delta)\} > 0$ means that some initiated cells are present. These initiated cells do not enhance the control response $R_o(\tau)$ over the study time, but may be potentiated into neoplasia by additional promotion. It appears that the value of I above some threshold (perhaps one or a few cells per site of neoplasia) may not result in increased carcinogenesis. The fraction of animals with tumors resulting from treatment with a chemical initiator generally seems to be independent of the dose level of the initiation stress (at high initiating doses)--provided that those initiating doses do not contribute significantly to promotion. Obviously, very low initiating doses or even certain chemicals are less effective at inducing DNA repair, DNA synthesis, and cell replication. Treatment with these doses or weak chemical initiators may result in suboptimal initiation. Thus, the observed tumor frequency expressed in percent of animals with neoplasia may be different for low initiating doses or for different initiating chemicals followed by identical potentiating schedules.

For most practical applications, we have doubts that the measured neoplastic frequency following a promotion schedule is a sensitive function of $I\{D(\Delta)\}$ other than a Kronecker delta type response. Hence in our modeling studies to date, we have assumed $I\{D(\Delta)\} = 1$ which corresponds to optimal initiating dose treatments. Even if the tumor rate can be shown to be dependent upon the dose level of an initiating chemical as opposed to our speculation of this paragraph, then the equations of this paper should still be valid because the initiation response function $I\{D(\Delta)\}$ becomes contained in the factor "A".

The time-profile of neoplasia in $P\{D(\Delta)\}$ and $R_o(\tau)$ will be estimated by T^N (Druckrey, 1967) in order to index the zero-dose homeostatic-driven monoclonal expansion of neoplastic growth from a single initiated cell. The onset of neoplasia is hastened by potentiating insults that accelerate cellular proliferation rates, so that treatment with 2-AAF results in cell-growth kinetics that we simulate by the product of T^N and a factor which approximates the dose-induced compensatory cellular-proliferation at the pathological site of interest. T^N should be quite consistent with compensatory cell proliferation resulting from a single treatment, from an undosed protocol, or from a continuously and evenly dosed study. However, numerical values of N should differ for each treatment protocol. T^N is simple, easy to evaluate, and leads to a high degree of utility (but not complete generality as in the series of experiments implementing nonuniform $D(\Delta)$ schedules). Also, the T^N approximation (when extrapolated to very long time periods) can cause the response probability function to exceed unity, but this problem is more academic than practical.

The competing--risk factor $C\{D(\Delta),T,\tau\}$ describes the survival of the animals at risk. For low doses, no significant mortality is observed so that $C\{D(\Delta),T,\tau\}$ is unity. $C = 1.0$ was used in this application to the ED_{01} study, but elsewhere C has been estimated from biological mechanisms (Jones, 1984).

P{D(Δ)} is a model of compensatory cell proliferation. There is significant evidence from both in vitro and in vivo studies that the growth of a single neoplastically altered cell is controlled strictly by intercellular communication with normal cells until the single cell has divided into many daughter cells. For example, the work of Farber (1980) suggests that in liver a hyperplastic nodule of 2^{16} (or 10^5) cells or smaller is growth inhibited by normal organ tissue, but larger foci can break into autonomous growth. Furthermore, our empirical correlations over several years (Jones et al., 1983) indicate that the degree of potentiation appears to be linearly proportional to the quantity (1-L), where L is the survival probability of stem cells in the tissue-at-risk. The quantity (1-L) is an index of the amount of compensatory cellular proliferation necessary to maintain tissue homeostasis.

For the ED_{01} study, the general response function was simplified to

$$y(D,\Delta,T,\tau) = T^N \cdot [A \cdot \{1-\exp(-\alpha D-\beta D^2)\} + R_o(\tau)] \qquad (2)$$

where D = D(Δ). Coefficients N, α, and β may be positive or zero. Although the algebra resembles the usual two-stage model, this model is not the traditional two-stage model and usually will not be used or evaluated in any similar manner (Jones et al., 1983; Jones, 1984). The two-stage model can only be evaluated with regression analysis techniques, whereas the α and β in this model are often taken directly from biological measurements.

If the pharmacodynamic processes or rate constants do not change with different degrees of dosing (Hoel et al., 1983), then the model can be evaluated according to the pharmacological dose rather than the dose which actually reaches the intracellular molecules at risk. In other examples it seems necessary to develop a time- or concentration- threshold formulation which describes the transfer of pharmacological dose to dose to the cells-at-risk (Hoel et al., 1983). For example, low doses given at low-dose rates may be detoxified by a certain metabolic (e.g., blood --> liver --> excretory) pathway (Russell et al., 1982). At higher doses the primary metabolic route may not be able to completely adjust to high-dose rates. Thus, a new metabolic pathway (e.g., urinary) could be activated, resulting in exposures to cells previously protected from weaker insults.

Evaluation of the Model when 2-AAF was Fed to Sacrifice

The Model of Eq. (2) was evaluated from the sacrifice-only data of the ED_{01} study (Littlefield et al., 1980). Because the animals were only 3-4 weeks of age at the start of the study and then were placed on a continuous diet of 2-AAF up to the time of sacrifice, the three measures of time are essentially equal in this part of the ED_{01} study.

Thus, (although τ is the age of the animals when they are removed from the study, T is described as a time measure from the start of the experiment to sacrifice or time-to-tumor, and Δ is the time on 2-AAF), all three indexes were taken to be equal and as published in units of months. In Eq. (2), R_o represents the probability of spontaneous neoplasia before age τ. It is important to use the fitted model precisely according to the dose and time definitions upon which the model coefficients were evaluated, and to not change the defined concepts once the numerical coefficients have been determined. Such changes can be made, but only if one has a clear understanding of the delicate interrelationships of the different experimental and biological parameters.

The constants in the models were evaluated from a nonlinear regression procedure of weighted least-squares estimates. Statistical weights were assigned initially according to the inverse of the variance of the binomial frequency distribution. After the first iteration, P's were replaced by P values which were calculated from the (i-1)th function in order to assign weight values in the ith iteration of the regression analysis.

In Eq. (2), the natural rate of liver neoplasia, $R_o(\tau)$ seemed to be best evaluated by the regression analysis routine, and was also assumed to have the T^N appearance rate; hence, the time-dependent $R_o(\tau)$ term for liver neoplasia would be $(9.59 \times 10^{-9})T^N$. Other constants were estimated to be $A = 2.55 \times 10^{-7}$, $\alpha = 5.72, \times 10^{-5}$, $\beta = 4.62 \times 10^{-8}$, and $N = 4.66$. The resulting fit is illustrated in Table 1. This fit was derived on the assumption that pharmacokinetic rate processes did not change with insult level or time-on-diet, i.e., the actual dose to cells of the liver was proportional (but not equal) to Haber's dose.

In subsequent computer runs, possible thresholds for liver neoplasia were offered by the transformation of $D = D-D_o$. Coefficients in the equation changed slightly, but the least squares fits did not improve. Thus, it seemed to us that the liver could not invoke harmless-detoxification of 2-AAF even at low concentrations or short times.

We could not fit the entire three-dimensional surface of bladder neoplasia according to $D = C \cdot \Delta$, nor have we found other modelers' fits to be accurate. A strong threshold behavior seems evident (p. 127 of SOT, 1981). To us, the perceived threshold seemed to vary across concentrations and with different times on the diet. Thus, we used a modified Haber's dose computed according to $D = C(\Delta-\Delta_o)$ or $D = \Delta(C-C_o)$. Δ_o was taken to be $a \cdot e^{-bC}$, where a and b were numerical coefficients evaluated by regression analysis, and C was concentration of 2-AAF (ppm). Likewise, C_o was chosen to be $a'e^{-b'\Delta}$ where a' and b' were numerical coefficients and Δ was time on 2-AAF (months). These assumptions gave great accuracy and were able to predict the outcome of the feeding-recovery-sacrifice protocol of the ED_{01} study.

For bladder neoplasi, the T^N factor was not needed or used. The value of A for bladder neoplasia includes the same time considerations as the product of $T^N \cdot A$ for the liver neoplasia. Thus, it appears that although both concentration and time-on-diet were important to bladder neoplasia, neither were significant in time-to-tumor.

The mathematical model resulting from the fits to the dose- and time-dependent bladder neoplasia is $A = 0.89$, $\alpha = 0.0$, $\beta = 4.99 \times 10^{-7}$, $R_o(\tau) = 0.0031$, and either $D = C(\Delta-\Delta_o)$ or $D = \Delta(C-C_o)$ when $\Delta_o = 42.9\ e^{-0.010 \cdot C}$ and $C_o = 168\ e^{-0.038\Delta}$. The results of the fitted model are compared with actual experimental data in Table 2.

Because no significant differences were noted (as a function of concentration or time-on-diet) for the latency period of bladder neoplasia, it appears that another promting action may possibly be dominant over the promoting action gained from 2-AAF products in the urine. The other model would have to be of constant magnitude in all treatment and control groups--at any particular sacrifice time.

Some natural component(s) of urine is (are) involved in promotion (Cohen et al., 1982; Rowland et al., 1980; Oyasu et al., 1981). Thus, it

Table 1

Incidence of liver neoplasms in sacrificed mice which were fed 2-AAF until time-of-sacrifice. Values inside parentheses are the number of predicted tumors for comparison with the actual pathologically counted tumors (i.e., numerator) in the sacrificed population (i.e., denominator). Values such as 0.1 mean that one tumor would be expected as predicted by the model if 10-fold more mice were sacrificed.

Dose (ppm)	Sacrifice Interval (Months)								
	9	12	14	15	16	17	18	24	33
0	(0.0) 0 / 142	(0.1) 0 / 140	(0.2) 0 / 113	(0.3) 0 / 88	(0.7) 1 / 183	(0.7) 0 / 128	(2.7) 1 / 401	(9.9) 9 / 383	(2.6) 8 / 23
30							(23) 17 / 1573	(63) 55 / 900	(38) 44 / 92
35						(4.7) 2 / 389	(13) 7 / 792	(51) 55 / 639	(22) 20 / 45
45					(2.9) 2 / 272	(3.9) 5 / 264	(7.7) 7 / 383	(45) 57 / 445	(7.6) 5 / 12
60	(0.2) 0 / 279	(0.7) 2 / 268	(1.4) 0 / 224	(1.7) 1 / 182	(3.6) 4 / 265	(4.0) 6 / 206	(7.1) 7 / 268	(58) 71 / 415	(9.8) 7 / 11
75	(0.1) 0 / 139	(0.5) 0 / 137	(0.9) 0 / 110	(1.1) 1 / 94	(3.0) 5 / 174	(3.3) 5 / 134	(9.0) 6 / 267	(56) 62 / 311	8[a] / 12
100	(0.1) 1 / 142	(0.6) 3 / 138	(1.3) 1 / 117	(1.5) 4 / 90	(2.1) 1 / 90	(2.3) 1 / 67	(6.2) 6 / 131	(41) 47 / 160	8[a] / 10
150	(0.2) 0 / 140	(1.0) 1 / 141	(2.0) 1 / 114	(2.4) 1 / 90	(3.3) 3 / 86	(3.6) 0 / 65	(9.3) 7 / 121	(53) 56 / 130	0[a] / 0

[a]Model invalid.

Table 2

Incidence of bladder neoplasms in sacrificed mice which were fed 2-AAF until time-of-sacrifice. Values inside parentheses are the number of predicted tumors for comparison with the actual pathologically counted tumors (i.e., numerator) in the sacrificed population (i.e., denominator). Values such as 0.1 mean that one tumor would be expected as predicted by the model if 10-fold more mice were sacrificed.

Concentration (ppm)	Sacrifice Interval (Months) 9	12	14	15	16	17	18	24	33
0	(0.4)[a] 1/142	(0.4) 0/140	(0.4) 0/113	(0.3) 0/88	(0.6) 0/183	(0.4) 0/127	(1.3) 1/400	(1.2) 1/384	(0.08) 0/24
30							(4.9) 4/1573	(2.8) 10/900	(0.3) 1/92
35						(1.2) 1/389	(2.5) 1/796	(2.0) 2/638	(0.4) (0.1) 0/45
45					(0.8) 1/271	(0.8) 1/264	(1.2) 1/383	(1.4) 1/445	(0.4) (0.04) 0/12
60	(0.9) 0/279	(0.8) 0/268	(0.7) 0/221	(0.6) 0/181	(0.8) 0/265	(0.6) 0/206	(0.8) 3/269	(1.9) (1.3) 3/415	(1.6) (0.8) 1/11
75	(0.4) 0/139	(0.4) 0/137	(0.3) 0/110	(0.3) 1/94	(0.5) 0/175	(0.4) 0/134	(0.8) 1/267	(15) (5.6) 3/311	(4.2) (3.6) 4/12
100	(0.4) 0/142	0.4) 0/138	(0.4) 0/117	(0.3) (0.5) 0/89	(0.5) (1.1) 1/90	(1.1) (1.5) 4/67	(4.8) 5/131	(46) (38) 25/160	(7.1) (6.9) 8/10
150	(0.4)[b] (5.1)[c] 1/140	(12) (16) 9/141	(24) (23) 21/114	(26) (23) 28/90	(32) (27) 36/85	(29) (25) 36/65	(64) (54) 62/121	(106) (99) 100/130	0/0

[a]Concentration-threshold and time-threshold model values are the same.

[b]Concentration-threshold model value.

[c]Time-threshold model value (i.e., time on 2-AAF).

Table 3

The model derived from data in Table 1 was used in order to predict the incidence of liver noeplasia in a different study, where mice were fed 2-AAF for 9, 12, and 15 months and then sacrificed at 18 and 24 months.

Δ = Time on 2-AAF (months)	T = Time to Sacrifice (months)	C = Concentration of 2-AAF (ppm)	Mice Sacrificed	(Haber's Dose D = CΔ) →	Model (Mice with Liver Neoplasia)	Experiment (Mice with Liver Neoplasia)
9	18	60	186	540	2.7	1
9	18	75	128	675	2.2	3
9	18	100	64	900	1.4	1
9	18	150	63	1350	2.1	4
9	24	60	108	540	6.0	13
9	24	75	66	675	4.3	10
9	24	100	35	900	3.0	5
9	24	150	33	1350	4.2	9
12	18	60	190	720	3.4	1
12	18	75	132	900	2.9	6
12	18	100	65	1200	1.9	0
12	18	150	63	1800	3.0	2
12	24	60	118	720	8.2	11
12	24	75	74	900	6.3	14
12	24	100	33	1200	3.7	5
12	24	150	29	1800	5.2	9
15	18	60	196	900	4.3	4
15	18	75	130	1125	3.6	5
15	18	100	64	1500	2.4	1
15	18	100	65	2250	4.0	4
15	24	60	114	900	9.6	15
15	24	75	86	1125	9.1	14
15	24	100	35	1500	5.1	6
15	24	150	28	2250	6.6	6

Table 4

The time-threshold and concentration-threshold models derived from data in Table 2 were used in order to predict the incidence of bladder neoplasia in a different study, where mice were fed 2-AAF for 9, 12, and 15 months, and then sacrificed at 18 and 24 months.

Δ = Time on 2-AAF (months)	T = Time to Sacrifice (months)	C = Concentration of 2-AAF (ppm)	Mice Sacrificed	(Δ_o = Time Threshold on 2-AAF (months)) →
9	18	60	(186)	19.7
9	18	75	128	16.2
9	18	100	64	11.7
9	18	150	63	6.1
9	24	60	108	19.7
9	24	75	66	16.2
9	24	100	35	11.7
9	24	150	33	6.1
12	18	60	190	19.7
12	18	75	132	16.2
12	18	100	65	11.7
12	18	150	63	6.1
12	24	60	118	19.7
12	24	75	74	16.2
12	24	100	33	11.7
12	24	150	29	6.1
15	18	60	196	19.7
15	18	75	130	16.2
15	18	100	64	11.7
15	18	150	65	6.1
15	24	60	114	19.7
15	24	75	86	16.2
15	24	100	35	11.7
15	24	150	28	6.1

Table 4 continued

Modified Haber's Dose $D = C(\Delta-\Delta_o)$ →	Model	Experiment (Mice with bladder neoplasia)	Model	← Modified Haber's Dose $D = \Delta(C-C_o)$ ←	C_o = Concentration Threshold of 2-AAF (ppm)
0	0.6	-0-	0.6	0	119
0	0.4	-2-	0.4	0	119
0	0.2	-1-	0.2	0	119
434	5.2	-4-	2.3	276	119
0	0.3	-1-	0.3	0	119
0	0.2	-0-	0.2	0	119
0	0.1	-0-	0.1	0	119
434	2.7	-6-	1.2	276	119
0	0.6	-0-	0.6	0	106
0	0.4	-0-	0.4	0	106
30.8	0.2	-1-	0.2	0	106
884	18	-14-	7.3	522	106
0	0.4	-0-	0.4	0	106
0	0.2	-1-	0.3	0	106
30.8	0.1	-1-	0.1	0	106
884	8.4	-7-	3.4	522	106
0	0.6	-1-	0.6	0	95
0	0.4	-0-	0.4	0	95
331	3.2	-0-	0.4	75	95
1330	34	-22-	17	825	95
0	0.4	-0-	0.4	0	95
0	0.3	-0-	0.3	0	95
331	1.8	-1-	0.2	75	95
1330	15	-11-	7.3	825	95

appears that this mode of action is the other and dominant mode of promotion. This hypothesis could be tested readily through a urinary diversion experiment designed according to methods of Rowland et al. (1980).

Our second speculation is that if "bladder dose" could be estimated (instead of using pharmacological doses), then the pharmacological threshold may persist but the tumor-threshold behavior of bladder neoplasia might disappear and both the probability of occurrence and time-to-tumor for bladder neoplasia would behave like comparable quantities in liver neoplasia (NCTR studies) and a T^N type consideration would be needed. Thus, we suspect that the unmodified form of Eq. (2) could fit both liver and bladder neoplasia data accurately if pathological doses were know.

Furthermore, if a more accurate dose-indexing system were used, then the values of α and β might be similar for different organs because these constants would no longer have to span pharmacological processes of transport and metabolism (Hoel et al., 1983).

Predictions of the Model When Treatment and Sacrifice Times are Different

The fitted equations for liver and bladder neoplasia resulting from continuous feeding of 2-AAF to time of sacrifice were evaluated in a predictive mode to test the fidelity of the mechanistic concepts (and of the algebraic analogue) when treatment time is different from sacrifice time, i.e., $T \neq \Delta$.

A second component of the NCTR study was to feed 2-AAF at concentrations of 60, 75, 100, and 150 ppm for 9, 12, and 15 months with sacrifice at 18 and 24 months. The comparison of forecasts of the models are compared with the actual experimental data in Tables 3 and 4.

CONCLUSIONS

Biological interpretation and mathematical modeling of the ED_{01} study has been a strong challenge to all (SOT, 1981). Most investigators usually consider tumors resulting from the 0-, 30-, 35-, 45-, 60-, 75-, 100-, and 150-ppm treatments at either 24 or 33 months. Thus, only one simple dose-response curve from a complex dose-time-response surface is studied. Some of the usual dose-response models can give reasonable fits to this one trace in time. The same model may not fit a different trace in time. Certainly, coefficients in the model must be redetermined. Only a few modelers have attempted to analyze the entire surfaces defined by eight treatment groups monitored at nine points in time, i.e., a surface as defined by 72 grid points. Results to date have been very unacceptable (SOT, 1981). But, we believe that the model of this study describes bladder and liver neoplasia accurately for continuous feeding and for a feed-wait-sacrifice treatment schedule.

ACKNOWLEDGMENTS

The helpful suggestions of M.D. Morris of the Mathematics and Statistics Research Department contributed greatly to the preparation of this manuscript.

REFERENCES

Cohen, S.M., Muraski, G., Fukushima, S., and Greenfield, R.E. (1982), Cancer Res. 42, 65.

Druckrey, H. 1967, UICC Monograph Series, Vol 7 Rene Truhaut, Ed., 60.

Farber, E., 1980, In Toxic Injury of the Liver, Farber, E., Fisher, M., eds., New York: M. Dekker.

Hoel, D.G., Kaplan, N.L., Anderson, M.W., 1983. Science 219, 1032.

Jones, T.D., 1984, "A Unifying Concept for Carcinogenic Risk Assessment: Comparison with Radiation-Induce Leukemia in Mice and Men" Health Physics 47(4), 533-558.

Jones, T.D., Griffin, G.D., Walsh, P.J., 1983, J. Theor, Biology 105, 35.

Littlefield, N.A., Farmer, J.H., Gaylor, D.W., 1980, J. of Envir. Pathology and Toxicology, 3, 17.

Oyasu, R., Hirao, Y., Izumi, K., 1981, Cancer Res. 41, 478.

Rowland, R.G., Henneberry, M.O., Oyasu, R., Grayhack, J.T., 1980, Cancer Res. 40, 4524.

Russell, W.L., Hunsicker, P.R., Raymer, G.D., Steele, M.H., Stelzner, K.F., Thompson, H.M., 1982, Proc. Natl. Acad. Sci. USA 79, 3589.

Society of Toxicology (SOT), ED Task Force, 1981, Fundamentals of Applied Toxicology, 1.

4,4'-METHYLENEDIANILINE: RISK ASSESSMENT AND RISK MANAGEMENT

J. William Hirzy

U.S. Environmental Protection Agency
Washington, D.C. 20460

ABSTRACT

Risks associated with the production, use and disposal of 4,4'-methylenedianiline (4,4'-MDA) have been under study by the Environmental Protection Agency (EPA). 4,4'-MDA was the first chemical to trigger section 4(f) of the Toxic Substances Control Act (TSCA), and EPA and the Occupational Safety and Health Administration (OSHA) have been working to find the most effective way to control the chemical's risks. Limited information on dermal penetration rates and exposure levels have introduced uncertainty into the dose side of the dose-response equation. A cancer bioassay, on the other hand, has provided good data for estimating response, at least within the tested dose range.

A technique for estimating lifetime average daily doses to workers based on the limited dermal exposure data and inhalational exposure levels has been employed. This technique has been used to estimate the degree of risk reduction that protective clothing and equipment and improved industrial hygiene practices can produce.

The risk assessment information, complete with its uncertainties, is being made available to risk managers in EPA and OSHA, while additional work is planned to reduce uncertainty in the assessment. This work includes additional monitoring of workplaces and surface waters (the latter to assess possible contamination of drinking water supplies) and measurement of the chemical's skin penetration rate. Since skin penetration rates can vary over several orders of magnitute, and since the limited data available indicate that dermal exposures in the workplace may be significant, reduction of uncertainty in this area is crucial for cost-effective risk management decisions.

KEY WORDS: 4,4'-Methylenedianinline; Cancer; Regulation; EPA; OSHA; TSCA; Workplace; Risk Assessment; Risk Management

1. INTRODUCTION

There is a certain redundancy implicit in the theme of this meeting: Uncertainty in Risk Assessment, Risk Management and Decision Making. Risk implies un-certainty. How likely is an untoward event? What are the likely consequences? What can be done about it? What should be done about it? By whom?

The Environmental Protection Agency has been engaged in just such an analysis (EPA, 1984) on the cancer risks associated with 4,4'-methylenedianiline (4,4'-MDA), a 400 million pounds per year chemical used

RISK ASSESSMENT

How likely is the occurance of an untoward event, and what are the likely consequences if it occurs?

RISK MANAGEMENT

What can be done if the likelihood of the event and/or its consequences are unacceptable?

What should be done?

Was the action taken adequate?

Figure 1

- ca. 400 million pounds annual production.
- ca. 97% used to make methylene diphenyldiisocyanate, a polyurethane precursor.
- ca. 3% used as epoxy curing agent, as a component of wire enamels, as a polyurethane component *per se*, and as a specialty resins intermediate.

Figure 2

Production and Use of 4,4'-Methylenedianiline (4,4-MDA)

mostly as an intermediate in the polyurethane industry. The Agency is continuously involved in other such analyses, and is, in fact, in the process of developing internal guidelines for their conduct. The analysis for the 4,4'-MDA case has, of necessity, been underway for some time, and has been in a rather evolutionary condition as the Agency's work on risk assessment guidelines has evolved and as information on 4,4'-MDA has been received from the public.

So, in addition to being an exposition of a particular case (4,4'-MDA) of how the Agency has dealt with uncertainty in risk assessment and risk management decisions, this presentation provides some insight into the evolving decision-making process at EPA.

2. HISTORICAL PERSPECTIVE: THE §4(f) DECISION

In March of 1983, EPA invoked section 4(f) of the Toxic Substances Control Act, because the hazard and limited exposure information available at that time indicated that there may be a reasonable basis to conclude

Table 1

Excerpts From Hazard Information for
Section 4(f) Decision (NIP, 1983, Draft)

Tumor	Incidence Control	Low Dose[(a)]	High Dose[(b)]
Male Rat			
Liver Neoplastic Nodules	1/50	12/50	25/50
Thyroid F-cell Carcinoma or Adenoma	1/49	4/47	10/48
Female Rat			
Thyroid F-cell Carcinoma or Adenoma	0/47	4/47	19/48
Thyroid C-cell Carcinoma	1/47	5/47	7/48
Male Mouse			
Adrenal-Pheochromocytoma	2/48	12/49	14/49
Female Mouse			
Liver Hepatocellular Carcinoma or Adenoma	4/50	15/50	23/50
Thyroid F-cell Carcinoma or Adenoma	0/50	1/47	13/50

(a) 150 ppm 4,4'-MDA·2HCl in drinking water.

(b) 300 ppm 4,4'-MDA·2HCl in drinking water.

4,4-MDA presents a significant risk of serious harm from cancer to certain people.

The hazard information included an NTP bioassay (NTP, 1983) on the dihydrochloride salt of 4,4'-MDA which showed statistically significant increases in tumors in both sexes of Fisher 344 rats and B6C3F mice at multiple sites. The data from the bioassay were sufficient to establish a clear dose-response relationship in the dose regions close to those studied in the bioassay, namely 40 to 80 mg/kg/day in mice and 8.5 to 17 mg/kg/day in rats. Using the body surface area extrapolation method, these dose regions translate to human doses of about 2.6 to 5.2 mg/kg/day.

There were available at the time of commencement of EPA's priority review other pieces of hazard information whose combined weight, when added to that of the NTP study, provided further impetus for the S4(f) decision -- and for the subsequent finding that exposures to 4,4'-MDA

June 1979 - Interagency Testing Committee (ITC) recommends consideration of testing 4,4'-MDA for teratogenicity, mutagenicity, carcinogenicity, other chronic effects, environmental effects, and epidemiology.

Mid 1982 - EPA makes 4,4'-MDA subject to reporting requirements of sections 8(a) and 8(d) of the Toxic Substances Control Act (TSCA). The Chemical Manufactures Association (CMA) establishes a 4,4'-MDA project panel.

Early 1983 - An NTP bioassay and anecdotal exposure information prompt the first invocation of section 4(f) of TSCA, the priority review section.

Figure 3

Historical Perspective

- McGill and Motto (1974) report that workers who breathe 0.1 ppm of 4,4'-MDA in air and handle a raw epoxy resin compound containing the amine suffered hepatotoxicity, while workers whose only exposure is inhalational at 0.1 ppm did not.
- CMA (Cox, 1982) reports adherence to 0.1 ppm 8-hr TWA exposure standard and notes analytical problems in measuring 4,4'-MDA levels.
- NIOSH (1976) reports that 8-hr TWAs of 0.03 to 0.4 ppm were reported by a former manufacturer, and that dermal exposure, while not quantified, appears to be a significant potential problem.

Figure 4

Exposure Information for Section 4(f) Decision

should be controlled. This information included reports of acute intoxication in the workplace that resulted in hepatotoxicity, among other effects.

Anecdotal exposure information available early in 1983, indicated that while dermal exposure was probably important, there were no data on which to estimate possible human doses. So only inhalational dosing was considered when we made the estimates of human doses and resultant risks in early 1983. Inhalation dose estimates were based on anecdotal reports that the recommended ACGIH TLV of 0.1 ppm (0.8 mg/m^3) was adhered to my manufacturers of 4,4'-MDA, and on assumptions that in certain use/processing settings this level might be exceeded by a factor as high as 20.

Table 2

Risk Estimations for Section 4(f) Decision

Conditions	LADDs[(a)]	Added Lifetime Risk Based on Tumor Types: FRTC[(b)]	FRTCA[(c)]	MRLNN[(d)]
0.01 ppm, 7 hr/day, 10% absorption	0.0010	1×10^{-9} 2×10^{-5}	1×10^{-5} 3×10^{-5}	2×10^{-5} 1×10^{-4}
0.01 ppm, 7 hr/day, 50% absorption	0.0049	3×10^{-8} 8×10^{-5}	7×10^{-5} 1×10^{-4}	1×10^{-4} 5×10^{-4}
0.1 ppm, 4 hr/day, 50% absorption	0.028	8×10^{-7} 5×10^{-4}	4×10^{-4} 7×10^{-4}	7×10^{-4} 3×10^{-3}
0.1 ppm, 7 hr/day, 50% absorption	0.049	3×10^{-6} 8×10^{-4}	7×10^{-4} 1×10^{-3}	1×10^{-3} 5×10^{-3}
2.0 ppm, 7 hr/day, 50% absorption	0.98	—	—	9×10^{-2}

(a) Lifetime Average Daily Dose in mg/kg/day

(b) Female rat, thyroid C-cell carcinoma. Higher value is upper 95% confidence limit, lower value is maximum likelihood estimate.

(c) Female rat, thyroid F-cell carcinoma or adenoma.

(d) Male rat, liver neoplastic nodules. This tumor type is not used in the current draft of the risk assessment (Aug. 1984).

There is uncertainty, of course, in these assumptions. We were uncertain as to what actual levels were in the 4,4'-MDA manufacturing setting - detailed reporting of those levels had not been made to EPA at that time by the manufacturers. We were uncertain about the levels that could be present in use/processing settings. And further, we were uncertain about the amount of inhaled 4,4'-MDA that was absorbed.

With all these uncertainties, we decided to set up a table of risk estimates that used 0.5, 0.1, 0.01 and 0.001 ppm as possible exposure

- o 4,4'-MDA exposures are not regulated by law.
- o Inhalational exposures, alone, at the voluntary ACGIH standard of 0.1 ppm (0.8 mg/m^3) may pose added cancer risks of 5×10^{-3}.
- o Exposures in 4,4-MDA processing facilities might pose risk as high as 9×10^{-2}.
- o About 2,500 workers may be exposed.

Conclusion - While widespread harm is unlikely, there is a reasonable basis to conclude that significant risk of serious harm from cancer is presented by 4,4'-MDA to workers.

Figure 5

The Section 4(f) Decision

- o Using airborne exposure level data from manufacturers and standard assumptions, calculate lifetime average daily doses (LADDs) via inhalational exposures.
- o Using data from hand-pad monitoring conducted by NIOSH and assumption described below, calculate LADDs (described below) via dermal exposures.
- o Combine LADDs calculated above and, using the Global 82 multi-stage model with body surface area extrapolation, estimate risks to low-dosed humans from tumor incidences in high-dosed test animals.

Figure 6

Risk Estimation Method

levels, absorption efficiencies of 20, 50 and 80% of inhaled 4,4'-MDA, the usual inhalation rate and worker weight, duration and frequency assumptions, and work careers of 15 and 40 years. With the resulting matrix of calculated lifetime average daily doses, the linerized multistage model of Crump (1980) was used with the bioassay data on 13 tumor types to calculate a matrix of risk estimates, both maximum likelihood estimates (mle) and upper 95% confidence limits (U95CL).

We extracted from that matrix risk estimates for 0.1 and 0.5 ppm exposure levels over a 40 year work career and 50 percent absorption. We also made the same kind of calculations for 1 and 2 ppm exposure levels using several of the more responsive tumor types from the bioassay. These individual lifetime added risks ranged from 6×10^{-7} (mle for least

Table 3

Subsequent Exposure Data Excerpts

Inhalational

Source	Average 8-Hour TWAs (ppm) <8 hrs/week	9-20 hrs/week	>20 hrs/week
CMA			
High	0.07	0.06	0.07
Low	<0.005	<0.005	0.004

Dermal

	Deposition Rate (ug/cm²/hr)					
	4,4-MDA Manufacturing				4,4-MDA Processing	
	Operator[a]		Flaker-Bagger[b]		Resin mixer[c]	
Source	palm	other[d]	palm	other[d]	palm	other[d]
NIOSH	9.0	2.5	4.2	0.7	250	27

(a) Wearing cotton gloves. Deposition measured after 6 hours, and was assumed to be at a uniform rate.

(b) Wearing mid-forearm latex gloves. Deposition measured after 6 hours, and was assumed to be at a uniform rate.

(c) No gloves. Exposure was for about 10 minutes.

(d) "Other" was measured on the back of the hand, and was designed to measure general environmental exposure

responsive tumor type) to 9 x 10^{-2} (U95CL for most responsive tumor type).

It was on the basis of these numbers that the 4(f) designation was made.

3. RISK ASSESSMENT DEVELOPMENTS

Following the 4(f) designation, we were able to obtain very detailed exposure information from 4,4'-MDA manufacturers, including number of workers exposed in different job classifications, the duration of exposures broken out into 8 hours per week or less, 9-20 hours per week and more than 20 hours per week, and exposure levels for each group of

$$LADD = C \times V \times A \times D \times 250/365 \times 40/70 \div W$$

Where C = concentration of 4,4-MDA in mg/m^3

V = inhalation rate, 1.2 m^3/hr

A = absorption rate, assumed to be 50%

D = exposure duration in hr/day

W = worker mass, 70 kg

250/365 averages exposures during 250 work days over a full year.

40/70 averages exposures during a 40 year work career over a full 70 year lifetime.

Figure 7

Inhalational Ladd Calculation Method

$$LADD = A \times S \times R \times \int_0^Y tdt \text{ +/or } A \times S \times L \times Y$$

Where A = absorption rate as fraction of deposited material per hour, assumed 1%/hr.

S = surface area of exposed skin in cm^2.

R = deposition rate in $mg/cm^2/hr$.

t = time

L = load of 4,4'-MDA on skin in mg/cm^2.

Y = duration of exposure in hours.

The first term accounts for a continous deposition/absorption situation with a uniform deposition rate, while the second term accounts for absorption occuring after deposition has stopped leaving a fixed load of 4,4'-MDA on the skin.

Figure 8

Dermal Ladd Calculation Method

workers (CMA, 1983). Analytical methods employed were also reported. This information reduced uncertainty about exposures in the 4,4'-MDA manufacturing workplace, but there remained (and still remain) certain data gaps.

Inhalational LADD

$0.57\ mg/m^3 \times 1.2\ m^3/hr \times 8\ hr/day \times 0.5 \times 250/365 \times 40/70 \div 70\ kg$

$= 0.015\ mg/kg/day$

Dermal LADD

Daily dose via palms:

$2 \times 0.01/hr \times 0.009\ mg/cm^2/hr \times 410\ cm^2 \int_0^{3.5} t\,dt = 0.45\ mg$

Daily dose via rest of upper body

$2 \times 0.01/hr \times 0.0025\ mg/cm^2/hr \times 10{,}900\ cm^2 \int_0^{3.5} t\,dt + 0.0025\ mg/cm^2/hr \times$

$3.5\ hr \times 10{,}900\ cm^2 \times 0.01/hr \times 1\ hr$

$= 4.3\ mg$

Total dermal LADD = $4.8\ mg/day \times 250/365 \times 40/70 \div 70\ kg = 0.027\ mg/kg/day$

Total LADD = 0.042 mg/kg/day

Figure 9

4,4'-MDA Manufacturing Ladd Calculation Example
For a 40 Hours Per Week Exposure

Table 4

Risk Estimation Examples

LADD	Added Lifetime Risk Based on			
	MRLNN[a]		FMHCC/A[b]	
mg/kg/day	mle[c]	U95CL[d]	mle	U95CL
0.00062	1×10^{-4}	3×10^{-4}	1×10^{-4}	1×10^{-4}
0.010	2×10^{-3}	5×10^{-3}	2×10^{-3}	2×10^{-3}
0.020	5×10^{-3}	1×10^{-2}	3×10^{-3}	5×10^{-3}
0.042	1×10^{-2}	2×10^{-2}	8×10^{-3}	1×10^{-2}

(a) male rat, liver neoplastic nodules. This tumor type is not used in the current risk assessment (Aug. 1984).

(b) female mouse, hepatocellular carcinoma or adenoma.

(c) maximum likelihood estimate.

(d) upper 95% confidence limit.

Inhalational LADD

$0.38 \text{ mg/m}^3 \times 1.2 \text{ m}^3\text{/hr} \times 8 \text{ hr/day} \times 0.5 \times 250/365 \times 40/70 \div 70 \text{ kg}$

$= 0.010 \text{ mg/kg/day}$

Dermal LADD: Impact of Wash-up Delay

0.25 hr/day exposure (immediate wash-up)

$0.01\text{/hr} \times 43 \text{ ug/cm}^2 \times 0.25 \text{ hr/day} \times 410 \text{ cm}^2 = 0.044 \text{ mg}$[a]

$0.01\text{/hr} \times 4.6 \text{ ug/cm}^2 \times 0.25 \text{ hr/day} \times 2400 \text{ cm}^2 = 0.028 \text{ mg}$[b]

$\text{LADD} = 0.072 \text{ mg/day} \times 250/365 \times 40/70 \div 70 \text{ kg}$

$= 0.00040 \text{ mg/kg/day}$

6 hr/day exposure (delayed wash-up)

LADD = 0.010 mg/kg/day

(a) palms

(b) rest of hands, forearms, face and neck-V.

Total LADD

Immediate wash-up: 0.010 + 0.00040 = 0.010 mg/kg/day

Delayed wash-up: 0.010 + 0.010 = 0.020 mg/kg/day

If Inhalational Exposure is at 0.001 ppm 8-hr. TWA,

LADD = 0.00022 mg/kg/day

Then: Immediate wash-up: 0.00022 + 0.00040 = 0.00062 mg/kg/day

Delayed wash-up: 0.00022 + 0.010 = 0.010 mg/kg/day

Figure 10

4,4'-MDA Processing Ladd Calculation Example

We also began to receive information from users and processors of the chemical. Our Economics and Technology Division undertook a study of exposures in the workplace, of engineering and personal protective equipment controls, and of uses and possible substitutes for the chemical (EPA, 1983a, 1983b). We received information from the Department of Energy on the uses of 4,4'-MDA and the personnel protection programs employed by DOE contractors who use the chemical for making nuclear weapons. Some exposure-level information was also received from DOE (DOE, 1983).

We had no data on dermal exposure. So, involvement of the National

o 1970 Workers wear "divers' suits," yet urinalysis shows 15% of samples contain at least 200 ug/l of 4,4'-MDA.

o 1978 Awareness of dermal exposure increases, with subsequent improvements in hygiene. Now 4% of urine samples have at least 200 ug/l, 2% have 80-200 ug/l, and 3% have 20-80 ug/l of 4,4'-MDA.

o 1980 Cooperative program in effect, with further hygiene improvements. 0.9% of urine samples contain 20-80 ug/l, none higher.

Estimate of Exposure Reduction

.15 x 200 ug/l = 30 ug/l "average" concentration.

.009 x 20 ug/l = 0.18 ug/l "average" concentration.

Reduction factor = 0.006

* M. Vaudaine, et al. Diamino-Diphenyl Methane: An Example of Toxic Agent Monitoring in the Workplace at Rhone-Poulenc Industrie. J. Toxicol. Medicale 2, 207 (1982).

Figure 11

An Industrial Hygiene Report*
*M. Vaudaine, et al. Diamino-Diphenyl Methane: An Example of Toxic Agent Monitoring in the Workplace at Rhone-Poulenc Industrie. J. Toxicol. Medicale 2, 207 (1982).

Institute of Occupational Safety and Health (NIOSH) to help us in the general area of industrial exposure and in the particular case of dermal exposure was begun. NIOSH visited one manufacturer who makes ca. 99% assay 4,4-MDA, and one processor who uses that form (NIOSH, 1984a, 1984b). Measurement of airborne levels of the chemical was attempted, but with limited success, and dermal exposures were measured for the first time, using a hand-pad technique pioneered by the California Department of Food and Agriculture.

With the information obtained in this segment of the investigation, we were able to make estimates of risks based on both dermal and inhalational exposures, lowering the level of uncertainty to some degree, but with recognition that the number of work situations in which dermal exposure is possible makes drawing broad conclusions based on two plant visits risky.

These two plant visits by NIOSH, however, did open a lead to an analysis of dermal exposure reduction. At the 4,4'-MDA manufacturing plant, a golf-glove-like hand-pad device was placed on a worker's hand inside a latex glove, and the worker operated a flaker-bagging machine, packaging 99% assay a mine for 6 hours while wearing gloves. At the 4,4'-

MDA-using facility, a hand pad was placed on a worker who wore no gloves, while weighing the pure a mine for a resin batch over a period of about 10 minutes. Both palm and back-of-the-hand areas were monitored to get both direct hand-contact and general ambient exposure data.

The results of this work showed that under the latex glove deposition rates were 4.2 ug cm hr and 0.70 ug cm hr for direct hand-contact and general ambient exposure, respectively (NIOSH, 1984a); and without gloves the respective rates were 250 and 27 ug cm hr (NIOSH, 1984b). Thus, deposition rates appear to be reducible by factors of about 40 to about 60 by the simple expedient of wearing latex gloves.

4. RISK MANAGEMENT DEVELOPMENTS

Confirmation of the results of the NIOSH studies -- reduction of uncertainty, if you will -- can have substantial impact on risk management decisions on this chemical.

- Was the wearer of latex gloves experiencing break-through? or leakage around the glove cuffs?
- If break-through was involved, when did it happen? could protection be enhanced even more by changing gloves at mid-shift?
- Do the data from back-of-the-hand monitor pads accurately reflect exposures to face, neck, upper body, etc?
- Can Tyvek© or similar protective outerwear eliminate dermal exposure?

Other aspects of EPA's investigation also affect risk management decisions. Substitution of other chemicals for 4,4'-MDA has been studied. Replacement of 4,4'-MDA in some application would require extensive, expensive research and development programs. Can workers be adequately protected by the use of gloves and impervious clothing so that R and D expenses can be avoided? How expensive will such R and D programs be? The cost of gloves and protective clothing can be easily obtained; the cost of R and D is much harder to quantify in advance. Here is a major uncertainty in risk management.

5. THE FUTURE

The authorities of OSHA and EPA intertwine; EPA can order a ban -- require substitution for 4,4'-MDA in any product subject to TSCA; both OSHA and EPA can set workplace standards, including use of protective equipment and setting of a Permissible Exposure Limit. The private sector is free to act prior to Federal regulatory action. It can impose work rules requiring use of protective clothing and put engineering controls in place to limit exposure; it might find substitutes and, if it produces quality health effect data, it could market the substitutes as lower risk chemicals.

All of these factors involve risks in the workplace, and their complexity, revealed in limited degree in this, the 4,4'-MDA case, affords you a picture of the difficulties that EPA, OSHA, industry and labor will continue to face in dealing with uncertainty in risk assessment and risk management in other cases.

It is EPA's hope that the completion of its risk assessment guidelines will improve the robustness of its risk assessments and, as a result, give risk management decision-makers, both inside and outside the Agency, better tools for finding rational solutions to some of the problems we encounter in our society, which is so dependent upon chemical technology for its well being. Better risk assessment opens the door to better risk management and more efficient use of our social resources.

ACKNOWLEDGMENT

The work reported here could not have been done without the highly professional contributions of David Eberly, Chemical Control Division; John Helm, Jeanette Wiltse and Paulette Grimes, Existing Chemical Assessment Division; Brion Cook, Gary Grindstaff, Karen Hammerstrom, Dan Heggem, James Remmers, Janet Springer and Sandra Strassman-Sundy, Cheryl Siegel-Scott, Exposure Evaluation Division; Roman Kuchkuda, Craig Matthiessen and Sammy Ng, Economics and Technology Division; Angela Auletta, Harry Milman, Greg Thies and Bill Thomson, Health and Environmental Review Division, all of the Office of Toxic Substances.

REFERENCES

CMA, 1983. Response of the Chemical Manufacturers Association Methylenedianiline Panel to EPA's Section 4(f) Designation for MDA. June 27, 1983.

Cox, J.V., 1982. Letter of Steven D. Newburg-Rinn, Chief, Test Rules Development Branch, Office of Toxic Substances, U.S. EPA. Re: 4,4'-Methylenedianiline. November 4, 1982.

Crump, 1980. An Improved Procedure for Low-Dose Carcinogenic Risk Assessment for Animal Data. J. Environ. Path. and Toxicol. 5, 675 (1980).

DOE, 1983, Department of Energy. Response of the Albuquerque Operations Office (AL) and Contractors to the Environmental Protection Agency (EPA) Request for Information for the Review of 4,4'-Methylenedianiline (MDA). Draft. Received by EPA. August 31, 1983.

EPA, 1983a. Exposure to and Control of 4,4'-Methylenedianiline. Draft report prepared under contract #68-02-3935. Work assignment 1-5. PEDCo Environmental, Inc. October, 1983.

EPA, 1983b. 4,4'-Methylenedianiline (MDA) Use and Substitutes Analysis. Draft Report prepared under contract #69-02-3935. Task order #6. ICF, Inc. September, 1983.

EPA, 1984, Docket No. OPTS 64000 A.

MCGill, D.B. and Motto, J.D., 1974. An Industrial Outbreak of Toxic Hepatitis Due to Methylenedianiline. N.Engl. J. Med. 291, 278 (1974).

NIOSH, 1976. Current Intelligence Bulletin 8. 4,4'-Diaminodiphenylmethane (DDM). January 30, 1976.

NIOSH, 1984a. Industrial Hygiene Survey Report of Olin Corporation, Moundsville, WV. Dr. Mark Boeniger, Industrial Hygiene Section, National Institute for Occupational Safety and Health, Cincinnati, OH. March 1984.

NIOSH, 1984b. Industrial Hygiene Survey Report of A.O. Smith/Inland Corporation, Little Rock, AR. Dr. Mark Boeniger, Industrial Hygiene Section, National Institute for Occupational Safety and Health, Cincinnati, OH. February 1984.

NTP, 1983. Xarcinogenesis Studies of 4,4'-Methylenedianiline Dihydrochloride in F344/N Rats and B6C3F Mice. National Toxicology Program Technical Report, Series No. 248. NTP-81-143. NIH Publication No. 83-2504. June 1983.

Vaudaine, M., Lery, N., Diter, J.N., Droin, M. and Charnaillard, C., 1982. Diamino-Diphenyl Methane: An Example of Toxic Agent Monitoring in the Workplace at Rhone-Poulenc Industrie. J. Toxicol. Medicale 2, 207 (1982).

FUZZY HIERARCHICAL ANALYSIS

J.J. Buckley* and V.R.R. Uppuluri**

University of Alabama in Birmingham*
Birmingham, Alabama 35294
Oak Ridge National Laboratory**
Oak Ridge, Tennessee 37830

ABSTRACT

This paper extends hierarchical analysis to the case where the participants are allowed to employ fuzzy ratios in place of exact ratios. If a person considers alternative A more important than alternative B, then the ratio used might be approximately 3 to 1, or between 2 to 1 and 4 to 1, or at most 5 to 1. The pairwise comparison of the issues and the criteria in the hierarchy produce fuzzy positive reciprocal matrices. The geometric mean method is employed to calculate the fuzzy weights for each fuzzy matrix, and these are combined in the usual manner to determine the final fuzzy weights for the alternatives. The finally fuzzy weights are used to rank the alternatives from highest to lowest. The highest ranking contains all the undominated issues. The procedure easily extends to the situation where many experts are utilized in the ranking process. Two examples are presented showing the final fuzzy weights and the final ranking.

KEY WORDS: Hierarchical Analysis; Fuzzy Numbers; Priority Theory; Ranking Alternatives; Multi-criteria Decision Analysis.

1. INTRODUCTION

This paper develops Saaty's hierarchical analysis[1,2,3], when the experts (judges,...) are allowed to use fuzzy ratios in place of exact ratios. In Saaty's hierarchical analysis a person (expert, judge,...) is asked to supply ratios a_{ij} for each pairwise comparison between issues (alternatives, candidates,...) $A_1, A_2, \cdots, A_m$ for each criterion (objective) in a hierarchy, and also between the criteria. For some specific criterion, if a person considers A_1 more important than A_5, then a_{15} might equal 3/1, or 5/1 or 7/1. The numbers for the ratios are usually taken from the set $\{1,2,\cdots,9\}$ so a_{15} could be s_1/s_5 for s_1, $s_5 \in \{1,2,\cdots,9\}$ and $s_1 > s_5$. The ratios a_{ij} indicate, for this expert, the strength with which A_i dominates A_j. If a_{15} is equal to 5/1, then a_{51} is taken as 1/5. That is, $a_{ji} = a_{ij}^{-1}$ and $a_{ii} = 1$ for all i. Let A be the m x m matrix whose entries are the ratios a_{ij}. A is called a positive reciprocal matrix. Saaty's procedure uses the pairwise comparison matrices A for each criterion, and also the pairwise comparison matrix for the criteria, to compute a final set of weights w_i ($w_i > 0$, $w_1 + \cdots + w_m = 1$) for the alternatives which can be used to rank the issues from highest to lowest.

In this paper the a_{ij} can be fuzzy numbers[4,5,6,7]. Fuzzy numbers can express an expert's feeling that a ratio a_{ij} is approximately 5 to 1 instead of exactly 5/1, or that a ratio is between 6 to 1 and 8 to 1 enstead of exactly 7/1. We all recognize that it is difficult for people to always assign exact ratios when comparing two alternatives. When comparing A_1 and A_5 a person might feel that A_1 is much more important than A_5. Does this mean that a_{15} equals 7/1, or 8/1 or 9/1? Using fuzzy numbers the expert can respond that a_{15} is between 7 to 1 and 9 to 1. Also, a person could feel that A_1 is a little more important than A_5. An appropriate fuzzy ratio in this case might be approximately 3 to 1. Fuzzy numbers automatically incorporate the vagueness of these replies.

In the next section we introduce the fuzzy numbers which experts will be allowed to use in pairwise comparisons and we discuss consistency of the data. The following section investigates methods of deriving the final fuzzy weights for the issues and discusses a procedure for ranking the alternatives given their fuzzy weights. The fourth section presents various properties of fuzzy hierarchical analysis. The following section extends the technique to the situation where input from many experts is desired. The relationship of this paper to previous research is discussed in the sixth section. Two examples are worked out in the following section and the last section contains a summary and our conclusions.

This paper only sketches the results of fuzzy hierarchical analysis. The complete details, plus proofs of all the theorems, may be found in the longer paper[8].

2. FUZZY RATIOS AND CONSISTENCY

The type of fuzzy numbers used[4] by the experts in paired comparisons is described by $(\alpha/\beta,\gamma/\delta)$ where $0 < \alpha \leq \beta \leq \gamma \leq \delta$. The graph of the membership function μ is determined by these four numbers as follows: zero to the left of α, a straight line segment from $(\alpha,0)$ to $(\beta,1)$, a horizontal line segment from $(\beta,1)$ to $(\gamma,1)$, a straight line segment from $(\gamma,1)$ to $(\delta,0)$, and zero to the right of δ. From now on we will place a bar over a symbol if it represents fuzzy numbers. If an expert believes alternative A_i is more important that A_j, then the fuzzy ratio $\bar{a}_{ij} = (\alpha_{ij}/\beta_{ij},\gamma_{ij}/\delta_{ij})$ has $\alpha, \beta, \gamma, \delta \in \{1,2,\cdots,9\}$ and $\bar{a}_{ji}$ is taken as $(\bar{a}_{ij})^{-1} = (\delta_{ij}^{-1}/\gamma_{ij}^{-1}, \beta_{ij}^{-1}/\alpha_{ij}^{-1})$. For the reciprocal of a fuzzy number see Dubois and Prade[3].

If two of the numbers α, β, or β, γ or γ, β are equal in a fuzzy number $\bar{a}_{ij}$, then the corresponding line segment doesn't exist. For example, (4/5, 5/6) is a triangle over the interval [4,6] and (2/2, 2/4) is a line segment from (2,1) to (4,0) and zero elsewhere. Any real number n is equal to (n/n,n/n). Therefore, if a person feels that A_i and A_j are equally important, then $\bar{a}_{ij} = (1/1,1/1)$. Let A be the m x m fuzzy matrix of all-paired comparisons for the issues $A_1,A_2,\cdots,A_m$. The elements in A are $\bar{a}_{ij}$ where $\bar{a}_{ji} = (\bar{a}_{ij})^{-1}$ and $\bar{a}_{ii} = (1/1,1/1)$ all i. A is called a fuzzy positive reciprocal matrix.

The fuzzy numbers are intuitively easy to use and have various interpretations which would be explained to a person before they would be asked to supply the fuzzy positive reciprocal matrices A. For example, approximately 5 to 1 might be (4/5,5/6) and between 6 to 1 and 8 to 1 could be (6/6,8/8) or (5/6,8/9). Also, one might interpret (3/3,3/5) as being at least 3 to 1.

We now need to introduce a method of comparing fuzzy numbers that will be used in the definition of consistency and in a later section when we want to rank the alternatives from highest to lowest. Let $\bar{M}$ and $\bar{N}$ be two fuzzy numbers with membership functions $\mu_m(x)$ and $\mu_n(x)$, respectively. An arbitrary fuzzy number M is also determined by the numbers $(m_1/m_2,m_3/m_4)$ but the graph of $\mu_m(x)$ need not be a straight line segment on $[m_1,m_2]$ and $[m_3,m_4]$. The membership function $\mu_m(x)$ is continuous monotonically increasing from zero to one on $[m_1,m_2]$ and continuous monotonically decreasing from one to zero on $[m_3,m_4]$. Define

$$v(\bar{M} \geq \bar{N}) = \sup_{x \geq y}(\min(\mu_m(x),\mu_n(y))).$$

We will say that $\bar{M}$ is greater than $\bar{N}$, written $\bar{M} > \bar{N}$, if $v(\bar{M} \geq \bar{N}) = 1$ and $v(N \geq M) < \theta$, where θ is some fixed positive fraction less than or equal to one[6,7]. Values like 0.7, 0.8, 0.9 might be appropriate for θ. If M is not greater than N and N is not greater than M, we will say M and N are approximately equal which is written as $M \approx N$. Therefore, if

$$\min(v(\bar{M} \geq \bar{N}),v(\bar{N} \geq \bar{M})) \geq \theta,$$

then $\bar{M} \approx \bar{N}$. We will write $\bar{M} \geq \bar{N}$ when $\bar{M} > \bar{N}$ or $\bar{M} \approx \bar{N}$. We will also need to add and multiply fuzzy numbers. Let $\oplus$ and $\odot$ be the standard fuzzy addition and multiplication, respectively[5,6,7].

<u>Definition 1.</u>

A fuzzy positive reciprocal matrix $\bar{A} = [\bar{a}_{ij}]$ is consistent if and only if $\bar{a}_{ik} \odot \bar{a}_{kj} \approx \bar{a}_{ij}$.

This definition is the direct extension to fuzzy positive reciprocal matrices of the definition of (ordinal) consistency for positive reciprocal matrices[1,3]. If $A = [a_{ij}]$ is a positive reciprocal matrix, then A is consistent if and only if $a_{ik}a_{kj} = a_{ij}$.

Much of fuzzy hierarchical analysis will reduce to Saaty's hierarchical analysis if all the fuzzy numbers are triangular ($\beta_{ij} = \gamma_{ij}$) and $\theta = 1$. A is consistent if and only if $\beta_{ik}\beta_{kj} = \beta_{ij}$ when $\beta_{ij} = \gamma_{ij}$ and $\theta = 1$. If some flat fuzzy numbers are employed ($\beta_{ij} < \gamma_{ij}$) and/or $\theta < 1$, then fuzzy hierarchical analysis is an extension of classical hierarchical analysis.

If a fuzzy positive reciprocal matrix is not consistent we may wish to revise the estimates of the ratios so that the estimates of the fuzzy weights may be improved. Saaty[1] states:

> Note that improving consistency does not mean getting an answer closer to the "real" life solution. It only means that the ratio estimates in the matrix, as a sample collection, are closer to being logically related than to being randomly chosen.

<u>Theorem 1.</u>

Let $\bar{A} = [\bar{a}_{ij}]$ where $\bar{a}_{ij} = (\alpha_{ij}/\beta_{ij},\gamma_{ij}/\delta_{ij})$ and let $\beta_{ij} \leq a_{ij} \leq \gamma_{ij}$ all i,j. If $A = [a_{ij}]$ is consistent, then $\bar{A}$ is consistent.

3. THE FUZZY WEIGHTS

In Saaty's hierarchical analysis one first computes the weights w_i ($w_i > 0$, $w_1 + \cdots + w_m = 1$) for each positive reciprocal matrix A. Then these weights are combined, depending on the hierarchical structure, to obtain the final set of weights for the alternatives. Therefore, we must first specify how to obtain the fuzzy weights $\bar{w}_i$ given any fuzzy positive reciprocal matrix A.

We will begin with Saaty's λ-max procedure for determining the weights. For a given A, λ-max is the largest (real) eigenvalue of A and the weights w_i are the components of the normalized (sum equals one) eigenvector corresponding to λ-max.. Now consider a fuzzy positive reciprocal matrix $A = [\bar{a}_{ij}]$ where $\bar{a}_{ij} = (\alpha_{ij}/\beta_{ij}, \gamma_{kj}/\delta_{ij})$. Generalizing the λ-max method we would consider

$$\bar{A} \odot \bar{w} = \bar{\lambda} \odot \bar{w}$$

where $\bar{w}^T = (\bar{w}_1, \cdots, \bar{w}_m)$ and the $\bar{w}_i$ and $\bar{\lambda}$ are fuzzy numbers. It is known[8] that the above equation does not have a normalized solution for $\bar{w}$. Other types of solutions for $\bar{w}$ may be considered but we shall not pursue this direction in this paper.

Uppuluri's geometric mean technique[9,10] for computing the weights w_i is easily extended to fuzzy positive reciprocal matrices A. Given a positive reciprocal matrix $A = [a_{ij}]$, he first computes the geometric mean of each row as

$$r_i = \{\prod_{j=1}^{m} a_{ij}\}^{1/m},$$

and then $w_i = r_i/(r_1+\cdots+r_m)$. If A is consistent, then Uppuluri's method always produces the same weights as Saaty's λ-max technique. Uppuluri showed[9] that for m = 3, both methods compute the same weights (A need not be consistent) and when m > 3 the numerical results for the weights in the two procedures are close to each other.

For $\bar{A} = [\bar{a}_{ij}]$ define

$$\bar{r}_i = (\bar{a}_{i1} \odot \cdots \odot a_{im})^{1/m}, \text{ and}$$

$$\bar{w}_i = \bar{r}_i \odot (\bar{r}_1 \oplus \cdots \oplus \bar{r}_m)^{-1}.$$

Roots of fuzzy numbers are discussed in Dubois and Prade[7]. In the remainder of this paper we will employ this technique for computing the fuzzy weights $\bar{w}_i$.

Based on Buckley[4] we will determine the membership function μ_i for $\bar{w}_i$. Let

$$f_i(y) = [\prod_{j=1}^{m} ((\beta_{ij} - \alpha_{ij})y + \alpha_{ij})]^{1/m},$$

$$g_i(y) = [\prod_{j=1}^{m} ((\gamma_{ij} - \delta_{ij})y + \delta_{ij})]^{1/m},$$

for $0 \leq y \leq 1$. Define $\alpha_i = [\prod_{j=1}^{m} \alpha_{ij}]^{1/m}$ and $\alpha = \sum_{i=1}^{m} \alpha_i$. Similarly, define β_i and β, γ_i and γ, δ_i and δ. Finally, let

$$f(y) = \sum_{i=1}^{m} f_i(y),$$

$$g(y) = \sum_{i=1}^{m} g_i(y).$$

The fuzzy weights $\bar{w}_i$ are determined by $(\alpha_i\delta_i^{-1}/\beta_i\gamma_i^{-1}, \gamma_i\beta_i^{-1}/\delta_i\alpha_i^{-1})$ where the graph of μ_i is zero to the left of $\alpha_i\delta_i^{-1}$, $x = f_i(y)/g(y)$ on the interval $[\alpha_i\delta^{-1}, \beta_i\gamma^{-1}]$, a horizontal line from $(\beta_i\gamma^{-1}, 1)$ to $(\gamma_i\beta^{-1}, 1)$, $x = g_i(y)/f(y)$ on the interval $[\gamma_i\beta^{-1}, \delta_i\alpha^{-1}]$, and zero to the right of $\delta_i\alpha^{-1}$. We have assumed the x-axis is horizontal and the y-axis is vertical so the graph of $x = f_i(y)/g(y)$ starts at $(\alpha_i\delta^{-1}, 0)$ for $y = 0$ and monotonically increases to $(\beta_i\gamma^{-1}, 1)$ as y grows from 0 to 1. One might wish to multiply each $\bar{w}_i$ by a normalizing constant so that the support of the $\bar{w}_i$ (where $\mu_i(x) > 0$) lies in the interval [0,1].

<u>Theorem 2. (Preservation of Ordinal Consistency[1]).</u>

Suppose $\bar{a}_{ij} > \bar{a}_{kj}$ for $j = 1,\cdots,m$. Then $\bar{w}_i \geq \bar{w}_k$.

The procedure outlined above determines the fuzzy weights for any fuzzy positive reciprocal matrix A. To obtain the final fuzzy weights, and the final ranking of the issues, we need to consider a specific hierarchy. To illustrate our method consider the structure in Figure 1. For each criterion C_k (aspect, characteristic,...) we obtain a fuzzy positive reciprocal matrix A_k of pairwise comparisons. We also obtain a fuzzy positive reciprocal matrix E for the pairwise comparison of the criteria. Fuzzy weights $\bar{w}_k = (\bar{w}_{1k},\cdots, \bar{w}_{mk})$ are computed for each A_k and fuzzy weights $\bar{e} = (\bar{e}_1,\cdots,\bar{e}_K)$ are derived from E. The final fuzzy weight for issue A_j is

$$\bar{f}_j = (\bar{w}_{j1} \odot \bar{e}_1) \oplus \cdots \oplus (\bar{w}_{jK} \odot \bar{e}_K).$$

The membership functions for the $\bar{f}_j$ are easily found from the membership functions for the $\bar{w}_{jk}$ and $\bar{e}_j$. Again, one might multiply each $\bar{f}_j$ by a suitable constant so that all the $\bar{f}_j$ have their support in [0,1].

<u>Definition 2.</u>

Alternative A_i dominates issue A_j, written $A_i > A_j$, if and only if $\bar{f}_i > \bar{f}_j$. If A_i doesn't dominate A_j and A_j doesn't dominate A_i, we write $A_i \approx A_j$.

Define the sets $H_1, H_2, \ldots$ as follows. H_1 contains all the undominated issues. H_2 has all the undominated alternatives after deleting all the A_i in H_1. Similarly, we define $H_3, H_4, \ldots$ with the last non-empty set called H_d. Clearly, $H_1 \neq \emptyset$, and as long as H_1 doesn't contain all the alternatives, H_2 is non-empty, etc. All the issues in a H_j are at the same level in the ranking. All the A_i in H_1 have the highest ranking, the issues in H_2 are all at the second highest level, etc. The final ranking of the $A_1, A_2, \ldots, A_m$ is accomplished by the sets $H_1, H_2, \cdots, H_d$.

4. PROPERTIES OF FUZZY HIERARCHICAL ANALYSIS

We write $A_i \geq A_j$ if $A_i > A_j$ or $A_i \approx A_j$. The following properties of $>$, $\geq$ and $\approx$ are easily verified and so the proofs are omitted[4].

a. $>$ is transitive, $\approx$ is not transitive.
b. Given any two issues A_i and A_j, then one of the following holds: $A_i > A_j$, or $A_i \approx A_j$, or $A_j > A_i$.

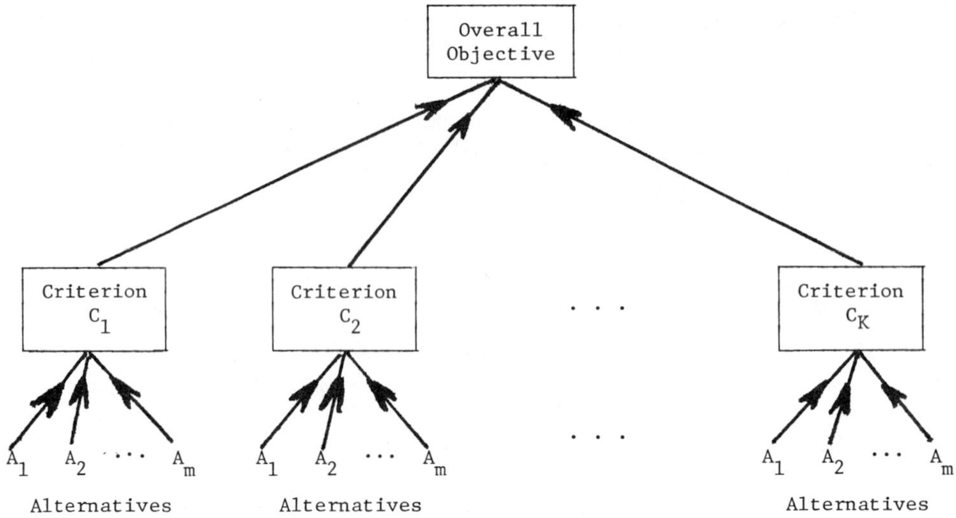

Figure 1

Hierarchical Structure

c. If A_i and A_j belong to H_k, then $A_i \approx A_j$.
d. Given any $A_i \in H_k$, $k > 1$, there is an $A_j \in H_{k-1}$ so that $A_j > A_i$.
e. $\geq$ is transitive, anti-symmetric and reflexive (a partial order).
f. If $A_i \in H_k$, then $A_i \geq A_j$ for all $A_j \in H_\ell$, $\ell \geq k$.
g. $\geq$ is a total order (partial order plus b above).

We next observe that fuzzy hierarchical analysis "contains" standard hierarchical analysis when the geometric mean procedure is employed to calculate the weights. As before $A_k = [\bar{a}^k_{ij}]$, $\bar{a}^k_{ij} = (\alpha^k_{ij}/\beta^k_{ij}, \gamma^k_{ij}/\delta^k_{ij})$, are the fuzzy positive reciprocal matrices for the criteria C_k, $k = 1,2,\cdots,K$ and $E = [\bar{e}_{ij}]$, $\bar{e}_{ij} = (\pi_{ij}/\rho_{ij}, \sigma_{ij}/\tau_{ij})$, is the pairwise comparison matrix for the criteria. Let $a^k_{ij} \in [\beta^k_{ij}, \gamma^k_{ij}]$ all i, j, k and let $b_{ij} \in [\rho_{ij}, \sigma_{ij}]$ all i,j. Form the matrices $M_k = [a^k_{ij}]$ and $M = [b_{ij}]$. Use the geometric mean method to find the weights for the matrices M_k and M and then combine them in the usual way to obtain the final weights f_i for the alternatives. If $\bar{f}_i = (\phi_i/\chi_i, \psi_i/\omega_i)$ are the fuzzy weights obtained from the A_k and E, then it is not difficult to see that $f_i \in [\chi_i, \psi_i]$ all i. That is, the (non-fuzzy) weights f_i belong to the interval where the membership function for $\bar{f}_i$ equals one.

Suppose in a pairwise comparison of the issues $A_1,\cdots,A_m$ the "true" weights are $w^*_1, w^*_2, \cdots, w^*_m$, where $w^*_i > 0$ and $w^*_1 + \cdots + w^*_m = 1$. If an expert knew the w^*_i, then all the a_{ij} in the pairwise comparison matrix A would equal w^*_i/w^*_j and A would be consistent. In practice, the $\bar{a}_{ij}$ used are considered estimates of w^*_i/w^*_j wince the true weights are usually now known.

<u>Theorem 3</u>

Let $\bar{A} = [\bar{a}_{ij}]$, $\bar{a}_{ij} = (\alpha_{ij}/\beta_{ij}, \gamma_{ij}/\delta_{ij})$ and suppose $w^*_i/w^*_j \in [\beta_{ij}, \gamma_{ij}]$ all i,j. Then $\bar{A}$ is consistent and $w_i^* \in [\xi_i, \eta_i]$ and i, where $\bar{w}_i = (\varepsilon_i/\xi_i, \eta_i/\theta_i)$.

Theorem 2 states that if the true ratios belong to the intervals where the membership function for $\bar{a}_{ij}$ is one, then the true weights belong to the intervals where the membership functions for $\bar{w}_i$ equal one.

It is of interest to speculate when all the fuzzy weights will be the same. Suppose an expert believes that A_1 is more important than A_2, A_2 is more important than $A_3, \cdots$, and A_m is more important than A_1. The expert's preferences are not transitive and the fuzzy weights $\bar{w}_i$ will depend on the other comparisons between A_1 and A_3, A_2 and A_4, etc. When there are only three issues we can show the fuzzy weights must all be equal. This situation is sometimes called the voter's paradox, or Arrow's paradox.

Theorem 4.

Suppose $m = 3$ and $\bar{a}_{12} = \bar{a}_{23} = \bar{a}_{31}$. Then $\bar{w}_1 = \bar{w}_2 = \bar{w}_3$.

5. MULTIPLE EXPERTS

Multiple experts are now employed in the hierarchical analysis in order to rank the alternatives. Uppuluri[10] has used the geometric mean to aggregate, or pool, all the data across the experts before computing the weights from the "average" matrix.

Suppose the experts (judges,...) are called $J_1, \cdots, J_n$. Each judge J_ℓ supplies a fuzzy positive reciprocal matrix $A_{k\ell}$ for each criterion C_k in the hierarchy, and J_ℓ also produces a matrix $\bar{E}_\ell$ of paired comparisons between the criteria. Let $A_{k\ell} = [\bar{a}_{ij}^{k\ell}]$ and $E_\ell = [\bar{e}_{ij}^{\ell}]$. The average fuzzy positive reciprocal matrices $\bar{A}_k = [\bar{a}_{ij}^{k}]$ and $\bar{E} = [\bar{e}_{ij}]$ are determined as follows:

$$\bar{a}_{ij}^{k} = [\bar{a}_{ij}^{k1} \odot \cdots \odot \bar{a}_{ij}^{kn}]^{1/n},$$

$$\bar{e}_{ij} = (\bar{e}_{ij}^{1} \odot \cdots \odot \bar{e}_{ij}^{n})^{1/n}.$$

It may be checked that $(\bar{a}_{ij}^{k})^{-1} = \bar{a}_{ji}^{k}$ and $(\bar{e}_{ij})^{-1} = \bar{e}_{ji}$.

Theorem 5.

If $\bar{A}_{k1}, \cdots, \bar{A}_{kn}$ are consistent, then $\bar{A}_k$ is consistent.

6. PREVIOUS RESEARCH

In a recent paper van Laarhoven and Pedrycz[11] attempted to extend Saaty's hierarchical analysis to fuzzy hierarchical analysis also using fuzzy numbers. They used Lootsma's results on log least-squares to extend Saaty's hierarchical analysis to the case of multiple estimates for the ratios and to the situation of missing data (no estimates for certain ratios).

For each ratio w_i^*/w_j^*, $i < j$, assume that we have n_{ij} estimates a_{ijk}, $k = 1,2,\cdots,n_{ij}$ where some of the n_{ij} could be zero (missing data case). The log least-squares estimate of the true weights w_i^* is a solution w_i to

$$\min \Big(\sum_{i<j} \sum_{k=1}^{n_{ij}} \big(\ln a_{ijk} - \ln(w_i/w_j)\big)^2 \Big).$$

When $n_{ij} = 1$, for $i < j$, Uppuluri[9] showed that the solution $\hat{w}_i$ is just the normalized geometric row mean discussed earlier. If $n_{ij} = n$, $i < j$, which is the multiple expert case discussed above, it can be shown the solution $\hat{w}_i$ is the one indicated in Section 5. That is, one first obtains the geometric average across all the judges to obtain the "average" positive reciprocal matrix and then computes the normalized geometric row means.

Van Laarhoven and Pedrycz assume the $\bar{a}_{ijk}$ are triangular fuzzy numbers and solve the normal equations for triangular fuzzy numbers $\bar{w}_i$ which are the fuzzy estimates of the true weights. The normal equations are obtained by taking the partials with respect to w_i and setting the result equal to zero. Our method is to substitute the fuzzy ratio $\bar{a}_{ijk}$ into the solution of the normal equations. The solution to the normal equation is of the form

$$\hat{w}_i = F(a_{ijk}),$$

for some function F which depends on the n_{ij}. Our method is simply

$$\bar{w}_i = F(\bar{a}_{ijk}).$$

The paper by van Laarhoven and Pedrycz is subject to two main criticisms. First, there are situations where the normal equations do not have a unique solution for the triangular fuzzy numbers $\bar{w}_i$. We saw a similar situation when we attempted to employ Saaty's λ-max method to generate the fuzzy weights. Algebraic equations with fuzzy variables sometimes do not yield unique solutions. Secondly, they insist on obtaining triangular fuzzy numbers for their weights. Since algebraic operations on triangular fuzzy numbers do not necessarily produce a triangular fuzzy number, they are forced to employ approximate methods to preserve the shape of the fuzzy number.

Our method does not suffer from either of these problems and it always produces a unique fuzzy number for the weight. It is equally applicable to the cases of missing data and multiple estimates.

7. APPLICATIONS

In any application the value of θ, the number used in comparing fuzzy numbers, must be decided on first. There is no set rule which dictates the value of θ and we have suggested values like 0.7, 0.8, 0.9. A θ value of one implies fuzzy number $M = (m_1/m_2, m_3/m_4)$ is greater than another fuzzy number $N = (n_1/n_2, n_3/n_4)$ if and only if $n_3 < m_2$. The ordering of fuzzy numbers is accomplished by comparing the intervals where their membership functions equal one when $\theta = 1$. In the following examples we have set $\theta = 0.8$. Notice as θ increases from 0.7 to 0.8 to 0.9 fuzzy numbers become more spread out. That is, we may have $M \approx N$ for $\theta = 0.7$ but $M > N$ for $\theta = 0.8$.

Assume $\theta < 1$ and consider comparing two fuzzy numbers $\bar{M}$ and $\bar{N}$ when $n_3 < m_2$. Then $M > N$ if and only if

$$\max(\min(\mu_m(x),\mu_n(x))) < \theta,$$

where $\mu_m(\mu_n)$ are the membership function for $\bar{M}(\bar{N})$. $\bar{M}$ is greater than $\bar{N}$ if the intersection of the graphs of μ_m and μ_n on $[n_3,m_2]$ lies below the horizontal line $y = \theta$. If this intersection lies on $y = \theta$ or above, then $M \approx N$. When $[n_2,n_3] \cap [m_2,m_3]$ is nonempty we have $M \approx N$ also.

Example 1

A government agency wishes to rank chemicals A_1, A_2, A_3 from most harmful to least harmful to the environment. The hierarchy is shown in Figure 1 with criterion C_1 = aquatic life, C_2 = agriculture and C_3 = timber. In a real study there would be more than three chemicals and possibly four or five criteria. The agency employs the testimony of a group of experts who supply the fuzzy positive reciprocal matrices $A_{k\ell}$ for each criterion C_k. The agency also collects data on the pairwise comparisons of the criteria to obtain the fuzzy positive reciprocal matrices E_ℓ for the criteria. Suppose the pooled information is given in Table 1. Each fuzzy positive reciprocal matrix is consistent.

Consider the fuzzy positive reciprocal matrix for criterion C_3 in Table 1C. The fuzzy ratios have the following interpretations: (1) (1/2,2/3) means that chemical A_1 is approximately twice as harmful to timber as chemical A_2; (2) (7/8,8/8) means that A_1 is at most eight times as harmful as A_3; and (3) (3/3,4/4) means A_2 is between 3 to 4 times as harmful as A_3.

It is not difficult to program a personal computer to determine the final fuzzy weights f_i and using graphics have their membership functions displayed on one coordinate system. The membership functions for the final fuzzy weights are shown in Figure 2. Therefore, chemical A_1 is the most harmful, and chemicals A_2 and A_3 are approximately equally harmful. That is, $H_1 = \{A_1\}$ and $H_2 = \{A_2,A_3\}$.

Example 2

A government wishes to rank various energy sources from most important to least important for the nation over the next ten years. The alternatives are A_1 = nuclear, A_2 = hydroelectric, A_3 = fossil, and A_4 = solar. The hierarchical structure is the same as in Figure 1. There are two criteria in the study. C_1 represents economical and political considerations including cost, balance of payments, etc. The other criterion C_2 represents military and defense considerations including self sufficiency, availability, etc. Expert opinion is solicited from energy experts, government officials, military personnel,... and the pooled data is presented in Table 2. Each fuzzy matrix is consistent.

The final fuzzy weights $\bar{f}_i$ for the alternatives A_i are shown in Figure 3. It is clear from Figure 3 that $H_1 = \{A_2,A_3\}$ and $H_2 = \{A_1,A_4\}$. This study has ranked hydroelectric and fossil fuel highest and approximately equally important. If it is desired to have H_1 contain only one alternative, then a second study comparing only A_2 and A_3 would be needed in order to try to differentiate between these two energy sources.

8. SUMMARY AND CONCLUSIONS

This paper investigates the possibility of allowing participants in a hierarchical analysis to give vague, or imprecise, replys when comparing

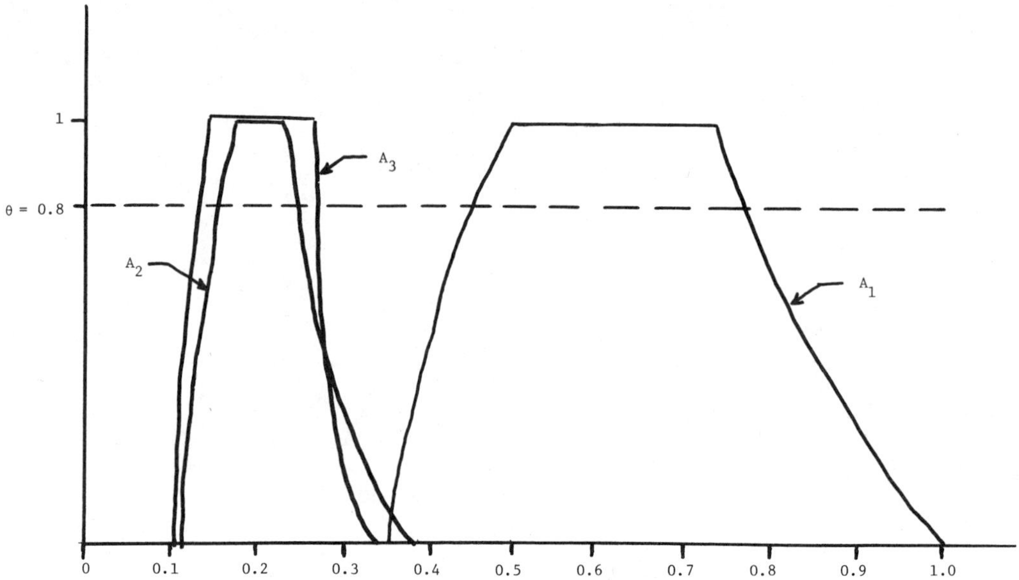

Figure 2

Membership Functions for the Final Fuzzy Weights in Example 1

Table 1a

Fuzzy Positive Reciprocal Matrix for Criterion C_1 in Example 1

	A_1	A_2	A_3
A_1	1	(1/4/1/3,1/3/1/2)	(1/2/1/2,1/2/1/2)
A_2	(2/3,3/4)	1	(1/1,2/2)
A_3	(2/2,2/2)	(1/2/1/2,1/1)	1

Table 1b

Fuzzy Positive Reciprocal Matrix for Criterion C_2 in Example 1

	A_1	A_2	A_3
A_1	1	(6/6,6/7)	(2/2,4/4)
A_2	(1/7/1/6,1/6/1/6)	1	(1/2/1/2,1/1)
A_3	(1/4/1/4,1/2/1/2)	(1/1,2/2)	1

Table 1c

Fuzzy Positive Reciprocal Matrix for Criterion C_3 in Example 1

	A_1	A_2	A_3
A_1	1	(1/2,2/3)	(7/8,8/8)
A_2	(1/3/1/2,1/2/1)	1	(3/3,4/4)
A_3	(1/8/1/8,1/8/1/7)	(1/4/1/4,1/3/1/3)	1

Table 1d

Fuzzy Positive Reciprocal Matrix for the Criteria in Example 1

	C_1	C_2	C_3
C_1	1	(1/7/1/6,1/6/1/5)	(1/3/1/2,1/2/1)
C_2	(5/6,6/7)	1	(3/3,3/3)
C_3	(1/2,2/3)	(1/3/1/3,1/3/1/3)	1

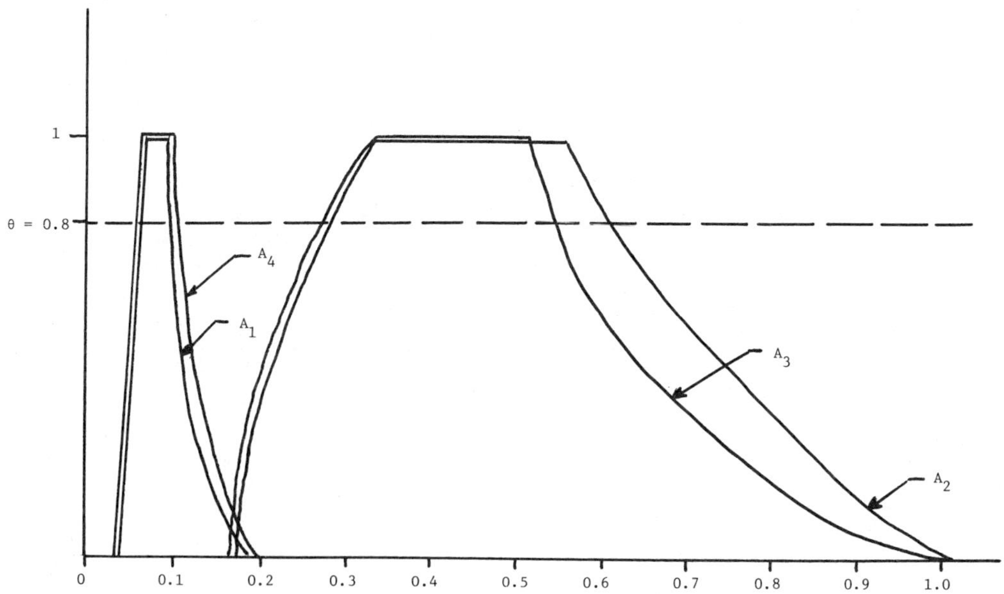

Figure 3

Membership Functions for the Final Fuzzy Weights in Example 2

Table 2a

Fuzzy Positive Reciprocal Matrix for Criterion C_1 in Example 2

	A_1	A_2	A_3	A_4
A_1	1	(1/7/1/6,1/6/1/5)	(1/6/1/5,1/5/1/4)	(1/1,1/1)
A_2	(5/6,6/7)	1	(1/1,2/2)	(4/4,6/6)
A_3	(4/5,5/6)	(1/2/1/2,1/1)	1	(3/4,5/6)
A_4	(1/1,1/1)	(1/6/1/6,1/4/1/4)	(1/6/1/5,1/4/1/3)	1

Table 2b

Fuzzy Positive Reciprocal Matrix for Criterion C_2 in Example 2

	A_1	A_2	A_3	A_4
A_1	1	(1/5/1/5,1/3/1/3)	(1/6/1/6,1/6/1/5)	(1/2/1/2,3/2/3/2)
A_2	(3/3,5/5)	1	(1/2/1/2,1/1)	(6/6,6/7)
A_3	(5/6,6/6)	(1/1,2/2)	1	(8/9,9/9)
A_4	(2/3/2/3,2/2)	(1/7/1/6,1/6/1/6)	(1/9/1/9,1/9/1/8)	1

Table 2c

Fuzzy Positive Reciprocal Matrix for the Criteria in Example 2

	C_1	C_2
C_1	1	(1/2,2/3)
C_2	(1/3/1/2,1/2/1)	1

two alternatives. If a person is comparing two alternatives A and B as to their relative importance (or weight, or brightness, etc.) and believes A is more important than B, then he/she may respond by saying A is approximately twice as important as B, or A is between 3 to 5 times as important as B, etc. Fuzzy numbers are used to capture the vagueness of these statements. Saaty's hierarchical analysis is generalized where fuzzy numbers are employed in place of exact ratios.

Much of classical hierarchical analysis may be accomplished with fuzzy numbers. This paper shows that the concept of consistency generalizes to fuzzy matrices. Saaty's λ-max method for determining the weights from a positive reciprocal matrix does not readily extend to fuzzy matrices. More research is needed on fuzzy eigenvalues and vectors of fuzzy positive reciprocal matrices. Uppuluri's geometric mean procedure is easily applied to a fuzzy matrix to obtain the fuzzy weights. The fuzzy weights are then combined in the usual way, depending on the hierarchical structure, to calculate the final fuzzy weights for the alternatives. The final fuzzy weights are utilized in partitioning the issues into sets $H_1, H_2, \cdots$. The set H_1 contains all the highest ranked alternatives, H_2 has those ranked second, etc. All the issues in H_1 are undominated and judged approximately equal. All the alternatives in H_k, $k > 1$, are approximately equal and are dominated by some issue in the next highest ranking H_{k-1}. Fuzzy hierarchical analysis is shown to possess a number of other desirable properties. The techniques readily extend to the situation where multiple experts are employed in the ranking process, or to the case of missing data.

If there are not too many alternatives and criteria fuzzy hierarchical analysis is easily programmed on a personal computer which would show graphically the membership functions for the final fuzzy weights. Then one could immediately pick off from the display of these membership functions the ranking $H_1, H_2, \cdots$.

REFERENCES

1. T.L. Saaty, A scaling method for priorities in hierarchical structures, J. of Mathematical Psychology 15 (1977) 234-281.
2. T.L. Saaty, Exploring the interface between hierarchies, multiple objectives and fuzzy sets, Fuzzy Sets and Systems 1 (1978) 57-68.
3. T.L. Saaty, The Analytic Hierarchy Process (McGraw-Hill, New York, 1980).
4. J.J. Buckley, Ranking alternatives using fuzzy numbers, Fuzzy Sets and Systems 15 (1985) 21-31.
5. D. Dubois and H. Prade, Operations on fuzzy numbers, Int. J. Systems Sci. 9 (1978) 613-626.
6. D. Dubois and H. Prade, Decision-making under fuzziness, in: M.M. Gupta, R.K. Ragade, R.R. Yager (eds.), Advances in Fuzzy Set Theory and Applications (North-Holland, Amsterdam, 1979) 279-302.
7. D. Dubois and H. Prade, Fuzzy Sets and Systems (Academic Press, New York, 1980).
8. J.J. Buckley, Fuzzy Hierarchical Analysis, Fuzzy Sets and Systems, 17 (1985) 233-247.
9. V.R.R. Uppuluri, Logarithmic least-squares approach to Saaty's decision problems, in: W.E. Lever, D.E. Shepherd, R.C. Ward, D.G. Wilson (eds.), Mathematics and Statistics Research Department Progress Report (Oak Ridge National Laboratory, Oak Ridge, TN 37830, 1978) 19-21.
10. V.R.R. Uppuluri, Expert opinion and ranking methods, Oak Ridge National Laboratory, NRC FIN No. B044, 1983, Oak Ridge, TN 37830.
11. P.J.M. van Laarhoven and W. Pedrycz, A Fuzzy Extension of Saaty's Priority Theory, Fuzzy Sets and Systems 11 (1983), 229-241.
12. F.A. Lootsma, Performance Evaluation of Non-Linear Optimization Methods via Multi-Criteria Decision Analysis and via Linear Model Analysis, in: M.J.D. Powell, Ed., Nonlinear Optimization 1981 (Academic Press, London, 1982) 419-453.

LIQUID METAL FAST BREEDER REACTOR (LMFBR) RISK ASSESSMENT

G. Sherwood

Office of Nuclear Energy (NE-52)
U.S. Dept. of Energy
Washington, D.C. 20545

ABSTRACT

A risk assessment has been performed for the potential future LMFBR fuel cycle, based on the results of several recent environmental reviews of the LMFBR program and of the Clinch River Breeder Reactor Plant (CRBRP), as well as applicable Light Water Reactor (LWR) data. The LMFBR fuel cycle consists of fuel fabrication, reactor operation, spent fuel reprocessing and associated waste management and transportation. No mining, milling, or enrichment of uranium will be required because the depleted uranium tails (0.2% U-235) already in existence will be sufficient for many, many years.

This risk analysis was performed by determining the characteristics and impacts of model facilities in the LMFBR fuel cycle. The following key assumptions were used to quantify these impacts:

1. There would be 200 cancer deaths per million man-rem of total wholebody radiation exposure. (This is used to convert calculated radiation exposures to estimated health effects and is consistent with estimates given elsewhere.)
2. There would be 30 metric tons per year of spent fuel per 1000 MWe LMFBR power plant. (This is used to establish fuel cycle needs and impacts and is consistent with previous estimates.)
3. Radiation releases would be essentially the same as those analyzed for the CRBRP and its supporting fuel cycle. (This is used to establish the general population radiation exposures and is consistent with previous estimates.)
4. LMFBR power plants would be at least as safe as current U.S. LWR power plants. (This is used to estimate the risks of catastrophic accidents and is consistent with other analyses.)

The resulting LMFBR fuel cycle risk was estimated to be 0.1-0.3 deaths per reference reactor year (0.8 GWe-yr). This is a factor of 2 or so less than comparable LWR fuel cycle risks and an order of magnitude or more less than comparable coal fuel cycle risks.

KEY WORDS: Risk; LMFBR; Fuel Cycle; Nuclear Power Plant; Occupational; Radiation; Accidents.

LIQUID METAL FAST BREEDER REACTOR (LMFBR) RISK ASSESSMENT

Description and Assumptions

The LMFBR fuel cycle consists of fuel fabrication, reactor operation, spent fuel reprocessing and associated waste management and transportation (see Figure 1). No mining, milling, or enrichment of uranium will be required because the depleted uranium tails (0.2% U-235) already in existence will be sufficient. In fact, the existing U.S. stockpiles would be enough to supply the equivalent of the entire U.S. electricity needs for 700 years with LMFBRs[1]. Plutonium for LMFBRs will initially be supplied from Light Water Reactor (LWR) spent fuel and later from LMFBR spent fuel. No credit will be taken for the surplus plutonium that will be produced in LMFBRs (i.e., with doubling times in the range of 10 to 20 years). This surplus plutonium could eventually be consumed in LWRs, which could reduce calculated LWR risks since LWR mining, milling, and enrichment requirements would be reduced if plutonium from LMFBRs forms part of the LWR fuel supply.

The following key assumptions were made for purposes of this risk assessment.

1. There would be 200 cancer deaths per million man-rem of total wholebody radiation exposure. (This is used to convert calculated radiation exposures to estimated health effects and is consistent with estimates given in BEIR I [2], BEIR III [3], UNSCEAR [4], EPA [5], etc.)
2. There would be 30 MT/yr of spent fuel per 1000 MWe LMFBR power plant.* (This is used to establish fuel cycle needs and impacts and is consistent with previous AEC [6], NRC [7], ERDA [8], and DOE [9] estimates).
3. Radiation releases would be essentially the same as those analyzed for the CRBRP and its supporting fuel cycle. (This is used to establish the general population radiation exposures and is consistent with previous DOE [10] and NRC [7] estimates).
4. LMFBR power plants would be at least as safe as current U.S. LWR power plants. (This is used to estimate the risks of catastrophic accidents and is consistent with CRBRP [11] and LDP [12] analyses).

Other Assumptions:

1. LMFBR power plants would have 30 year operating lifetimes. (This is used to establish impacts associated with radioactive waste management and is consistent with DOE [13] and NRC [7] projections).
2. LMFBR fuel reprocessing occupational doses and accidental fatalities would be the same as those for LWRs [14].
3. LMFBR transportation risks would be consistent with current U.S. experience for all types of transportation (by mode and commodity, especially truck and rail for coal and LWR fuel cycles).
4. There would be recycle of both plutonium and uranium in a closed LMFBR fuel cycle, as shown in Figure 1. Surplus plutonium would be used either for initial cores for other LMFBRs or for recycle in LWRs.
5. Transportation distances would be as follows:
   Depleted uranium stockpile to fuel fabrication plant - 1,000 miles.
   Fuel fabrication plant to LMFBR power plant - 1,000 miles.

---

*Operating at an 80% capacity factor.

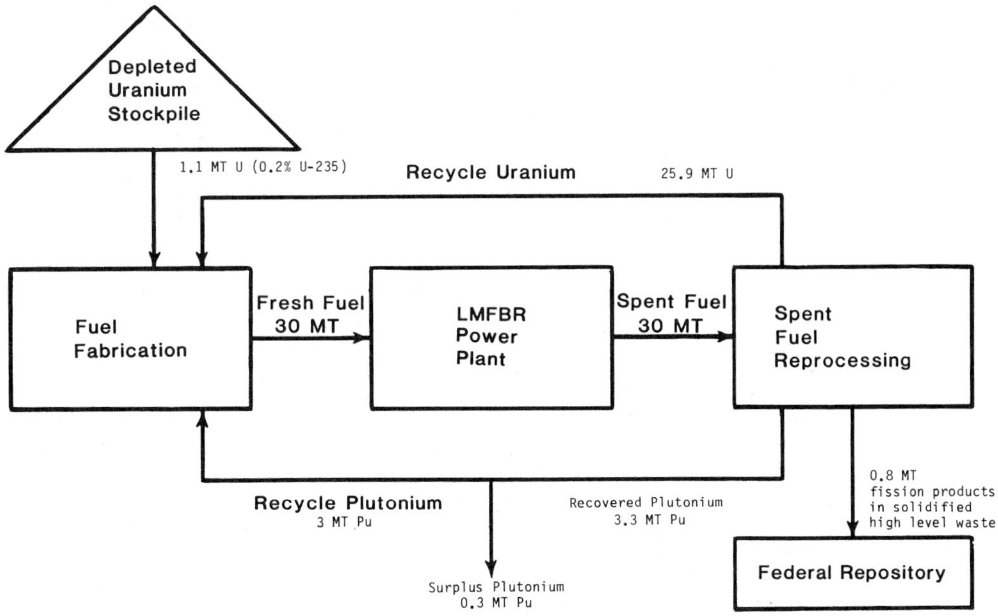

Figure 1

LMFBR Fuel Cycle
(Showing Annual Material FLows for a 1000 MWe Plant)

LMFBR power plant to spent fuel reprocessing plant - 1,000 miles.
Spent fuel reprocessing plant to fuel fabrication plant and to federal repository - 2,000 miles.
Fuel fabrication plant to federal repository - 2,000 miles.
Fuel fabrication plant, spent fuel reprocessing plant and LMFBR power plant to low level waste burial ground - 1,000 miles.

6. Radioactive waste quantities would be consistent with those given in the Environmental Impact Statement for the LMFBR Program [6] (except for low level wastes from the fuel fabrication plant).

ANALYSIS

Transportation

Transportation risks were determined using the data given in previous environmental reviews of the LMFBR program [6,8] and the CRBRP [7,10] as well as other sources [9,15,16]. First, the total quantity of spent and fresh fuel required annually was estimated, as shown in Table 1. Note that fuel cycle losses, typically 1/2% each for fuel fabrication and fuel reprocessing, were not included in this analysis, since they aren't important here. Then quantities of fuel materials and wastes were determined using the above data sources. The results are shown in Table 2. Next, shipping modes (truck or rail) were selected, using normal assumptions[6,7]. Finally, total shipping distances were determined using the shipping distances given in Figure 2 and the quantities to be shipped. The results are given in Table 2 and shown in Figure 3.

For truck shipments, there were an average of 0.033 deaths per million vehicle miles, based on data from 1974 to 1981 [16]. Thus the risk for truck shipments is

Table 1

Mass Flow Data (MT/YR)****
(1000 MWe LMFBR, 80% CF, Annual Average)

Fuel Cycle Component	NASAP**	CRBRP*	Source Brewer	NASAP***	WASH-1535	This Study (Assumed)
Depleted Uranium	2.5 MT	33 MT	19 MT	2.5 MT	17.3 MT	1.1 MT
Fresh Fuel	28.2 MT	36 MT	20.9 MT	35.4 MT	18.8 MT	30 MT
Spent Fuel	28.2 MT	36 MT	20.1 MT	35.4 MT	18.8 MT	30 MT
Uranium Recycle	24.8 MT	31.5 MT	18 MT	31.4 MT	16 MT	25.9 MT
Plutonium Recycle	1.8 MT	3 MT	1.9 MT	2.3 MT	1.7 MT	3 MT
Plutonium Storage	0.2 MT	0.3 MT	0.2 MT	0.3 MT	0.2 MT	0.3 MT
Solidified High Level Waste (Fission Products)	0.8 MT	0.8 MT	1 MT	0.7 MT	0.7 MT	0.8 MT
Losses	0.6 MT	0.4 MT	--	0.7 MT	0.2 MT	--

* Scaled up (times 3).

** Homogeneous.

*** Heterogeneous (spiked).

**** Per RRY (Reference Reactor Year - 0.8 GWe-yr)
(Data sources - NASAP [9], CRBRP [7], Brewer [15], and WASH-1535 [6]).

$$70{,}000 \text{ miles/RRY} \times \frac{0.033 \text{ deaths}}{1{,}000{,}000 \text{ miles}} \times 2 = 0.005 \text{ deaths/RRY.}$$

This is probably split about equally between occupational workers and members of the general public. The multiplication by 2 is to account for empty return trips.

For rail shipments, there were 4.4×10^{-10} deaths per ton-mile, based on 1979 data [16]. Tonnages were estimated as follows:

Spent fuel (30 MT) -
2000 miles x 30 rail cars x 100 tons/rail car = 6,000,000 ton-miles.

High Level Waste (0.8 MT) -
2000 miles x 1 rail car x 100 tons/rail car = 200,000 ton-miles.

Transuranic Wastes -

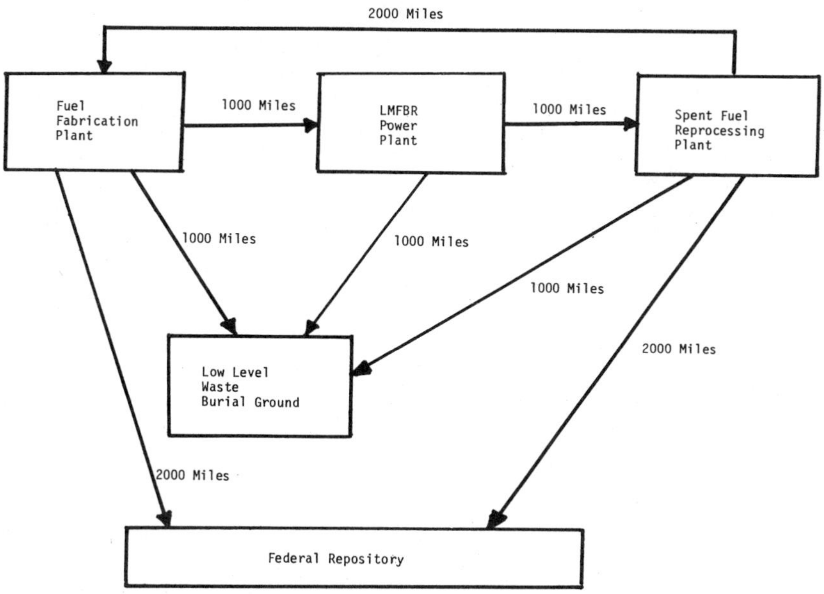

Figure 2

Distances Between LMFBR Fuel Cycle Facilities

2000 miles x 16 rail cars x 100 tons/rail car = 3,200,000 ton-miles. Thus the risk for rail shipments is:

$$9.4\times10^{6}\ \frac{\text{ton-miles}}{\text{RRY}} \times 4.4\times10^{-10}\ \frac{\text{deaths}}{\text{ton-mile}} = 0.004\ \text{deaths/PRY.}$$

About 10 percent of this number would be railroad workers [16].

Risks due to radioactivity releases following transportation accidents have been evaluated previously [6]. These risks are much smaller (about 2×10^{-8} deaths/RRY) than the normal risks due to transportation accidents that were evaluated above (about 0.01 deaths/RRY). The fuel cycle data selected in Table 1 ("This Study") are believed to envelope nearly all practical U.S. 1,000 MWe LMFBR designs. Note also that if uranium is not recycled, transportation requirements would be decreased slightly (see Figure 3 and Table 2), since about four 1,000 mile shipments of depleted uranium would then replace six 2,000 mile shipments of recycle uranium.

Occupational

There are two major types of occupational risks for the LMFBR fuel cycle - occupational radiation exposures and occupational accidents. These risks were estimated using existing data from a variety of sources [6,7,14,15]. Occupational radiation exposures are given in Table 3. The LMFBR power plant exposures span a very wide range. The LMFBR fuel reprocessing plant occupational exposures range from 6-7 man-

Note - RRY stands for reference reactor year (0.8 GWe-yr.)

Table 2

Transportation Requirements And Quantities
(1,000 MWe LMFBR and Supporting Fuel Cycle - Annual Average Amounts)

Material	Quantity	Mode	Mileage
Depleted Uranium	1.1 MT	Truck	1,000 miles (1 trip)
Fresh Fuel	30 MT	Truck	30,000 miles (30 trips)
Spent Fuel	30 MT	Rail	3,000 miles (3 trips)
High Level Waste	0.8 MT ($60\ ft^3$)	Rail	2,000 miles (1 trip)
Recycle Uranium	25.9 MT	Truck	12,000 miles (6 trips)
Recycle Plutonium	3 MT	Truck	12,000 miles (6 trips)
Low Level Waste (Fuel Fabrication)	$800\ ft^3$	Truck	2,000 miles (2 trips)
Low Level Waste (Power Plant)	$2,000\ ft^3$	Truck	5,000 miles (5 trips)
Transuranic Waste (Fuel Fabrication)	$15,000\ ft^3$	Rail	2,000 miles (1 trip)
Transuranic Waste (Fuel Reprocessing)	$1,000\ ft^3$	Rail	2,000 miles (1 trip)
Low Level Waste (Fuel Reprocessing)	$3,000\ ft^3$	Truck	8,000 miles (8 trips)
	Totals:	Rail Truck	9,000 miles 70,000 miles

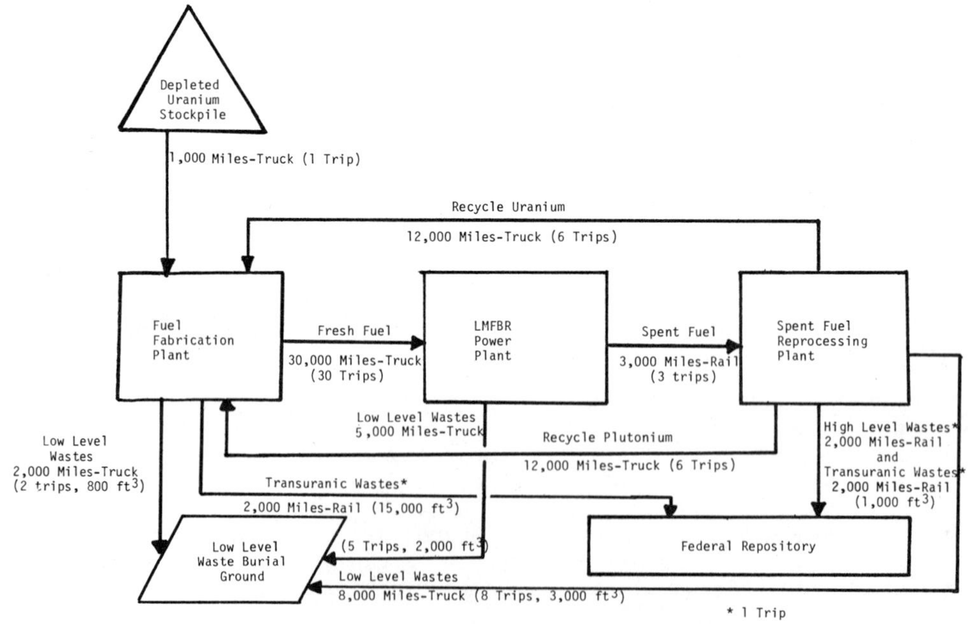

Figure 3

LMFBR Fuel Cycle - Transportation Data

rem per metric ton of fuel reprocessed, based on NFS experience[14], to about 0.3 man-rem per metric ton, based on Barnwell estimates[33]. The estimated LMFBR fuel reprocessing plant occupational exposures are based upon an average of 1 man-rem per metric ton of fuel reprocessed[8]. The LMFBR fuel fabrication plant exposures were assumed to be similar, based on estimates for LWR mixed-oxide fuel fabrication[19]. Waste management occupational exposures were very small[13].* Selected occupational exposures and resulting risks are given in Table 4, along with occupational accident risks. As can be seen, these estimates also span a

Table 3

LMFBR Annual Occupational Exposures
(Total Man-Rem, Whole-Body)

Fuel Cycle Segment	Exposure	Source of Estimate
LMFBR Power Plant	1000 800 250 14 8 7.5 17-55 23,34	NRC Estimate [7] LWR Experience [17] CRBRP [18] EBR-II Experience [31] Foreign Experience [17] Phoenix Experience (1981) [32] Rapsodie Experience (1981) [32] FFTF Experience (1981,1982) [37]
Spent Fuel Reprocessing	200 20-30 10	Nuclear Fuel Services [14] DOE [6] and NRC [19] Estimates Barnwell Estimate [33]
Fuel Fabrication	100 30 10,13 48	B&W (Apollo) Experience [39] LWR (mixed oxide) [19] 308 Building Experience (1981,1982) [37] SAF Line Estimate [38]
Transportation	25 9 (5-20)	CRBRP [7] LMFBR [6] (based on distances in Table 2)
Radioactive Waste Management	<1	GEIS [13]

*See also footnote on page 416.

Table 4

LMFBR Annual Occupational Risks

- Occupational accidents*
  - LMFBR Powerplant (0.01) [20,**]
  - Fuel cycle facilities (0.02) [20,**]
  - Transportation (0.01) [15]
- Occupational radiation exposures
  - LMFBR Powerplant (10-1000 man-rem)
  - Fuel cycle facilities (60 man-rem)
  - Transportation (10-25 man-rem)
  - Total (80-1085 man-rem or 0.02-0.22*)
- Total risk is 0.06 - 0.26*

---

* Deaths for 1000 MWe LMFBR powerplant and supporting fuel cycle on annual basis (with 80% capacity factor).

** Alternate estimates

Spent fuel reprocessing: $\frac{3 \text{ workers}^{33}}{\text{RRY}} \times \frac{12 \text{ deaths}^{35}}{100{,}000} = 0.0004$ deaths/RRY

LMFBR powerplant: $\frac{246 \text{ workers}^{18}}{\text{RRY}} \times \frac{12 \text{ deaths}^{35}}{100{,}000} = 0.03$ deaths/RRY

Fuel fabrication: $\frac{85 \text{ workers}^{36}}{\text{RRY}} \times \frac{12 \text{ deaths}^{35}}{100{,}000} = 0.01$ deaths/RRY

wide range. LMFBR operating experience will be needed to narrow this range.

Population Exposure

The major routine releases of radioactivity from the LMFBR fuel cycle occur during spend fuel reprocessing operations. Releases from all other parts of the fuel cycle, including reactor operation, are much smaller and lead to doses that are minor in comparison to those associated with spent fuel reprocessing plant releases. For example, estimated annual whole-body population doses for the CRBRP power plant radioactivity releases are less than 1 man-rem [7]. The major contributors to population exposures are krypton-85, carbon-14 and tritium released during fuel reprocessing. Krypton-85 is a fission product. Carbon-14 is generated primarily via the

$^{14}N(n,p)^{14}C$ reaction [21] where the nitrogen is an impurity in the mixed oxide fuel (about 20 ppm [22]) and in the stainless steel cladding and other structural materials (about 45 ppm [22]). The tritium is produced during fission and from activation of lithium and boron impurities in the sodium coolant. Table 5 shows local population (one million people within 50 miles of a spent fuel reprocessing plant), U.S. population (300 million people), and worldwide population (constant world population of 6 billion people) exposures for the complete release of tritium, carbon-14 and krypton-85 from reprocessing spent fuel from 1 year of operation of a 1000 MWe LMFBR. These exposures would be lower, of course, if control technology is applied at fuel reprocessing plants to reduce releases of H-3, C-14 and Kr-85. In addition, H-3 releases and resulting exposures would be be reduced if H-3 is not retained in the fuel, but is instead captured in powerplant cold traps.

The U.S. population exposure due to the release of transurancies (from reprocessing spent fuel due to 1 year of operation of a 1000 MWe LMFBR) is shown in Table 7 [23]. This table is the result of an analysis of the health effects of routine plutonium releases for a 1000 MWe LMFBR power plant and its supporting fuel cycle [17]. The risks due to LMFBR fuel cycle radioactivity releases are given in Table 6.

It should be noted that, except for releases of transuranics during spent fuel reprocessing, LMFBR population exposures are primarily whole-body ones.

As previously stated, releases from all parts of the LMFBR fuel cycle, and resulting doses, are minor compared to those associated with spent fuel reprocessing plant releases. Annual average U.S. population whole body doses due to LMFBR power plant releases have generally been estimated to be less than 1 man-rem [7,24]. Doses due to fuel fabrication radioactivity releases have likewise generally been estimated to be less than 1 man-rem, on a reference reactor year (0.8 GWe-yr) basis [7,24]. Doses due to high level radioactive waste management are also estimated to be less than 1 man-rem per reference reactor year [13],*.

However, the total U.S. population exposure consists of both doses due to radioactivity releases and doses due to direct exposure to radiation. This direct exposure for the LMFBR fuel cycle comes from transportation. Estimates of the annual average total-body U.S. population direct radiation exposure due to LMFBR fuel cycle transportation range from 2 to 5** man-rem/RRY [6,7]. Assuming 5 man-rem/RRY as a reasonable upper limit, the total annual risk to the U.S. population from LMFBR fuel cycle transportation is 5 man-rem times 200 cancer deaths per million man-rem or 0.001 deaths/RRY. Thus the total annual risk due to all sources of radiation exposure for the LMFBR fuel cycle is 0.019 (Table 6) plus 0.001 which equals 0.02 deaths per reference reactor year.

Catastrophic Accidents

The risk associated with LMFBR fuel cycle catastrophic accidents is basically the early and late fatalities associated with very large LMFBR

*2800 man-rem, worldwide, for 500 GWe by 2040, LWR once-through fuel cycle = 0.2 man-rem/RRY.

**1-3 man-rem/RRY based on distances in Table 2.

Table 5

LMFBR Fuel Cycle Population Exposures [16, 8]
(Total Man-Rem, Whole-Body, Environmental Dose Commitments, for 1 Year of Operation)

Isotope	Local Population	World Population	U.S Population
H-3	0.6	30	15
C-14	0.5	1300	65
Kr-85	<0.1	100	10

Table 6

LMFBR Fuel Cycle Annual Average Radioactivity Release Risks

- o Cancer Risks (World)
  - H-3 (0.006)
  - C-14 (0.26)
  - Kr-85 (0.02)
- o Cancer Risks (U.S.)
  - H-3 (0.003)
  - C-14 (0.013)
  - Kr-85 (0.002)
  - Transuranics (0.0012)

o Total = 0.019 (U.S. only)

power plant radioactivity releases following a major, but highly unlikely, accident. Accidents associated with other LMFBR fuel cycle facilities don't make a significant contribution to this risk [6]. It is believed that LMFBRs can be designed, built, operated and regulated to be at least as safe as current LWRs. That being the case, LMFBR power plant accident risks should be less than or equal to LWR power plant accident risks on an annual average basis, as shown in several recent analyses [11,12].

It should be noted that the potential for release of any radioactivity exists only for accidents of very low probability. For example, the most likely such accident was estimated to have a probability of occurrence of about 1 in 50,000 per year of CRBRP operation [11]. The radioactivity release associated with that event would be quite small and would cause an insignificant public health effect. On the other hand, an extremely improbable accident that could affect a large number of people was estimated to have a probability of occurrence of 1 in 200 million per year of CRBRP operation [11].

Table 7

LMFBR Plutonium Release Health Effect Estimates

	WASH-1535[6] (1974)	Supplement[17] (1982)
Dose Equivalent Commitment (rem)	(70 Yr Commitment, Q = 10)	(50 Yr Commitment, Q = 20)
Lung	4	9
Bone	26	86 (Surface)
		7 (Marrow)
Liver	10	18
Gonads	0.4	1
Risk Factors (per 10^6 organ-rem)		
Lung	16 - 110*	35
Bone	2 - 17*	5 (Surface) 25 (Marrow)
Liver	1 - 7*	15
Gonads	60 - 1500*	1000
Excess Cancer Deaths		
Lung	.000064 - .00044*	.00032
Bone	.000052 - .00044*	.00043 (Surface) .00018 (Marrow)
Liver	.000010 - .00007*	.00027
Total	.0001 - .001*	.0012
Genetic Effects		
Total	.0002 - .002*	.0006

* Lower estimate based on absolute risk model, higher estimate based on relative risk model.

The assumption that LMFBR and LWR power plant catastrophic accident risks are essentially the same is a reasonable and necessary one. On this basis, the annual average risk associated with LMFBR power plant accidents would be less than 0.02 deaths per reference reactor year [25].

RESULTS

The results of this LMFBR risk analysis are given in Table 8 below.

TABLE 8
LMFBR ANNUAL ELECTRICITY PRODUCTION RISKS (per RRY)

Occupational Risks	0.06 - 0.26
Transportation Risks	0.01
Catastrophic Accidents	0.02
Public Risks (due primarily to radioactivity releases)	0.02
Total	0.1-0.3

Most of the uncertainty shown in this estimate is due to the LMFBR power plant component of LMFBR fuel cycle occupational radiation exposure risks. As previously stated, this uncertainty will only be reduced by obtaining operating experience with large (1000 MWe) commercial U.S. LMFBR power plants. The total public risk for the LMFBR fuel cycle is quite low - only 0.05 deaths per reference reactor year (not counting the fact that a portion of the transportation risk should be assigned to transportation workers).

The major portion of the LMFBR fuel cycle risk is due to radiation exposure - 0.02 to 0.22 deaths/RRY for occupational radiation exposures*, 0.02 deaths/RRY for doses due to radioactivity releases following catastrophic accidents and 0.02 deaths /RRY for public radiation exposures* - a total of 0.06 - 0.26 deaths per reference reactor year. Non-radiation related risk is only 0.05 deaths per reference reactor year (occupational accidents and transportation accidents).

There is a minor amount of "double counting" in these estimates. First, there is undoubtedly some overlap between transportation accident risks and occupational risks due to transportation (see page 412 - Table 4). Also, the accident risk due to rail transportation includes accidental deaths at grade crossings, which are probably also included in the number of deaths due to truck transportation. However, this "double counting" probably isn't significant since the transportation component of the total risk is small - 0.01 deaths/RRY for both transportation accident risks and for occupational risks due to transportation.

There is also some uncertainty that is not shown in these estimates. The assumption of 200 cancer deaths per million man-rem of radiation exposure was used for calculating both public and worker deaths due to cancer following exposure to ionizing radiation. This number could change by roughly plus or minus a factor of 2, as shown in Table 9. Thus, the risk due to radiation exposure might range from 0.03 - 0.13 deaths/RRY to 0.12 to 0.52 deaths/RRY instead of the 0.06 - 0.26 deaths per reference reactor year previously mentioned. However, this assumption (200 cancer deaths per million man-rem of exposure) is believed to be quite reasonable. In addition, there is no need to worry about variations in estimates for the different types of cancer since only total cancer

*These are primarily whole-body exposures.

Table 9

Comparison of Various Estimates of Cancer Deaths Per Million MAN-REM

Type of Cancer	BEIR I Report (NAS-NRC 1972)[2] Absolute Risk Model 30-Year Plateau	BEIR I Absolute Risk Model Life Plateau	BEIR I Relative Risk Model 30-Year Plateau	BEIR I Relative Risk Model Life Plateau	EPA[27]	Reactor Safety Study (NRC 1975)[25] Upper Bound[(b)]	Reactor Safety Study Central Estimate[(c)]	UNSCEAR Report (1977)[4]	ICRP-26[26]	BEIR III Report (1980)[3] Probable Risk[(d)]	BEIR III Maximum Risk[(e)]
Leukemia	-26-[(a)]	-	-37-[(a)]	-	54	28	5.6	15-25	20	20[(f)]	45[(f)]
Non-leukemic	60	74	122	417	-	106	42	-	-	-	-
Lung	16	19	-	-	60	22	4.4	25-50	20	28	100
Bone	2.4	3.0	-	-	16	7	1.4	2-5	5	-	-
Thyroid	-	-	-	-	-	13	2.6	5-15	5	6.9	25
Total Cancers	86	100	159	454	200	134	48	100	100	120	400

(a) 10-year risk plateau following in utero exposure, otherwise 25 years.
(b) Lower bound is zero for all cancers.
(c) Calculated on the assumption that no individual dose will exceed 10 rem.
(d) Linear-quadratic average of absolute and relative models.
(e) Linear, relative model only.
(f) Leukemia and bone cancers combined.

estimates were used and because nearly all of the risk was due to whole body exposures in the LMFBR fuel cycle.

Also, there is considerable uncertainty in the risk associated with catastrophic accidents. This uncertainty would be roughly an order of magnitude [34]. Thus the risk due to catastrophic LMFBR power plant accidents would range from 0.002 to 0.2 deaths/RRY, instead of the 0.02 deaths/RRY previously mentioned.

Potential changes in LMFBR fuel cycle parameters, such as the amount of spent fuel discharged annually (which is related to reactor core design and fuel burnup), would not lead to significant changes in these risk estimates. For example, variations in the annual fuel requirement of plus or minus 20 percent would affect primarily the fuel reprocessing, fuel fabrication and transportation annual occupational exposures and risks - only about 0.05 deaths/RRY, which would increase or decrease by 0.01 deaths/RRY for this 20 percent variation. Or, if control technology is used to reduce C-14 releases by a factor of 5, the resulting U.S. cancer risk due to this release would be reduced from 0.013 to 0.003 deaths/RRY, a decrease of 0.01 deaths/RRY. It should be noted that changes in fuel burnup would not change radioactivity releases or resulting doses significantly.

As can be seen in Figures 2 and 3, it is unlikely that total shipping distances would increase by more than a factor of 2. Thus the transportation risk could increase from the 0.01 deaths/RRY in this analysis to 0.02 deaths per reference reactor year at most.

DISCUSSION

In comparing the results of this LMFBR risk assessment with the results of risk assessments for other energy technologies, it should be noted that no risks were included for LMFBR fuel cycle facility construction or decommissioning. However, it can be assumed that the risks associated with LMFBR fuel cycle facility construction and decommissioning are the same as those for LWR fuel cycle facility construction or decommissioning. For example, a risk of about 0.006 deaths per reference reactor year could be assigned to LMFBR fuel cycle facility decommissioning [28] and a risk of about 0.06 deaths per reference reactor year could be assigned to LMFBR fuel cycle facility construction [*,29,30].

---

* $$\frac{9 \text{ man-hrs.}^{29}}{\text{kWe}} \times 10^6 \ \frac{\text{kWe}}{\text{GWe}} \times \frac{1}{30 \text{ yrs.}} = 3\text{x}10^5 \ \frac{\text{man-hrs}}{\text{RRY}}.$$

and $$\frac{3\text{x}10^5 \text{ man-hrs/RRY}}{2000 \text{ hrs/worker/yr.}} \times \frac{40 \text{ deaths/yr}^{30}}{100{,}000 \text{ workers}} = 0.06 \ \frac{\text{deaths}}{\text{RRY}} .$$

(This figure, 0.06 deaths/RRY, should probably be multiplied by a factor of 2 or 3 to account for increased construction efforts now estimated for nuclear powerplants, 18-27 man-hrs/kWe.)

Genetic effects due to radiation exposures were not included either. This is because fatal genetic effects have not been shown to result from exposures to low levels of ionizing radiation. Also, such effects could only occur many years after LMFBR fuel cycle operations, unlike accidental deaths, which would occur during LMFBR fuel cycle operations, and cancer deaths, which would be expected to occur within 3 to 50 years after exposure to ionizing radiation. Furthermore, potential genetic effects of other energy technologies are generally not included in risk assessments.

Lastly, only effects that could be readily quantified were included in this LMFBR risk assessment. Thus, no risks were calculated for sabotage of LMFBR fuel cycle facilities, since probabilities for such sabotage aren't established. Also, there was no attempt to quantify the potential risks associated with the diversion of LMFBR fuel cycle plutonium to the production of nuclear weapons by either terrorist groups or nations that don't possess nuclear weapons yet. In this case there are many other ways to obtain nuclear weapons or nuclear weapons materials. This include purchase and theft, and the construction of facilities dedicated to that purpose (as opposed to diversion from commercial systems) as well as diversion from non-LMFBR commercial systems. The particular approach that a nation might employ in securing nuclear weapons depends upon desired objectives, the size and development time of the weapons force desired and a variety of characteristics of the country itself such as its technological capability, the availability of other nuclear facilities, sensitivity to detection, vulnerability to sanctions, and political relationships. Given these various factors, it is extremely difficult to establish any probabilities concerning the existence of a particular commercial fuel cycle and the proliferation of nuclear weapons.

In summary, it is believed that the estimate of 0.1 to 0.3 deaths per reference reactor year developed in this LMFBR risk assessment is a reasonable one. This estimate can be compared with results of risk assessments for other energy technologies, provided that differences in what is included in these other risk assessments are properly accounted for.

REFERENCES

1. DOE/NE-0041, "Program Summary for the Civilian Reactor Development Program", U.S. Department of Energy, July 1982, page 14.
2. U.S. National Academy of Sciences - National Research Council, "The Effects on Populations of Exposure to Low Levels of Ionizing Radiation," Report of the Advisory Committee on the Biological Effects of Ionizing Radiations, Washington, D.C., 1972.
3. U.S. National Academy of Sciences - National Research Council, "The Effects on Populations of Exposure to Low Levels of Ionizing Radiation: 1980," National Academy Press, Washington, D.C. 1980.
4. U.N. Scientific Committee on the Effects of Atomic Radiation, "Sources and Effects of Ionizing Radiation," United Nations, New York, N.Y. 1977.
5. EPA 520/7-79-006, "Radionuclide Impact Caused by Emissions of Radionuclides into Air in the United States," U.S. Environmental Protection Agency, August 1979.
6. WASH-1535, "Proposed Final Environmental Statement, Liquid Metal Fast Breeder Reactor Program," U.S. Atomic Energy Commission, December 1974, Volume II, Sections 4.4, 4.5 and 4.7.
7. NUREG-0139, Supplement No. 1 "Final Supplement to Final Environmental Statement, Clinch River Breeder Reactor Plant," U.S. Nuclear Regulatory Commission, October 1982.
8. ERDA - 1535, "Final Environmental Statement, Liquid Metal Fast Breeder Reactor Program," U.S. Energy Research and Development Administration, December 1975.
9. DOE/NE-0001/9, "Nuclear Proliferation and Civilian Nuclear Power,"

Report of the Nonproliferation Alternative Systems Assessment Program, Volume IX, Reactor and Fuel Cycle Description, U.S. Department of Energy, June 1980.
10. "Environmental Report, Clinch River Breeder Reactor Plant," Project Management Corporation, Amendments XIV and XVI, May and October, 1982.
11. CRBRP-1, "CRBRP Safety Study--An Assessment of Accident Risks in the CRBRP," March 1977.
12. "Single Plant Risk Model Development and Application," GEFR-00573(DR), Advanced Reactor Systems Department, General Electric Company, August 1981.
13. DOE/EIS-0046F, "Final Environmental Impact Statement: Management of Commercially Generated Radioactive Waste," U.S. Department of Energy, October 1980.
14. ORNL/Sub-81/31066/1, "Review of the Operating History of the Nuclear Fuel Services, Inc., West Valley, New York Irradiated Fuel Processing Plant, September 1981. See also WASH-1327 (GESMO), August 1974, Volume III and Volume V of WASH-1535 (Reference 6).
15. "Quantification and Comparison of External Costs of Nuclear and Fossil Electrical Power Systems," S.T. Brewer, Energy Research and Development Administration, Washington, D.C. 1975.
16. "Statistical Abstract of the United States, 1982-83", U.S. Bureau of the Census, Washington, D.C., 1983, Tables No. 1047, 1061 and 1088.
17. DOE/EIS-0085-FS, "Final Environmental Impact Statement Supplement, Liquid Metal Fast Breeder Reactor Program," U.S. Department of Energy, May 1982.
18. Based upon operational workforce (246 people) from "Environmental Report, Clinch River Breeder Reactor Plant," Project Management Corporation, Volume 3, p. 8.2-7 and assuming about 1 man-rem occupational exposure per person per year.
19. NUREG-0002, "Final Generic Environmental Statement on the Use of Recycle Plutonium in Mixed Oxide Fuel in Light Water Cooled Reactors," Volume 3, U.S. Nuclear Regulatory Commission, August 1976.
20. R.L. Gotchy, "Health Risks from the Nuclear Fuel Cycle," in "Health Risks of Energy Technologies," AAAS Selected Symposium, C.C. Travis and E.L. Etnier, Editors, Westview Press, Boulder, Colorado, 1983.
21. ERDA-1535, Volume I, op. cit., page III D-27.
22. ERDA-1535, Volume I, op. cit., page III D-23.
23. DOE/EIS-0085-FS, op. cit., page 26.
24. WASH-1535, Volume II, op. cit., pages 4.2 - 73 and 4.3 - 61.
25. WASH-1400 (NUREG-75/014), "Reactor Safety Study," Executive Summary, U.S. Nuclear Regulatory Commission, October 1975.
26. International Commission on Radiological Protection, "Recommendations of the International Commission on Radiological Protection," ICRP Publication 26, Pergamon Press, Oxford, 1977.
27. EPA-520/9-73-003-B, D, "Environmental Analysis of the Uranium Fuel Cycle," Part I - Fuel Supply, Part III - Nuclear Fuel Reprocessing, U.S. Environmental Protection Agency, 1973.
28. IAEA Bulletin, Vol. 22, No. 5/6, "Comparative Risks from Different Energy Systems: Evolution of the Methods of Studies", L.D. Hamilton, October 1980, page 57.
29. WASH-1535, op. cit., page 10.2-4 (Volume 4).
30. "Statistical Abstract of the United States, 1982-83," op. cit., page 412 (construction, 1981 data).
31. ANL-W-RE-82-65, "EBR-II Project Exposure History: 1970-1981," July 29, 1982.
32. "The Development of Fast Neutron Reactors in France," L. Vautrey, at 15th Annual Meeting, International Working Group on Fast Reactors, Obninsk, 30 March - 2 April 1982.

33. "Final Safety Analysis Report, Barnwell Nuclear Fuel Plant Separations Facility," Allied-Gulf Nuclear Services, October 1973, Volume IV, Section 12.
34. NUREG/CR-0040, "Risk Assessment Review Group Report to the U.S. Nuclear Regulatory Commission," Ad Hoc Risk Assessment Review Group, September 1978.
35. "Statistical Abstract of the United States, 1982-83," op. cit., page 412 (industry average, 1981 data).
36. "Environmental Analysis of the Secure Automated Fabrication (SAF) Facility," Supplement to Environmental Assessment for the Fuels and Materials Examination Facility, U.S. Department of Energy, August 1981.
37. Letter Report, "Exposure Usage at 308 Building Laboratory and FFTF for 1981 and 1982," July 15, 1983.
38. Letter Report, "Annual Man-rem Estimated for CRBR Fuel Pin Reduction," July 20, 1983.
39. Based upon data from 1/75 to 3/78 (3 FFTF cores), personal communication from Michael Austin, Manager, Technical Control, Nuclear Power Generation Division, Babcock & Wilcox, June 17, 1983.

EXAMINING THE REALITIES OF RISK MANAGEMENT

B. John Garrick

Pickard, Lowe and Garrick, Inc.
Newport Beach, CA 92660

ABSTRACT

Quantitative risk management in nuclear power plants has been demonstrated on a very limited basis. While not yet receiving the general support of plant owners and operators, the results are impressive and encouraging. It is clear that risk can be quantified, contributors pinpointed, and cost-effective changes can be made for controlling and, if appropriate, reducing risk. It is also clear that an integrated safety model of the plant is necessary to be successful. Experience indicates the extreme importance to risk of support systems such as electric power, room cooling, and secondary heat removal systems. External events are also an important class of contributors to risk. There is still the regulatory mind set that impedes rapid progress toward meaningful risk management based on costs, risks, and benefits. Utility executives and decision makers need to embrace the concept for it to be a realistic success. This has not happened although there is a beginning.

KEY WORDS: Risk, Safety, Management

INTRODUCTION

Sufficient experience now exists, especially in the nuclear industry, to consider the progress that has been made toward meaningful tools or aids for the control and, hence, management of risk. The considerable activity in the field of probabilistic risk assessment (PRA) suggests a high level of interest and application. It is the purpose of this paper to examine our own experience in this regard and to offer some observations and opinions about current practices in risk management and the requirements for success.

RISK MANAGEMENT - WHAT IS IT AND WHY DO WE NEED IT?

One of the truly remarkable observations about the nuclear power industry is the enormous amount of effort given to management issues and procedures, and yet only recently has the idea of an integrated risk management model begun to take hold--and then only ever so slightly. Rather, the U.S. Nuclear Regulatory Commission (NRC) and the industry have chosen to rely almost entirely on regulations to insure safety. This is not a very satisfactory approach. First, it is not satisfactory because it tends psychologically to take responsibility for safety off the designers and operators of the plant and transfer it remotely to people in Washington, D.C. Designers and operators then have their attention on satisfying regulations rather than on creating safety. That is risky;

that is dangerous in my opinion. Second, it is not satisfactory because once the idea takes root that regulations produce safety, the natural next idea is that more regulations mean more safety. Of course, this is erroneous. There comes a point when more regulation means less safety, simply out of the complexity of it, let alone the shifting of responsibility that takes place and the fact that the regulations are often poorly conceived and based on somebody's pet idea rather than on a solid quantitative and plant-specific appraisal. This brings us to a third point.

Relying on the licensing process is not satisfactory because risk is often driven by matters not explicitly addressed by regulations. That is, risk is a matter of hardware, software, and people, and the very detailed way in which they interact. Such detail often transcends licensing coverage. It is the detail with respect to specific plants that must be the basis for meaningful risk management. It is specific, not generic, plants that fail. Thus, what is meant by risk management, is a means of managing risk by exercising an integrated safety model of a specific plant and weighing the costs, benefits, and risks of all available options for achieving risk control.

This rather basic approach has the potential for extraordinary payoffs. First, there is the safety incentive, the improvements that result from a quantitative and real-time knowledge of the risk rather than just of the hazards. A byproduct of the risk model is an identification of options for controlling and maintaining safety. The risk model--that is, the integrated safety model of the plant-provides a quantitative basis for seeing beyond regulatory compliance to the actual bottom-line risk and to the sources of that risk. Second, there is the economic incentive that comes from being able to choose cost-effective options to meet safety objectives. With the risk model, a logical basis exists to minimize plant unavailability due to regulatory requirements that are often imposed intuitively and arbitrarily. A well-done risk model instead provides a quantitative basis on which to take alternative steps that may achieve safety goals while allowing the plant to continue to operate. Other direct economic benefits include (1) the ability to reduce the plant owners' investment risk by having the information necessary to reduce and control the core damage frequency, and (2) the ability to delay and possibly eliminate costly plant modifications resulting from ill-advised suggestions or mandates by the regulatory agencies. In other words, the owner is in the position of dealing with matters of safety and economics from his position of strength--a detailed knowledge of the causes and effects associated with his specific plant. He is liberated from having to depend totally on sometimes arbitrary licensing rules and is, instead, free to make logical and responsible decisions about the safe and economical operation of his plant. Indeed, the risk management approach is the only way to quantitatively provide "assurance to the health and safety of the public."

THE WAY RISK MANAGEMENT HAS WORKED SO FAR

Although it is only a beginning, some very positive results have come out of the nuclear plant risk models completed to date. Generally, the observations made here will be based on the direct experience of the author. No attempt is made to represent other experiences although, obviously, they are a very important adjunct to the story being told here. The most significant result has been the increased understanding by the owners of the design of their plant and how to control its safety. In those risk models in which we have been involved, the approach has been to establish a PRA team made up of both consultant employees and technical staff from the plant owner. The benefits have been extensive and the potential is there for still more. With some exceptions, the benefits have been very rapid and obvious during the course of developing the risk model but have experienced a drop-off once the PRA is complete. There are many reasons for this. First, there is the excitement and level of interest that comes from developing the first integrated model of the plant. Suddenly, there is context and perspective with respect to what is and what is not important to safety. Indeed, there is an alternative to the often vague and intangible interpretations of the licensing requirements. Once such a perspective is developed, often the interest has peaked and a somewhat false sense of security sets in--a security that comes from believing the results of the PRA too much in the first place--and, in the second place, not recognizing the dynamics of risk; that is, the everchanging risk profile. A second reason for a drop-off in benefits following the completion of a PRA is that the PRA team loses its momentum. With the completion of the risk model, a PRA team that can effectively implement the model often ceases to exist. The technology transfer is insufficient, the owner's PRA team lacks clout or support, or the real essence of the risk management concept has failed to capture the owner's imagination. The attitudes and perceptions of the owners toward PRA will be addressed later. For now, some positive examples and benefits of some of the studies completed to date will be described.

Consideration has been given elsewhere (Reference 1) to the methodology of risk management. The methodology for developing a risk model is founded on an approach that accommodates the decomposition of risk into basic causes to facilitate the easy identification of candidate corrective actions. The underlying methodology was developed by Kaplan (Reference 2) and has been applied and expanded on elsewhere (Reference 3). Figure 1 conceptualizes the decompositions steps from effects to root causes.

The level 1 to level 2 decomposition makes direct use of the Kaplan matrix formalism. For level 3 and beyond, the process involves the examination of logic models, cause tables, and data packages--all part of the risk model. This process exposes the hardware, software, and human actions driving the risk and thus facilitates the identification of explicit corrective actions. To illustrate the process of risk management, two examples are discussed. Specific identification of the plants is withheld. The first example is a situation where an external event (examples of external events are wind, floods, external fires, earthquakes, etc.) was the major contributor to risk. In the second example, internal plant failures dominated the risk.

Seismic Example

The PRA for one specific plant identified seismic events as an important contributor to core melt frequency and latent offsite health effects. The failure mode of greatest importance involves the loss of plant control as a result of disabling the control room. The sequence of events included a strong motion earthquake and the subsequent interaction (bumping) of the control room building roof line with an adjoining structure. The result is a possible collapse of the control room roof and inaccessibility of plant controls.

An analysis of this scenario identified an effective fix. It involved increasing the gap between the two buildings and installing rubber bumpers. The cost was about $200,000. This simple modification increased the seismic capacity from 0.27g to approximately 1.00g. Figure

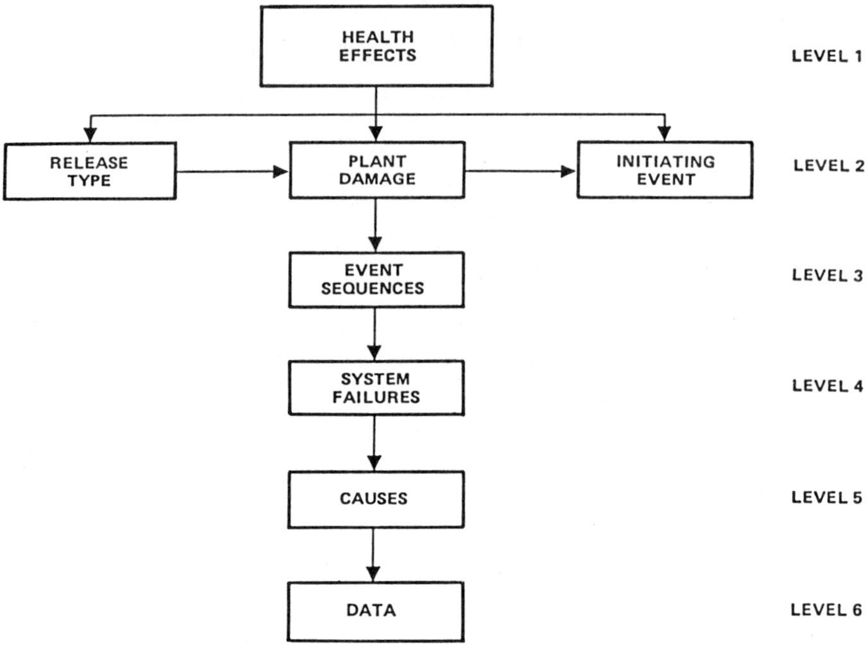

Figure 1

Risk Decomposition

2 shows the relative contribution to core melt frequency before and after the modification. In this particular instance, this was the only seismic modification made. Decomposing the contributors to the "before modification" risk curve showed that there was one major seismic contributor and that there was not much doubt about it. The PRA model also showed quantitatively the considerable risk benefit that could be obtained from this single fix. Further, the model showed that additional modifications to reduce the seismic risk do not exhibit a high cost-benefit ratio. There are two reasons for this. The first involves the fact that in the "after" state, seismic is no longer a significant contributor to the total plant risk. Second, no single seismic fix has much impact on the "after" curve. That is, the after curve derives from multiple failures, and no single failure is dominant.

Example for a Plant Under Construction

Unlike the previous example that involved a plant already in operation, this example illustrates the value of a risk management model during the design and construction phase. In this case, the opportunity existed for a high payoff of the PRA since fixes could be made in many cases without major costly hardware modifications. This example also illustrates the ability to expose the rank order of fixes in terms of risk reduction and cost. As soon as the PRA began to reveal the contributors to risk, it was employed in a risk management sense. The steps involved delineation of the rank and magnitude of the contributors, identification of options for reducing their impact on risk, recommendation of specific fixes, and, finally, a requantification of the risk including the impact of specific contributors. Over a period of 3 years, numerous iterations

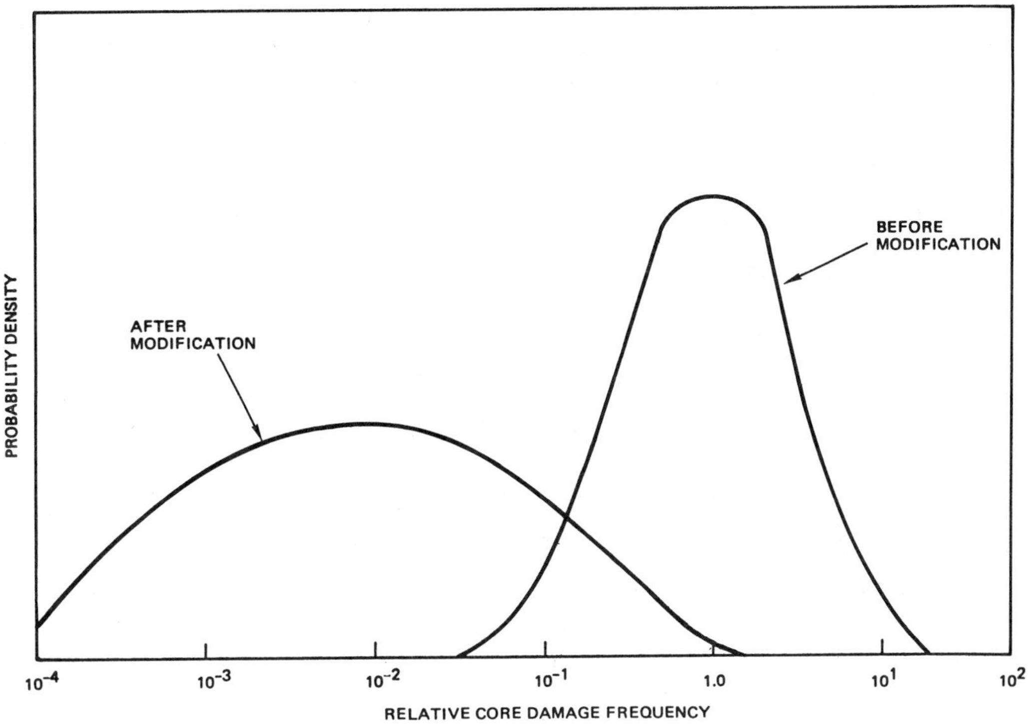

Figure 2

Seismic Risk

of the risk model were made as changes were made to enhance plant safety. The changes involved design; procedures; and in some cases, improved analyses. At each iteration, the risk management team made recommendations of changes based on results from exercising the risk model. The risk management team consisted of risk assessment practitioners, plant operations personnel, licensing representatives, and executives responsible for the plant. At each iteration, the team met to consider the recommendations made by the risk assessment practitioners. The actual decisions were made by the plant owner executives.

It took only a few iterations of the risk model to feel confident that the public risk was acceptably small, well understood, and amenable to control. Several more iterations of changes were required to achieve what the plant owner considered to be an acceptable core damage frequency and, hence, investment risk.

Results used for four of the several risk management iterations are presented in Table 1. These results are presented as a list of manually or automatically actuated systems in terms of their importance to core damage frequency. Percentage importance in Table 1 is a measure of the maximum potential reduction in core damage frequency associated with redesigning a particular system or improving the likelihood that a particular operator action will be performed correctly. The importance of a particular system is based on the ratio of the sum of the frequency of scenarios containing the system in its failed state and the total core

Table 1

Contributors to Core Damage for Four Phases of Risk Management

SYSTEM(s) OR OPERATOR ACTION	PERCENT REDUCTION IN CORE DAMAGE FREQUENCY IF THE INDIVIDUAL SYSTEM (OR OPERATOR ACTION) FAILURE FREQUENCY COULD BE REDUCED TO ZERO			
	FIRST ITERATION	SECOND ITERATION	THIRD ITERATION	FOURTH ITERATION
1. ELECTRIC POWER	11	65	43	52
2. AUXILIARY FEEDWATER	9	11	11	31
3. TWO TRAINS OF ELECTRIC POWER RECOVERED				21
4. LOW PRESSURE INJECTION/DECAY HEAT REMOVAL	4	3	8	19
5. FAILURE TO RECLOSE PORV/PSVs		5	20	17
6. ESFAS/ECCAS			14	15
7. HIGH PRESSURE INJECTION SYSTEMS	3	9	15	14
8. OPERATOR RECOVERY OF ELECTRIC POWER DURING STATION BLACKOUT		50	14	14
9. SUMP RECIRCULATION WATER SOURCE				11
10. COMPONENT COOLING WATER			3	8
11. THROTTLE HPI FLOW (OPERATOR ACTION)			1	4
12. FAILURE OF MAIN STEAM SAFETY VALVE TO RECLOSE			1	4
13. SERVICE WATER	32	<1	10	4
14. SAFEGUARDS CHILLED WATER	20	8	13	1
15. BWST SUCTION VALVE				1
16. CONTAINMENT ISOLATION			1	
17. STEAM GENERATOR TUBE RUPTURE	7	<1		
18. HIGH PRESSURE RECIRCULATION (OPERATOR ACTION)	5	2	11	
19. MAIN FEEDWATER		4	1	
20. FAST TRANSFER OF ELECTRIC BUS			7	
21. OTHER	4	6	6	$<<1$
RELATIVE CORE MELT FREQUENCY	**1.00**	**0.30**	**0.10**	**0.06**

damage frequency. Since fixing any system that fails in a particular scenario will reduce the frequency of that scenario, the reduction potential at each iteration can equal more than 100%.

At each iteration, the total core damage frequency was reduced and the relative risk reduction potential of each system/operator action was changed. After the iteration, each system or action represents a different fraction of the total core damage frequency. If the total goes down and the system remains unchanged, its contribution on a percentage basis will of course increase. The relative total core damage frequency at each iteration of the risk management process is presented in the bottom row of Table 1.

To illustrate the process, the first iteration shown on Table 1 indicates that service water system failures occurred in scenarios that contributed 32% of the total core damage frequency. At this point, this was the system that, if fixed, would result in the greatest reduction of risk. As shown in Figure 1, the next step in the risk management process was to "turn up the microscope" and find the specific failures responsible for the unavailability of the service water system. It turned out that service water unavailability was dominated by failure to switch between normal and emergency water cooling modes (from cooling tower to pond) following initiating events that produced an emergency safety feature actuation signal. One set of values had to close while another opened. An inexpensive fix involved installing an interlock that prevented the normal valves from closing before the emergency valves opened. This fix netted a full 30% reduction in core damage frequency and a very favorable benefit-to-cost ratio. Furthermore, service water failure ceased to be an important contributor to risk in any subsequent interactions.

In the second iteration, the electric power system, which contributed only 11% to the core damage frequency in Iteration 1, becomes relatively more important even though its design has not been changed. As the contribution to risk from the service water and safeguards chilled water system decreases, resulting in a decrease in the core damage frequency, the relative contribution of the electric power system increases. Fixes on the electric power system were not as effective for reducing risk and it remained as an important contributor throughout the risk management process. The contribution to risk from electric power was dominated by the unavailability of diesels. There were few simple procedural or hardware fixes with a favorable cost benefit. One fix that was favorable is discussed later.

An important adjunct of the risk management process, also shown in Table 1, is the gradual leveling of system importance as the process continues. In the fourth iteration, the system importance showed a continuum more characteristic of a balanced design. The core damage frequency is relatively low and no single system is so poorly configured that it dominates the risk. Of course, this characteristic makes it increasingly difficult to determine which system to attack next if significant further reductions in risk are required.

Table 2 delineates some of the specific recommendations and results that are background to the iterations noted in Table 1. The risk management activity was deliberate, painstaking, and systematic. It covered a period of 3 years and was dependent on having computational tools specifically designed to allow the decomposition of risk into the important contributors at each step along the way.

In addition to the two examples of risk management summarized above,

Table 2

Examples of Risk Management Recommendations Derived from a Risk Model

SYSTEM OR SCENARIO	ITERATION	CAUSES CONTRIBUTING TO SYSTEM FAILURE	RECOMMENDATION	CORE DAMAGE FREQUENCY REDUCTION	ESTIMATED COST ($1,000)
1. SERVICE WATER SYSTEM	FIRST	LOSS OF SERVICE WATER PUMPS DUE TO CLOSED DISCHARGE PATH	• INSTALL INTERLOCK TO ENSURE THAT THE DISCHARGE VALVES TO THE COOLING POND ARE OPEN BEFORE THE DISCHARGE VALVES TO THE COOLING TOWER CLOSE	31	5
2. SAFEGUARDS CHILLED WATER	FIRST	DEPENDENCY OF AUXILIARY FEEDWATER PUMP ON CHILLED WATER COOLING (DURING REACTOR TRIP A CHILLED WATER FAILS TO CONTINUE TO RUN AND TRAIN B FAILS TO START)	• REEVALUATE REQUIREMENT FOR CHILLED WATER COOLING FOR AUXILIARY FEEDWATER PUMP OPERATION	10	15
			• CONTINUOUSLY OPERATE BOTH TRAINS	2	
			• CROSSTIE SAFEGUARD CHILLED WATER TRAINS	10	600
3. AUXILIARY FEEDWATER	FOURTH	FAILURE OF AUXILIARY FEEDWATER AFTER LOSS OF OFFSITE POWER (BOTH ESSENTIAL BUSES) 81% – TURBINE PUMP HARDWARE 15% – TURBINE PUMP MAINTENANCE	• INSTALL THIRD AUXILIARY FEEDWATER PUMP AND CROSSTIE SAFEGUARDS CHILLED WATER TRAINS	25	2,600
		AFTER LOSS OF ONE ESSENTIAL BUS 32% – MOTOR-DRIVEN PUMP HARDWARE 32% – INJECTION VALVE HARDWARE 25% – MOTOR-DRIVEN PUMP MAINTENANCE 8% – DEPENDENT VALVE FAILURES	• INSTALL THIRD AUXILIARY FEEDWATER PUMP	20	2,000
4. ELECTRIC POWER RECOVERY	FOURTH	DIESELS FAIL TO START AND THE OPERATORS FAIL TO RECOVER OFFSITE POWER	• ADD ONE OR TWO ADDITIONAL EMERGENCY DIESEL GENERATORS	40	—

there have been many other cases of successfully impacting risk in a quantitative way. Some highlights are as follows:

- Essential Electric Power Systems. Early results from the PRA of one unit of a two-unit plant indicated that several station blackout scenarios were important contributors to the frequency of occurrence of core damage. All together, these scenarios contributed approximately 60% of the core damage frequency. From these results, it was possible to identify specific modification options in the design of the AC power systems that would increase the likelihood that AC power was recovered, thus reducing the core damage frequency. For the most attractive options, the impact in core damage frequency was computed with the cost of implementation. Based on this analysis, a decision was made to establish the procedures for manually cross-connecting the same train (A to A or B to B) essential buses of Unit 1 to those of Unit 2, thus enhancing the availability of at least one operating diesel generator to both units. Incorporation of this new procedure reduced the frequency of station blackout scenarios by approximately 55%. The total core damage frequency was reduced by about 33%.
- Third Auxiliary Feedwater Pump. Following the accident at Three Mile Island Unite 2, the NRC asked all utilities with operating or soon to be operating nuclear plants to perform a detailed calculation of the reliability of their auxiliary feedwater (AFW) system. The calculation showed that when used in conjunction with the other systems available for removing heat from the reactor coolant system (RCS), the ability of the plant to remove heat from the RCS satisfied the reliability criteria provided by the NRC. Nevertheless, the NRC mandated the utility to install an additional auxiliary feedwater pump at each unit of the two-unit plant. The PRA for this plant put this action in its proper perspective. As a result of the PRA, it was clear that the type, location, and required support systems (not the absolute number of pumps) were the keys to determining the importance of the auxiliary feedwater system to risk.
  The third pump proposed by the utility and accepted by the NRC was a nonsafety-grade, motor-driven pump installed in the turbine building and connected to either of the plant's other two AFW pump discharge lines. The other two pumps were installed in fire and flood protected rooms with tight doors and thick concrete walls surrounding them as required to qualify them as safety grade. In order to operate during most scenarios, these two rooms required cooling to protect the electronic pump controls. Because the third pump was not in such a room, it did not require similar cooling. Adding this pump, which is independent from the cooling system, significantly reduced the overall contribution of the AFW system to the frequency of severe core damage. PRA analysis and risk decomposition showed that more reduction in risk came from lessening the dependence of the AFW system on the room cooling systems than from adding a third pump.
- Service Water. The probabilistic risk assessment for this large pressurized water reactor indicated high frequencies of failure for the service water system. The importance of this system is that its failure could lead to failure of the diesel generators (emergency power) and the fan coolers (containment heat removal capability) due to lack of cooling water. The scenarios of concern in this case involved the

mispositioning of a pump group selector switch and the resulting unavailability of service water to essential safety systems. A simple change in procedure to assure a daily check of the switch position reduced the mean frequency of failure of the service water system on demand by a factor of approximately 300. While not nearly by as big a factor, there was also a reduction in the core melt frequency of 3 to 10.

- Emergency Operations Support. A recent event at a BWR with an ongoing PRA involved the inability to open a shutdown cooling line valve on demand. The PRA provided information to the emergency response center regarding probable failure modes of the valve and several alternate cooling configurations. The latter included potential "off normal" use of systems, such as heat removal via the reactor water cleanup system and a feed and bleed option.
- Emergency Electric Power. This was a case where a support system dependency compromised the redundancy of a main line safety system. The scenario involved the loss of the long-term fuel supply to a diesel generator as a result of no power to the fuel oil transfer pump. In particular, because of the alignment among the fuel oil transfer pumps, their power supplies, and the diesel generators, there resulted an asymmetric set of fuel oil failure state dependencies. Although the design included three diesel generators and three fuel oil transfer pumps, the net effect of the asymmetric power loading was to make one diesel generator dependent on the status of the other two and to make the entire system very sensitive to the status of one particular diesel generator. The PRA revealed these dependencies and identified them as the major contributor to failure of electric power and an important contributor to core melt frequency.

  A modification was proposed in the power supply logic for the fuel oil transfer pumps. The modification results in a more reliable supply of fuel to the three diesel generators under normal ad degraded power conditions and eliminates the dependency of one diesel generator on the operability of the other two. The impact of this proposed design change is to reduce the frequency of fuel damage by over a factor of 50.
- Fires. As a result of a plant-specific PRA, it was discovered that fire was an important contributor to risk. In this case, two fire locations dominated: the switchgear room and the electrical tunnel entrance area. Analysis indicated that large fires in these areas could lead to reactor coolant pump seal failure [small loss of coolant accident (LOCA)], loss of all components necessary to mitigate core melt, and loss of the safety systems necessary to prevent containment failure. The risk assessment revealed a number of options for reducing the impact of fire on risk. Three options were considered in detail: (1) self-contained seal injection charging pumps not receiving power through critical fire zones; (2) addition of alternate cable feeds and contingency cables to emergency equipment; and (3) additional fire barriers to increase time available to extinguish fires.

  The analysis based on the PRA indicated that type 1 fixes lead to core melt frequency reduction factors of 3 and 8 and release frequency reduction factors of 2 and 15. The corresponding reductions for type 2 fixes are 3 to 10 and 10 to 100, respectively. Type 3 fixes result in generally smaller reduction factors varying from 3 to 10 for both core melt frequency and release.

- Break Between Isolation Valves. A high pressure coolant injection (HPCI) system or reactor core isolation cooling (RCIC) system steam line break at the drywell penetration (between isolation valves) was previously identified as a BWR risk concern. An ongoing BWR PRA explicitly addressed this question. To do so required the construction of a more comprehensive event model than has been previously attempted as well as an empirical pipe-break model. The model addresses those fractions of HPCI or RCIC steam line breaks that occur specifically between the drywell wall and the outboard isolation valve. Further, successful isolation for these cases would depend solely on the inboard valve (one out of one). The preliminary results suggest that such breaks are not significant risk concerns.

All of the above are real examples of risk management. Other examples include the decision by a plant owner not to implement a costly offsite power transmission modification, a decision to modify the battery room to increase its resistance to fire, and a quantification of the effectiveness of specific safeguard systems. The significance of all these examples is that the initiative was the plant owner's desire to control and, where appropriate, to reduce the risk to the plant and to the public. This is in marked contrast to only responding to a requirement of the NRC.

Even with the successful examples cited above, it appears that the nuclear industry is still a long way from generally embracing the risk management idea. It is important to ask why and to determine if anything can be done about it.

WHAT WILL IT FINALLY TAKE FOR RISK MANAGEMENT TO BECOME A REALITY?

Before risk management can become a generally accepted concept, several very important things must happen. Probably the most important requirement is convincing the industry of the need for it. A close second requirement is the need for greater confidence in the concept of risk management itself. The thought processes of plant owners are sometimes so regulatory-oriented that to do something that is not a specific regulatory requirement, especially when it involves safety, is almost outside their mind set. Of course, there are reasons for this attitude. One is that they believe their plants are safe and that further expenditure of resources is not in their best interest. Another is that they have confidence in the licensing process and truly believe that if they are in compliance, safety is assured. There are serious flaws with both of these views. First, compliance without regard to cost and real risk will inevitably destroy the industry. The runaway costs associated with some of the plants recently starting up are a clear manifestation of the weakness of the "compliance at any cost" approach. Meanwhile, there is often no idea of what the real risk is in those plants absent an integrated safety model. Second, there is risk in these plants--perhaps not to the health and safety of the public--but certainly the investment risk is not inconsequential. In other words, the comprehensive risk studies performed to date have pretty well confirmed the high degree of safety of nuclear power plants with respect to public health. On the other hand, the likelihood of costly plant damage is considerably greater than most observers had believed. This kind of risk protection has not been, and perhaps should not be, the emphasis of the NRC. However, the absence of such emphasis has left the plant owners vulnerable to severe financial setbacks. Thus, the need for comprehensive risk management with the owners as prime movers for its implementation seems obvious. The failure to recognize this, except in selected cases, seems to be a reflection of the lack of management interest.

To change the situation will require a change in the consciousness of utility management on the matter of risk. Some of the actions that might help are as follows:

- Senior management of power plant owners must break way from the compliance state of mind and embrace a more fundamental and responsible notion of risk management than can be provided through regulations.
- A more visible commitment needs to be made by senior management to a meaningful risk management activity not encumbered with day-to-day compliance activities. The plant owners must be more eager to lead--not wait to see what "the NRC is going to do."
- There needs to be within the plant owners organization a dedicated champion of the risk model--somebody who has some real clout with the decision makers and who is very knowledgeable about how power plants work. The risk manager must have a program for deliberate and systematic application of the risk model and be in the path of the decision process.
- Plant owners must overcome their communication difficulties between different internal organizations, such as engineering and operations. Operations must embrace, apply, use, and have some control over the concept of risk management.

Of course not all the initiative for successful risk management must come from the plant owners. Others, such as the consultants, have to get their act together as well. Foremost, the methodology has to become more "user friendly" and not make the plant owners hostage to the consultants for effective application. There needs to be more packaging of the expertise required to make risk management a reality. That is, implementation needs to be more automated. Consultants have not done a particularly good job of apprising the plant owners of the commitments necessary for meaningful risk management. There is too much attention on developing the model in the first place and not nearly enough on the requirements for making the PRA a "living" model. The steps being taken by the NRC with their safety assurance programs are a constructive beginning. Again, it is the NRC providing leadership here when it should be the plant owners.

In spite of the difficulties and the lack of confidence by some in quantitative risk management, there is reason for hope. Quantitative and user friendly risk models are being developed. The NRC is utilizing the PRA thought process more in its licensing activities. Plant owners are beginning to see not only the safety value of PRA but other spinoffs, such as economic benefits and a basis for improving plant availability. Finally, the greatest benefit of all is the very good prospect of an enhanced public image--something nuclear power desperately needs and deserves.

ACKNOWLEDGMENT

The author wishes to thank Frank R. Hubbard, John G. Stampelos, and Stan Kaplan of Pickard, Lowe and Garrick, Inc., for their assistance in preparation of this paper.

REFERENCES

1. Garrick, B.J., "Experience and Advancements in Risk Assessment," Nuclear Power Reactor Safety Course, Massachusetts Institute of Technology, July 16, 1984.
2. Kaplan, S., "Matrix Theory Formalism for Event Tree Analysis--Application to Nuclear Risk Analysis," Risk Analysis, Vol. 2, No. 1, March 1982.
3. Kindinger, J. P., F.R. Hubbard, "Application of the Midland PRA to Plant Design Decision," ANS-Sponsored International Nuclear Power Plant Thermal Hydraulics and Operations Topical Meeting, Taipei, Taiwan, October 1984.

INTEGRATED ECONOMIC RISK MANAGEMENT IN A NUCLEAR POWER PLANT

R.N.M. Hunt* and M. Modarres**

Baltimore Gas and Electric Company*
Baltimore, Maryland 21203
University of Maryland**
College Park, Maryland 20742

ABSTRACT

Goal trees have been found to be a powerful tool which can be employed as the integrating foundation for a comprehensive industrial economic risk management program. The paper discusses the rules for development of these goal trees and provides an example which demonstrates their usefulness within the context of a nuclear power plant economic risk management program.

KEY WORDS: Economic Risk; Nuclear Power Plant; Goal Trees; Risk Management.

INTRODUCTION

When risk management is discussed within the context of a nuclear power plant, the issue which immediately comes to the fore is the one associated with public health and welfare. This is clearly very important to the plant owner, but represents only one of the many sources of risk to which he is exposed and from which he wishes to optimize exposure. It is to the integral risk of all contributions that risk management programs must address themselves, since it is this factor which governs the extent to which a nuclear plant contributes to the owner's economic strength. Development of a comprehensive risk management program requires that an integrating framework be devised to provide not only identification of all the discrete contributors to economic risk, but also an understanding of the ways in which they contribute. It is this understanding which enables the owner to effectively evaluate all potentially feasible schemes to be employed in maintaining overall risk at an acceptable or optimal level.

A tool which has proven useful in providing this required degree of understanding is called a "goal tree". The remainder of this paper will detail how such a goal tree can be rigorously constructed and describe an example which will briefly demonstrate how such a tree can provide a means for prioritization within the complete spectrum of risk contributors.

Goal Tree Development

To develop a goal tree, the first step involves definition of the top event or "objective" for the tree. This top event, which for a nuclear plant must reflect the economic objective, should be explicitly defined in terms which make it a single unambiguous statement. It is from this definition that the analyst will identify and relate all the different plant goals and subgoals which must be achieved to attain the overall economic objective. Typically the objective and its supporting goals are displayed in tree format so that their interrelationships can be clearly represented and easily recognized, as demonstrated graphically in figure (1).

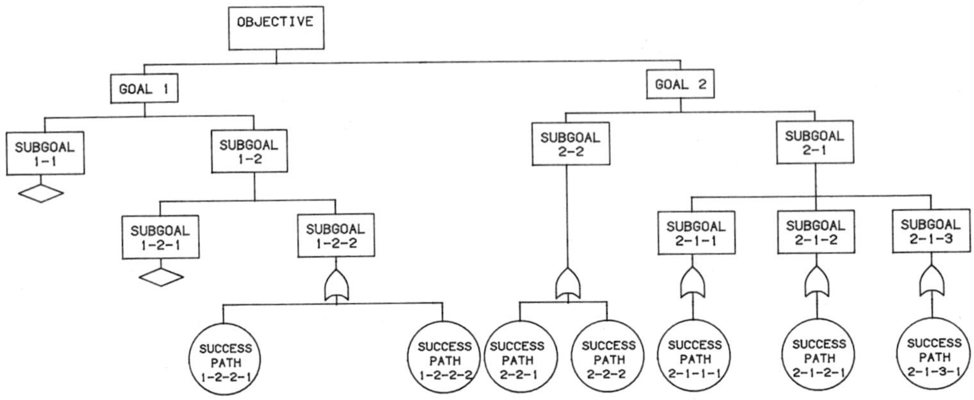

Figure 1

Typical Goal Tree Structure

Rules for Goal Tree Development

The goal tree is built vertically downward from the objective in levels, wherein the analyst subsequently decomposes each identified goal to its necessary and sufficient set of dependent subgoals. The prime attribute required of the analyst during this task is that he have a broad and expert understanding of the physical processes involved, since it is the achievement of these physical processes by the hardware that will allow their optimal performance and maximization of the objective function. It is this composition of fundamental processes which will be represented in graphical form by the goal trees.

As the vertical detailed development of the tree increases, it is necessary that tests by applied to ensure its accuracy and completeness, and that hierarchy between goals and subgoals is rigorously maintained. This latter requirement is crucial to the tree's being capable of providing the needed framework for future cause-consequence analyses.

The rules or tests which are to be applied at every level of the developed tree to ensure adequacy are:

- Upon looking upward from any subgoal towards the objective or tree-top, it is possible to explicitly define why the specific goal or subgoal must be satisfied.
- Upon looking downward from any goals towards the bottom of the tree, it is possible to explicitly define how the specific goal or subgoal is satisfied.

Failure of the tree to pass either of these tests at any level implies that:

There is a lack of completeness, and intermediate subgoals have been omitted, or
the hierarchy of goals or their interdependencies have neither been rigorously nor completely defined.

The process of downward development of the tree continues until the analyst is unable to further define subgoals without recourse to using references to actual hardware. As soon as hardware is explicitly mentioned, the tree becomes one which describes "success paths," not goals. It is important that this boundary between goals and success paths be recognized because this is the point at which the logical structure of the tree changes. Within the goal tree, all goals must be connected by logical "AND" gates, whereas the success paths must be connected to the goals they serve by logical 'OR' gates. In this latter case, it should be recognized that a single input 'OR' gate will be used if there is only one success path for any specific goal.

Use of the fact that all connectors in the goal tree are logical 'AND's' can be made in a convention which allows the analyst to provide a means for easy distinction between the goal tree and the hardware success paths. The goal tree is drawn with no gates explicitly defined. It a gate is shown, success paths are being described.

An understanding of the logical structure of the tree is necessary to recognize that the nature of the goal tree does not in any way change the boolean expression representing logical relationship of components to achieve the top event for the plant, which is clearly a function of hardware configuration not the individual goal structure. The implication of the above statements is that if the tree is to be used as an explicitly quantifiable analytic tool, its structure below the goal level is adequate. The goal portion of the composite tree merely provides a visual display of the analyst's understanding of how the hardware interacts, and the affect of its performance on the plant objective. It is this understanding, or intelligence within the tree which makes it so valuable in a risk management capacity. Since it is possible to look at the composite goal-success path tree and identify the exact relationship between any piece of hardware and the plant objective, the tree can become the desired cause-consequence framework so necessary for a comprehensive risk management program. Through this mechanism it is possible, conceptually, to establish hardware performance requirements which are commensurate with the degree of their contributing risk.

Hardware Support Goal Trees

The plant description provided by the combination of success paths and the goals which they support, can be used to both understand the plant and provide a base for quantification of the probability of successful achievement of each goal. To do this, it is necessary to know the frequency with which the success paths are demanded and their individual reliabilities upon demand. Empirical estimates can readily be assessed for both individual component reliabilities and success path demand rates so that estimates of goal success probabilities can be made. The weakness in this approach lies with the fact that the assessment originates with acturial data, without regard as to why the data has the magnitude that it does. In reality, the reliability or availability of any component is determined by the design, construction, operation, or maintenance activities to which it is subjected. These are, in a way, directly analogous to the support systems or success paths which support the overall plant objective. In fact, it is possible to construct analogous goal tree and success path presentations which are of essentially the same structure as the whole plant tree, but with the prime difference that there is a distinct goal for each system, equipment, or component which provides an understanding of the various subgoals and shows how the hardware must perform to meet its objective. The four goals to be satisfied by this tree are:

- energy use of the equipment is maximized (efficiency)
- time required to start the equipment upon demand is minimized (demand failure rate and time to restore)
- time between failures is maximized (operating failure rate)
- time required to return failed hardware to service, given the occurrence of a failure is minimized (time to restore)

All activities by the human are in some way attempting to address and affect one of the above goals. The goal tree concept is used to decompose each of the identified goals to provide the necessary understanding of the minutiae of individual goals needed, and show how and why they are needed, just as they were in the main tree. An example of a partial hardware goal tree is provided in figure (2). Connected to the bottom of this goal tree will be the success paths which, however, do not represent hardware but will describe explicit human activities. As these success paths are examined in further detail, it becomes apparent that they represent plant programmatic activities. This allows the analyst to physically relate all plant activities to the overall economic plant objective, and recognize that the complete tree provides the necessary framework to integrate all risk management activities in an optimal fashion.

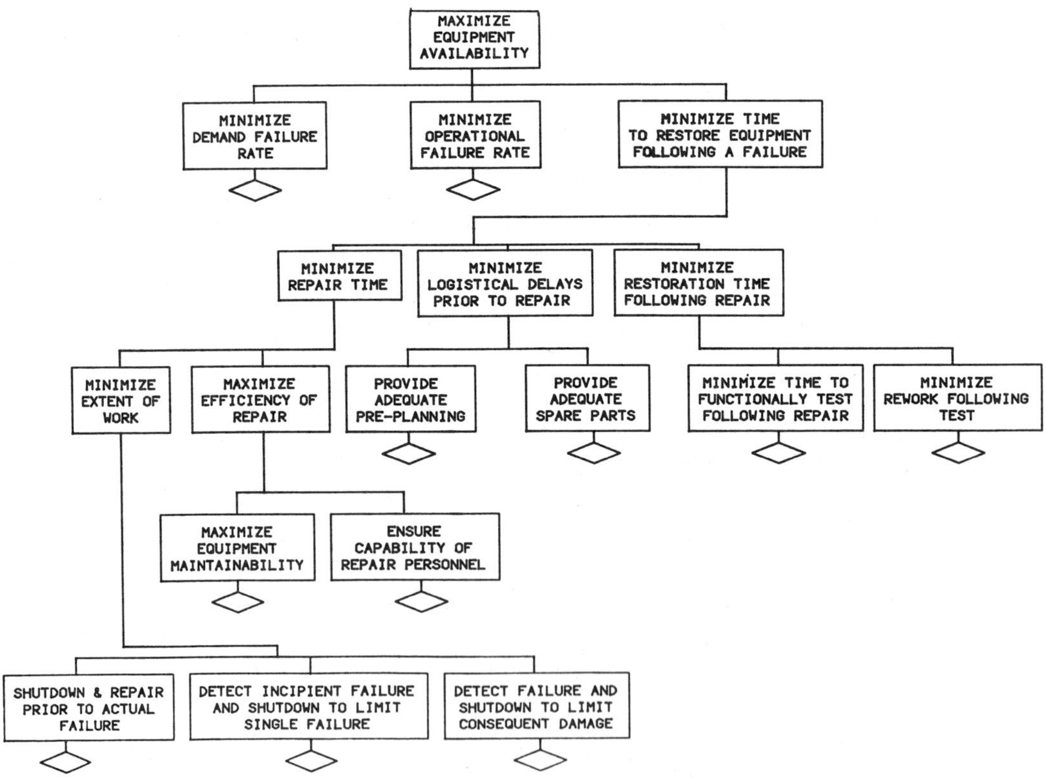

Figure 2

Example of a Partial Equipment Goal Tree

Comments on the Tree

As the tree is constructed, a strict hierarchy is maintained, so that the relationships between goals are held true. The level of the goals in the tree do not, however, indicate their importance. It is the degree o effect on the objective that determines their importance, in fact, it is possible to visualize the entire tree as a "mobile" handing in space with each of the elements connected or suspended by elastic whose length is determined by the inverse of its importance. As the plant moves through its spectrum of possible states, the goals would move relative to each other, so that those which have the higher importance at any point in time will be closer to the objective, conceptually.

Clearly, as the reader visualizes the complete plant tree with all its various branches moving around as the plant state changes, he is bound to wonder at the practicality of the approach. The practicality arises from the analyst having the basic understanding of how, conceptually, the plant is put together and then use this profound understanding to reduce the tree to a point at which it is useful. The following brief example of a plant goal tree and the way in which it can be examined to understand the concept of importance for a goal which is dependent on plant state is provided in the hopes that the merits of this extremely powerful approach can be clearly recognized.

Nuclear Power Plant Goal Tree - An Example

To demonstrate the principles outlined in the previous sections of this paper, a brief example of a goal tree has been developed. This goal tree, typical of a nuclear power plant, is shown in figure (3).

From this figure, it is easy to not only identify the many concerns associated with nuclear generation of electric power, but to understand how they interact with each other. The right side of the tree represents the "productivity" concerns and the left branch represents those associated with "safety". This tree perhaps reflects the problems within the nuclear industry, wherein there have tended to be adversarial roles played by both the regulators and owners. The regulatory agencies have been concerned primarily with the left hand side of the tree, since this is the side where risk contributions to the general public originate, and the owners have focused their own attentions primarily to the right hand side. What can easily be recognized from the tree is that there are many issues common to both sides. These issues are clearly recognized by the industry and the regulators, but had there originally been more integration of both productivity and safety concerns during the plant design, both would have been more effectively addressed.

Reduction of trips or plant transients which have a large impact on economic performance can also be recognized as safety issues if the tree is developed fully, since they appear as initiators which require actions to protect the core, which is a necessary step towards maximizing safety. Analogously, protection of the core is easily identified as a productivity concern since it severely impacts the time needed to restore the plant, should this goal not be achieved.

By closely examining the tree, it is possible to understand all the issues associated with plant operation and place them into their proper context. Both competing and complementary issues can be recognized so that when decisions are made their full ramifications are obvious. The tree does not explicitly contain consequence definitions, but it is easy to use the tree as the framework for a full cause-consequence model for the plant.

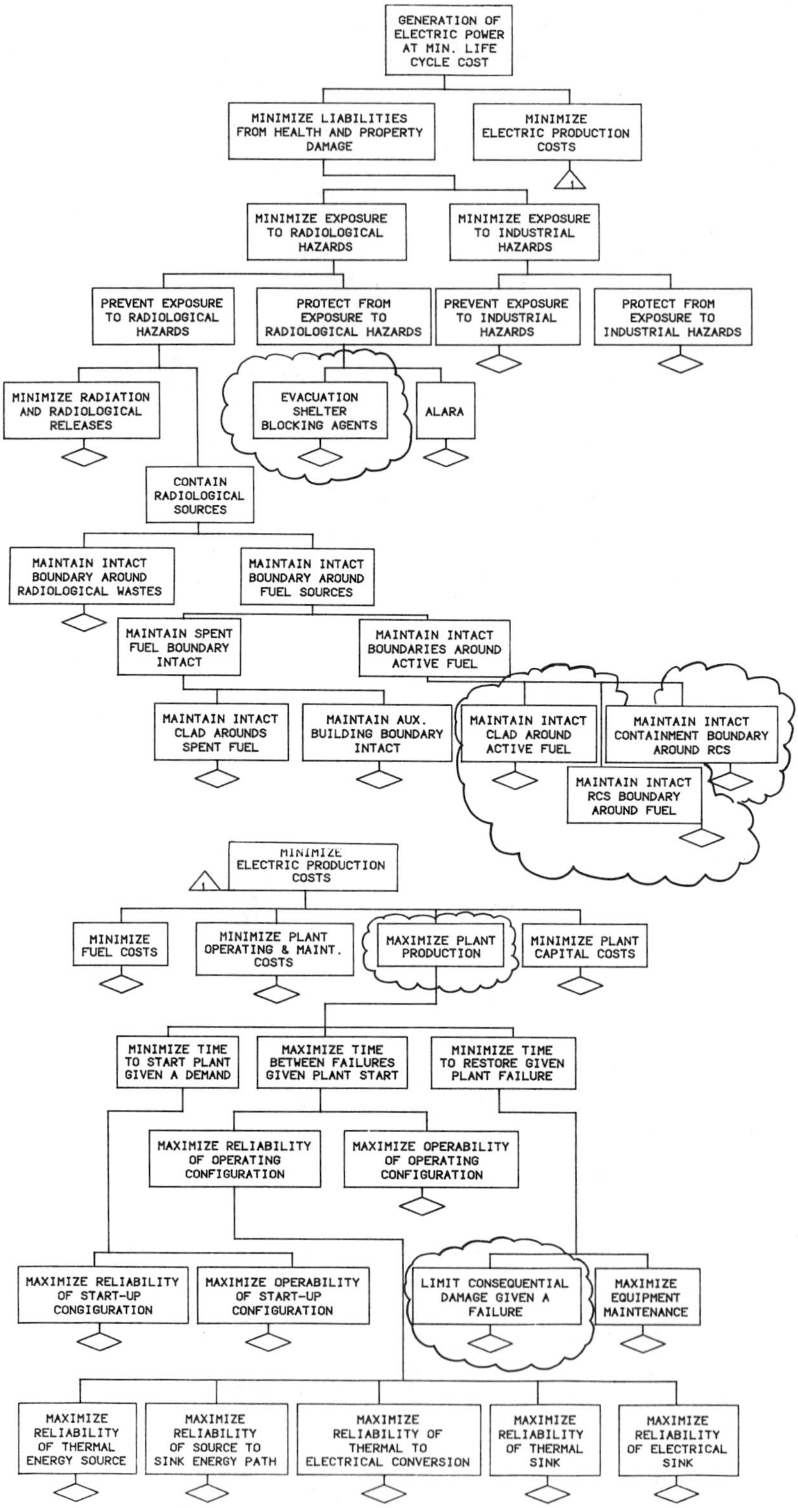

Figure 3

A Typical Goal Tree for Nuclear Power Plants

Pragmatic Application of the Tree

Following the description of the tree provided above and in figure (3), clearly the reader must have some concerns as to the practicality of its application even though many of its inherent capabilities have been hinted at.

First, can the tree be used to establish the relative importance of the identified goals in an operating industrial plant environment? Secondly, can absolute measures of these identified importances be calculated?

Demonstration of the pragmatic qualities of the tree for this paper requires that some simplifying assumptions be made, but which do not invalidate the results. The simplification first requires pruning of the tree so that it represents only a limited number of the issues. For the example presented, only the goals within the "clouds" on figure 3 will be assumed to be of concern. These goals are easily recognized by industry analysts as representative of the "defense-in-depth" philosophy, and with which most are well familiar. They are:

- power production
- core and plant protection
- containment integrity
- site emergency preparedness

Below each of these goals, if the tree were fully developed, would eventually appear the hardware required to support the goals, logically combined to provide the required success paths. Using this hardware model, it would be possible to calculate a conditional probability for the successful achievement of each goal, together with an assessment of the consequences should the goal not be achieved; such an analysis utilizes the cause consequence nature of the tree.

If the events which result in a demand for successful achievement of a goal are identified and their frequencies known, clearly the risk associated with each goal can be calculated.

"Importance" of a goal clearly can be represented as the risk contribution which it makes toward the overall objective. This risk contribution is dependent upon the likelihood of a demand of successful demand for successful achievement of a goal during any particular plant state, so the risk contribution of each goal is a variable and its "importance" so measured becomes a function of plant state. (Picture the goal tree elastically connected) This is entirely consistent with intuition since:

- When the plant is starting up, the important goals are associated with the reliability and operability of the start-up configuration.
- When the plant is operating, the important goals are associated with the reliability of power production.
- When the plant fails (trips), the importance of both the restoration goals and those goals associated with "safety" increase. Following a plant trip, it is necessary to not only restore the power generation capability, but also ensure that the core protection systems respond effectively and ensure minimum chances of core damage.

Visual display of the importance of various goals and their

relationship to plant state can be provided as a composite risk curve, as shown in figure (4). This curve, as presented in the paper, does not represent the product of expensive calculations, but is provided to show how it can be used to demonstrate the concepts and provide the basis for a discussion on its future potential uses.

The Use of the Composite Risk Curves

The actual cost vs. plant state curve (figure 4) can be quantified using simplifying, but documented, assumptions and conventional reliability, availability, and consequence analytical and modeling techniques. Any postulated change in plant character can be evaluated within the context of its effect of plant performance over the complete spectrum of operating states and thus the true economic cost or benefit of the changes assessed. Quantification of this net benefit is provided by the integration of the area between the original and the newly calculated risk curves.

CONCLUSIONS

The foregoing descriptions of the goal tree methodology and its application in the development of a process suitable for integrating the effects of all individual plant risk contributors, seems to fulfill the requirements of the fully comprehensive risk management program briefly outlined in the introduction to this paper.

The use of the methodology is by no means limited to nuclear power generation. When the concepts and approach were applied in other areas, they showed themselves to be entirely suitable in developing the required understanding of any physical or industrial process, and to be a powerful tool which can be applied in the definition and solution of problems which involve complex decision making elements.

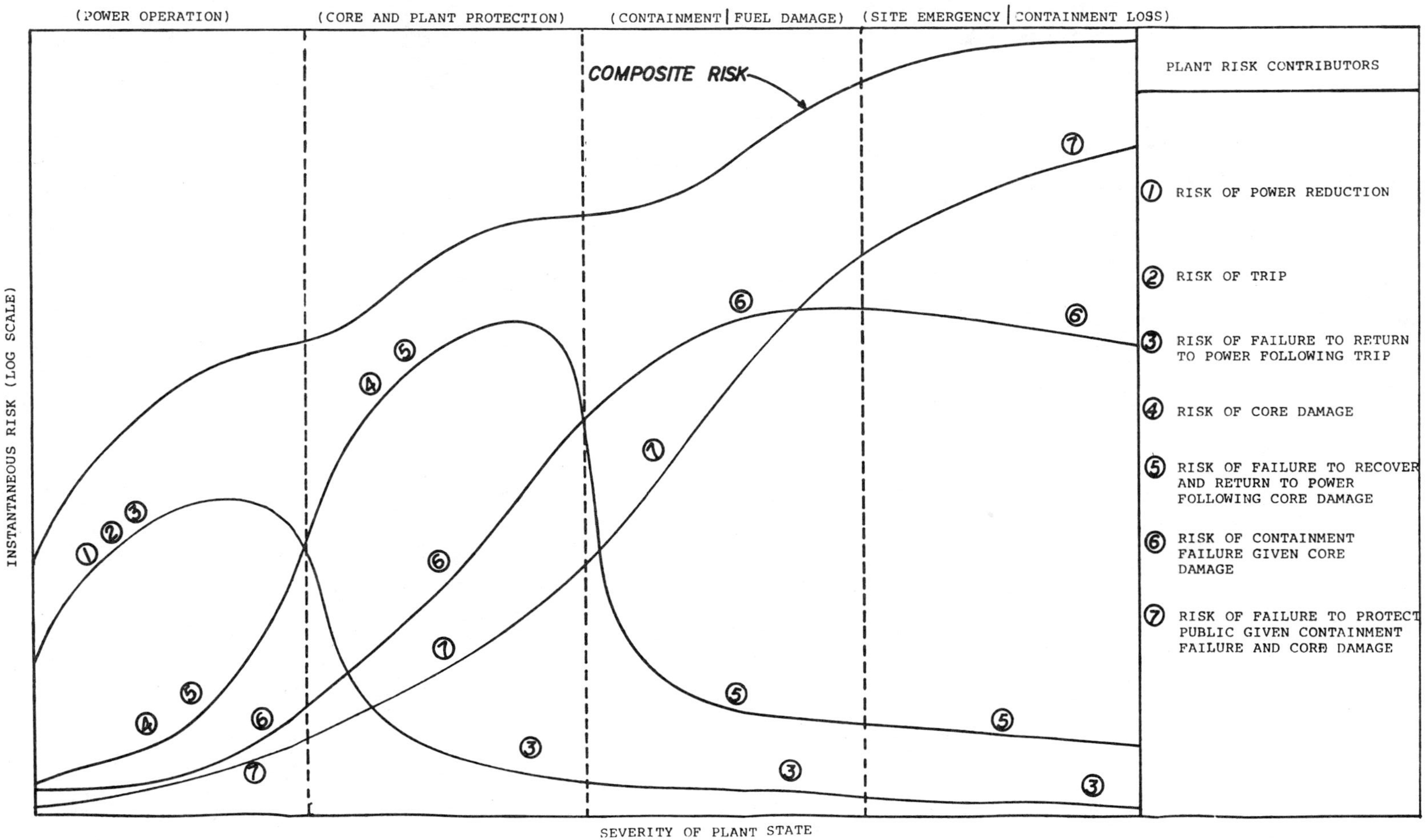

Figure 4
Conditional Risk Vs. Plant State
(Not Based on Calculated Values - For Demonstration Only)

THE OREDA HANDBOOK AND ITS ROLE IN OFFSHORE RISK ANALYSIS

Torkell Gjerstad

Elf Aquitaine Norge A/S
Stavanger, N-4001, Norway

ABSTRACT

Probabilistic risk assessment and production availability studies are being applied extensively to offshore development projects in Europe. The associated need for reliability data has led eight major oil companies to cooperate in the development of the OREDA Handbook - an offshore reliability data reference to be published in 1985.

KEY WORDS: Reliability Data; Offshore Oil and Gas Exploitation; Risk Assessment; Production Availability.

1. INTRODUCTION

Safety, reliability and availability of offshore drilling and production platforms are of considerable concern to the oil industry. Analyses are applied to an increasing extent as a means of providing a basis for decision making where risk or reliability are important properties. Despite the considerable interest in risk and reliability engineering studies for offshore systems, no authenticated source of failure information exists for offshore installations. In previous studies it was necessary to resort to generic data from onshore petroleum plants and data from other industries.

A number of offshore platforms have now been in operation in Europe for a significant length of time, and the OREDA Handbook project - Offshore REliability DAta, was established to compile a comprehensive basis of reliability information from failure and repair records already existing in company files and records.

The offshore operators participating in the OREDA Project have responded to this situation by agreeing to collect, analyze and merge reliability information to be presented in the OREDA Handbook.

2. SAFETY AND AVAILABILITY OFFSHORE

2.1 Background

During the past few years, probabilistic risk assessment has gained widespread application in the oil and gas exploitation industry in the North Sea. Major accidents like the Bravo blow-out, several helicopter crashes and the Alexander L. Kielland platform capsizing, have made the industry and the authorities focus on safety in offshore operations.

Several safety research programmes have been successfully completed over the last years, dealing with every facet of offshore safety. The lack of relevant reliability data and the associated uncertainties in risk figures was extensively discussed in the OREDA pre-project, which was part of these programmes.

Risk assessment has been recognized as valuable means in the design and decision-making process. Many operators have set up within their organization expert teams in this area. This has lead to a better understanding of the practical benefits and limitations of probabilistic assessment, and a more widespread discussion of the application of such techniques and the results thereof.

Parallel to these efforts within the safety area, a strong interest has developed within operating companies in the application of availability analysis techniques within the production area. Development projects in the North Sea typically requires financial investments in the order of 1 - 2 billion US dollars, often put into single platform concepts. Consequently, space and weight restrictions are extremely tight, and cost-benefit optimal design is of paramount importance in order to minimize financial penalties.

2.2 Offshore Design and the Need for Data

Regulatory bodies involved in North Sea operations appear to be moving from detail design regulations to regulations based on functional criteria. Having experienced a number of safety problems in some development projects involving large integrated platforms (accommodation, utilities, drilling and production facilities on the same deck), the Norwegian Petroleum Directorate (NPD) set out to develop a set of "Guidelines for Safety Evaluation of Platform Conceptual Design". These Guidelines were issued in 1981. The Concept Safety Evaluation (CSE) is to be included in the Main Plan which forms the basis for the authorities approval of the proposed concept.

The primary objectives of CSE is to demonstrate an acceptable safety level and to identify Design Accidental Events in terms of heat radiation, blast overpressure and structural impact. The CSE is thus a design tool. The Guidelines introduces a probabilistic cut-off limit with regards to Design Accidental Events:

> "In practical terms, it may be considered necessary to exclude the most improbable accidental events from the analysis. However, the total probability of each type of excluded situation should not by best available estimate, exceed 10-4 per year for any of the main functions (escapeways, sheltered area and man support structure).
> This number is meant to indicate the magnitude to aim for, as detailed calculations of probabilities in many cases will be impossible due to lack of relevant data".
>
> (para 4.2.2)

The latter statement was probably included to comfort the opponents of probabilistic risk assessment and to allow for the really few cases when no relevant data can be found. Because, in practice, the CSEs which have been submitted to NPD over the last few years have all included a comprehensive application of probabilistic assessment.

The introduction of CSE Guidelines was initially met with some scepticism within the industry, fearing that the methodology and the acceptance criteria suggested by NPD might escalate cost without necessarily improving the safety level significantly, experience with a number of CSE studies has, however, demonstrated that application of risk assessment techniques in the early phases of a platform development project is indeed beneficial. Potential safety problems have been identified at an early stage of design, permitting modifications to be implemented without major cost increase. Furthermore, the CSE has proved to be an excellent basis for risk management programmes throughout a development project and into the operational phase.

The introduction of probabilistic risk assessment studies in development projects highlighted the lack of an authenticated source of reliability data for offshore safety and production systems. Recourse had to be made to data from onshore process plants and the nuclear industry (typically WASH 1400 and IEEE Std 500), which inevitably led to an undesirable level of uncertainty in some instances. During this period of time the OREDA pre-project, which is discussed in section 3.1 of this paper, was completed, and the proposal to develop the OREDA Handbook caught the interest of the industry.

The application of the CSE technique only partly explains the growing industry interest in establishing an offshore reliability data reference. Most of the companies who joined the OREDA project expressed an equally or even stronger interest from the production availability point of view. As mentioned previously, weight and space restrictions are very tight in an offshore development project, leading to a strong incentive for optimal design. In the development of smaller, marginal oil and gas fields, this incentive is further amplified by the fact that profit margins may become tighter.

Availability studies are typically applied to problems such as selecting the number and capacity of process trains, gas compressor and utility systems configurations, etc. Several offshore operators today apply availability modelling techniques extensively for such purposes. Equally important is the area of new design concepts, for which existing regulations and standards may lead to technical solutions which are both impractical and associated with unacceptable cost penalties.

In such cases, reliability and availability techniques may be applied to demonstrate that design options which do not satisfy a regulation or a standard, nevertheless may provide an adequate safety level. In a recent case, cost savings in the order of 50 million US dollars were made possible by applying computer technology rather than mechanical devices on a safety trip function (thus violating an API Recommended Practice). The basis for the authorities' approval was an availability study demonstrating the superiority of the computer-based design.

In developing the first issue of the OREDA Handbook, priority has been given to safety related items, but a number of process and utility items have also been included. It is anticipated that any future updates of the Handbook will include a major extension of the latter parts.

3.THE OREDA PROJECT

3.1. Pre-Project

In 1980, a research team was established under the sponsorship of the Safety Offshore programmes and began a study to determine the feasibility of establishing a base of reliability data from the maintenance records of the companies involved in offshore operations, specifically exploration, drilling, and production in the Norwegian Sector of the North Sea.

The pre-project period commenced from Autumn 1981 and consisted of

(1) studying existing reliability data banks, including some which were successful and some which failed:

(2) the development of an approach, including methods of collection and analysis of data, and the format for its presentation; and

(3) discussion/meetings with the industry.

The pre-project phase included a number of visits to organization in the USA witch are engaged in reliability data activity. The aim was to study the organization, documentation and analytical methods used in data collection projects carried out in the USA and relevance of these to the preparation of a offshore reliability handbook. the approach adopted for the development of the IEEE Std 500 has been a particularly useful reference for the OREDA project.

In addition to the three tasks identified as part of the pre-project, modification were made to the original plan based on the discussions with the oil industry. Experience so far indicated that the OREDA approach was highly feasible. Once the project moved into an operational phase in the beginning of 1983, the management and funding was transferred to the OREDA Steering Committee which is composed of members representing operating companies. Hence, the project changed from a governmental research project to an industry sponsored and controlled project.

3.2 Scope

The scope of this project included the collection, analysis and presentation of both quantitative and qualitative reliability information in a handbook format which was intended to address a restricted number of items.

For each item covered, the quantitative generic information consist of:

- failure modes
- failure rate for each failure mode with associated uncertainty limits
- supportive information, e.g. number of events, time in service, population.

In addition to the quantitative information, the Handbook presents the following qualitative reliability information for each class of items:

- item boundary descriptions
- offshore application
- environmental and operations conditions
- failure causes/mechanisms
- data sources.

The OREDA Handbook covers the following types of equipment:

- safety systems
- process systems
- electrical systems
- cranes
- drilling equipment.

3.3 Purpose of the Handbook

The Handbook was developed in order to provide safety and reliability/availability studies of offshore systems with a sound base of generic reliability data gathered from maintenance systems, test records, operational logbooks and other sources of technical information.

The methods and techniques for parametre estimation and the evaluation of data gathered from various sources are well documented in the Handbook. This enables the user of the Handbook to make individual evaluations of the validity of the information included.

The Handbook should provide a basis for further data collection and analysis within each participating company. Subsequent editions of the Handbook may appear when intercompany data collection has progressed far enough to suit republication.

The data base to be included in the first edition of the published Handbook provides a unique source of authenticated data for many other areas of interest, including design audits, maintenance planning, safety system testing, plant availability and cost-benefit studies.

The OREDA Handbook is not published with the expectation or intention that any individual or organization would necessarily apply all of the principles presented, but is prepared in a comprehensive way to provide an immediate source of reliability data. Such a reliability data base must include applicable and significant failure modes as well as failure rate/repair data for systems and equipment used offshore. This Handbook represents an effort to establish this data base. It is expected that the data presented will be a source of information for reliability analysts throughout the industry, thus providing a common base for their use.

The individual event information which is a basis of the estimates is in most cases gathered from two or more installations or companies, and consequently the figures in the handbook reflect a weighted average of the experience within the industry, thus enhancing the credibility of the data.

As the OREDA project is limited in time and restricted to data gathering on hardware components and systems, human reliability information was not included. Nevertheless, component failures caused by human errors will be included in the estimate of the failure rates.

The pre-project discovered that the type and details of the information recorded varied widely for different types of equipment and from company to company. In order to produce a generic data base, a form containing failure events and inventory was required prior to starting the data collection. In addition, it was found necessary to predefine failure modes and specify the component boundaries. The Handbook will include data on a restricted number of items, and due to this restriction the taxonomy does not give a general classification of components and systems on offshore platforms. However the taxonomy does reflect the items for which historical failure data could be made available within the frame of the project.

Figure 1 illustrates the data processing procedure from the extraction and pre-analysis of data from the individual company files to the confidential treatment of it, and then on to the pooling of the data and the estimation of generic reliability parameters to be published in the Handbook.

3.4 Project Organization

Extensive efforts were put into designing the project organization in a way which would ensure an acceptable level of confidentiality for the companies which would join the Project. Following the pre-project, a number of operators had by now expressed an interest in OREDA, and did

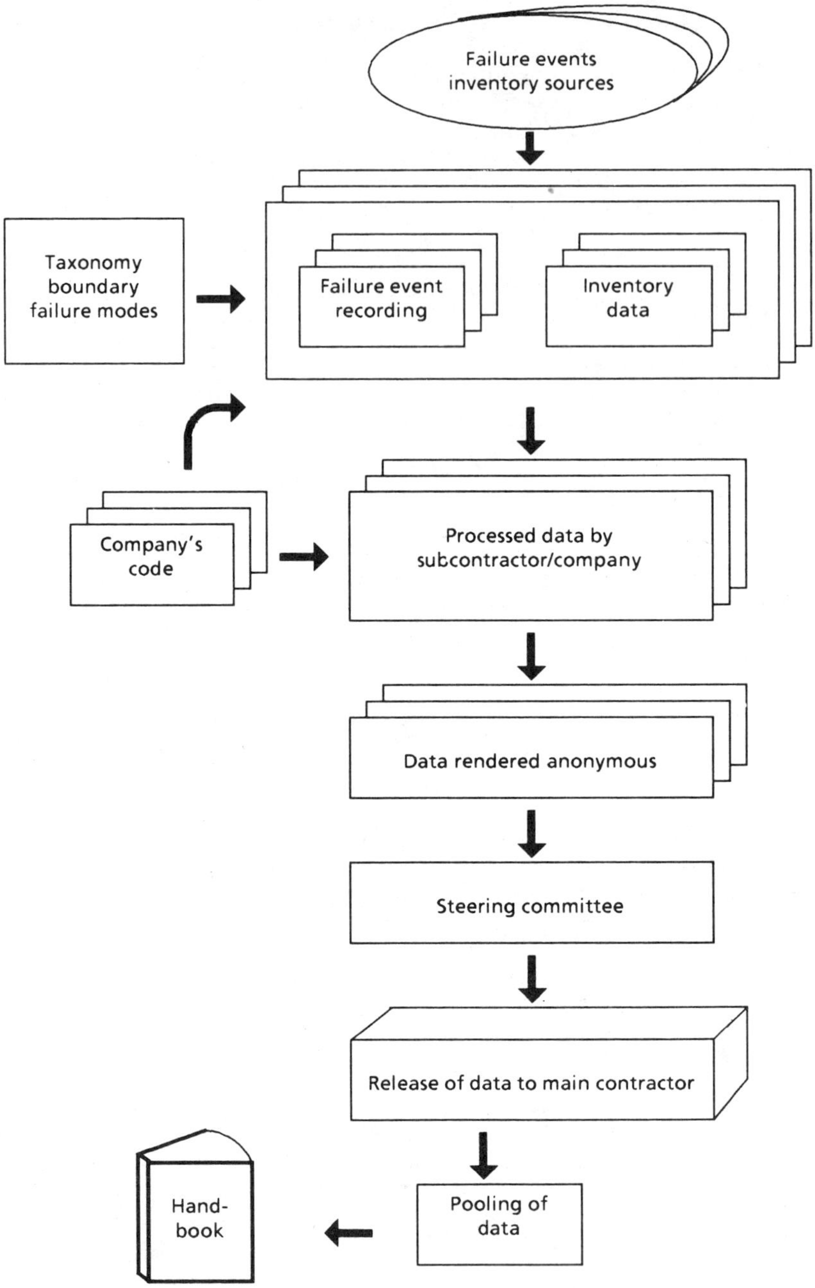

Figure 1

Data Processing Procedure in the OREDA Project

actively contribute to find a satisfactory solution to this problem.

A main contractor was selected by the participating companies and was given the responsibility for administering the development and production of the Handbook.

Each participating company was given the choice to select himself a subcontractor to be responsible for the data collection and processing within the company. The financial support provided by the companies has been put into a central fund, and the main contractor and the subcontractor expenses were paid from this fund. In this way, participating companies which supplied limited amounts of data shared the cost of collecting data within those companies which provided more extensive input. Still, in each instance the site specific input, such as detailed failure histories of equipment, has only been made available to the originating company.

Each participant selected the set of components on which data was to be collected from their files. Hence, the total effort has been controlled by each company.

Eight operating companies joined OREDA:
- Norsk AGIP A/S
- BP Petroleum Development Ltd., Norway
- Elf Aquitaine Norge A/S
- Norsk Hydro A/S
- Saga Petroleum
- A/S Norske Shell
- Statoil, Den norske Stats Oljeselskap A/S
- Total Oil Marine.

The present OREDA budget is USED 650.00, paid through equal shares by the participants. The OREDA Project today is an exclusive industry project, with no involvement from the Norwegian Authorities. It is a 2-year project planned to be completed by the end of 1984, but it is likely that the project will continue beyond this first phase.

Experience data are collected from both the Norwegian and British sectors of the North Sea and from the Adriatic Sea. By being involved in the project, each participant will have the opportunity to compare their own past experience with that of other operating companies. Finally, the whole offshore industry will gain from the publication of this Handbook, which is believed will become a standard reference for risk and reliability data for offshore installations.

3.5 Data Collection

The basic information necessary for the production of this book is the property of the individual oil companies participating in the project. Traditionally, this kind of information has not been released to outside parties. The data collection approach chosen therefore ensured that the data were treated in a confidential manner throughout the study. Information made available from the individual company records has been treated on two levels:

On a strictly confidential level where the raw data from company records being analyzed are fed back to the company for their own information and approval before release, and on a second level where pooling of data from several companies has taken place on a generic basis and provided the main output for the Handbook. The first confidential level was obtained by each individual oil company appointing a subcontractor. The subcontractor alone or together with personnel from the oil company formed a data collection team within that company.

The extent of the data collection within each company, i.e., number of generic items to be covered by the exercise, was agreed upon on the basis of the expected availability of failure data within the company and a consideration of the importance of the items in terms of safety and reliability for offshore platforms. The number of items covered was adjusted during the data collection according to the experience gained about the availability of failure data information within each company.

Failure and inventory information has mainly been extracted from

- maintenance files/reports
- inspection and test reports
- work orders
- equipment histories
- daily drilling reports

Only a very few cases has the basic failure information been obtained from sources other than written historical files, i.e., by interviewing experienced personnel. In most cases, the main data sources have been made available onshore at company offices, but visits to offshore platforms have also been necessary in some instances.

The way the data collection was organized in the OREDA project implies that several teams were used for the collection of data within the oil companies. For this reason it was of vital importance to have a specification of the interface boundaries for the items in the taxonomy, so that the data collection teams had the same understanding of which failures were to be considered within the boundary of a particular item. The boundary specification is also of vital importance to the user of the estimated reliability data because it is the only mechanism by which the user of the data can judge whether or not a failure has been included in the reliability data source.

An item may have several interfaces with other items. Prior to the data collection the boundary specification was given in a tubular form with corresponding schematic figures of each generic item. These figures illustrates what was included within the boundary of the item. It was interesting to note that the figures alone in most cases gave sufficient information to the data collection teams to determine the boundaries.

The failure modes are also categorized for each item according to failure severity class, i.e., critical, degraded or incipient, and segregated according to time related and demand related modes. The failure modes are further divided into submode classes which are intended to provide more knowledge of the nature of the failure or of the portion of the item affected.

The following changes in item operational conditions have been used as a base for the subdivision of the failure modes:

- change in operation, e.g., failure to operate once started, low output, high output
- change without command, e.g., spurious operation
- no change on demand, e.g., failure to start, partially open.

Data evaluation and verification were undertaken within each company and between companies and are described further below. In cases where such evaluation and verification made it necessary, additional studies were initiated, e.g., when estimated repair times for a given generic item showed a large deviation between two or more companies, or when part of the data base has contained unsufficient data for estimation of reliability parameters.

Failure and event data collected for each platform after being rendered anonymous, transferred to a data transmission form, and were approved by each company before being submitted to the Main Contractor for further analysis, pooling and presentation in the Handbook.

Because of the requirement of strict confidentiality placed on the data collection and presentation, important background information necessary for supplementing the data forms with qualitative information was also processed at company level and rendered anonymous.

3.6 Data Verification

Although the basic data collection approach was applied for all exercises, the detailed data collection had to be adapted to suit the structure and operation of the engineering information system within the individual oil company. Consequently, the data verification also had to be adjusted to the data collection within each company. In general the data verification was carried out on four different levels.

Level 1:

Routine physical checks were carried out on the data during the initial inventory data extraction, during the transfer of failure data from historical records to data processing forms, and when included in the final database. Random checks on population, failure frequencies etc. were also made against the original data processing form.

Level 2:

Sampling exercises were carried out to compare the output from one data source with the output of another source covering the same items. Estimates of failure rates were carried out in order to check the completeness of the information recorded in that particular file.

Level 3:

Maintenance and reliability engineers within the companies were confronted with failure, repair and inventory data extracted from the historical records in some of the cases. In all cases failure histories on drilling equipment were sent back to the rig operators for their verification and comments.

Level 4:

A comparison of failure and repair data between the companies involved was carried out. In some cases detection of wide discrepancies between companies initiated extra checks of data involved.

3.7 The Handbook

The OREDA Handbook will present reliability data on approximately 150 items, many of which are unique to offshore drilling and production installations. The book will be published by the OREDA Project and distributed in cooperation with an international publishing house. The intention is to have the Handbook available on the market during spring 1985.

Copies of the OREDA Handbook will be available from:

OREDA
P.O. Box 370
N-1322 HOVIK

The price is estimated to USED 150,-.

The reliability data will be presented in standard format tables as shown in figure 2. Each data table will be accompanied by a free format table giving the following information:

- A technical description of the item
- The type of application (service)
- Operational modes (e.g. continuous, stand-by, etc.)
- Internal environment (e.g. fluid passing through a valve)
- External environment (e.g. ambient temperature and air humidity)
- Testing and maintenance routines
- Supplementary information on failure courses and failure modes
- Description of data sources.

The pairs of reliability data tables and free format tables will make the OREDA Handbook a practical and easy-to-use reference. The amount of information given for each item is very comprehensive and probably unique in the area of reliability data sources.

The OREDA Project covers a wide range of components and systems. The following main areas will be included:

1. Safety Systems
   - 1.1 Gas and fire Detection System
   - 1.2 Process Sensors
   - 1.3 Firefighting Systems
   - 1.4 Emergency Shut-down Systems
   - 1.5 Pressure Relieving Systems
   - 1.6 General Alarm and PA Systems
   - 1.7 Evacuation Systems
2. Process Systems
   - 2.1 Vessels
   - 2.2 Valves
   - 2.3 Pumps
   - 2.4 Heat Exchangers
   - 2.5 Compressors
   - 2.6 Pig Launching/Receiving Systems
3. Electrical Systems
   - 3.1 Power Generation
   - 3.2 Power Conditioning
   - 3.3 Power Protection
   - 3.4 Control Instrumentation
4. Utility Systems
   - 4.1 Slop and Drain Systems
   - 4.2 Ventilation and Heating Systems
   - 4.3 Hydraulic PA Supply Systems
   - 4.4 Pneumatic Power Supply Systems
5. Crane Systems
6. Drilling Equipment
   - 6.1 Drawworks
   - 6.2 Hoisting Equipment
   - 6.3 Diverter Systems
   - 6.4 Drilling Riser
   - 6.5 BOP Systems
   - 6.6 Mud Systems

3.8 OREDA Phase II

The OREDA Project is an ongoing exercise run, funded and controlled by international offshore operating oil companies. It is the intention of

Taxonomy no	Item
1.4.4.1.1	Safety Systems ESD (Emergency Shut-Down) Systems ESDV (Emergency Shut-Down Valves), (excl. XMV and AFV) Hydraulically Operated, Globe/Gate/Ball Hydrocarbon, Oil, Less than 24"

Population	Samples	Aggregated time in service (10^6 hours)		No of demands
		Calendar time*	Operational time+	
18	2	.5016	.3291	

Failure mode	No of failures	Failure rate (per 10^6 hrs)			Active repair (hours)	Repair (manhours)		
		Lower	Mean	Upper		Min	Mean	Max
Critical	**2 ***	**0.71**	**4**	**13**	-	**4**	**7**	**10**
	2 +	**1.1**	**6.1**	**19**				
Failed to close	2 *	0.71	4	13	-	4	7	10
	2 +	1.1	6.1	19				
Degraded	**2 ***	**0.71**	**4**	**13**	-	**2**	**6**	**10**
	2 +	**1.1**	**6.1**	**19**				
External leakage	1 *	0.1	2	9.5	-	10	10	10
	1 +	0.16	3	14				
Internal leakage	1 *	0.1	2	9.5	-	2	2	2
	1 +	0.16	3	14				
Incipient	**7 ***	**6.5**	**14**	**26**	-	**2**	**5.9**	**10.5**
	7 +	**12**	**23**	**42**				
Seepage	4 *	0	7.5	22	-	2	6.8	10.5
	4 +	4.2	12	28				
Faulty indication	3 *	0	6.4	20	-	2.3	4.8	6
	3 +	0	11	36				
All modes	**11 ***	**12**	**22**	**36**	-	**2**	**6.1**	**10.5**
	11 +	**21**	**35**	**57**				

Comments

Figure 2

Example of Reliability Data Table in the OREDA Handbook

these companies to continue their joint efforts beyond the publication of the first issue of the OREDA Handbook.

An intermediate phase following the publication of the Handbook is foreseen, in which the experience from the first phase will be reviewed together with feedback from Handbook users. The objective of the intermediate phase will be to define the scope and procedures for OREDA phase II.

It is foreseen that the second phase will include both a refinement within the areas covered in the first issue of the Handbook, and a broadening of the scope to include additional items and new areas.

The OREDA Project is today an international project. The Steering Committee will welcome additional participation in the future in order to broaden the experience basis which is the key element in any reliability data collection exercise.

WARNING SYSTEMS: RESPONSE MODELS AND OPTIMIZATION

M.E. Paté-Cornell and C. Pe Benito-Claudio

Stanford University
Stanford, California 94305

ABSTRACT

Probabilistic analysis of warning systems can be used to make technical, strategic, and tactical decisions regarding the signals to be observed, the warning to be issued, and the response to alerts. Response to alerts can be either a decision variable or a parameter of the problem.

A modeling of the response function depends on the frequency and the uncertainty of the hazardous event. Two cases corresponding to different response situations are presented here. Fires in students' dormitories are frequent events: the response to a firm alarm is determined by the system's past performance; false alarms create the crying-wolf effect. Accidents in nuclear power plants are rare events for which we have only limited knowledge and experience: the response to emergency warning is influenced by the people's perception of the risk and the sociological context of the warning. Different models of response to warnings are developed and integrated into the probabilistic analysis of warning systems for decision-making purposes.

KEY WORDS: Warning Systems; Response Models; Probability; Fire Alarms; Nuclear Power Plants.

1. INTRODUCTION

The probabilistic analysis of warning systems, including assessment of response to alerts, can be used as a support for three types of decisions: technical, strategic, and tactical. Technical decisions, which are treated here separately for emphasis, involve the choice of phenomena to be monitored and the choice of an observation system; for example, what are the characteristics (temperature, pressure) of a sub-system in a nuclear power plant to be observed and by what means. Strategic decisions include the identification of the signalor set of signals that will constitute an alert and the definition of the kind of response to be considered or recommended given the occurrence of the signal. This can be, for example, a threshold of particle densities for a fire alarm calling for evacuation. Tactical decisions involve response decisions to be made at the time of an alert on the basis of previously decided strategies, and of the specific circumstances of the warning (lead time, additional information, local constraints).

Given that most warning systems can issue false alerts or miss the signals altogether, a trade-off has to be found in most cases between Type

I errors (defined here as failure to warn) and Type II errors (defined here as false alerts). Optimization of the system can be done if the consequences of all types of scenarios can be assessed and compared (true alert, false alert, missed prediction) (1).

The focus in this paper is on technical and strategic decisions. The emphasis is on the response model, either from a descriptive or from a normative perspective, and its integration into a probabilistic analysis of a warning system. For frequent events (for example, fires in students' dormitories), the parameter of interest is the memory of past performance of the system and the "crying-wolf" effect. For rare events (for example, accidents in nuclear power plants), the emphasis is on the uncertain behavior of the population and its willingness to respond, given the perception of the risk and the sociological context of the warning.

2. FREQUENT EVENTS: THE CRYING-WOLF EFFECT

Assume that a fire alarm in a student dormitory has to be set (or located), i.e., a level of sensitivity has to be chosen. A tradeoff has to be found between a high sensitivity and therefore a short lead time (or no lead time at all), for response (see Figure 1). Optimization in this case can be performed as a maximization of the number of people saved, within reasonable cost constraints. The dominant effect that determines the response level is the memory of past performance of the warning system (1).

Three models can be used to describe the effect of memory on the response to warnings:

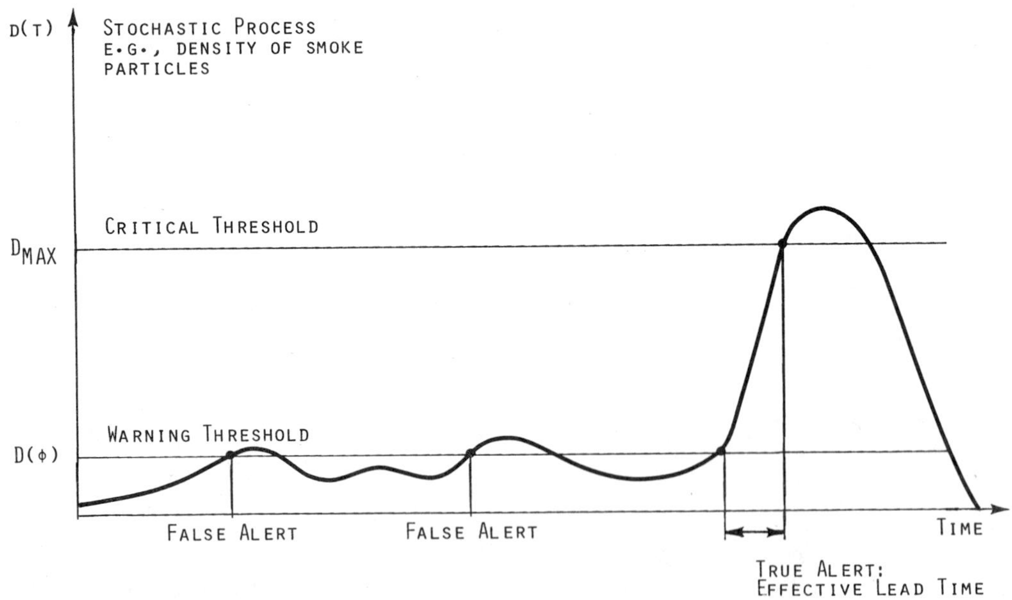

Figure 1

Stochastic Precursor: Monitoring an Underlying Phenomenon
From Ref. 1

- a nonparametric descriptive Markov model in which the response level is a function of the quality of the n past alerts (true or false) that people remember;
- a parametric descriptive model, in which the response level is determined by the impact of past true and false alerts on people's willingness to respond, given memory decay;
- a normative model in which each individual makes a decision to respond or not, given: (1) the probability that he assigns to the occurrence of the event following a warning; (2) his probability of being a victim of this event; (3) the costs to him of responding to the alert; and (4) the value that he places on his own life.

The following example illustrates these three models; the details of the model can be found in reference 1. The notations the submodels are the following:

$EV(\Delta L)$: expected value of the number of people saved per time unit.

ϕ: sensitivity parameter (decision variable).

$\lambda_1(\phi)$: rate of irrelevant false alerts (not linked to the occurrence of events).

$\lambda_2(\phi)$: rate of alerts (true or false) due to actual exceedence of the ϕ threshold (e.g., cigarette smoke under a fire alarm).

N_E: loss model (expected number of deaths or injuried per event assuming no warning).

i: index of different patterns of true (T) and false (F) alerts (e.g., TFFT, the most recent event being the last one in the sequence).

{k}: set of alert patterns ending in a true alert.

α_i: rate of response to a new alert given a past pattern of true and false alerts.

$p_i^*(\phi)$: steady-state probability of each pattern of true and false alerts given a sensitivity level.

$v(\alpha_i)$: speed of evacuation given the number of people willing to respond. It is assumed that below a certain response rate there is a free flow, then a gradual engorgement (linear decrease of v with α), then total jam and v = 0 above a certain level of response.

D_{MAX}: threshold of the monitored variable (density of smoke) above which people are hurt.

$D(\phi)$: threshold of the monitored variable above which the signal occurs (decision variable).

$\bar{\nu}$: mean rate at which the system deteriorates over time once the threshold $D(\phi)$ ex exceeded (e.g., rate of increase of the smoke density).

It is shown in reference 1 that the benefits of the warning system per time unit can be written as

$$EV(\Delta L) = [\lambda_1(\phi)+\lambda_2(\phi)]N_E \times \left[\sum_{i\in k} \alpha_i \, v(\alpha_i)\, p_i^*(\phi)\right] \times [D_{MAX} - D(\phi)]\,\bar{\nu}. \quad (1)$$

The rate of response α can be modelled in the three different ways described above.

The nonparametric descriptive model requires the assessment of the number of past warnings n that are remembered and affect the response (e.g., n = 4), and the determination of the rate of response for each pattern. For example, α(TTFT) equals 0.7. The value of the rate of response (e.g., 0.7) can be estimated either on the basis of past statistical data, or by use of a questionnaire. This model has the advantage of its simplicity: in the long run, what matters is the steady-state probability of each sequence of true and false alerts relative to true alerts. A large number of steps thus presents no computational problem. This model presents both the flexibility and the clumsiness of such nonparametric description of behavior: one needs to assess the rate of response of the population for each of the 2^{n+1} patterns of true/false alerts involved in an n-step memory model. It has one major drawback, however: it ignores the severity of past events (which may be critical to memory and response) and the time elapsed since they occurred (except indirectly through the rate λ of events E per time unit).

The parametric, descriptive exponential model has the advantage of including the effects on memory of events, severity, and elapsed time. It assumes exponential decay with time of memory of past events, and requires adjustment of the different parameters to reflect observed or revealed rates of response. The response factor α is of the following form:

$$Q(t) = \int_{-\infty}^{t} \{\exp[-m(t-\tau)]\} \times [k_A M(\tau) - k_B M(\tau) - k_C(\tau)]\, d\tau$$

$$\alpha(t) = \Phi\left[\frac{Q(t) - \beta}{\gamma}\right]. \tag{2}$$

$\alpha(t)$ is the proportion of people willing to respond to a warning at time t, given the past history of true alerts (A), Type I errors (b), and Type II errors (C) at all times τ preceding t.

$Q(t)$ is an intermediate function in the computation of α.

$\Phi(\cdot)$ is a scaling function, for example, a Gaussian cumulative, in which β (≥ 0) and γ are scaling factors. Φ insures that α remains between 0 and I.

m is the memory-discounting factor.

$M(\tau)$ is the magnitude of the events that occurred at different times

k_A is the factor measuring the positive effect on future response rates of a true alert.

k_B is the factor measuring the negative effect on future response rates of a missed prediction (Type I error).

k_C is the factor measuring the negative effect on future response rates of a false alert (Type II error).

This model has the advantage of treating explicitly the length of the time interval between warnings, and between events and warnings. A disadvantage, however, is the assumption of additivity of these effects on the response rate, which may not correspond to reality. Also, one may question in some cases the reality of exponential decay. A very long time after the last earthquake, for example, there is a popular perception that chances are increasing that another one may occur, and there may be a great willingness to respond to a warning. In other cases, one may argue that events that have occurred a long time ago have left a stronger print in some people's minds than more recent ones. In general, however, memory decay is a reasonable assumption.

The parametric normative model starts from the individual point of view instead of the collective one. For each potential victim, the problem is to know whether he will respond or not to the warnings, assuming rationality of choices (2). This model includes (1) Bayesian updating of information every time a new event of type A, B, or C occurs, and (2) a study of individual decisions to take protective measures, given individual risk attitudes, costs, and values. Assuming that all people interpret in the same way each new piece of information, and that they are all equally likely to be victims if they do not respond to the warning, the willingness or not to respond is determined by individual risk attitudes, cost of events, and costs of response to alerts. A warning having been issued, the individual decision can be modelled by the decision tree shown on Figure 2. Assuming that $p_t(E|W)$ is the same for all individuals, and that different people value lives differently according to the perceived risk of death, the criterion EV(u(R)) > EV(u(R)) (maximization of expected utility) determines whether or not the individual in question decides to respond to a new warning. Given a joint distribution of risk attitudes, costs of response and value of one's own life in the considered population, one can, therefore, compute the proportion of people who are willing to respond to warnings. This approach has the advantage of examining the roots of individual reactions, and the inconvenience of assuming rationality of individual choices in the evaluation of a parameter of the model, which might turn out to be unrealistic.

The following numerical application is presented for illustrative purpose. Assume that 3 fires occur per year on the average, exposing 100 students and that the number of people injured (or killed) per event is 5 (mean value), that the sensitivity of the system is defined on a scale of 0 to 100. The sensitivity is the decision variable and determines the probability of an event conditional on the signal--which increases as sensitivity decreases--and the probability of a signal conditional on an event--which decreases as sensitivity increases. The nonparametric rate of response was assumed to vary from a low 20% for a system that tends towards 100% false alerts to a high 90% for a system that delivers 100%

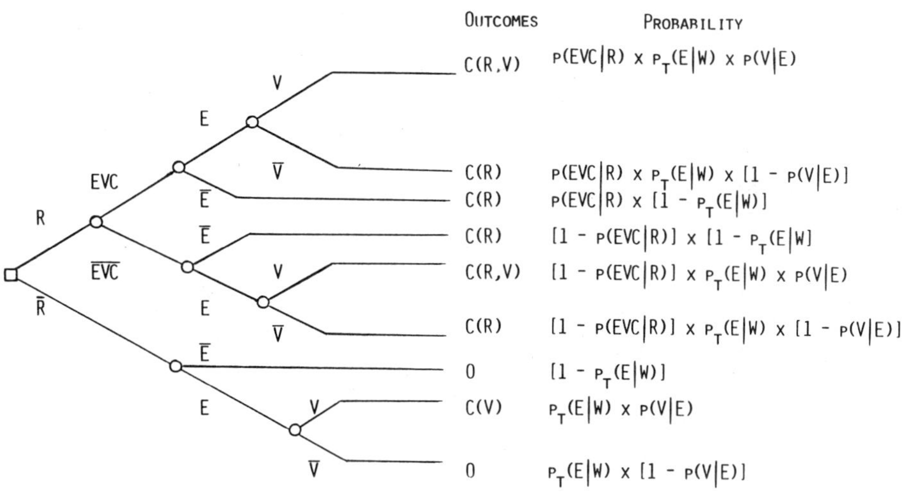

LEGEND:

R: RESPONSE
W: WARNING
EVC: EVACUATION
E: EVENT OCCURRENCE
V: BEING A VICTIM
‾: NEGATION OF EVENT

C: COST
$P_T(E|W)$: PROBABLITY AT TIME T OF EVENT E CONDITIONAL ON WARNING W, AS A FUNCTION OF PAST HISTORY OF WARNINGS QUALITY.

ASSUMPTION: IT IS ASSUMED THAT THE COST OF RESPONSE DOES NOT DEPEND ON SUCCESS OF EVACUATION ATTEMPT.

NODES:
EVENT NODE
DECISION NODE

Figure 2

Decision Tree for the Individual Response to an Alert

true alerts (but is prone to missing events). This very imperfect system is also assumed to issue irrelevant alerts at the rate of 2 per year. Figure 3 shows the variations of the expected number of avoided injuries as a function of ϕ. A maximum of 5.5 avoided injuries (expected value) is reached for a high sensitivity of 30 (from 0: maximum sensitivity, to 100: minimum sensitivity) at an average low cost of $3,000 per life saved.

The parametric descriptive model assumes coefficients k_A = 1.1, k_B = -1.1, and k_C = -1.2, therefore slightly higher negative effect of false alerts than missed events. The memory decay factor is 10% per year and the factor is renormalized through a Gaussian cumulative curve. Figure 4 shows the variation of the expected number of avoided injuries as a function of ϕ. A maximum of 7.5 is reached for a high sensitivity of 38. The discrepancy with previous results i not of major significance given that thee was not attempt to match entirely the parameters of both models.

Finally, for the normative model of response, the example of a group of ten persons was considered. To each of them a different value was attributed to the probability of being a victim if an event were to occur (from 0.1 to 0.99). The risk attitude, the cost of evacuation and the value of life were assumed to be uniform among individuals. The decision of each person to evacuate or not is analyzed assuming a rational decision model displayed in Figure 2. The rate of response α is calculated in this manner for each considered value of ϕ Figure 4 shows the variation of the expected number of avoided injuries as a function of ϕ. A maximum of 10 is obtained for a high sensitivity of 24.

These three models illustrate clearly the memory effect and the variations of the benefits of a warning system between two extreme values of sensitivity: in all three cases one observes zero or very low benefit for extremely sensitive or extremely insensitive systems and a maximum benefit at an intermediate point that depends on the physical characteristics of the system and the description of the "crying-wolf" effect assumed in the response model.

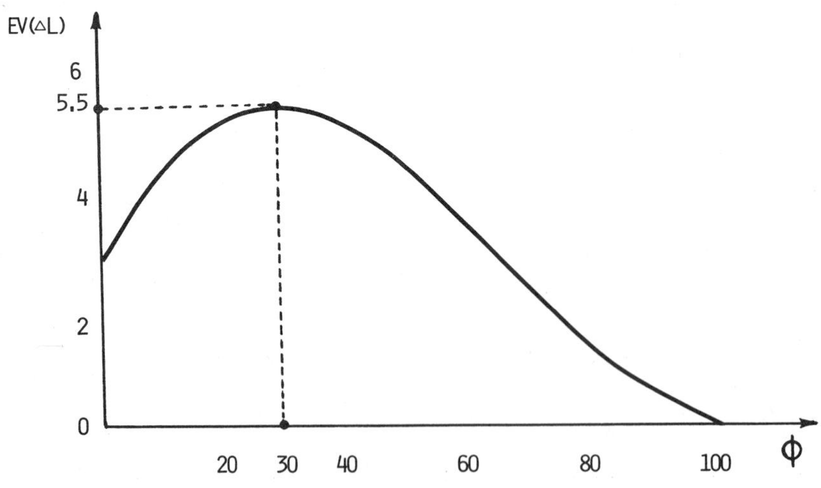

Figure 3

Optimum Sensitivity for the Non-Parametric Descriptive Response Model

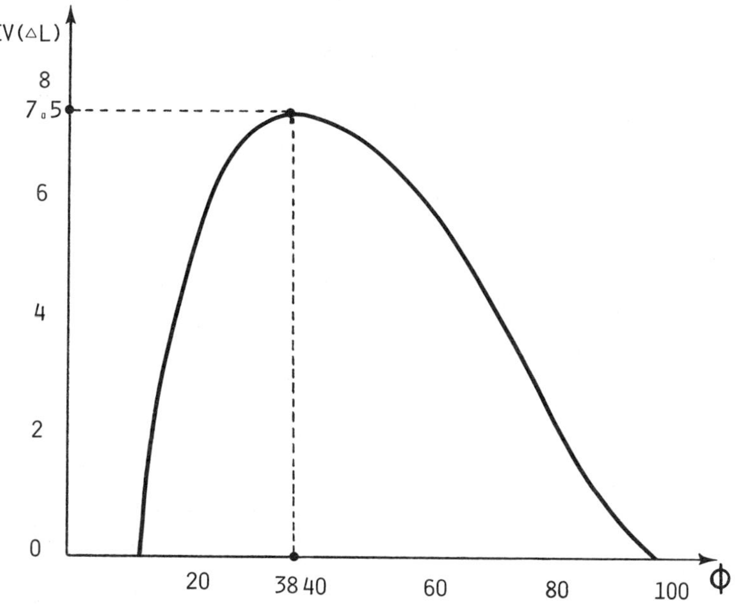

Figure 4

Optimum Sensitivity for the Parametric Descriptive Response Model

3. RARE EVENTS: PERCEIVED RISK AND SOCIOLOGICAL CONTEXT

In the case of rare events or events for which even comparable occurrences are rare, the memory effect is no longer the key element. This is the situation encountered when doing emergency planning for nuclear power plants (3). Following an initiating event, the plant state evolves in a manner that can be observed through the monitoring of different elements (example: temperature or pressure). A first choice has to be made as to which of these characteristics constitutes a signal on-site, and above what threshold the signal should appear. Operator intervention following the in-plant signal can either accelerate, leave unchanged, slow down, or avoid the evolution of the accident towards radioactive release. A second threshold has to be set above which an off-site warning is issued and a recommendation is made to the public to either evacuate, shelter, take other protective measures, or do nothing. The lead time available for response, the public's willingness and capability to respond to the warning, and the possible release of radioactivity are the three variables linking the dynamic evolution of an accident on-site and the public response to off-site warning (see Figure 6).

The probabilistic analysis of alternative warning systems is done using influence diagrams such as that shown in Figure 7 (18 and 19). An important element of the diagram is the response model, which includes the public's willingness to respond and the capability to respond (for example, an evacuation model). A simplified behavioral response model has

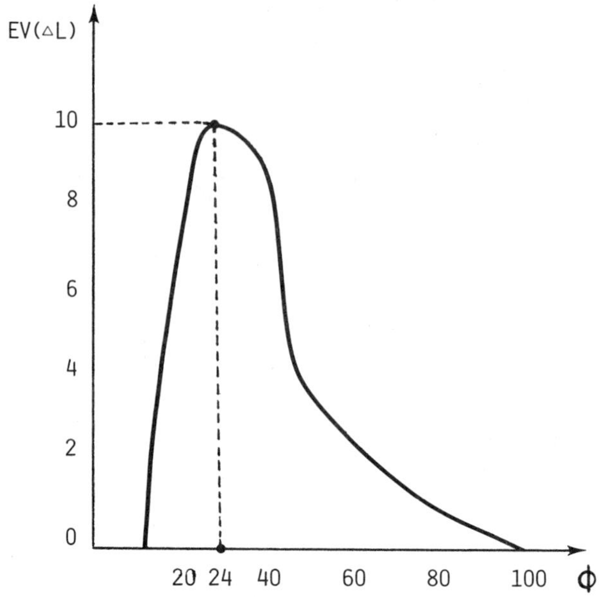

Figure 5

Optimum Sensitivity for the Normative Response Model

been developed to assess the public's willingness to respond to an emergency warning, using previous work in the field of sociology and psychology.

Two different theoretical approaches in the study of behavioral response to warning can be found in the literature. The first one focuses on individual decision-making processes of warning respondents in a given psychological context. it gives importance to one or more of the following factors: cognitive abilities, conflict, learning and feedback, memory, and motivation. This includes, for example: Janis's conflict model of emergency decision-making (4), Fishbien's theory of reasoned action, which was used in recent works on nuclear-reactor disasters (5), and models developed recently by researchers coming from the engineering discipline (6 and 1). Another approach is the sociological one. It emphasizes the role of social and cultural variables in shaping response, and the actions of individuals and groups (7, 8, 9, 10, and 17).

Some recent works combine aspects of these two approaches. The resulting theoretical framework is a crisis decision-making model which explains how some important psychological and sociological variables, and the warning itself, influence the individual's response (Examples: (7 and 9)). Crisis decision-making models range from the stimulus-response model, which assumes "an individual captured," to various derivatives of the traditional decision-making model, which treat individuals as capable of "evaluating and differentially acting upon informational inputs" (7 and 10).

These models may be integrated into a framework that describes the

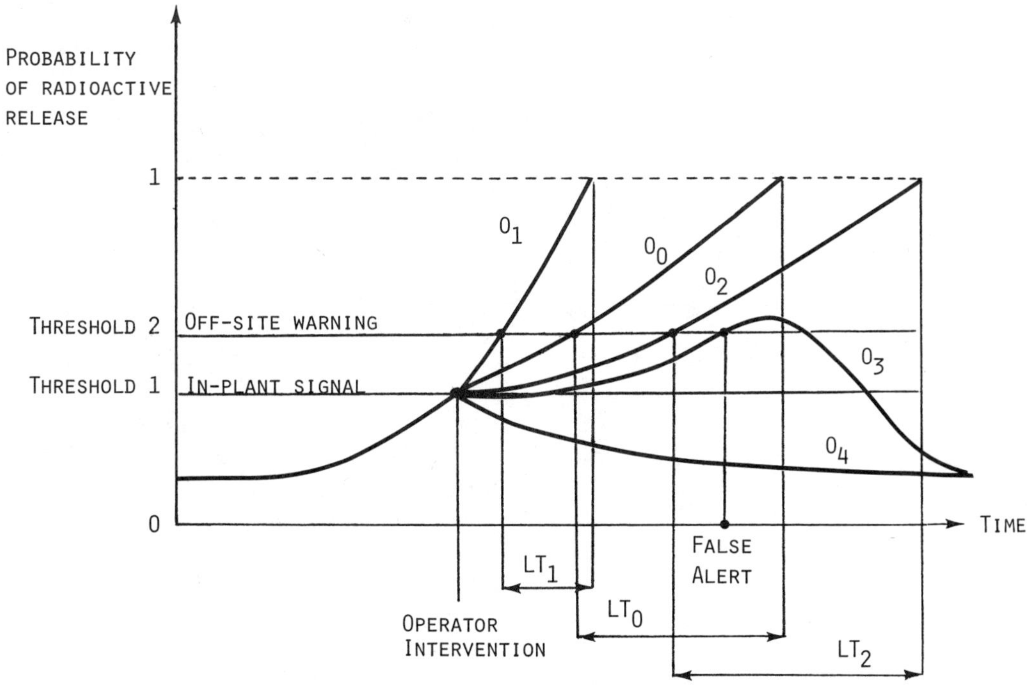

Figure 6

Warning System for Nuclear Power Plants - Thresholds and Lead Times

crisis behavior of individuals in two key phases: definition of the warning situation and decision to take protective action. To analyze the first phase, Janis emphasizes the psychological conditions (conflict, hope, and time) and the individual's coping pattern; sociologists focus on the sociological conditions--the changing social norms during the warning situation, and the individual's coping behavior in such environment (9). After the individual has defined the warning situation (hopefully in a state of vigilance, as in Janis's model), he goes on to the second phase and decides to take a protective action. The concepts of rationality and optimality apply to this phase of the response process. Two models are available: the subjective expected utility model, which assumes that the individual--a rational economic man--maximizes his expected utility, and the bounded rationality model, which asserts that the individual tries to achieve a satisfactory, though not maximal, level of achievement (9 and 11). Perry (1983) developed a qualitative model, which combines the emergent norm approach and bounded rationality, within Rowe's risk analysis framework (9 and 12).

Many psychological and sociological variables influence the behavior of individuals in times of crisis (13, 14, 15, 16, 17). Which variables are most important is subject to contention (5, 8, 9, 10, 17). The qualitative model that we used in the analysis of warning systems for nuclear power plants has the following variables: (1) prior and situational risk perception; (2) social confirmation; (3) warning belief; (4) adaptive plan; and (5) distance from the plant (see Figure 8).

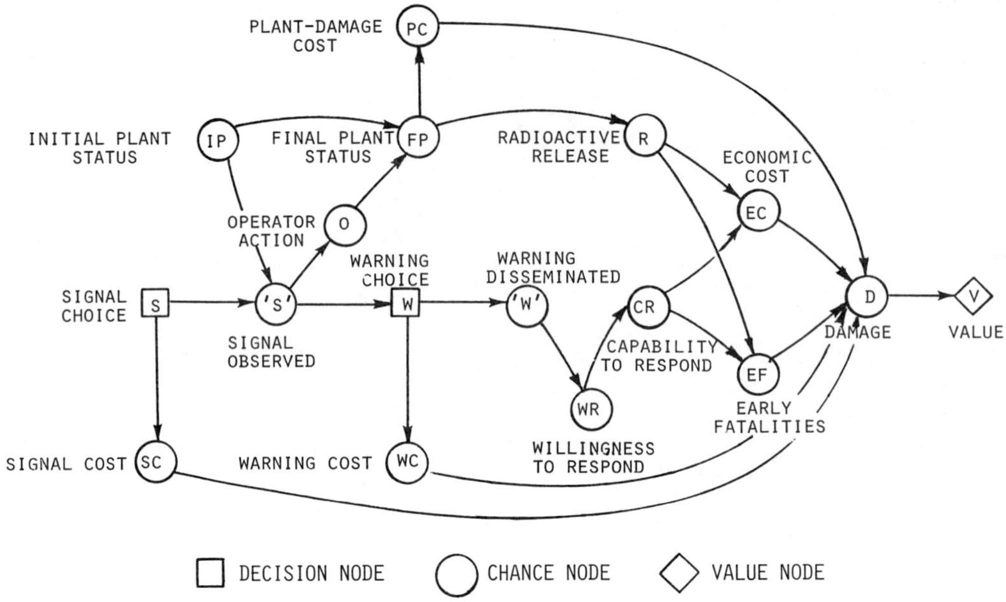

Figure 7

Analysis of a Warning System for a Nuclear Power Plant
Simplified Influence Diagram

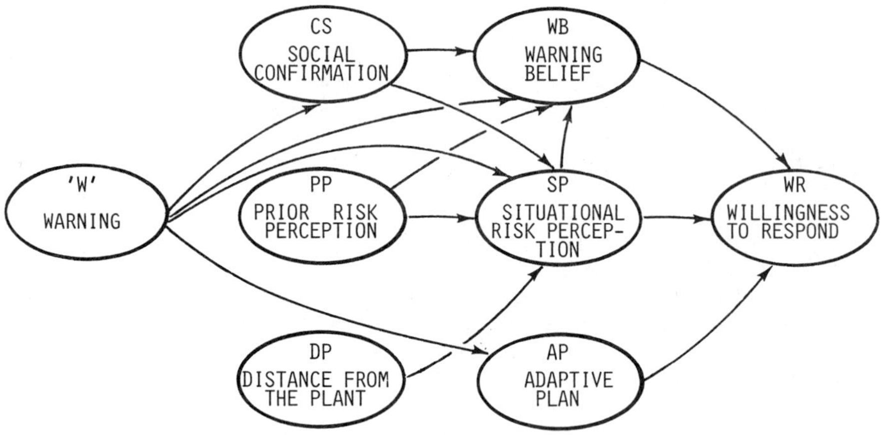

Figure 8

Behavioral Response Model for Nuclear Power Plant Emergency

For a quantitative probabilistic analysis, one can model the behavioral response by treating explicitly the two key phases described above, the variables that affect them, and the crisis decision-making process of individuals. In the present study, a simplified model is used that focuses on the effect of five key variables--warning, social confirmation, prior risk perception, distance from the plant and adaptive plan--on the willingness to respond, and consequently on the outcome of an accident. The probability distribution of the willingness to respond of a given population is calculated based on empirical findings, and integrated into the influence diagrams (e.g., Figure 7).

Preliminary results of this study suggest that, during the early stage of an accident, when both the final plant state and the release of radioactivity are uncertain, it is better to wait for more information before taking action. This is due to the negative effects of false alerts, and to the low probability and generally slow evolution of events that could lead to off-site damage. Sensitivity analysis shows the effect of different levels of response on the choice of a warning system.

5. CONCLUSIONS

Warning systems can be a powerful way to reduce risks. They may also create risks themselves. The best warning systems are not necessarily the most sensitive ones. The sensitivity that gives optimum benefits (i.e., maximum risk reduction) depends on the expected response to signals, the reliability of the system, and the values that are involved in the trade-off between the consequences of Type I and Type II errors.

7. ACKNOWLEDGEMENT

This work was funded in part by the Center for Economic Policy Research (Stanford University) and by the Electric Power Research Institute.

REFERENCES

Paté-Cornell, M.E., Probabilistic Assessment of Warning Systems, Paper No. 13, Center for Economic Policy Research, Stanford University, Stanford, California, 1984.

Raiffa, H., Decision Analysis, Reading, Mass.: Addison-Wesley Publishers, 1968.

Claudio, C.P. and M.E. Paté, "Theory, Models, and Analysis of Warning Systems with Applications to Nuclear Power Plants," Presentation to the annual ORSA-TIMS conference, San Francisco, California, 1984.

Janis, I.L. and L. Mann, Decision-Making: A Psychological Analysis of Conflict, Choice and Commitment. New York: The Free Press, 1977.

Ziegler, D.J., S.D. Brunn, and J.H. Johnson, Jr., "Evacuation from a Nuclear Technological Disaster," Geographical Review, Vol. 71, No. 1, January 1981.

Ferrell, W.R. and R. Krzysztofowicz, "A Model of Human Response to Flood Warnings for System Evaluation," Water Resources Research, Vol. 19, No. 6, December 1983, pp. 1467-1475.

Leik, R.K., T.M. Carter, and J.P. Clark, Community Response to Natural Hazards Warning. Springfield, Virginia: National Technical Information Service, 1981.

Mileti, D.S., T. Drabek, and J.E. Haas. Human System Environments: A Sociological Perspective. Boulder, Colorado: Institute of Behavioral Science, University of Colorado, 1975.

Perry, R.W., Comprehensive Emergency Management: Evaluating Threatened Populations, (Draft book) 1983.

Quarantelli, E.L., People's Reactions to Emergency Warnings. A Preliminary Paper, Department of Sociology, Disaster Research Center, The Ohio State University, June 1982.

Simon, H.A., Models of Bounded Rationality. Cambridge, Mass.: The MIT Press, Vol. 2, 1982.

Rowe, W.D., An Anatomy of Risk. New York: John Wiley & Sons, Inc., 1977.

Dynes, R.R., Organized Behavior in Disaster. lexington, Mass.: Heath Lexington Books, 1970.

Dynes, R.R. and E.L. Quarantelli, "Group Behavior Under Stress: A Required Convergence of Organizational and Collective Behavioral Perspectives," Collective Behavior - A Source Book, Ed. M.D. Pugh. New York: West Publish Co., 1980.

Drabek, T.E., "A Methodology of Studying Disasters, American Behavioral Scientist, Vol. 13, No. 3, January-February, 1970, pp. 331-343.

Clark, J.P. and T. Carter, Response to Hurricane Warnings as a Process: Determinants of Household Behavior, Natural Hazard Warning Systems Report Series 79-08, Department of Sociology, University of Minnesota, 1979.

Sorensen, J. and B. Richardson, "Evacuation Behavior at TMI: Review and Examination," (draft paper). Oak Ridge, Tenn.: Oak Ridge National Laboratory, 1983.

Howard, R.A. and J.E. Matheson, "Influence Diagrams," in The Principles and Applications of Decision Analysis," R.A. Howard and J.E. Matheson (eds.), Vol. II, Strategic Decisions Group, Menlo Park, California, 1984.

Shachter, R., "Evaluating Influence Diagrams," Department of Engineering-Economic Systems, Stanford University, Stanford, California, May 1984.

UNCERTAINTY ANALYSIS AS APPLIED TO PROBABILISTIC RISK ASSESSMENT

V.M. Bier

Pickard, Lowe and Garrick, Inc.
Newport Beach, California 92660

ABSTRACT

There are three basic stages in any uncertainty analysis. The first is data analysis, including the use of expert opinion. This task is the development of probability distributions for the basic quantities of interest, such as equipment failure rates and repair times. These distributions can be either generic or plant-specific, depending on the purpose of the analysis and the availability of plant-specific information. The second stage involves propagating the distributions for the basic quantities of interest to obtain distributions for the desired overall performance measure (e.g., the frequency of a hazardous release from a chemical plant). This can be done using either Monte Carlo simulation or the method of discrete probability distributions. Finally, the results of the uncertainty propagation must be interpreted and their implications for decision making clarified. This stage involves significant interaction between the analysis and the decision makers.

KEY WORDS: Uncertainty Analysis; Data Analysis; Bayesian Methods; Uncertainty Propagation

1. INTRODUCTION

The approach to uncertainty analysis adopted by Pickard, Lowe and Garrick, Inc. (PLG), for use in probabilistic risk assessments is based on the subjective, Bayesian theory of probability and the theory of decision analysis. This approach yields rigorous answers to many questions about risk and can also aid in clear problem formulation.

The rationale for doing an uncertainty analysis is to help in decision making. All decisions involve uncertainty. The role of an uncertainty analysis is therefore to quantify the existing uncertainty in order to help the decision maker choose among the available options. These options may include strategies for reducing the uncertainty.

The three stages of an uncertainty analysis are data analysis (including the use of expert opinion), uncertainty propagation, and the interpretation of results. These three tasks are discussed in the following sections.

2. DATA ANALYSIS

In general, there are three types of information available for use in a plant-specific risk analysis: general engineering knowledge about the design, manufacture, and operation of the equipment being analyzed; the historical performance of that equipment at plants other than the one under consideration; and the performance of the equipment at the specific plant being analyzed. The results of an uncertainty analysis should include all three types of evidence. That is the reason for doing a Bayesian data analysis, since it allows for the appropriate weighting of all types of evidence and for the explicit use of judgment.

As an example, consider the decision regarding which type of pump to buy. If the salesman for one particular type of pump says that its failure rate is equal to two per year, you could then develop a subjective probability distribution for the pump failure rate based on such factors as your engineering knowledge about pumps in general and your degree of trust in the salesman. Thus, if you believe that the pump salesman is more likely to underestimate the failure rate than to overestimate it, your probability distribution might span a range from about one to five failures per year, as shown in Figure 1.

This distribution could then be updated as additional evidence became available. For example, if you found out that a neighboring plant with the same type of pump had consistently experienced only two or three failures per year for a large number of years, this would probably convince you that the failure rate at your own plant was very unlikely to be as high as five per year. Your updated subjective probability distribution might then look something like the one in Figure 2.

The process for performing this updating can be described more rigorously by Bayes' theorem:

$$p(\lambda|E) = \frac{p(\lambda)L(E|\lambda)}{L(E)}$$

In this notation, λ is one possible value of the unknown quantity of interest (e.g., a failure rate); E is the new evidence that has become available; $p(\cdot)$ is the prior distribution for λ; $p(\cdot|E)$ is the posterior distribution for λ in light of the new evidence E; $(L(E|\lambda)$ is the likelihood of observing the evidence E given the specified value of λ; and

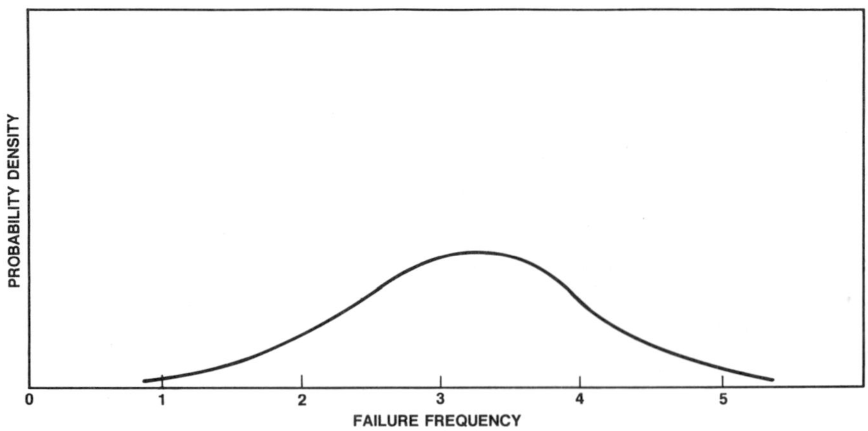

Figure 1

Probability Distribution for Pump Frequency Based on Salesman's Statement

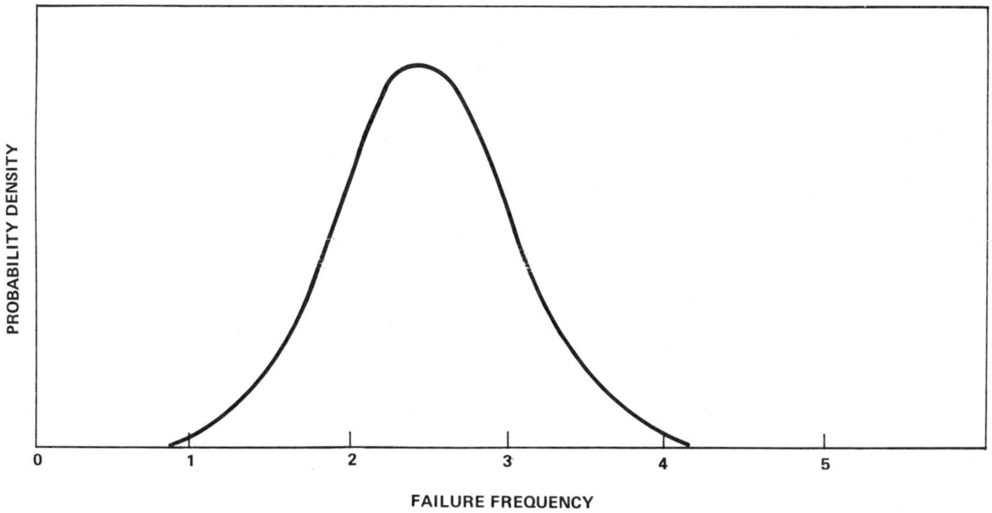

Figure 2

Probability Distribution for Pump Failure Frequency after Learning Neighbor's Experience

L(E) is the overall likelihood of observing the evidence E based on the prior distribution for λ. Thus, the relative likelihood assigned to any particular value of λ after the evidence E has been obtained is proportional to two factors: how likely that value of λ was thought to be the right one before the evidence E was obtained (as given by the prior distribution of λ); and how likely the evidence E would be if that value of λ were the right one (as given by $L(E|\lambda)$.

There are two basic models for use in combining generic and plant-specific data. The simpler, uniform-population model assumes that all elements of the population share the same values of all important parameters (e.g., failure rates or mean repair times). The data can then be pooled as if all came from the same item of equipment, since there is assumed to be no population variability.

In the more complicated, variable-population model, each element of the population is assumed to have a unique set of parameter values (e.g., failure rates and mean repair times). The population can then be described by a median value and a measure of variability for each parameter. These median values and variabilities are generally treated as unknown quantities, and they can be assessed by Bayesian analyses just like those performed for the actual failure rates and mean repair times at particular plants.

Figure 3 indicates how data on observed failure rates for similar items of equipment at different plants can be used to develop a population variability curve. This curve shows how the actual failure rate, λ, of the equipment in question varies from plant to plant. In this hypothetical example, observed failure rates of approximately 5 per year were found at 3 plants, observed failure rates of approximately 10 per year were found at 8 plants, and observed failure rates of approximately

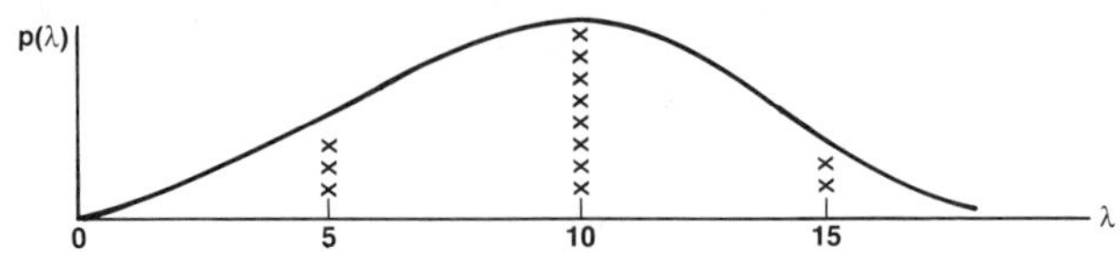

Figure 3

The Variable-Population Model: Population Variability Curve

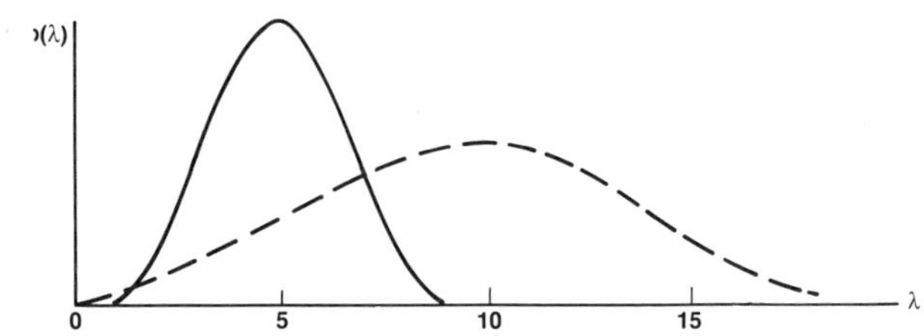

Figure 4

The Variable-Population Model: Plant-Specific State-of-Knowledge Curve Based on 10 Failures in 2 Years

15 per year were found at 2 plants. On this basis, a population variability curve ranging from nearly 0 to a little under 20 failures per year was constructed. This curve can be thought of as a probability density function for the failure rate of the type of equipment in question at a randomly chosen plant from the population.

Figure 4 shows now this population variability curve can be updated to yield a probability distribution for the failure rate of the type of equipment being considered at one specific plant. This example might correspond to a case in which roughly 10 failures had been observed at the plant in question over about 2 years, so one would be quite confident that the failure rate at that plant was about 5 per year. By contrast, it only 2 or 3 failures had been observed over a period of half a year, the plant-specific failure rate distribution would look more like the generic distribution or population variability curve, since the evidence would not distinguish the plant in question from many of the other plants in the population.

For a more realistic example, Table 1 presents actual data on motor-operated valve (MOV) failures on demand at several plants. To begin the analysis of these data, a population variability curve for the failure rate was developed. Based on general engineering knowledge, it was assumed that the median failure rate for the population of similar MOVs at all plants could be anywhere between 4.0×10^{-4} and 1.5×10^{-2} per demand. It was similarly assumed that the range factor of the population variability curve (i.e., the ratio between the 95th percentile and the median or between the median and the 5th percentile) could be anywhere between 1.5 and 11.9.

Table 1

Data on Motor-Operated Valve (MOV) Failures on Demand

SOURCE	DATA		
	NUMBER OF FAILURES	NUMBER OF DEMANDS	FAILURE RATE ESTIMATE
GENERIC			
PLANT A	10	1.65+4	6.06-3
PLANT B	14	1.13+4	1.24-3
PLANT C	7	1.73+3	4.05-3
PLANT D	42	6.72+3	6.25-3
PLANT E	3	1.26+3	2.38-3
PLANT F	31	9.72+3	3.19-3
PLANT–SPECIFIC	15	5.32+3	2.82-3

NOTE:

EXPONENTIAL NOTATIONS ARE IN ABBREVIATED FORM;
i.e., 1.65 +4 = 1.65×10^4; 6.06 – 3 = 6.06×10^{-3}

The joint prior distribution for the median failure rate and the range factor was then updated using the generic data in Table 1. The results indicated that the most likely value of the median failure rate for the population of MOVs was 3.0×10^{-3}. Similarly, the most likely value of the range factor describing the population variability was found to be between 2 and 3. Finally, the generic distribution or population variability curve resulting from the previous step was updated using the plant-specific data from Table 1.

The results of both the generic and the plant-specific analyses are shown in Figure 5. As can be seen from that figure, the plant-specific distribution is much narrower than the generic one, since it reflects only the uncertainty about the failure rate of MOVs at the particular plant in question, not the variability between the plants in the population as a whole.

3. UNCERTAINTY PROPAGATION

The purpose of techniques for uncertainty propagation is to combine the probability distributions for the various input quantities in an analysis (e.g., equipment failure rates) to yield a probability distribution for the desired output quantity (e.g., the frequency of a hazardous chemical release). The distributions for the input quantities are typically obtained as the result of a data analysis.

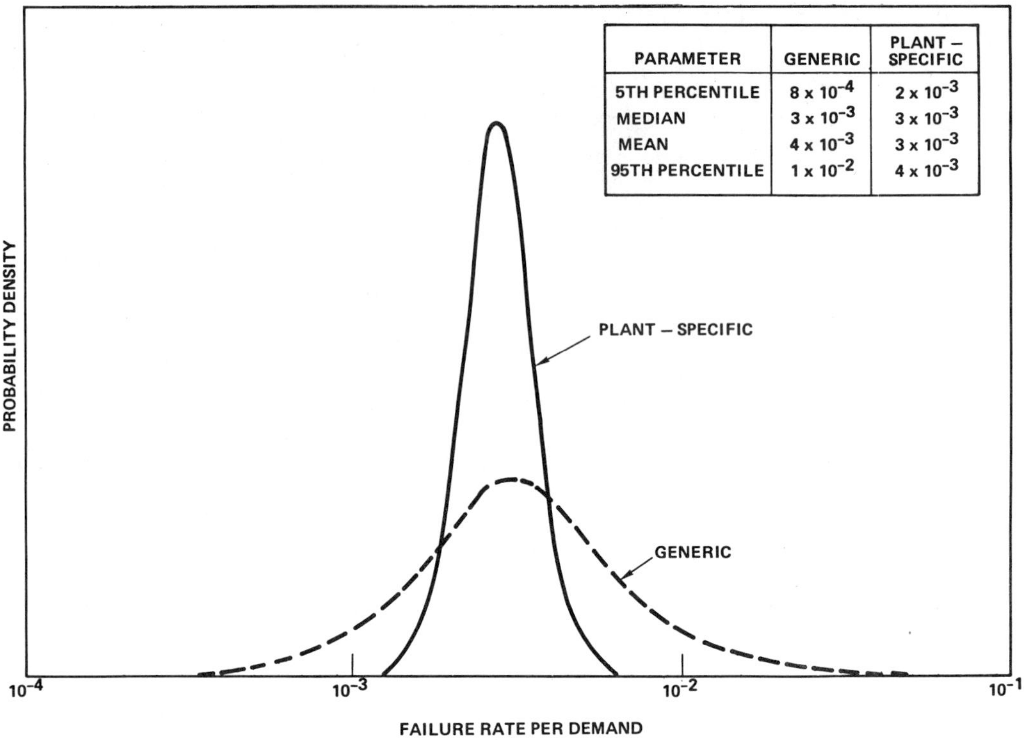

Figure 5

Generic and Plant-Specific Distributions for MOV Failure Rates on Demand

Three basic methods of uncertainty propagation are available: the method of moments, Monte Carlo simulation, and the method of discrete probability distributions. The method of moments propagates the means and variances of the input quantities through the risk model to yield the mean and variance of the output quantity. Monte Carlo simulation is based on random sampling from the probability distributions for the input quantities, whereas the method of discrete probability distributions is based on systematic sampling.

The method of moments is quite simple in the case where the input quantities are all mutually independent. For example, if A and B are mutually independent and

$$U = A + B$$

is the desired output quantity,

$$\text{Mean } (U) = \text{Mean } (A) + \text{Mean } (B)$$

and

$$\text{Variance } (U) = \text{Variance } (A) + \text{Variance } (B)$$

Similarly, if the desired output quantity is given by

$$U = A \bullet B$$

then

$$\text{Mean (U)} = \text{Mean (A)} \cdot \text{Mean (B)}$$

and

$$\text{Variance (U)} = \text{Variance (A)} \cdot \text{Variance (B)}$$
$$+ \text{Variance (A)} \cdot [\text{Mean (B)}]^2$$
$$+ \text{Variance (B)} \cdot [\text{Mean (A)}]^2$$

In the case where the input quantities are not mutually independent, the expressions for the mean and variance of the output quantity are more complicated. However, for many expressions and many types of input distributions, the mean and variance of the desired output quantity can still be evaluated analytically.

The advantage of the method of moments is that in many cases it yields closed-form analytic results. This makes it possible to see exactly how the results depend on the means and variances of the input quantities without having to reevaluate the entire model every time an input quantity is changed. However, the method of moments yields only the mean and variance of the desired output quantity, not its entire distribution. Therefore, some form for the distribution must be assumed, introducing a possible source of inaccuracy.

In Monte Carlo simulation, a value is sampled at random from the distribution for each input quantity. The sample values of the various input quantities are then combined to yield a sample value of the desired output quantity. The entire process is repeated many times, until enough sample values of the desired output quantity have been obtained to yield a good approximation of its distribution.

One of the primary advantages of Monte Carlo simulation is that the accuracy of the results can be determined rigorously. In addition, any desired degree of accuracy can be obtained simply by increasing the number of samples taken from the distributions for the input quantities. However, to obtain highly accurate results, an extremely large number of samples may be required, which could make this method costly.

Finally, in the method of discrete probability distributions, each continuous input variable is approximated by a discrete variable with a finite number of possible values. Each value is associated with a probability indicating the likelihood of the actual input quantity taking on a value near the specified one. Each possible combination of values of the discrete input variables is then associated with the appropriate probability to form one point in the discrete probability distribution for the desired output quantity. The number of points in the output distribution can be reduced by grouping points with similar values.

One advantage of the method of discrete probability distributions is that it tends to be inexpensive. Relatively few points are needed in the distribution for the desired output quantity, since each point is already associated with a probability. This is unlike simulation, in which information about the relative likelihood of any given value can be inferred only from the density of points in the vicinity of that value, so a large number of points is needed to yield meaningful results.

However, the accuracy of the results obtained using the method of discrete probability distributions generally cannot be rigorously determined. In addition, using this method can make it difficult to properly account for dependencies in large or complex problems, (e.g., the dependencies arising when the same input quantity appears at several different places in the expression for the desired output quantity).

A simple example illustrating the process of uncertainty propagation using the method of discrete probability distributions involves the ignition of a flammable chemical. Figure 6 presents a three-point discrete probability distribution for the frequency of leaks from a hypothetical storage tank, measured in events per year. Similarly, Figure 7 presents a three-point distribution for the conditional frequency of ignition given that a leak does occur.

Figure 8 presents a discrete probability distribution for the product of the leak and ignition frequencies, which is the actual fire frequency measured in events per year. Note that some of the points in Figure 8 appear very close together. These pairs of points can be grouped for greater ease of interpretation, as shown by the dotted lines.

4. INTERPRETATION OF RESULTS

The results of uncertainty analyses can be difficult to interpret, so risk analysts must interact extensively with the decision makers for whom an analysis was performed. Examples of areas in which risk analysts can make important contributions to the decision process are weighing the merits of an assured small benefit against the merits of a small chance of a larger benefit; identifying the dominant contributors to the uncertainty about each option; and identifying strategies for reducing those uncertainties.

Figure 9 presents the results of a hypothetical uncertainty analysis comparing the magnitude of risk reduction provided by a new design against that provided by the use of diagnostic instrumentation with an existing design. This example shows that the new design is likely to yield a reduction in risk significantly greater than that associated with the use of the diagnostic instrumentation, but that there is also a small chance that the new design will actually result in an increase in risk (shown in Figure 9 as a negative risk reduction).

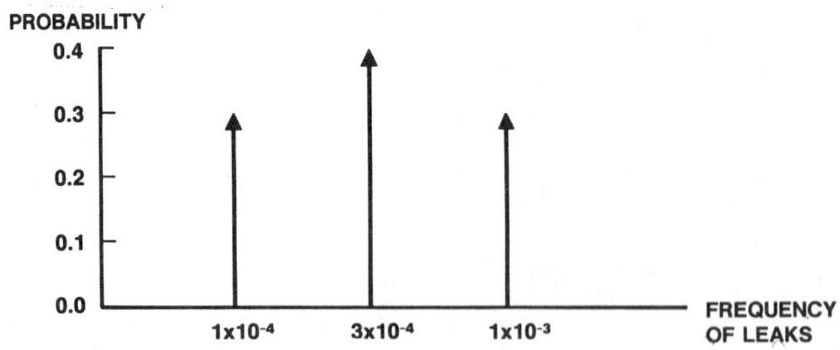

Figure 6

Discrete Probability Distribution for the Frequency of Leaks

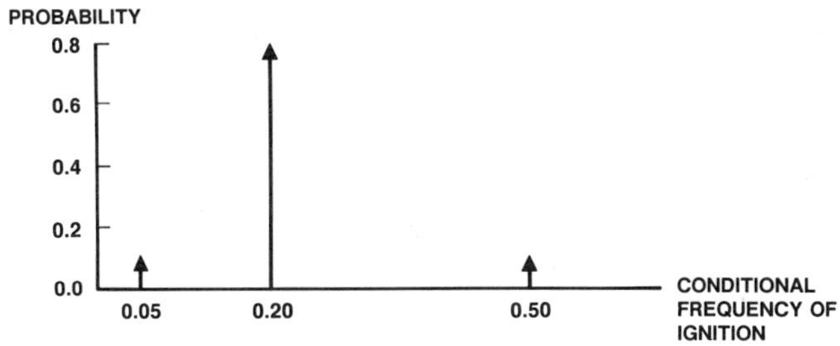

Figure 7

Discrete Probability Distribution for the Conditional Frequency of Ignition Given a Leak

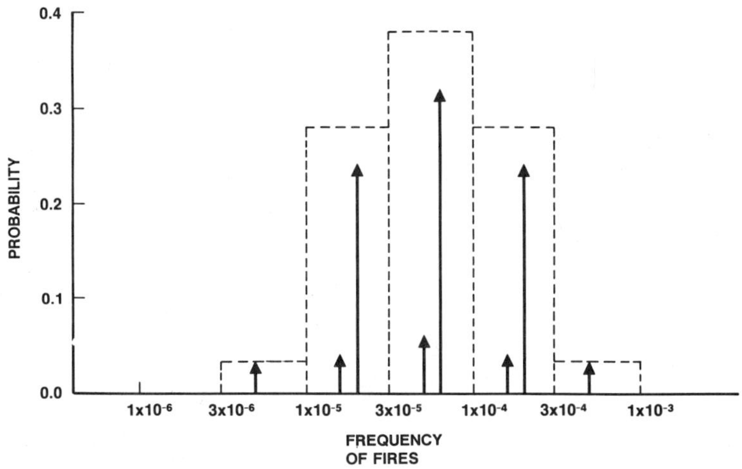

Figure 8

Results of Example

The large uncertainty associated with the new design is likely to be unacceptable to many decision makers. However, the chance of an increase in risk associated with the new design may be due to its potentially greater vulnerability to only one or two failure modes. Either a redesign of the equipment to eliminate that potential vulnerability or research to determine whether it is in fact a problem could enable a decision maker to realize the potential benefit associated with the new design without an undue chance of poor performance.

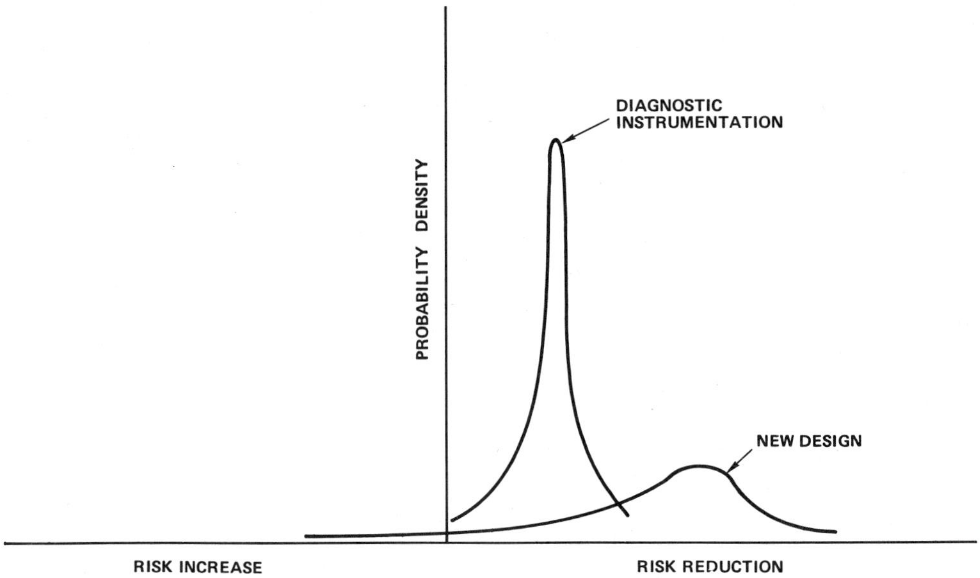

Figure 9

Example - Net Benefits of Corrective Actions

ANALYSIS OF RISKS OF DIVERSION OF PLUTONIUM OR HIGHLY ENRICHED URANIUM

E.L. Zebroski

Electric Power Research Institute
Palo Alto, California 94303

ABSTRACT

The production of a nuclear explosive device requires a range of specialized skills and facilities and a sufficient quantity of concentrated fissionable material. Significant amounts of such materials exist in hundred of locations, in a variety of forms including mixtures with nonfissionable materials and with radioisotopes. Nuclear weapons start with pure metallic materials of predominantly a single isotope (uranium-235 or plutonium-239). Chemically or isotopically diluted materials either require larger amounts of fissile isotope or else cannot be made to produce an explosive reaction without extensive processing involving sophisticated techniques and facilities. Obstacles to diversion of fissionable materials include combinations of physical security, guards and defense forces, accountancy, diluents, limited amounts of material in any one location, and timely detection and forceful reaction to attempts at diversion or theft.

A generalized risk assessment model is developed, incorporating the main factors that influence the probability of successful diversion. A roughly inverse relationship is involved between the amount of material potentially seized or diverted and the levels, skills, facilities, and resources required to make a workable nuclear device. Absolute probabilities are indeterminate. Experience indicates an upper bound of about 1.6×10^{-6}/year/weapon-equivalent amount of material. Inequality relationships can be used to develop relative rankings and semiquantitative magnitudes of diversion risks.

Small university research reactors appear to contribute less than 0.1% of the total diversion risk of U.S. materials. Limiting the amounts of material at any one university location is practical so that thefts from several--perhaps half a dozen--reactors would be required to make a single explosive device plausible.

A related, nonnuclear risk appears relatively likely, arising from the likelihood of unwarranted public overreactions to a variety of actions from vandalism to attempted thefts. These reactions are likely, given the present levels of discrimination on this topic.

Cost/benefit relationships are also derived. These exhibit disparities of several orders of magnitude in the ratios of commitments made or planned to the size of risk exposure reductions attainable.

KEY WORDS: Diversion; Weapons Material; Nuclear Bombs; Research Reactors; U-235; Plutonium.

Probabilistic Formulation of Risk Exposure

The main factors which affect the likelihood of successful diversion, or theft of material to the end result of a workable nuclear device are exhibited in equations (1) through (4).

$$P_{tot} = \sum_{i,t} P_{i,t} T_i \qquad \text{(Eqn. 1)}$$

$$P_{i,t} = \frac{d_{i,t} A_{i,t}}{M_b} (1-r_{i,t})Q_{i,t} \qquad \text{(Eqn. 2)}$$

P_{tot} = Total probability of successful diversion of material from all i facilities in time interval t.

$P_{i,t}$ = The probability of successful diversion of material to produce n workable nuclear explosive devices from location i, in time interval t.

$A_{i,t}$ = The amount of weapons-useable material contained in location i, during time interval t.

M_b = The mass of fissile material required for a nuclear explosive device of type b.

$d_{i,t}$ = The probability of access for theft or diversion from location in in time interval t, of amount $A_{i,t}$.

$r_{i,t}$ = The probability of detection and protection or recovery of material $A_{i,t}$, given that access has occurred or been attempted.

$Q_{i,t}$ = The probability of success in assembling a workable nuclear device given mass M_b.

T_i = Time of exposure of amount $A_{i,t}$ at location i.

There is a further important relationship affecting the value of M_b for a given diversion attempt

$$M_b = \frac{M_0}{f(R,S,D,E,G,T)} \qquad \text{(Eqn. 3)}$$

where M_0 = just critical mass of material without efficient reflector and at nominal density

$f(R,S,D,E,G,T)$ = a complex relationship generally proportional to the product of the availability of resources R, Scientific and technological skills S, specialized data D, specialized equipment and facilities E, specialized materials G, and time T, available to accomplish the design, processing, testing, and assembly.

$$\text{Also } Q_{i,t} \propto f(R,S,D,E,G,T) \qquad \text{(Eqn. 4)}$$

And for $M_b \ll M_0$; $Q_{i,t} \cong R \times 10^{-9}$

where R = resources in \$

The expressions given can not be quantitatively evaluated in closed form. However, they permit several of order-of-magnitude qualitative or ranking comparisons that can be useful in decisions on risk management activities.

Assumption of a First Law of Risk Management

The value of these inequalities assumes their use in the ideal mode of rational risk management to optimize the total amount of safety which is obtained for the total effort available or expended. This might be referred to as the "first law of risk management," which can be formulated as follows: "the relative effort applied to limit or control a given risk should be in proportion to the product of the relative size of the risk and the fractional reduction in that risk achievable for the effort applied."

It is recognized that such a first law is very limited in real world application for several reasons. The discipline for establishing relative risks within a given field of or type of risks is now often practical, but is used only sporadically in practice. The problem of developing relative risks between different areas or functions of society, is still in its infancy, is overlaid with subjective values and customs, and has virtually no impact in public risk management and policy. Exogenous factors commonly drive the priority and resources devoted to risk management activities of various types. The extended literature on the influence of risk perception highlights the large gaps between rational deployment of effort in risk management and actual current practices. This analysis adds to the literature of such gaps.

Recognizing all of the foregoing limitations, nevertheless, it is assumed that awareness of the relative rankings of risk exposures to a common set of hazards can be useful to policy makers and in drafting specific rules, regulations, and procedures which indicate the type and intensity of efforts to be applied to various risk management activities.

Effects of the Controlling Factors

The general relationship equation (1) can be interpreted as a measure of the annual probability that significant amounts of fissionable material might be seized, stolen, or diverted, from locations holding such materials, and in amounts sufficient to produce one or more nuclear explosives. Recent attention on improvement in security measures in certain weapons storage facilities are aimed at reducing the probability of success of access ($d_{i,t}$). However, a little reflection will suggest that vaults or passive measures can always be breached if the incentive is large enough, and that the factor ($r_{i,t}$) of detection, recovery and reaction forces is at least equally important in preventing successful theft or diversion.

Another factor, ($Q_{i,t}$), is extremely important in evaluating risks at locations at which only small amounts of material are present. Given a relatively large amount of material, an explosion with some degree of nuclear yield might be produced, perhaps not as a garage operation, but certainly with the materials and equipment available in almost any modern university with science and engineering departments. For this reason, much more intensive protective measures (affecting $d_{i,t}$ and $r_{i,t}$) are appropriate for facilities which routinely contain many times the amount of material (M_b). Given 50 or more kilograms of highly enriched U-235, it becomes plausible that an adequately educated but inexperienced team of people could product a single nuclear explosive. To be sure, the reliability of the device might be uncertain, the yield might range anywhere from ability to destroy several city blocks to a nuclear "fizzle" with a nuclear yield hardly more than the chemical explosive used. Nevertheless, even such a primitive device could have major public and governmental impacts. For perspective, it should be noted that similar capacity for blackmail, terror, or damage can be obtained with much smaller initial resources and using much more readily available materials and means.

With smaller quantities of highly enriched U-235, say 20 to 30 kg, (or more to allow for processing losses), somewhat higher levels of skills and facilities are required, and there is greater need for specific physical data and calculations and process information for which the essential know-how is not available in the open literature. Unless it is further assumed that a sizeable team of people with extensive experience in a national weapons program are available, the time required to attain a plausibly workable device, and the probability of failure, both increase substantially.

For still smaller amounts of material, say 10 to 20 kilograms of U-235, net after processing losses, the requirements are sharply increased for specialized skills, data, facilities, and materials. As far as can be deduced from the historical record, this stage of development in each of the weapons countries has required an all-out national effort with essentially unlimited financial and material resources. Total times on the order of three years or more were required for a first successful effort in each of the countries which are currently weapons producers. The focus on the production of fissile materials tends to overlook the parallel, and in some respects more difficult developments required to produce the actual device.

The replication of such results by a team with all of the relevant data and experience, obviously would take less time and effort. However, it is highly implausible that a limited team of people with subnational levels of skills, resources, and facilities could accomplish this task in a short periof of time. Three to five years can plausibly be regarded as being a minimum to achieve an appreciable probability of success.

The well-publicized assertions by a former Los Alamos employee that small amounts of materials could readily be converted to nuclear weapons given readily available unclassified data, skills, materials, and facilities, appears to have hidden within it several interesting assumptions. The most basic assumption required (to give a significant probability success) is that a substantial team of people with a wide range of theoretical, engineering, chemical, metallurgical, and mechanical skills <u>and directly applicable experience</u> are available--presumably recruited from a national weapons laboratory such as Los Alamos--and that they willingly cooperate in a clandestine subnational or terrorist effort. A second assumption is that such a team would bring with it a large volume of detailed physical data, and nuclear, chemical, metallurgical, and specialized know-how, including explosives and hydrodynamic technology. A further hidden assumption appears to be that resources, and facilities--often estimated to be in the range of several hundred million dollars--are available to such a team, and that no obstacles in procurement, or hazards of detection of such activities are operative. For the case of limited amounts of fissile material requiring highly sophisticated design, fabrication, assembly, and packaging, the probability $Q_{i,t}$ can be represented as a judgment approximately by the product of resources available in dollars times 10^{-9}, and three to five years' time.

<u>Alternate Paths and Diversion Scenarios</u>

For plutonium-239 as the target material for theft or diversion, many of the same considerations apply, except that the reference mass of material M_b is less than 1/3 as large as for U-235. This highlights an interesting strategic question on the motives and tactics of a would-be diverter. If weapons of medium or high technology are desired, able to work with relatively small masses of material, then diversion of <u>small</u>

amounts of material (say less than 50 kg of U-235) is unprofitable, relative to the overall effort required. A more direct, secure, and expandable route is the development of dedicated facilities to produce U-235 by any of a considerable number of proven enrichment processes. Equally simple, at this level resources, is the production of plutonium in a low-temperature, low-technology production reactor. The scenarios which lead to substantial prospects of success in diversion can be qualitatively ranked as follows:

(1) A national decision to "go for nuclear weapons" by seizure of substantial stockpiles of existing nuclear weapons already produced by one of the present nuclear weapons countries, and stored and deployed in several hundred locations throughout the world.

(2) A national level decision to "go nuclear" by clandestine development of as much as possible of the theory and technology of the device in a secret manner, followed by seizure of major amounts of fissile material (for U-235 ranging from several hundred to several thousand kgs) which may be available in aggregate as raw material, fabricated fuel, and spent fuel associated with research and testing reactors. Under similar circumstances, an alternate decision may be to seize substantial tonnages of spent fuel from power reactors, with the intention of reprocessing to extract plutonium, using previously built and tested facilities. Both of these options involve large risks of international interventions.

(3) A national, but more likely subnational, faction decision to seize relatively small amount of material (for example, 25 to 100 kgs of U-235) which may be available in research, testing, or prototype reactors.

(4) Subnational factions or terrorist groups' decision to steal or divert relatively small amounts of material from fabrication stockpiles, or research and test reactors for criminal, terrorist, or ideological ends.

(5) Diversion or theft of material from U.S. medium or high-powered research or test reactors.

(6) Diversion or theft from low-powered U.S. research reactors commonly at universities.

We note that each scenario involves somewhat different objectives, requirements, and resources. It is important in assessing risks of various scenarios to use self-consistent sets of assumptions. For example, a highly skilled, highly resourced group with near--national capabilities would have little motivation for thefts of small quantities of material, since this would rate poorly relative to a self-production approach or relative to seizure from locations with large amounts which could be expanded to a significant offensive or defensive deterrent capability.

Limiting Value Approach to Aggregate Historical Exposure

A large number of nuclear weapons are deployed in several hundred locations within the United States. All but three states are known to have several such locations. In addition, weapons are deployed in over 100 locations in Europe, and several hundred locations in Southeast Asia, not counting ships at sea. Given the buildup of production of nuclear warheads (including aircraft-carried bombs, artillery shells, and warheads for ballistic and cruise missiles), something in excess of 30,000 such devices have been produced in the United States and stockpiled or deployed. Allowing for the buildup over time since 1945, we can estimate a cumulative exposure of approximately $4x10^5$ weapon-years has occurred. We can take account of raw material and material in production and rework pipelines by dividing the amount of such materials by the average mass required for a weapon. This gives a further estimate of $2x10^5$ weapons-equivalent years of exposure. Similarly, we can estimate the exposure for non-weapons materials in several categories. For example, highly enriched materials supplied for research in test reactors overseas are approximately $1.5x10^4$ weapon-years equivalent. Domestic research and test reactors (medium and high power) constitute about $1.5x10^3$ weapon-year equivalent. Small low-powered university type research reactors give a cumulative exposure of less than $4x10^2$ weapon-year equivalents.

The overall cumulative exposure is roughly $6x10^5$ weapon-years exposure to date. Given that no successful diversion resulting in a nuclear device has occurred, the current limiting value for diversion or theft is about $1.6x10^{-6}$ per weapon or weapon equivalent, per year.

Derived in the same fashion, the current annual exposure is approximately 4×10^{-4} weapon-years per year. If the historical cumulative value is taken as an approximation to the expected value, then the overall exposure of the entire inventory leads to a current value of $P_{tot} = 0.06$ per year. This number declines slowly with time assuming the favorable experience to continue. Nevertheless, it is not entirely comforting, since it leads to a 50% expectation of a successful diversion in about 12 years.

The foregoing discussions of the historical record is neutral to the differences in the values of the controlling variables as exhibited in equation 2. For practical decision-making purposes it is necessary to examine at least the order-of-magnitude differences in the controlling variables between the different segments of exposure.

Factors Controlling Weapons Diversion Exposures. It is evident that the preferred target for a would-be diverter, whether national, subnational, or terrorist, would be an actual weapon. Actual weapons represent over 96% of the weapon-years exposure to date, yet there apparently has been no actual diversion, nor even a competently mounted threat of such diversion evident.

Weapons storage locations are protected by vaults and guards. If the statistics of successful breaches of bank vaults applied, one would expect to have seen several dozens of occurrences. Evidently the factor of detection and reaction ($r_{i,t}$) is quite large. The presence of armed guards and barriers of the military installation are certainly significant obstacles. However, under some circumstances such obstacles could be relatively minor to a determined band of attackers, as for example, when weapons or weapons material are being transported by truck, train, or aircraft. The dominant protective factor operating, we may surmise, is the high assurance of detection and reaction, with indefinitely increasing force if necessary.

The probability factor of achieving the successful fabrication of a nuclear device ($Q_{i,t}$) is relatively high for weapons, but not unity as might be supposed. Weapons have defensive features such as separate keys of unique materials, and certain self-destruct capabilities, which prevent them from being useful to an adversary who might capture them in a battlefield situation. For the same reason, even a successful peacetime diversion would still require a high degree of technical sophistication, and preferably a variety of detailed technology knowledge accessible only to "insiders", if $Q_{i,t}$ is to attain values as high as 0.1 to 0.5, and given limited periods of time, of the order of several months.

Overseas Exposure of U.S.-Supplied High Enriched Uranium

Approximately 18,000 kilograms of this material have been supplied to countries overseas (1). For various reasons, the United States did not establish a pattern of tracking this material point by point and moment and moment but relies on either bilateral agreements or IAEA accountability and inspection. At any given time, up to 2,000 kg. of such material is present overseas as raw material (uranium metal, uranium hexafluoride, or uranium oxide) or as material undergoing fabrication processes and including fabricated fresh fuel.

Something in the range of a half of a ton of such material is present at any given time in operating reactors, where it has associated with it the high levels of radioactivity of operating nuclear fuel. By far the largest part of this material exists in spent fuel pools scattered in roughly 100 locations, in 40 countries, with the bulk of it in Western European countries and Japan. Allowing for the elimination of some U-235 by burnup in the fuel, and the relatively small amount which has been returned to the United States for reprocessing, a rough estimate of 6,000 to 10,000 kgs. of highly enriched U-235 remains.

From a diverter's standpoint, the resources, skills and facilities required are substantially higher for radioactive fuel which might be taken from an operating reactor than for unirradiated fuel or raw materials. The highly radioactive fuel from the reactor core has the further disadvantage that attempts to remove or transport it can be detected at long distances by conventional radiation detectors, so that the likelihood of effective detection, reaction and recovery can be very large (provided that such functions are planned and implemented in a given region).

Spent fuel from fuel storage pools still requires substantially more difficult and complex processing than unirradiated fuel or raw material. However, the level of radiation emitted falls off inversely with time (approximately) so that detection at a distance after 1 year and 10 year storage is orders of magnitude less sensitive than for fuel recently removed from an operating reactor.

Medium and High-Powered Test Reactors in the United States. Most of these reactors are operated by the Department of Energy and are located at national laboratories or weapons laboratories and associated facilities. In aggregate, these reactors involve similar amounts of material (within factors of 2) to the medium and high-power test reactors overseas. However, the effective exposure is an order of magnitude smaller than for the overseas reactors, since the spent fuel is routinely returned for processing, and returned to locations with military grade levels of protection and reaction forces. There appear to be no plans to convert domestic research and test reactors operated by the Department of Energy to lower enrichments.

Research and Test Reactors - Low Power

Over 50 low-power research and test reactors are operable within the United States, and roughly 30 of them involve some high enriched U-235 fuel. Most of these reactors have "lifetime cores" since the power level and duty cycle will take at least 20 years to reach the physical burn-up limits of the fuel. Many of these reactors are licensed to carry as inventory substantial additional amounts of unirradiated fuel, either as fresh fuel or as spent fuel awaiting shipping and storage. Augmented security procedures brought into effect in 1983 have largely eliminated the inventory of unirradiated fuel (or else provided for its storage at a DOE facility). With these measures, the amount of inventory at a typical university research reactor facility is less than 5 kg. and for some reactors less than 2 kg. A would-be diverter would have to remove and recover the material from at least three reactors to achieve some probability of producing a "high technology" device, and from at least half a dozen research reactors in order to have some chance of producing a "low technology" device. Nevertheless, the NRC has promulgated a rule that requires the conversion of all remaining non-DOE research reactors to low enrichment, under 20% U-235 (2).

A decision was reached in the late 1970s that the continued shipment of tonnage quantities of highly enriched of U-235 overseas constituted an undesirable risk, particularly since assurance of low values of $d_{i,t}$, or high values of $r_{i,t}$ could not be assured. Quantities of material $A_{i,t}$ at single locations range from the equivalent of several to many "low technology" nuclear devices. The process of conversion to low enrichment for exports is underway. However, the technical obstacles to producing high performance fuel at 20% enrichment (without substantial redesign of the reactor facility), are delaying implementation of this change to the late 1980s.

The order for conversion of domestic university research reactors to low enrichment was taken for the nominal reasons of further reduction of the risks of diversion, and to "set an example" for overseas research reactors. However a possibly more compelling reason is also wroth noting. Vandalism or break-ins at university research or computing facilities have occurred occasionally in recent years. The possibility exists that such a future incident might be followed by threats or rumors, based on allegations that material usable for a nuclear explosive was stolen. There is an evident advantage for NRC officials to be able to announce that no nuclear explosive was possible, regardless of whether some material was actually taken. In such a case, the regulatory agency is assured of being perceived as highly responsible. Nevertheless the reduction in the actual risk of a diversion leading to a bona fide nuclear explosive is largely illusory. It begs the question of proper attention to the other, much larger and very real diversion risks listed in Table I.

Measures of Cost Effectiveness of Risk Management Options

Substantial increments in physical security and presumably in detection and recovery forces have been implemented for military stores of weapons and weapons materials in recent months. These actions appear to be in response to the impacts of the terrorist bombings in Lebanon and elsewhere, and a general White House directive to augment physical security measures for facilities of all kinds. This appears to be a cost-effective approach, since nearly all of the people involved are already federal or contractor employees, so the main effect is assure that effective plans and communications are organized.

COST EFFECTIVENESS OF CONVERSION TO LOW ENRICHMENT

The federal costs (mostly R&D) for conversion of overseas reactors are of the order of $30 to $50 millions. By the late 1980s this action will reduce the continuing inventory of high enrichment uranium in reactors and in the production pipeline by more than two tons.

The cost of this improvement to the U.S. is about $20/gram of material removed from service (not counting the procurement and conversion costs incurred overseas).

This strategy avoids increasing the continuing diversion risk exposure from the tonnage of spent fuel in storage pools, but does nothing to decrease it. This remains by far the largest store of weapons-useable U-235 not in U.S. custody. Overseas holders of this material are not commonly returning it to the United States since it costs them more than retaining it in storage pools. (The United States is prohibited by the NNPA of 1978* from expending any funds for receiving, storing, or processing any kind of spent fuel from overseas, but a year-to-year exemption for some high enrichment fuel has been decreed).

On the current course of events, the cost vs. risk reduction benefits in vogue are shown in Table 2. The U.S. costs committed per gram removed from service (ideally a rough measure of assigned risk-reduction value), are actually in reverse order of the size of risk exposure. The most curious disparity is the effective assignment of a negative value to the dominant residual diversion risk of this type, namely the spent fuel tonnages overseas, which contain several hundred weapon equivalents.

The diversion risk ($d_{i,t}$) refers to the probability of removal of a weapons-useable amount from a given location (see equation 2). The likelihood of theft or diversion would seem obviously smaller in a military facility than in an civilian facility--say at a university. However the risk-controlling factor at a given location is not ($d_{i,t}$) but the entire expression $P_{i,t}$ in Equation 2. The value of $A_{i,t}/M_b$ is much smaller than 1.0 for low power research reactors, while it can take values in excess of several hundred at military locations.

Table 1

Ranking of Diversion Exposures

	Category	Estimated Value $P_{i,t}$ Per Year (1)
I.	WEAPONS MATERIALS AND FACILITIES	.05
II.	LARGE TEST RECTORS USING HEU ABROAD	
	A. Unirradiated fuel materials	.005*
	B. Irradiated fuel	.005
III.	LARGE TEST REACTORS - DOMESTIC	.0005
IV.	SMALL RESEARCH REACTORS AT UNIVERSITIES	.00005

*Subject to gradual reduction 1985-1990 by displacement with low enriched uranium.

(1) Using equations (1) and (2) to partition the risk contributions; with $P_{tot} = 0.06$

*Nuclear Nonproliferation Act of 1978.

The assumption that diversion risk exposure per kilogram is not widely different, between the different materials (because of the offsetting effects of the factors controlling $P_{i,t}$) enables a semi-quantitative ordering of the rank of diversion risks for the categories shown in Table 1. Taking the historical value to date as the annual expected value $P_{tot} = .06$, the values in the right hand column are derived. On this scale, the conversion of overseas reactors to lower enrichment has appreciable overall value in reducing P_{tot}.

This leaves the spent fuel overseas as the dominant risk other than military. On this scale the small domestic reactor exposure is about two orders of magnitude smaller than the residual risk from spent fuel overseas.

Reduction in the diversion risk exposure of U-235 in spent fuel overseas has at least equal value, per unit retired, as for small domestic research reactors. If $100 per gram retired from service domestically is a proper valuation, then instead of charging for the return of overseas U-235, the U.S. should be willing to pay for its return. If the values are equated, the U.S. should be willing to pay at least $0.6 billion for comparable value in risk reduction.

Table 2

Relative Cost vs. Risk Reduction Benefit of Current Commitments

Location	Type of Material At Risk	Weapons Equivalents Amounts Per Location	$ gram committed to reduce exposure
Domestic	Fuel for DOE Reactors	Several to Many	~0
Domestic	Fuel for non-DOE Reactors	0.1 to 0.2	>> $100*
Overseas	Future Fuel for Medium and high power Research & Test Reactors	Many	~$20
Overseas	Spent Fuel Containing Highly Enriched U-235	Many dozens	< 0**

*Based on ref. (1), if promulgated and implemented without changes.
**U.S. requires substantial payment to accept the return of such fuel.

REFERENCES

(1) "Proposed Rule for Limiting the Use of High-Enriched Uranium (HEU) in Research and Test Reactors" NRC document Federal Register June 30, 1984.

(2) "Obstacles to U.S. Ability to Control and Track Weapons-Grade Uranium Supplied Abroad" U.S. General Accounting office Report GAO/ID-82-21, August 1982.

AN ASSESSMENT OF THE RISK TO PLANT PERSONNEL AND THE PUBLIC FROM PETROLEUM REFINERY ACCIDENTS

J.S. Arendt, D.J. Campbell, M.L. Casada and D.K. Lorenzo

JBF Associates, Inc.
Knoxville, Tennessee 37932

ABSTRACT

Petroleum refinery and chemical plant accidents represent significant risks to the health and safety of plant personnel and the surrounding public. Though large amounts of time and money have been spent over the last 15 years to assess the risk of nuclear power plant accidents, until recently very little work had been done to assess the health and safety impacts associated with accidents at conventional industrial plants.

This paper presents an overview of an assessment of the safety risk to workers and to the public at a large petroleum refinery located near Toronto, Ontario. This study estimated the expected frequencies and consequences of potential loss of containment accidents (e.g., leaks and ruptures of major process equipment and storage vessels). The personnel risk analysis was confided to the effects of random hit events--a situation when a worker happens to be in an area (hazard zone) that is exposed to the effects of an accident and the worker is injured or killed. The public risk analysis considered meteorological, occupancy, and evacuation effects that can influence the expected casualties from an accident.

This paper described methods and results of the work and discusses some of the general engineering recommendations suggested for reducing the estimated risk of operating the refinery.

KEY WORD: Refinery; Loss of Containment Events; Risk; Casualties; Constant Risk; Concern Risk; Random Hit; Hazard Zone; Fire and Explosion effects; Casualty Probability.

INTRODUCTION

Large amounts of time and money have been spent over the last 15 years to assess the safety risks of operating nuclear power plants. But until recently, very little work had been done to assess the safety impacts associated with accidents at conventional industrial plants. Reliability engineering and risk assessment techniques now offer new opportunities to reduce the risks involved in operating a facility.

This reliability technology is appealing to plant managers because it systematically assesses the important causes of events that threaten the health and safety of plant personnel and the public. This paper presents an overview of a pioneering application of reliability engineering and risk assessment techniques to a full-range petroleum refinery.

Purpose

This work identified ways to improve availability and safety at a full-range petroleum refinery. The project was divided into two phases. First, a preliminary hazards analysis (Phase I) identified areas that were potentially important contributors to the risk associated with loss of containment events at the refinery.[1] Second, a risk-based analysis (Phase II) was performed on those refinery areas identified in the PHA.

Scope

This analysis estimated the expected frequencies and consequences of potential accidents caused by loss of containment events in major sections of process equipment in the refinery. A loss of containment event is a leak or rupture of major process equipment or piping; it results in a release of hydrocarbon or other process material. For example, a heat exchanger flange that separates because of differential expansion due to a temperature transient is considered a "leak" event. A "blow-out" of the seals of a crude tower bottoms pump could release a significant amount of burning oil; this failure is considered a "rupture" event.

Our analysis considered the possible failures of engineered systems and refinery personnel to mitigate the effects of loss of containment accidents. Each process unit at the refinery was divided into process sections (collections of connected equipment). We estimated the risk associated with process sections and then compared these estimates to an acceptance criterion provided by the refinery. The major contributors to the risk associated with these sections indicate where improvement is needed and where efforts to reduce risk may be most cost-effective.

We considered rupture events to be more severe than leak events because they usually caused significant changes in process conditions (pressure, temperature, flow, etc.). These process deviations could be quickly recognized by alert operators from existing control room indications and alarms. Leak events, on the other hand, were assumed small enough that they registered no noticeable effect on process instruments. However, the leak events considered in this study were larger than minor leaks, which occur on a daily basis in the refinery.

The project team selected 12 refinery areas for the risk-based analysis. For convenience, we classified the selected areas according to the following groups:

- Group 1 - nine process units including the Hydrogen, 1st-Stage and 2nd-Stage Hydrotreater, Dewax, Crude, Condensate, Hydrobon/ Platformer, FCC, and Sat Gas units.
- Group 2 - ten LPG storage tanks
- Group 3 - three other propane vessels in the Propane Deasphalting (PDA) Unit and one propylene vessel in the Dewax Unit
- Group 4 - one Amine/Sulfur Unit
- Group 5 - ten bulk oil storage tanks located around the northern perimeter of the refinery property

Accidents involving process areas in groups 1-4 were analyzed in the personnel risk analysis. The public risk analysis considered accidents involving Groups 2 and 5.

Risk Analysis Concepts

Understanding certain concepts of risk is a prerequisite to understanding the results presented in this paper.[2,3] In this study, we developed estimates of the expected frequencies and consequences of refinery accidents caused by loss of containment events. The risk, or absolute risk, associated with an accident is then defined as the combination of the frequency of the accident and the consequence of the accident. The frequencies of loss of containment accidents are reported in units of events per year, and the consequences of these accidents are in units of total dollar cost or expected casualties (serious injuries or deaths of refinery personnel or members of the public).

The motivation for estimating the risk of a particular activity is often to judge whether the level of risk is acceptable. One example of a criterion for acceptable risk comparison is the constant risk criterion, or Farmer curve. A level of constant risk exists when the product of the frequency and consequence for each category of accidents is the same for all categories of accidents regardless of the level of consequence of the accidents. The constant risk criterion is a completely objective or rational criterion for risk comparison when the decision-maker has infinite resources with which to offset losses from accidents.

Often, however, the level of concern that a company has for some types of accidents is greater than that for other types of accidents. For example, a company may be willing to spend resources to reduce the products of expected frequencies and consequences for accidents that could totally destroy a plant (low-frequency, high-consequence accidents), while they may not be willing to spend more to prevent nuisance-type accidents (high-frequency, low-consequence accidents) even if the absolute risk associated with the two types of accidents is the same. In other words, the company expresses more concern for very high-consequence accidents and is willing to spend more than a fair share (from an absolute risk standpoint) to reduce the absolute risk of high-consequence accidents below the absolute risk associated with nuisance-type accidents.

In this study, we define concern risk as a measure of risk that indicates the refinery's desire to avoid accidents that have very high consequence potential, such as accidents that could disable one or more process units for extended periods of time. The concern risk effectively magnifies the contribution of the larger consequence categories to the total risk.

Technical Approach

For the personnel and public risk analyses we estimated the frequencies and consequences of loss of containment accidents in all five groups of refinery areas. The fault tree and event tree techniques[4-6] we used to identify the various accident sequences and to estimate their frequencies are similar to methods used to analyze the risk associated with nuclear power plant accidents. The methods to determine the safety consequences of refinery accidents included those developed and used in the Canvey Study.*

---

*The Canvey Study was a safety analysis performed in 1978 (and updated in 1981) by the United Kingdom Safety and Reliability Directorate to estimate the public risk from industrial activity in the Canvey Island-Thurrock area of England. See Reference 5.

Our project team successfully modified and extended these frequency and consequence assessment methods to product results useful for understanding the safety risks associated with operating a modern, full-range petroleum refinery. Thus, this pioneering project demonstrated a general methodology that can be efficiently used at other refineries to improve refinery availability and safety. The following sections discuss some of the methods and results of the personnel and public risk analyses.

OVERVIEW OF PERSONNEL RISK ANALYSIS

The personnel risk analysis involved a four-step procedure. In the first step, potential accidents that could affect the safety of refinery personnel were identified and possible accident sequences were modeled in event trees. In the second and third steps, we estimated the expected frequency and consequence of each accident sequence. Finally, we evaluated the various frequency/consequence combinations to estimate the risk to refinery personnel from these loss of containment accidents.

This analysis considered the immediate effects of loss of containment accidents on Groups 1, 2, 3, and 4. Personnel involved included:

- process unit operating personnel inside and outside the control rooms,
- maintenance personnel,
- outside contractor maintenance personnel, and
- engineering, administrative, and other refinery personnel.

This analysis did not consider the risk to refinery personnel who knowingly subject themselves to greater risk by responding to an emergency situation created by a particular accident. Specifically, this analysis did not consider risk contributions that could be attributed to:

- operators who remain on the scene and attempt to perform accident mitigating actions such as isolating failed equipment or shutting down a unit,
- fire-fighting personnel who converge on the accident location to contain the effects of the accident and prevent propagation to adjacent areas, and
- emergency team members or other personnel who return to the accident scene to rescue injured personnel.

The analysis determined major contributors to refinery personnel risk from random hit effects created by the loss of containment accidents. A random hit occurs when a worker happens to be in an area that is exposed to the effects of an accident. The area of exposure is called a hazard zone. A worker in the hazard zone for a particular accident will likely be seriously injured or killed if the accident occurs. The size of a hazard zone depends on the particular accident (i.e, a fire, an explosion, or a toxic material release). A hazard zone normally includes the area surrounding the process equipment that has failed. However, the hazard zones for some accidents can extend far beyond the actual location of the loss of containment event. For example, the hazard zone created by an LPG storage tank rupture that releases a drifting cloud of LPG vapor can be several square kilometers.

Local fire and explosion effects on nearby refinery personnel are the major risk contributors from accidents occurring in Group 1. Accidents occurring in Groups 2 and 3 can result in large releases of LPG, with both local and far-reaching fire and explosion effects. Loss of containment events in the Amine/Sulfur Unit (Group 4) can cause releases of gaseous hydrogen sulfide (H_2S), which is highly toxic. Inhaling low concentrations can cause unconsciousness, and inhaling high concentrations can cause respiratory failure and death.

We considered over 2,000 accident sequences for the personnel risk analysis of these four groups of refinery areas, but we will use only the accident sequence development of Group 2 accidents to illustrate our method. Results from all four groups will be presented, however.

To define the possible accident sequences following a rupture of an LPG storage tank, we used the event tree format in Figure 1. There are seven branching points listed across the top of the event tree; they represent events or conditions that can affect personnel safety. The branching points are:

1. local ignition,
2. explosion occurs,
3. limit damage,
4. general wind direction,
5. wind to location X,
6. weather type, and
7. delayed ignition.

Combinations of the events defined by these branching points determine the severity of accident consequences. If the LPG released from a rupture ignites locally (Branching Point 1), either an explosion or a fireball can

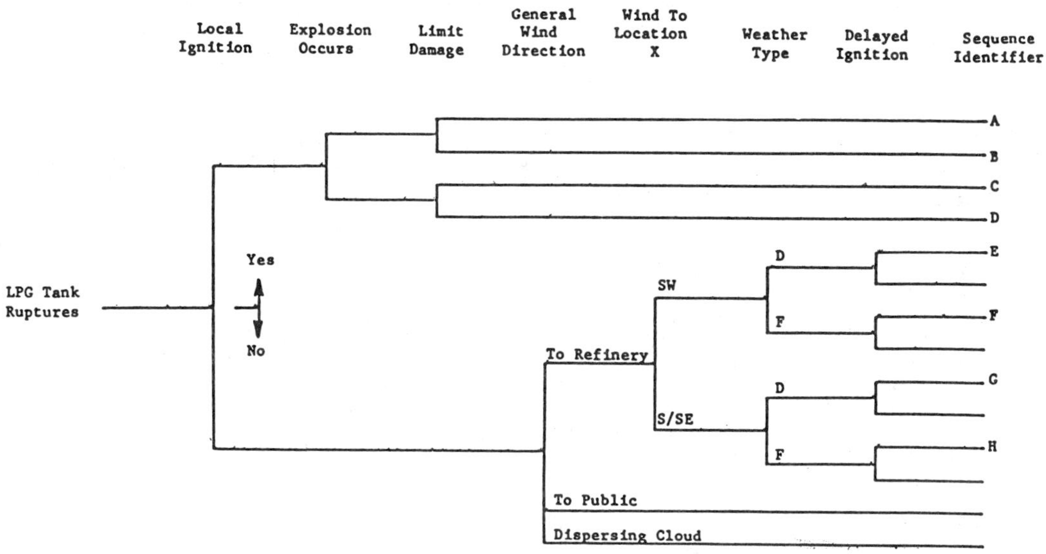

Figure 1

Personnel Risk Analysis Event Tree for the Rupture of an LPG Storage Tank occur (Branching Point 2). The "limit damage" branching point represents the capability of refinery fire-fighting teams to prevent the propagation of the fire to adjacent storage tanks. The four accident sequences resulting from the "limit damage" branching point can represent safety impacts on both refinery personnel and the public.

As indicated in Figure 1, an unignited cloud of LPG vapor (no local ignition) can drift in three general directions:

- toward refinery population centers located southwest and south/southeast of the LPG storage area (TO REFINERY)
- toward densely populated areas located west to the northeast of the LPG storage area (TO PUBLIC)
- toward the east where there are no densely populated areas (DISPERSING CLOUD)

Since this part of the risk-based analysis was concerned with the risk to refinery personnel only, and since the main refinery population centers are (1) the in-line blending control room and the marketing area to the southwest and (2) the main process areas to the south/southeast, only these two wind directions were pertinent to this analysis.

The sixth branching point discriminates between the two standard weather types assumed for an accident: Weather Category D, with an average wind velocity of 5 m/s; and Weather Category F, with an average wind velocity of 2 m/s. Generally, Category F conditions result in larger hazard zones since the inversion conditions tend to contain rather than disperse any released LPG vapor.

The last branching point in Figure 1 represents the case that a previously unignited vapor cloud drifts over a population area and then ignites.

The second step of the personnel risk analysis was to estimate the frequency of each accident sequence that can affect personnel safety. The accident frequency is the product of the initiating event frequency and the appropriate branching point probabilities along the sequence path. The branching point probabilities used for Group 2 event trees are based on meteorological data,[4] industry experience,[5] and the engineering judgment of project team members.

The local ignition and explosion probabilities, 0.2 and 0.1 respectively, are the same for each of the four sizes of storage tank releases. Their values are based on the results of previous studies.[5] The probabilities of limiting damage in all four trees range between 0.0 and 0.5 depending on the size of the release and whether the release ignites as an explosion or as a fireball.

If a release of LPG does not locally ignite, then it can form a drifting vapor cloud. The produce of the "general wind direction" (TO REFINERY) and the "wind to location X" (SW or S/SE) probabilities is the probability that a vapor cloud drifts to one of the two refinery areas of interest. These probabilities are based on wind direction data for the area around the refinery. Two standard weather conditions that could affect the atmospheric dispersion of a vapor cloud were also used to calculate accident sequence frequencies. We assumed that normal conditions (a 5 m/s wind) prevailed 90% of the time. The delayed ignition probabilities for LPG vapor clouds drifting over the main refinery areas range between 0.1 and 0.9 depending upon the cloud size and the specific wind direction.

The third step of the personnel risk analysis was to evaluate the impact to the safety of refinery workers from loss of containment accidents. An accident in any of the four groups creates a hazard zone. If a refinery worker happens to be in that hazard zone, he could be injured or killed. To determine the expected number of casualties for accidents, we estimated the hazard zone area for each accident and then determined the average number of refinery personnel (per m^2) in the hazard zone area. The product of the hazard zone area and the average number of refinery personnel in that area gives the average number of personnel exposed to the accident. This number, when multiplied by the casualty probability (the fraction of exposed personnel that are seriously injured or killed), gives the expected number of casualties for a particular accident.

For sparsely populated areas of the refinery, the average number of personnel (per m^2) can be a very small number, thus the total expected number of expected casualties for some accidents can be much less than one. On the other hand, accidents creating hazard zones in densely populated areas (e.g., near office buildings) can result in fairly large expected casualty estimates.

Comparing the refinery personnel distribution data to estimates of hazard zones for accidents, we determined the expected number of casualties for each accident and assigned the outcome to one of nine personnel safety consequence categories (Table 1). The category midpoints were used with the accident frequencies to determine the personnel risk for each of the accidents.

Table 2 summarizes the personnel risk analysis results for each refinery area in Groups 1-4. This analysis estimated the risk to refinery personnel from accidents in these areas to be 0.012 casualties per year,

Table 1

Refinery Personnel Safety Consequence Categories

Category	Range (Expected Casualties)	Midpoint (Expected Casualties)
1	Less than 0.01	0.0056
2	0.01 to 0.032	0.0178
3	0.032 to 0.1	0.056
4	0.1 to 0.32	0.178
5	0.32 to 1.0	0.56
6	1.0 to 3.2	1.78
7	3.2 to 10	5.6
8	10 to 32	17.8
9	Greater than 32	56.0

Table 2

Personnel Risk Summary for the Refinery by Area

Refinery Area	Absolute Risk (c/yr)	Contribution to Refinery Personnel Risk (%)
1st-Stage Hydrotreater	0.0021	18
Dewax	0.0020	17
FCC	0.0016	13
Sat Gas	0.0016	13
Amine/ Sulfur	0.0013	11
2nd-Stage Hydrotreater	0.00090	7.5
#3 Crude	0.00079	6.6
Hydrobon/ Platformer	0.00053	4.4
Hydrogen	0.00049	4.1
PDA	0.00025	2.1
Condensate	0.00025	2.1
LPG Storage Tanks	7.0×10^{-5}	0.58
Total	0.012	100

which is an average of 1 expected casualty from all accidents every 83 years. The 1st-Stage Hydrotreater Unit (18%) and the Dewax Unit (17%) have almost equal risk contributions. The 1st-Stage Hydrotreater is important because the hazard zones created by accidents there are normally much larger than for most other units because of the high pressure of the system. The Dewax Unit is important because the portions we analyzed contain LPG within enclosed areas. The high fire and explosion potential resulted in large hazard zones and higher casualty estimates.

The FCC, Sat Gas, and Amine/Sulfur units all have about equal risk contributions (13%, 13%, and 11%, respectively). The estimated personnel risk from accidents in all the other units is relatively low. The majority of these process unit accidents would create relatively small hazard zones, and because the population densities in the units are low, the casualty estimates for most accidents are small.

The final step in the personnel risk analysis was the evaluation of the frequency/consequence combinations for each of the 2,000 accident

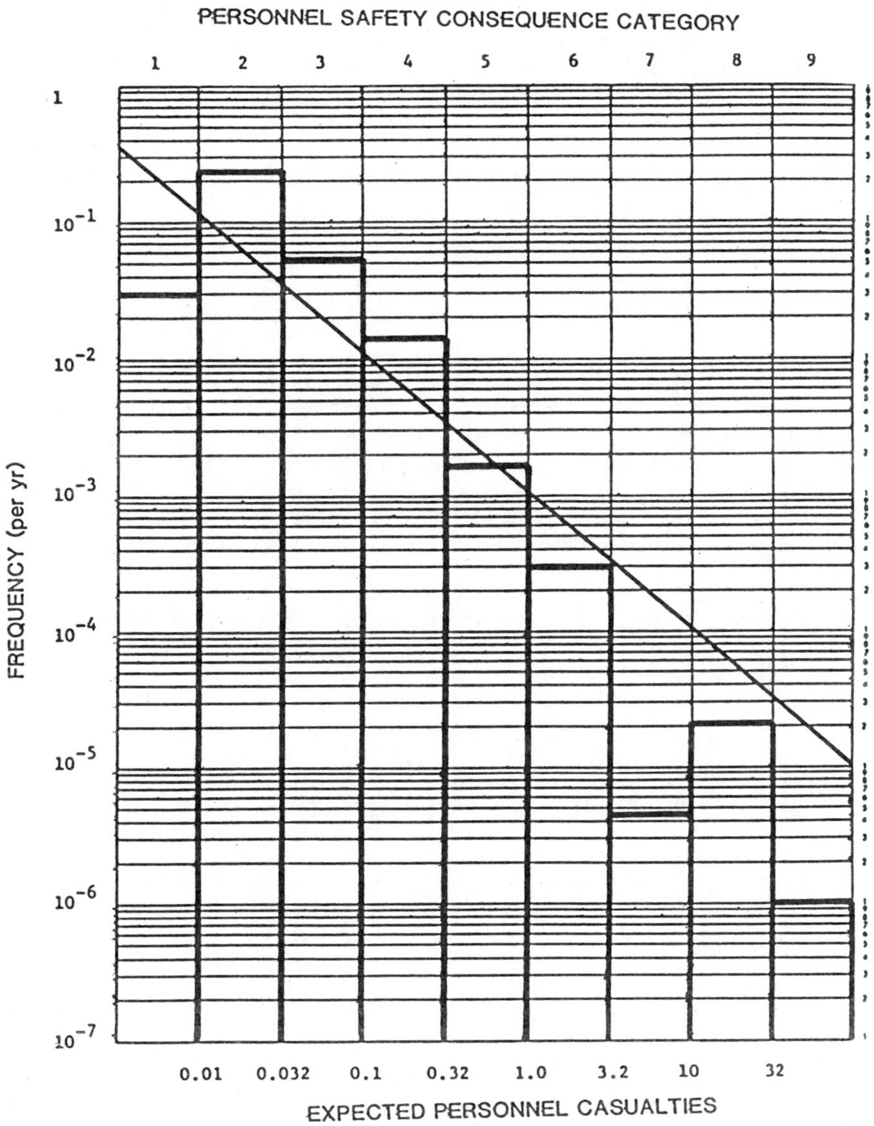

Figure 2

Personnel Risk Histogram for the Refinery

sequences considered for personnel safety consequences. Figure 2 is the risk histogram for all of the refinery areas included in this personnel risk analysis. The solid line is the acceptable risk criterion provided by the refinery. This line corresponds to an average risk of 0.01 expected casualties/yr or a total of 1 expected casualty every 100 years.

This histogram indicates three things:

1. The estimated frequency (therefore, the risk) for accidents in Category 1 is about a factor of 9 beneath the criterion line.

2. The risk for accidents in Categories 2-4 is slightly larger than the acceptable risk indicated by the criterion line.
3. The risk for accidents in Categories 7-9 is well below the acceptable risk indicated by the criterion line (by much as a factor of 40 for Category 7 accidents).

The low estimated risk from Category 1 accidents compared to the criterion line is not surprising. Category 1 accidents are the least hazardous of the events examined, and a study such as this one, which focused on relatively large accidents, can underestimate the risk from such events. It is also possible that some of the predicted Category 2-type accidents could actually result in smaller consequences.

The risk for accidents in Categories 7-9 is substantially lower than the acceptable risk. These categories represent accidents resulting in, on the average, about 4 casualties per accident (the midpoint of Category 7). The low estimates of personnel risk highlight the refiner's success in virtually eliminating the potential for these large accidents; no large groups of personnel are routinely located in hazardous areas of the refinery. The low frequency for events involving large numbers of casualties also reflects the "random hit" approach of this analysis.

Table 3 summarizes the information in Figure 2 and includes the personnel risk for accidents contributing to each consequence category as well as the category's importance to total refinery risk. This analysis estimated the risk to refinery personnel from accidents in Groups 1-4 to be 0.012 personnel casualties per year, or an average of 1 expected casualty from all accidents every 83 years. Categories 2-4, which are accidents resulting in from 0.01 to 0.32 expected casualties per accident, are the most important contributors to personnel risk. These categories contain accidents that pose 82% of the estimated personnel risk for the refinery.

Table 4 lists the major contributors to total refinery personnel risk. Although there are no exceptionally large contributions from any single process section, we can make several observations from these results.

The 1st-Stage and 2nd-Stage Hydrotreater units have two major contributors involving large process sections consisting of several vessels and long piping runs. These process sections are important because (1) accidents there create large hazard zones (possibly affecting many people) and (2) occurrence frequencies there are high due to the large numbers of components. The HTU compressors are major contributors because ruptures in the compressor building can result in explosions.

The Dewax Unit has three major risk contributors: two pumps and one compressor. Each of these components contains propylene, and leaks or ruptures of these components can generate local vapor clouds that are explosion hazards over relatively large areas. Leaks in the FCCU wet gas compressor can also create explosion hazards.

We concluded that providing adequate ventilation in compressor buildings will reduce the risk to refinery personnel in these areas. In addition, we offered two recommendations for reducing refinery risk based specifically on our personnel risk analysis results.

Table 3

Personnel Risk Summary for the Refinery by Consequence Category

Consequence Category	Frequency (per yr)	Absolute Risk (c/yr)	Refinery Absolute Risk Importance
1	0.029	0.00016	0.013
2	0.24	0.0042	0.35
3	0.055	0.0031	0.26
4	0.014	0.0025	0.21
5	0.0017	0.00095	0.079
6	0.00029	0.00052	0.043
7	4.4×10^{-6}	2.5×10^{-5}	0.0021
8	2.1×10^{-5}	0.00037	0.031
9	1.0×10^{-6}	5.6×10^{-5}	0.0047
Total	0.35	0.012	1.0

Table 4

Major Personnel Risk Contributors for the Refinery

Process Section Description	Refinery Personnel Risk Importance
HTU 1 section including recycle gas cooler, hot and cold high-pressure separators, recycle gas scrubber and piping	0.05
HTU compressors	0.04
Dewax refrigerant utility pump	0.04
Dewax refrigerant transfer pump (P-335)	0.04
HTU 2 feed/effluent heat exchanger train piping	0.03
Dewax refrigeration compressor	0.03
FCCU wet gas compressor	0.02

Recommendation 1

We recommended installing H_2S detection and alarm capability in and around the Amine/Sulfur Plant, which is the fifth largest contributor (11%) to refinery personnel risk. There are no exceptionally large risk contributors in this unit. However, providing adequate early warning capability for H_2S releases will reduce the risk associated with the Amine/Sulfur Unit.

Detectors should be strategically located in three places? (1) in the amine storage tank area by the area supervisor's office building, (2) along the pipe rack from the amine tank to the sulfur plant, and (3) surrounding the sulfur plant (except on the lake side). These detectors should provide readings and alarms in the Sulfur Plant control room, as well as alarms outside. They can be set at relatively high setpoints to minimize inadvertent alarms from other sources. This system can also be used to automatically actuate existing emergency shutoff valves in the event of a release.

Recommendation 2

We recommended that the refinery should consider defining special restricted access areas around certain equipment as a way of further minimizing the chance of personnel injury in the event of an accident. Areas that should be restricted include compressor buildings, the Amine/Sulfur Unit, and other areas where H_2S hazards exist (but which were not a part of this study). Limiting the number of people allowed in these areas is a cost-effective way of reducing the personnel risk at the refinery.

OVERVIEW OF THE PUBLIC RISK ANALYSIS

The public risk analysis used the same four-step approach as the personnel risk analysis. We identified accidents with the potential to affect the safety of the public living near the refinery, and we estimated the frequency and the consequence of each accident based on industry-average data, best-estimate consequence models, meteorological data, and engineering judgment. Finally, we evaluated the various frequency/consequence combinations to estimate the risk to the public from refinery loss of containment accidents.

The refinery is bounded on the west to the northeast by neighborhoods with an average population density of 2,400 persons/km^2. To the southwest these are industrial plants with an average population density of 7.4 persons/km^2. This study considered the risk to the public located only in the densely populated areas west to the northeast of the refinery.

Our analysis estimated the risk to the public from Group 2 and Group 5 accidents. Group 2 accidents are releases of LPG that can ignite locally or form drifting vapor clouds that can ignite over populated areas. Accidents considered in group 5 are fires from ten large gasoline and crude oil storage tanks, which can result in thermal radiation hazards to the nearby population. These particular tanks were selected for analysis because they are the only tanks close enough to the refinery boundary to present any significant thermal radiation hazard to the public.

Figure 3 is the event tree format we used to model Group 2 accident sequences that are potential contributors to public risk. (Only ruptures

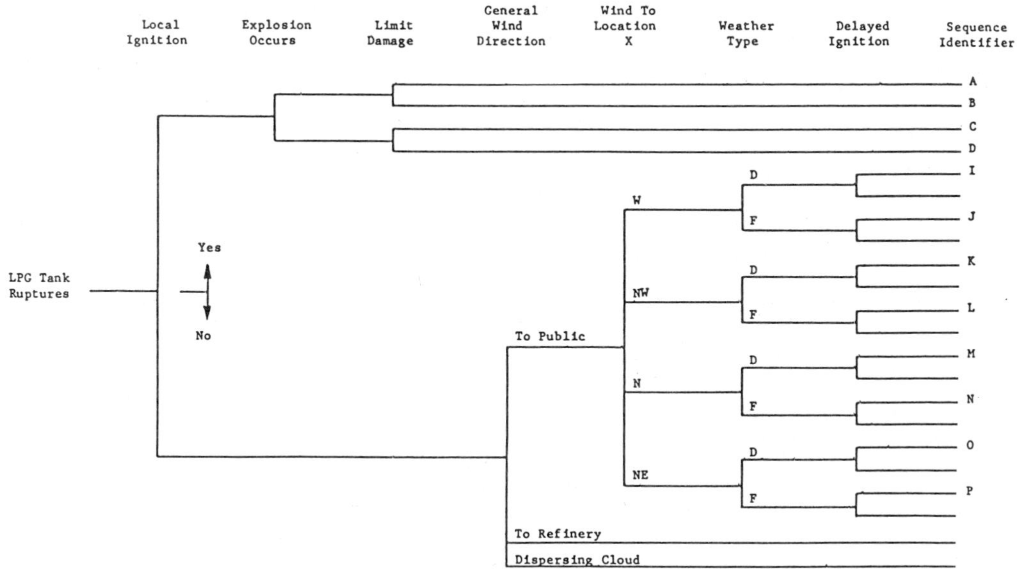

Figure 3

Public Risk Analysis Event Tree for the Rupture of an LPG Storage Tank

of these tanks were modeled because we assumed that leaks of these tanks do not present significant hazards to the public.) The seven branching points across the top of the event tree are identical to those defined for the Group 2 personnel risk analysis. However, this tree incorporates four specific wind directions--for the four populations of interest. These specific "wind to location X" directions are the compass directions west, northwest, north, and northeast of the LPG storage tank areas to which an unignited vapor cloud can drift.

We did not develop an event tree model for accident sequences involving the ten perimeter oil storage tanks. The only accident sequence for each tank is the tank catching on fire and the fire being contained within the shell of the storage tank. Large fires emanating from the tops of the tanks would create thermal radiation hazards to the populations west to the northeast of the refinery. No accidents involving ground fires in the tank compounds were considered since the public would likely be protected from thermal radiation by the tail dikes between the tank compounds and the refinery boundary.

The most likely cause of ignition of these perimeter oil storage tanks, which are external and internal floating-roof tanks, is a lightning storm. Lightning can strike a tank, causing a ignition of any flammable gas-air mixtures. In some cases, the lightning can strike near the tank releasing electrostatic charges bound in the tank. The sudden release of a bound charge from the floating roof to the shell of a tank can cause arcs that will ignite the contents unless the tank has been properly equipped to prevent arcing.

The frequencies of the LPG storage tank accidents were calculated in the same way as for the Group 2 personnel risk analysis, using the same component failure data. Event tree models for the rupture events contained the initiating event frequencies, the assigned branching point probabilities (based on industry experience, meteorological data,[3] and the engineering judgment of project team members), and the calculated accident sequence frequencies.

The local ignition and explosion probabilities, 0.2 and 0.1, respectively, were taken from previous studies,[5] and these probabilities are the same for each of the four sizes of storage tank releases. The damage-limiting probabilities are the same as those used in the personnel risk analysis. The delayed ignition probability for LPG vapor clouds drifting over a population area is 0.2. This probability is based on the assumption that clouds fully expand to the 50% LFL before igniting.

The product of the initiating event frequency, the first three branching point probabilities, and the "general wind direction" and "wind to location X" probabilities is the probability that a vapor cloud drifts in one of the four compass directions of interest. The probabilities for general wind direction and wind to a specific location are based on local wind direction data.[4] We assumed that normal weather conditions (a 5 m/s wind) prevailed 90% of the time.

For the analysis of Group 5 accidents, we assumed that the frequency of a storage tank fire contained within the tank shell was 10^{3} per year. This frequency is based on industry experience and data from the American Petroleum Institute.[6] No other probabilities were necessary for estimating the frequency of these accidents.

The third step of the public risk analysis was evaluating the impact on public safety from accidents involving the LPG storage tanks (Group 2) and the ten selected perimeter storage tanks (Group 5). Therefore a hazard zone for each accident was defined as the area beyond the refinery boundary affected by the accident.

This analysis estimated (1) the average number of people in each hazard zone for the accident sequences of interest and (2) the probability that these people are seriously injured or killed. We assumed that the average population density of the nearby neighborhoods was 2,400 persons/km^2. The casualty estimate for each accident was assigned to one of nine public safety consequence categories (Table 5). The category midpoints in this table were used with the accident sequence frequencies to determine the public risk for each of the accidents.

In estimating the expected number of public casualties from Group 2 accident sequences, we considered the same four types of effects considered in the personnel risk analysis:

1. blast effects from local explosions
2. burns from thermal radiation from fireballs
3. blast, missile impact, and thermal radiation effects from BLEVEs
4. blast and thermal radiation effects from drifting vapor cloud explosions

As with the personnel risk analysis, methods for determining expected casualties from the public sector are dependent on the type of accident (i.e., explosion, fireball, BLEVE, and drifting vapor cloud explosion).

Table 6 summarizes the consequence analysis results for explosions of the LPG storage tanks. Similar calculations were performed to calculate the consequence to the public of the other types of Group 2 and Group 5 accidents.

Table 5

Public Safety Consequence Categories

Category	Range (Expected Casualties)	Midpoint (Expected Casualties)
1	Less than 0.1	.0056
2	0.1 to 0.32	.0178
3	0.32 to 1.0	.056
4	1.0 to 3.2	1.78
5	3.2 to 10	5.6
6	10 to 32	17.8
7	32 to 100	56
8	100 to 320	178
9	Greater than 320	560

The final step of the public risk analysis was to evaluate the frequency/consequence combinations for the 130 accident sequences considered. The procedure for combining the expected frequencies and consequences is similar to the one employed in the personnel risk analysis. The only difference is that we used the public safety consequence category midpoints (Table 5) when multiplying by frequencies to obtain the public risk for accidents.

Table 7 summarizes the estimated public risk as a function of public consequence category. Over 83% of the risk to the public is from severe accidents that could result in hundreds of casualties. Table 8 shows that 86% of the risk to the public is from accidents involving the LPG storage tanks.

Our estimate of the total risk to the public located west to northeast of the refinery is 0.0028 casualties per year, or a total of 1 expected public casualty from all accidents about every 357 years. The weighted-average risk to an individual (the probability per year of becoming a casualty) is 5.2×10^{-7}/yr. The average probability of a person being struck and killed by lightning is about 10^{-7}/yr.

The maximum risk to an individual located at the nearest point (500 m northwest of the storage tanks) is 8.8×10^{-5}/yr. This is the average probability per year of the individual at greatest risk becoming a casualty due to refinery accidents.

Table 9 lists the major contributing accident sequences to the total public risk. An accident involving a rupture and a fireball of a butane sphere-followed by the boiling liquid expansion vapor explosion (BLEVE) of the remaining butane sphere and one of the large propane bullets--is by far the dominant risk contributor (64%). Accident sequences involving the

Table 6

Public Consequence Analysis Summary for Local Explosion Effects for LPG Storage Tanks

Description	Hazard Zone Area (km^2) (1)	Average Number of People in Area (per km^2) (2)	Casualty Prob (3)	Average Number of Public Casualties (4)
small bullet explodes	no effect outside refinery			0.0
small bullet explodes, causes fireball of all bullets				
large bullet explodes	0.12	2,400	0.10	29
large bullet explodes, causes fireball of all bullets				
small sphere explodes, causes large sphere and 1 large bullet fireball	0.56	2,400	0.10	13
large sphere explodes, causes small sphere and 1 large bullet fireball	0.057 0.89	2,400 2,400	0.25 0.10	34 21

ignition of drifting vapor clouds from the two butane spheres and the two large bullets contribute 7%, 7%, 5%, and 5%, respectively.

All of these results indicate that the most effective way to reduce public risk is to reduce the frequencies and/or consequences of accidents involving the two large butane spheres. In particular, these results show that if the refinery can successfully prevent the BLEVE of any LPG tanks following the rupture and fireball of another tank, then it can reduce the total public risk by a factor of two.

Table 7

Public Risk Summary for the Refinery by Consequence Category

Public Consequence Category	Category Midpoint (Expected Casualties)	Absolute Risk (c/yr)	Contribution to Refinery Public Risk
1	0.056	1.1×10^{-5}	0.39
2	0.178	0.0	0.0
3	0.56	3.9×10^{-4}	14
4	1.78	0.0	0.0
5	5.6	1.8×10^{-6}	0.064
6	17.8	3.7×10^{-5}	1.3
7	56	6.2×10^{-5}	2.2
8	178	4.1×10^{-4}	15
9	560	1.9×10^{-3}	68
Total		2.8×10^{-3}	100

Recommendation 1

We recommended that the refinery install fixed water deluge systems on the two butane storage spheres to provide vessel cooling in the event of a nearby fire. This external cooling will reduce the chance of these spheres experiencing a BLEVE. This system must be weather-proofed (steam-traced or dry-type) to withstand freezing temperatures. This type of externally mounted system is vulnerable to accidents involving blast effects from nearby explosions. However, our analysis shows that the fireball of the contents of nearby vessels is much more likely than an explosion and it warrants preventive measures.

Recommendation 2

For the LPG storage facility, we recommended developing comprehensive emergency procedures and contingency plans for operators and fire department personnel. These procedures should be integrated in an operator and emergency team training program. A general fire procedure for the refinery already existed; however, it did not provide sufficient detail to guide personnel responsible for coping with specific LPG release emergencies.

The procedures should address generic loss of containment failures for each LPG facility area, and they should define appropriate operator

Table 8

Public Risk Summary for the Refinery by Area

Refinery Area	Absolute Risk (c/yr)	Contribution to Refinery Public Risk (%)
LPG storage tanks	0.0024	86
Perimeter storage tanks	0.0004	14
Total	0.0028	100

The weighted-average risk to an individual (probability per year of becoming a casualty) is 5.2×10^{-7}/yr.

The maximum risk to an individual (located at the nearest point 500 m northwest of the storage tanks) is 8.8×10^{-5}/yr.

Table 9

Major Public Risk Contributors for the Refinery

Description	Absolute Risk (c/yr)	Contribution to Refinery Public Risk (%)
large sphere fireballs, causes small sphere and 1 large bullet BLEVE	9.0×10^{-4}	32
small sphere fireballs, causes large sphere and 1 large bullet BLEVE	9.0×10^{-4}	32
large sphere release, drifting vapor cloud explosion	1.9×10^{-4}	6.9
small sphere release, drifting vapor cloud explosion	1.9×10^{-4}	6.9
large bullet release, drifting vapor cloud explosion	1.3×10^{-4}	4.5
large bullet release, drifting vapor cloud explosion	1.3×10^{-4}	4.5

actions and alternatives. For example, a contingency plan should be developed for dealing with minor leaks of each of the four types of pressurized storage tanks.

In addition, plans for contacting appropriate local authorities should be outlined in the event of a major LPG release. This is necessary because of the close proximity of the facility to the high-population-density area near the refinery. An evacuation plan should be developed in concert with the responsible local authorities.

CONCLUSIONS

Prevailing economic conditions have forced many companies in the process industries to find more effective ways of increasing plant productivity and safety. The results of properly applied and scoped reliability engineering and probabilistic risk assessment studies offer ways of ranking plant availability and safety improvement alternatives.

The key to far-reaching changes, however, is not the methods themselves, but the philosophy that makes the tools work. Whether you refer to this philosophy as "pre-active" or as a "reliability ethic," it is becoming a new way of life for the process industries. These high technology tools are not magic or infallible, but they are based on sound engineering principles. They can help companies achieve high levels of safety while making the most effective use of every dollar they spend on improving productivity.

REFERENCES

1. J.S. Arendt, et. al., "Qualitative Risk Assessment of Engineered Systems," AIChE 71st Annual Meeting Proceedings, Miami, November 1978.
2. W. Lowrence, Of Acceptable Risk, William Kaufmann, Inc., Los Altos, California, 1976.
3. S. Lichtenstein, et. al., Journal of Experimental Psychology: Human Learning and Memory 4, pp. 551-563.
4. A.D. Little, An Assessment of the Risks Presented by the HF Alkylation Unit in the Clarkson Refinery, Report No. 87428, Canada Limited, Toronto, Ontario, September 1982.
5. Canvey: A Second Report, A Review of Potential Hazards from Operations in the Canvey Island/Thurrock Area, UK Safety and Reliability Directorate, England, September 1981.
6. Safety Digest of Lessons Learned - Safe Operation of Storage Facilities, American Petroleum Institute, Publication 758, Section 6, Washington, D.C., 1982.

INCOMPLETENESS IN DATA BASES:

IMPACT ON PARAMETER ESTIMATION UNCERTAINTY

Gareth W. Parry

NUS Corporation
Gaithersburg, Maryland 20878

ABSTRACT

The estimation of parameter values from data is discussed. Sources of uncertainty other than statistical sources are identified. The impact of incompleteness of the data is addresses, with a particular example of two contrasting methods of estimating a common cause failure parameter.

KEY WORDS: Probabilistic Risk Assessment; Parameter Estimation; Uncertainty; Data Sources; Common Cause Failures.

1. INTRODUCTION

One of the most important tasks in a nuclear power plant probabilistic risk assessment (PRA) is the estimation of the frequencies of core melt accidents. This is done by firstly constructing logic models, event trees and fault trees, which determine those combinations of basic events which can lead to core melt. The basic events include the occurrence of a plant transient or accident, the failure or unavailability of a component of an ECCS system, and the failure of an operator to perform a required function. This breaking down into basic events has two great advantages; it provides an engineering understanding of the plant, and it also provides a way of using numerical data, or other information on the likelihood of the more common basic events to construct estimates of the frequencies of events (core melt accidents) for which no data exists or is expected to exist. The estimation of the frequencies of core melt accidents is achieved by converting the logic model of the event and fault trees into a probability model, so that the frequency of each accident sequence may be expressed in terms of combinations of an initiating event frequency and probabilities that components, subsystems, or operators will not perform their function successfully when called upon to do so.

The ideal situation would be one in which sufficient data existed to allow reliable estimates of the basic event probabilities to be made. This however is generally not the case; there are certain basic events for which data is relatively plentiful, but others for which there is virtually none. This lack of data has not deterred analysts from providing estimates based on engineering judgment where data are lacking.

Because of the fact that all estimates are subject to uncertainty it has become customary to perform an uncertainty analysis which propagates the uncertainty on the basic event probabilities (and initiating event frequencies) through the analysis to provide a measure of the uncertainty in the predicted accident sequence frequencies.

While there is an intuitive feeling that estimates based on data are much more reliable than those based on judgment, and hence should have smaller ranges of uncertainty, this may be true to a lesser degree than might be supposed. In any analysis of data there are many assumptions to be made, of which assumptions about the applicability of the data, of its completeness, about the appropriate probability model to be used to describe the data, are a few examples. These assumptions affect the estimates made and their associated uncertainties. While there are many assumptions commonly used by virtually all PRA practitioners, there are others which vary between analysts, and unless they are clearly stated and clearly understood, a comparison between different PRA results can be misleading. It is possible that estimates made under one set of assumptions can be significantly different from those made under another, so that their respective uncertainty ranges barely overlap.

The purpose of this paper is to explore the impact of modeling assumptions on the analysis and collection of data and to discuss some potential pitfalls in comparing different analyses. Section 2 is a brief discussion of the commonly used probability models and the influence their adoption has on data collection. Section 3 discusses some problems associated with the interpretation and analysis of data. Section 4 is an example of how the incompleteness of a data set has led the author of this paper and his colleagues to make assumptions in estimating a common cause failure probability that differ from one of the more traditional approaches. Finally, Section 5 presents some conclusions.

This paper will not address the recurrent issue of Bayesian versus classical methods of estimation. Indeed, in the area of concern of this paper--making estimates of parameters when data exist--the difference are probably more philosophical than numerical. The problems to be discussed here have to be faced equally by classical and Bayesian analysts.

2. PROBABILITY MODELS, PARAMETER ESTIMATION, AND DATA COLLECTION

In order to convert the logic model of the event trees and fault trees into an estimator of accident sequence frequencies, it is necessary, for each basic event, to define a probability model. This model is basically what determines what data are required and how they should be manipulated to provide the appropriate probability estimates. Typically in PRAs very simple models are adopted. These are discussed briefly below for each of the basic events in turn except for human error probabilities which, since they are mainly derived judgmentally, are outside the scope of this paper.

Initiating Events

Initiating events are assumed to have a constant occurrence rate f, so that initiating events are distributed in time according to a Poisson distribution with parameter f, i.e., the probability of r events in time t is given by $P(r|t) = \frac{(ft)^r}{r!}\exp(-ft)$. The data required to estimate f is the number N of events recorded in a time T.

The underlying assumption is that the occurrence rate does not vary with time; there is no degradation or improvement in plant performance.

Failures to Run

Failures to run are modeled in a similar manner to initiating events. Failures are assumed to have a constant failure rate λ. Again the data required to estimate λ is the number of failures recorded in the total operational time.

Failures to Start

Components are modeled as either having a constant probability of failure on demand or a constant failure rate while in standby. In the first case the number of failures n in a fixed number of demands N is distributed according to the binomial distribution i.e., the probability that n events have occurred in N demands is $P(n|N) = \binom{N}{n} p^n (1-p)^{N-n}$ where p is the constant failure probability. The data required to estimate the parameter of the model is the number of failures in a known number of demands.

If the standby failure rate model is preferred, then, as before, the number of failures in a measured time in standby is the data required to estimate the standby failure rate λ_s.

In most PRAs the use of the constant failure of probability on demand seems to be preferred for most cases. Adopting it does imply an assumption that the probability of failure is independent of the time spent in standby, and indeed that it is independent of the number of demands. The former assumption is not an unreasonable approximation for components which are tested at relatively frequent, regular intervals. However, if it is applied to a component which is not tested regularly and which may experience long periods between demands this assumption may not be valid since the component may suffer a gradual deterioration with time, as a result of corrosion for instance.

Unavailability of a Component or System due to Maintenance

The unavailability of a component or system due to maintenance is the probability that, when called upon, the component or system is undergoing maintenance and cannot perform its function. If there is sufficient data, a reasonable estimate of this component of the unavailability may be obtained by summing the times the component or system spent in maintenance, and dividing by the total plant operational time. This was the approach adopted in the Limerick PRA (PECo, 1981); the data was not specific to Limerick since the plant is not yet operational, but it was data from another BWR operated by the Philadelphia Electric Company.

Another approach has been to estimate the unavailability in two steps; firstly by estimating the probability that a maintenance act is required and secondly by estimating the average time spent in maintenance. The unavailability U_m is then given by

$$U_m = p_m \cdot \frac{\langle t \rangle}{T}$$

where p_m is the probability of a maintenance act being required per test interval, t is the time spent in maintenance, and T the interval between tests. Some analysts have preferred to use a constant rate of occurrence of maintenance acts. In either case a time independence is implied by the formalism. The data required to estimate the unavailability by either method is a complete record of all maintenance acts including their duration.

The constant failure rate and failure probability models have become the folklore of PRAs. They are the simplest models and have minimal data requirements for estimating their parameters. The term minimal is not intended to signify the quantity of data, but only the type of data required. Typically this is a number of events in a fixed number of demands or a fixed time. Since PRAs are generally intended to characterize the risk profile of a fixed moment in time, models that ignore time dependence of failure rates or probabilities are probably adequate. However, to ignore the potential for time dependence in the data can lead to results which are misleading. An example of this can be seen in the results in EPRI NP-2230 (McClymont and Poehlman, 1982) for the frequencies of anticipated transients. There a marked decrease is shown with time following the first year of operation.

Of course it is much easier to investigate time dependent effects in a situation where events occur relatively frequently (initiating events) than in those where events occur rarely (failure of a particular pump). However, at least in a global sense, it may be possible to demonstrate a time trend even for the parameters of these rare events. This has been shown in Drago et al., (1982). Plant modification, improved maintenance practices, and greater experience with running the plant are all factors which can be expected to produce a variability with time.

The estimation of the unavailability due to maintenance is difficult mainly due to the difficulty in extracting the required data from plant records. while for some plants the records may be complete enough that the analyst can have confidence in the data extracted, in others this is not so. This affects both the estimation of the frequency and average duration of the maintenance acts. If the data are incomplete, then the analyst will have to supplement what data he has with judgment. The IPRDS system (Drago et. al., 1982a) has provided some information on the frequency of maintenance acts for certain components, in relation to the frequency of catastrophic failures, as well as estimates of mean duration times. This is for specific plants, but may be judged to provide an adequate basis for use in other plants. If plant-specific information is preferred, the opinions of the maintenance staff may be solicited, although conversion of their opinions on, for instance, the extremes of the distribution of repair times into estimates of mean duration time will generally require some assumption of distribution of the times to repair. The choice of the distributional form (historically the lognormal distribution has been favored) will clearly influence the estimate of the mean.

The case of the unavailability due to maintenance just discussed is somewhat typical in that, despite the apparent minimal requirement for data, it is often very difficult to obtain even that minimal amount. Extraction of data from plant records is time consuming, costly, and may be more or less rewarding depending on the care with which records are kept. If the data are not complete in detail, then the analyst must supplement it with his judgment. Even if the data are complete, there are still problems for the data analyst as described in the next section.

3. SOME PROBLEMS AND DECISIONS IN DATA ANALYSIS

Having decided on what models to use to describe the basic events, the data analyst must next collect the data that is necessary to estimate the parameters of those models. The first task is to identify the sources of data. While many PRA studies have chosen to use plant specific data, as a general rule it is not plentiful, and in order to 'improve' the estimates made from that data, plant specific data is often supplemented by data from other nuclear plants. This data may be obtained from other PRAs or from some of the data compilations such as the EG&G reports (Hubble et. al., 1982; Sullivan and Poloski, 1980; Trojovsky, 1982) which analyze the events that are recorded in the Licensee Event Reports (LERs), and the IPRDS (Drago et. al., 1982a) reports which present results from a detailed analysis of a few nuclear power plants.

One of the first questions to be asked is what relationship does one set of data have to another--for instance what does the data on the HPCI pump mean in relation to that on the RCIC pump in a BWR. It has been assumed that each pump has a constant probability of failure on demand--but is that probability the same for both pumps? If there is felt to be no reason to doubt that this is the case then the data on both pumps can be combined to give a larger sample. A larger sample will result in a narrower confidence interval therefore this grouping has an obvious desirable feature. (A corollary to this is that if one wishes to become more specific, treating every component as having a different failure rate or probability, one has to accept the larger uncertainty that accompanies this.)

However, how far should this grouping be continued? Should it be extended to include all turbine driven pumps in all plants? Doing this would result in an extremely narrow confidence interval as can be seen in the EG&G reports (e.g., Trojovsky, 1982). There are many reasons why pumps in different plants would be expected to not necessarily have the same failure probability; they may be of different design, subject to different maintenance regimes, etc., so this particular grouping does not seem to be appropriate. In some recent PRAs the plant-to-plant variation has been incorporated into the measure of uncertainty on the failure rates and probabilities used in analysis. There may of course be other factors of importance than plant-to-plant variation. Recent work by Martz et. al. (1982) is an attempt to group data but take into account factors which can lead to systematic differences from one member of the generalized population (pumps for example) to another. The method results in narrower confidence intervals on the estimates of the relevant parameter for a particular sub-group of the population than if the data for that sub-group alone were considered (Martz et. al., 1984). This is bought at the expense of an assumption about how the various factors which can influence the failure probability actually manifest their influence. Nevertheless, the method has obvious potential for a more systematic evaluation of how grouping should be done. Application of the method clearly requires a fairly detailed breakdown and engineering understanding of the data to determine what factors should be considered as potential candidates.

Another important issue is the compatibility of different data sets. Is the definition of the component and its failure the same as one data compilation as in another? The IPRDS (Drago et. al., 1982a) reports categorize failures as catastrophic, degraded or incipient. Typically, only catastrophic failures are relevant to PRAs. Are the failures recorded in other data complication catastrophic or not? Both the IPRDs and EG&G data compilations differentiate between failures with and without command faults. These distinguish between different definitions of a component. In the case of 'with command faults,' the component boundary is extended from the actual physical boundary of the component to include certain portions of the control and power systems. It is not easy to determine from the brief narrative in the respective reports whether these definitions are compatible.

The evaluation of running times for standby components and the number of demands is difficult and time consuming. The EG&G data compilations estimated these numbers based on test schedules e.g. the assumption that pumps are tested once a month. However, there are additional demands on components, e.g., operational demands, demands due to interfacing maintenance, corrective maintenance; these may be determined from a detailed examination of plant records and procedures. In many cases these additional demands may not be important. However, as an example, in the case of diesel generators, an EPRI report (McClymont and McLagan, 1982) has demonstrated that the two assumptions made in the EG&G compilation, monthly testing and weekly testing, are not correct. The situation for diesel generators is complicated by the fact that the test schedule is set by the number of failures according to Regulatory Guide #1.108. This means that different plants may have a very different frequency of demands, as verified by the EPRI report.

It can be seen by these few examples that there are potentially many pitfalls in using ready-made estimates from various data sources without a detailed understanding of the assumptions that underly the bottom line numbers they present. Many of these assumption may be left unstated or may be somewhat obscure. In making assumptions such as the aggregation of data, the analyst is responding to a desire to reduce uncertainty. In making assumptions about the number of demands seen by valves in a particular plant the analyst is responding to the need to supplement incomplete data by judgment. A further example of the response to incompleteness is presented in the next section, where two different methods for estimating a parameter of a simple common cause failure model are contrasted.

4. AN EXAMPLE - COMMON CAUSE FAILURES

Common cause failures are potentially significant to system unavailability or accident frequency; for example, the common cause failure of diesel generators is an important event in the case that offsite power is lost. A common cause failure is one where the failure of nominally independent components become dependent, thus overriding the redundancy incorporated in the plant design. Several methods have been proposed for estimating the probability of common cause failures, these include the Beta factor method (Fleming and Raabe, 1978; Fleming et al., 1983), the binomial failure rate method (Vesely, 1977), and the c-factor method (Evans et. al., 1984). The simplest model for common cause failures is that used in both the beta and c-factor methods. In both these methods it is assumed that for a set of a like components in parallel the probability of failure on demand is given by

$$p^n + p_c$$

where p is the constant probability that a component fails on demand and p_c is the constant common cause failure probability. Similarly for running components the total failure probability of n components is

$$(\lambda t)^n + \lambda_c t$$

where λ is the constant single component failure rate and λ_c is the constant common cause failure rate.

Since this is the simplest model, it places the least stringent conditions on the data required to provide the estimates of the parameters. At face value, based on the numerical manipulations to provide parameter estimates, the required data is the number of single component and multiple component demands or a record of single component and multiple component running times, and the number of single component failures and multiple component failures. Moreover since component failure rates and probabilities are already relatively small, and common cause failure rates or probabilities should be considerably smaller, it is necessary to have a fairly extensive data base. This is unlikely to be available in a plant specific data base, and more extensive data compilation will have to be used. Probably the most comprehensive data compilations for this purpose are the summary reports prepared by EG&G Idaho (Sullivan and Poloski, 1980; Hubble et. al., 1982; Trojovsky, 1982) for diesel generators, pumps, valves. These data compilations are based on Licensee Event Reports (LERs). They contain a record of failure events, some of them multiple failure events, estimate of demands or running times where applicable and perhaps most useful of all, one line descriptions of the failures. However, one very important piece of information is missing, and that is, were the demands on components, or the requirement that they run, correlated in any way. In other words, were any of the demand to start or run, demands on complete systems, or on individual components in the systems. The number of demands, and the running times reported in the LER summary reports are estimated on the basis of the regular surveillance testing and do not include operational demands or demands due to maintenance and testing after maintenance. Thus, the demands are assumed in the reports to derive mainly from system tests. In many system tests, however, tests are performed on individual legs of redundant systems independently so while the demands are correlated, they are less so than if all legs were demanded simultaneously. In addition, demands on individual trains (demands due to maintenance for instance) could have resulted in some of the failures recorded. The number of these demands is probably most significant for valves. Furthermore, the data consist of compilations of data from systems with different degrees of redundancy--for example, the number of diesels is different at different plants. This is correctly accounted for in the estimated total number of demands or running times.

These data compilations, therefore, are somewhat ill-defined in detail and are non-homogeneous. The Beta factor and c-factor methods would use these data in different ways. In this paper, the discussion is restricted to the original Beta factor model, not the more recent version (Fleming et. al., 1983), since our purpose is to contrast methods of data analysis. The c-factor method was derived specifically in response to the incompleteness of specification of the data in the EG&G reports while the Beta factor method was derived under more general considerations. The application of the two methods to the one set of data however can be interpreted as showing a basic difference in assumptions about the data and the way it was collected.

The Beta factor analyst essentially would make the assumption that all the data is obtained by tests at the system (or multiple component) level so that a set of N system demands would give rise to 3N component demands if the system were 3-fold redundant. The Beta factor is a factor which relates the common cause failure probability p_c to the individual component failure probability, p_s, (which contains contributions from independent and multiple failures)

$$p_c = \beta p_s$$

The estimation of β is based on the assumption that the data are collected from N tests at the system level (i.e., N system demands in which the common cause events are such that all the redundant components fail). In our previous example the number of components failed in $3n_c$ where n_c is the number of common cause events. The Beta factor, in this case, is given by

$$\beta = \frac{3n_c}{n_i + 3n_c}$$

where n_i is the number of independent failure events. Since the single component failure probability, p_s, is estimated by $(n_i+3n_c)/3N$ it can be seen that under the assumptions stated, the Beta factor method predicts accurately the system failure rate from common causes since

$$\beta p_s = \frac{3n_c}{n_i + 3n_c} \cdot \frac{n_i+3n_c}{3N} = \frac{n_c}{N}$$

However, in the case that the data comes from systems with different degrees of redundancy, or that incomplete common cause failures occur (i.e., only 2 of the 3 components fail) the generalization of the formalism to define the Beta-factor as the ratio of the number of components failed in common cause events to the total number failed is only approximate. Furthermore, it is not clear that all the demands on components are as closely correlated as this method assumes.

The c-factor approach was developed as an alternative to the Beta factor and is equivalent to making a different assumption about the data and the way it was collected. In essence, it assumes that essentially all demands are independent demands on the components. This is clearly also an approximation since the multiple failures recorded could have been the result of multiple demands. What is different in the c-factor approach is that it is the narrative description of each failure which is regarded as the most important part of the data (this does not minimize its importance in the Beta-factor approach).

Each failure event is studied and classified as an independent failure, a true multiple failure event, or a potential common cause failure event i.e. one for which it is judged that all redundant components would have failed had they been demanded simultaneously. Then a thought experiment is performed. For each of the (assumed) independent component demands the question is asked, "What would have happened if all redundant components had been demanded simultaneously?" The results of the thought experiment are; if there was no single component failure the system did not fail; if there was an independent failure, the system could fail only because of independent failures of the other redundant components; and if there was a potential common cause failure the system would have failed. Thus, if there are n_{pc} potential common cause events, and n_m multiple failure events the data has been interpreted as indicating $n_{pc} + n_m$ occurrences of a common cause. The fact that, in the n_m cases, the common cause resulted in more than one component failure is not important--the model does not distinguish between numbers of components failed. What we are estimating is the probability of occurrence of a common cause event and then assuming that it fails all components demanded. The number of demands N in the thought experiment should be corrected for the fact that in these multiple failure events the demands are correlated--while all the components that failed in these multiple failure events were demanded, one component demand should have revealed the common cause nature of the failure. The other demands are merely confirmation that the common cause failure existed--they provide no new information. However, this correction, because of the fact that n_m is small, is minor. The system common cause failure probability is therefore approximately

$$\frac{n_{pc} + n_m}{N'}$$

where N' is the number of demands on components. A c-factor is defined, which relates the system to component failure probability on the same way as the Beta factor, i.e.,

$$p_c = cp$$

Since p is estimated by n_i/N' where n_i is the number of independent failure events in the data base

$$c = \frac{n_{pc} + n_m}{n_i}$$

In the c-factor approach, therefore, the common cause failure probability is estimated on an inferential approach rather than on the basis of assuming a very specific sampling scheme. It can be seen that, if the data were truly from N' system demands on a group of identical systems, and if n_{pc} were judged to be zero, the c-factor approach would be non-conservative by a factor equal to the redundancy (in our example-3). It is the inclusion of the potential common cause failures which is crucial in the method which may be more or less conservative when compared to the Beta-factor method depending on the analyst's judgment. The c-factor is an alternative and perhaps complementary method to the Beta-factor methods in that it makes more use of the failure descriptions than the numbers of observed common cause events. They are both approximations and neither is "correct" except under very specific circumstances. The assumptions which underly the use of the methods are a response of the analyst to the incompleteness in specification of the data.

In the case that the way in which the data were generated were known explicitly, it is not necessary to make assumptions about what the data mean, and the analysts would be able to estimate the factor which relates single to multiple component failures directly.

One further point to make is that the narrative description of the failures is useful in another sense. Some of the real or potential common cause failures identified may already be included in the systems modeling, e.g., failures resulting from failures of system such as A.C. power, service water, others may not be possible failure mechanisms at the plant in question. These should not be included in the estimate of the β or c-factors since they are already modeled explicitly.

5. CONCLUSION

In this paper, we have discussed some of the problems that arise in the estimation of the parameters of the models for basic events in PRAs. It is clear that, even when estimates may be labeled as being data based, engineering judgment plays a very strong role. Since different analysts may have a very different perspective, their judgment can affect the estimates, giving rise to a source of uncertainty that is not treated in a "standard" uncertainty analysis which generally treats parameter value uncertainty in the context of a single model or set of assumptions.

This situation is unavoidable for many reasons. For example, the events considered are generally rare, so to provide some reasonable basis for numerical manipulation, and to reduce uncertainties, some aggregation of data is desirable. There are many ways this aggregation could be performed; the analyst's task is to choose the one he feels most appropriate. The definition of failure is not clear-cut; what one analyst calls a failure, another might not. The "sampling scheme" which provides the data is not under the control of the data analyst so that the process does not necessarily conform to a statistical experiment. In particular, the data may not be complete in that not all the pieces of information required to estimate certain parameter values are accurately determined.

These issues all contribute to uncertainty but are generally not addressed in an uncertainty analysis. This may give the uncertainty analysis of the systems analysis part of the PRA an unwarranted air of accuracy. Since different analysts do make different assumptions, certain results may not be directly comparable; differences may be due to differences in the analysis rather than in real differences between plants. While these may not be major in the analysis of internal initiating events, for external initiating events where the role of judgment is enhanced, this may be a significant factor.

The emphasis on judgment places a much greater burden on the data analyst than a mere manipulation of numbers. He must have an appreciation of the causes of the failure, their consequences, and how the plant is managed. As such the data task should become more and more important in PRAs, not necessarily for the numbers it produces, but for the by-products, the insights that are gained into plant design and operation.

REFERENCES

Drago, J.P., R.J. Borkowski, D.H. Pike, and F.F. Goldberg. "The In-Plant Reliability Data Base for Nuclear Power Plant Components: Data Collection and Methodology Report," Oak Ridge National Laboratory report ORNL/TM-8271, NUREG/CR-2641, (1982a).

Drago, J.P., R.J. Borkowski, J.R. Fragola, J.W. Johnson, 1982. The In-Plant Reliability Data Base for Nuclear Plant Components: Interim Data Report-the Pump Component, NUREG/CR-2886, U.S. Nuclear Regulatory Commission, Washington, D.C.

Evans, M.G.K., G.W. Parry, and J. Wreathall, 1984. "On the Treatment of Common Cause Failures in System Analysis," Reliability Engineering, Vol. 9, No. 2.

Fleming, K.N., and P.H. Raabe, 1978. "A Comparison of Three Methods for the Quantitative Analysis of Common-Cause Failures," in Proceedings ANS Nuclear Reactor Safety Division on Probabilistic Analysis of Nuclear Reactor Safety, May 8-10, 1978, Los Angeles, CA.

Fleming, K.W., A. Mosleh, and A.P. Kelley. 1983, "On the Analysis of Dependent Failures in Risk Assessment and Reliability Evaluation," Nuclear Safety, Vol. 24, No. 6, p. 637.

Hubble, W.H., C.F. Miller, M. Trojovsky, and S.R. Brown, 1982. Data Summaries of Licensee Event Reports of Valves of U.S. Commercial Nuclear Power Plants, EGG-EA-5816, EEG G Idaho, Idaho Falls.

Martz, H.F., and D.E. Whiteman, 1984. A Statistical Analysis of Nuclear Power Plant Pump Failure Rate Variability - Some Preliminary Results, NUREG/CR-3650.

Martz, H.F., R.J. Beckman, and C.R. McInteer, "FRAC (Failure Rate Analysis Code): A Computer Program for Analysis of Variance of Failure Rates," Los Alamos National Laboratory report LA-9116-MS (NUREG/CR-2434) (1982).

McClymont, A., and G. McLagan, 1982. Diesel Generator Reliability at Nuclear Power Plants: Data and Preliminary Analysis, EPRI-NP-2433, Electric Power Research Institute, Palo Alto, California.

McClymont, A., and B. Poehlman, 1982. ATWS: A Reappraised part 3: Frequency of Anticipated Transients, EPRI-NP-2230, Electric Power Research Institute, Palo Alto, California.

Philadelphia Electric Company, 1981. Probabilistic Risk Assessment: Limerick Generating Station.

Sullivan, H., and J.P. Poloski, 1980a. Data Summaries of Licensee Event Reports of Diesel Generators at U.S. Commercial Nuclear Power Plants, NUREG/CR-1362, U.S. Nuclear Regulatory Commission, Washington, D.C.

Trojovsky, M., 1982. Data Summaries of Licensee Event Reports on Pumps at U.S. Commercial Nuclear Power Plants, NUREG/CR-1205, Rev. 1., Washington, D.C.

Vesely, W.E. "Estimating Common-Cause Failure Probability in Reliability and Risk Analyses: Marshall-Olkin Specilization," in Proceedings, International Conference on Nuclear Systems Reliability Engineering and Risk Assessment, Gatlinburg, Tennessee, June 1977.

RISK ASSESSMENT OF HAZARDOUS MATERIAL TRANSPORTATION

P. Hubert and P. Pages

C.E.P.N. (Centre d'étude sur l'Evaluation de la Protection dans le domaine Nucléaire)
F.92260 Fontenay-Aux-Roses

ABSTRACT

The purpose of this paper is to present the series of models to be implemented for a probabilistic risk assessment associated with the transportation of hazardous materials. All types of models (road environment, container failure, atmospheric dispersion, gas cloud explosion...) are considered. Emphasis is placed on the features specific of the transportation case and on the description of the alternative ways of analyzing accident sequences and tank or package failures. These analyses are illustrated with figures from french road data collected by the CEPN. The results of the assessments related to propane, uranium hexafluoride, and small radioactive package transport are presented. An analysis of the uncertainties associated with these models, which leads back to the analysis of their use, is carried out.

KEY WORDS: Hazardous Material, Risk Assessment, Transportation Risk Evaluation Criteria, Decision Making, Reliability Analysis, Uncertainties.

INTRODUCTION

The quantitative assessment of risk associated with hazardous material transportation calls on methods developed within the framework of reliability and safety analysis. Like the safety study concerning fixed installations, it is not yet currently used. So far a small number of exhaustive probabilistic evaluations have been carried out in the transportation field: the cases of nuclear material, of liquefied gases, of chlorine and gasoline have been studied, mainly in the United States (/1/ to /4/), but also in Finland /5/, and in France by the CEPN (/6/ to /10/). The subject was also partially treated in the case of global site studies as the one of Canvey Island /11/ or of the Rijnmond area /12/, and in the case of special routing problems /13/. The aim of this paper is to present the available methods for assessing these risks. The features specific to the case of road transportation will be emphasized: the difficulties of modeling the environment of a system which may be quite simple in itself, the co-existence among the risks of rare events of catastrophic character and of less severe but almost certain accidents. The results of some assessments carried out by the CEPN will be presented along with an analysis of their significance and of their use.

1. THE REASONS FOR IMPLEMENTING QUANTITATIVE RISK ASSESSMENT

Three relatively distinct types of incitement can motivate hazardous material transportation risk assessment studies.

First of all, this risk poses an "acceptability problem". Although the number of victims is low, averaging about 1 per year in France, hazardous material transportation is perceived as a high risk activity. Without going into the complicated analysis of acceptability and "risk perception" phenomena, we can see that the "catastrophic" aspect of this risk is a determining factor (cf Los Alfaques 1978: 200 dead, St Amand (1973): 12 dead, or Lievin (1968): 6 dead). If the quantitative risk assessment is not in itself the solution to the problem of public hostility towards certain industry activities, it is nevertheless a necessary element.

The development of safety studies in Europe is parallel to the one in the United States (cf the "SEVESO" directive of the EEC). This trend is associated with a growing awareness and with the training of managers and authorities in concepts and methods of probabilistic risk analysis. It also corresponds to a marked progress in the modelling of physical phenomena. We refer in particular to the improved knowledge in the domain of unconfined vapor cloud explosion (UVCE) (/14/ to /17/) since the Flixborough disaster in 1974.

The complexity of hazardous material transportation risk management is also a strong incitement to the development of quantitative assessments leading to rationalization criteria. Naturally, the complexity of the regulations is most often put forward.

Knowledge of the level of the risk associated with new transports would help to clarify the regulatory procedures. The number of diversity of the actors concerned by the transportation of hazardous material (producers and carriers, central administrations, local communities, public security) is also a source of extra burden in risk management. Discussion among all these parties implies a minimum of figured data. A precious tool in particular is the cost-effectiveness analysis of protection options that is often carried out together with safety studies (/8/ and /9/)...

2. THE PROBABILISTIC RISK ASSESSMENT PROCESS

As in every system safety assessment, identification of the studied system is a prerequisite. In the case of hazardous material transportation, the system is in general very simple and its functions limited in number. For a road transport we may for example distinguish two sub-system, the vehicle and the tank or the packaging. Besides transporting from one point to the other the vehicle's function is to avoïd the various hazards of the road environment. The tank's or the packaging's function is most often limited to confining the material although sometimes other functions must be taken into account such as refrigeration or protection against radiation. The packaging is designed to ensure these functions in either a "normal" or "accident" environment, the latter being defined by reference situations of collision, water immersion or fire. The specificity and also the major difficulty in assessing this type of risk is that the environment is essentially variable and that its description is therefore probabilistic. Two aspects must be distinguished in the system's environment, an "aggressor" aspect which is expressed as a certain level of constraint on the package and an

Figure 1

Incitements to Quantitative Assessment

"aggressed" or "target" aspect where the population density, the protection level available to individuals and, if necessary, weather conditions parameters must be considered. Furthermore the modification of the human environment following the accident must be included, that is the evacuation possibilities or, on the contrary, crowd phenomena.

Definition of the risk modes and delimitation of the object of the study must be carried out in interaction with the identification of the system. In this identification process, risks are found to be heterogeneous in quite a few aspects.

Besides lethal accidents that come immediately to one's mind, one must not forget that other events, such as a release of carcinogenic substance for instance, are only perceptible in the long term. There are also undesirable events, although their direct consequences on the population may be averted, as in the case of contaminated groundwater. Finally, economic damages cannot be neglected. The destruction of a factory by a fire initiated by a gasoline truck, in 1979, caused damages of FF 5 millions without resulting in any casualty. Large scale evacuations as the one in Toronto involving more than 200,000 persons, are also to be considered among the consequences of accidents caused by hazardous materials. So far, risk assessment studies are only concerned with the immediate effects on the health of the populations, and are often limited to a mortality index. An exception are the studies on radioactive material transportation that take into account long term health effects and sometimes economic impacts such as evacuation and decontamination costs /11/.

Other heterogeneities are to be considered such as those concerning the nature of the dangers. For the same product, liquefied propane for example, the accident scenarios may lead to several alternative physical events. Following excessive pressure or fragilization, the tank can explode with flying fragments without ignition of the propane. But in case of a small as well as a major containment failure there are both a liquid and a gaseous release. The liquid pool and the cloud may ignite. In this last case two scenarios are possible: formation of a fireball if the cloud is ignited soon, or deflagration producing a shock or blast wave.

Another problem makes the interpretation of the risk evaluation results more difficult. It is the need to aggregate or to compare the frequent events of low consequences with rare events or high consequences: the disasters of "major events". The mathematical expectation of the number of deaths is in general used for this, but it is not satisfactory

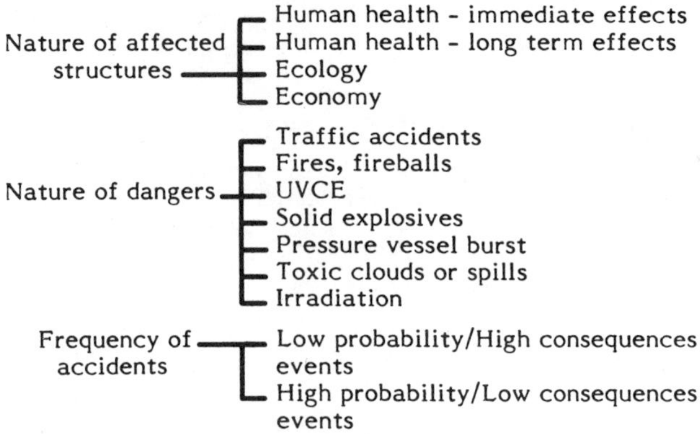

Figure 2

The Risk Heterogeneity

in many respects. Neither the public, nor the industry and administration authorities nor the insurance companies give the same importance to an event occurring once a year and causing one death and to an event occurring once every 100 years and causing 100 deaths. Thus, the assessment of risk associated with the transport of a given material, emphasizes a whole set of events from the almost certain road accident to the most unlikely unconfined vapor cloud explosions in urban zone, with all intermediate cases (tank explosion in rural zone for instance). The assessment must then show both risk dimensions: probability and consequence and the user of these studies must keep them in mind.

The system being known, the quantification process is always more or less the same. The case of propane (cf fig. 3) is a good example, but of course it is the nature of the danger that generally determines the type of model to be implemented (thermal transfer, atmospheric transport, external irradiation...).

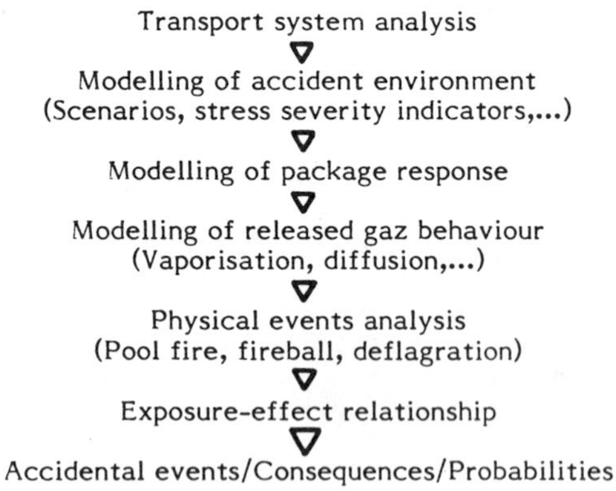

Figure 3

The Steps of a Probabilistic Assessment: The Case of Propane

The model's sophistication will also depend on the available information and on the level of safety offered by the packaging. Thus, the risk of a liquid hydrocarbon tank breaking open can be directly estimated from the available annual accident statistics, while the release of fission products after an accident involving a spent fuel cask will require the construction of a fault tree and the assessment of "rare events" among expected road accident environments.

In fact, these assessment practices lead to the use of the whole body of reliability analysis techniques, among which fault tree analysis occupies here a modest place mainly because of the simplicity of the systems.

3. MODELING THE ACCIDENT ENVIRONMENT

Availability of accident data, but also an a priori idea of the thresholds and the failure modes of the packages studied are the two guiding criteria when modeling accidents. Concerning road accidents, the CEPN used data provided by the "Sécurité Civile" on truck fires and by the CITMD (Commission Interministérielle pour le Transport des Matieces Dangereuses, the french equivalent of the National Transportation Safety Board) for truck accidents. Based on the latter, a data base has been constituted including a total of 1300 accidents. One interesting feature of this base is the presence of collision physical parameters (mass, velocity, collision angle, etc...) usually not found in ordinary accident files.

A first use of accident reports is qualitative. It allows us to form an idea of the reality of accident environment and to emphasize event sequences difficult to imagine a priori. We have thus noticed a fall from a bridge on railway tracks followed by impact with an oncoming train, a hydrocarbon liquid spill into a sewer followed by underground explosions in the network, the role of electricity lines in initiating fires, and the possibility of road accidents involving 500 vehicles. In practice, data collected on accidents are mainly used in the construction of event trees and assessment of the probabilities of the different scenarios: fire, fire + collision, collision against fixed objects (cf Figure 4).

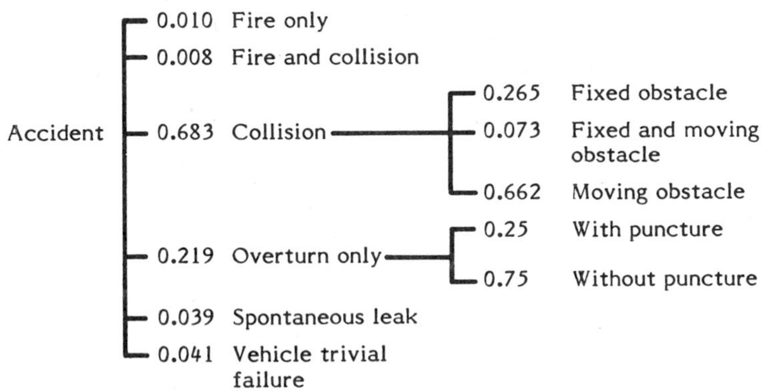

Figure 4

Event Tree for a Road Accident

In parallel, the statistical distribution of severity indicators for accident stresses is assessed: fire duration, occurrence of puncture, collision kinetic energy. One may then either be satisfied with the empirical distribution or fit a known distribution function as in the case of collision energy (of Figure 5). If the package failure thresholds lie within the range of observed values, the latter approach will be used. This is precisely the case for uranium hexafluoride and liquefied gas tanks.

On the other hand, other transports, because of their high level of potential risk involve packagings with much higher safety margins. It is then no longer possible to extrapolate from observed statistical distributions. The exceptional stress levels required to damage these packages would necessitate some kind of a combinatorial approach. We are thus led to quantify the "rare events": packages being squeezed in collision involving two other trucks, which would make possible a criticality accident for a shipment of an array of plutonium packages /18/, or packages catching fire after a collision with a gasoline truck (on the other of 5 per 10,000 accidents).

4. CONTAINMENT FAILURE ANALYSIS

Except for the situations just mentioned (potentially hazardous product and packaging with a high level of safety) the construction of a fault tree is not necessary. The elementary event probabilities are rarely measurable and when their quantification is considered the complex trees that have been worked out are reduced to very simple forms (1 or 2 branches). Failure mode identification can most often be made directly: collision, fire, collision followed by fire, puncture are in general the only sequences to be singled out. We can then associate the rupture probabilities with the severity indicators worked out in the preceding step.

Several estimation techniques are possible. In the case of a package which can be considered as behaving like a liquid hydrocarbon tank we may directly use tank truck accident statistics. The number of such accidents, about 150 per years in France, is in fact sufficient for estimating the occurrence probability of certain events. We observe thus the rate of containment failures (40 to 50%) or of fires (2% to 3%). When possible, the use of a similarity for instance between studied and reference (hydrocarbon) tanks is a simple solution for quantifying the

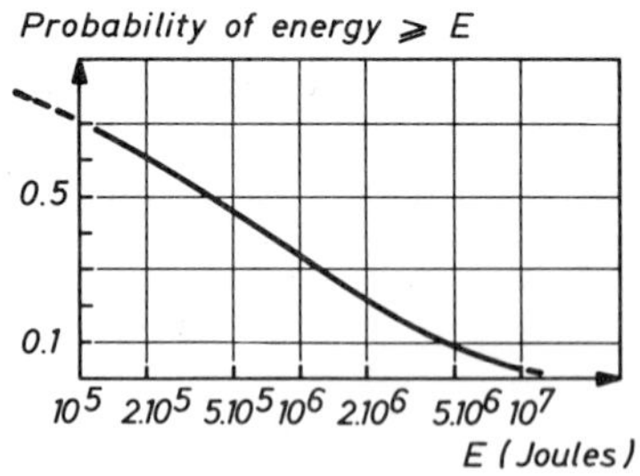

Figure 5

Energy Associated with Truck Accidents

failure. The failure rate for ordinary tanks as a function of a collision severity indicator (energy in our case) is shown on figure 6, based on accident file data.

In the case of tanks an exponential function is quite good in describing the failure as a function of energy (Figure 6). The scaling will then be expressed as a multiplying factor acting on the failure function parameter. To use this method one must know the phenomenon that causes the rupture, however the scaling ratio can be evaluated with he help of only a stress indicator since the severity distribution function was fitted on that same indicator and not on the real intensity of the stress. Another interesting point is that the resulting failure rate is an analytical expression which makes sensitivity or cost-effectiveness analysis easy /8/.

The most commonly used technique however, is the evaluation of a failure threshold in relation to each stress. Among the examples presented below, this method was adopted for the evaluation of the risk associated with the transport of propane (energy threshold of 2.10^6 J. /2/) and for the resistance to fire of uranium hexafluoride (60 mn in a 800° fire: /20/), the mechanical resistance having been evaluated by a scaling method. It must be noted though, that thresholds can be more or less realistic depending on whether they are estimated from regulatory design formulae, from more sophisticated and less conservative calculation codes, or even from destructive experiments.

5. THE ASSESSMENT OF PHYSICAL CONSEQUENCES AND HEALTH EFFECTS

The previous steps allow us to estimate the containment failure probability of a package under different accident scenarios. It is rarely possible to directly deduce the consequences on humans of these events. An exception is the bursting of a pressure vessel, for which the Pacific Northwest Laboratory /2/proposes considering a lethal surface 600 m long by the width of the tank corresponding to the zone swept by flying metal fragments. In general exposure-effect relationships (for chemical toxicity, radiations, overpressure and thermal effects) have to be used together with further developments in the analysis of accident sequences in order to model the behavior of the material released, which amounts in

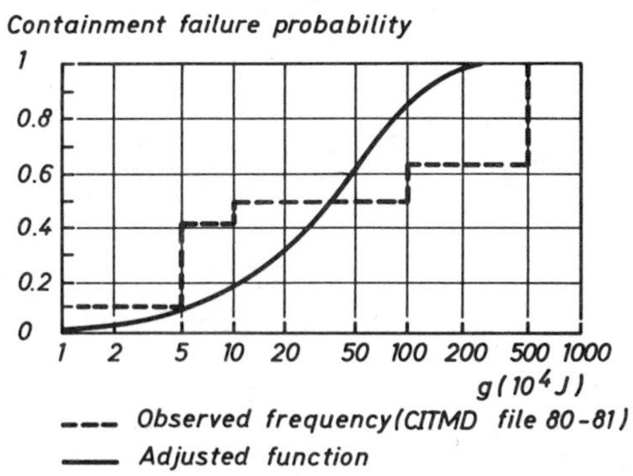

Figure 6

Adjustment for Ordinary Tank Trucks

fact, to extend the event tree of the accident scenarios.

The analysis of the physical phenomena should be further advanced up to the point of involving explicitly the parameters required by the exposure-effect relationship (concentration x time, thermal flux...). For the last few years, models simulating the above described phenomena have been made available (see Figure 7). It is the case of the formalization of the "source term": leakage rate at opening, flash evaporation, slow evaporation (cf /15/ for example). It is also the caseo of explosion in gaseous state that can be calculated in a more or less sophisticated way.

The most simple method uses a gross explosion yield (5%), based on observation, that allows us to assess the TNT equivalent of released mass. Other models take into account the formation and later expansion of the gas cloud (cf /15/, /14/, and /17/), and even its shape /21/. The A. Lannoy's approach has been used here to compute the effects of a hydrocarbon gas cloud explosion. It helped determine an average radius of 75 m with 100% mortality rate and of 370 m with 10% for a catastrophic spill from a tank truck and given Baker's exposure-effect relationship /14/.

The description of fireballs has also been the object of a certain number of studies that propose relatively converging practical rules as to evaluating their dimensions and the thermal flux at a given distance (/14/, /17/, /22/). Taking the previous example again, if the released gas ignites prematurely, the resulting phenomenon is not an explosion but a fireball with an estimated radius of 40 m and a 100% mortality rate. The dangerous zone radius (350 kJ in 6 seconds giving 10% mortality) is then 108 m. These two types of phenomena are still roughly modelled, in particular specific gas parameters rarely enter this empirical approach and often the same value is assigned to all fuels. But if we accept to rely on gross estimates their application is relatively easy. Another classic case is the atmospheric dispersion of toxic products.

With the exception of the carcinogenic and radioactive substances where each exposure is associated with a death probability, the effect of inhaling toxic products involves a threshold. Calculating the number of deaths then comes down to determining the zone in which the value of the product "concentration x time" exceeds that threshold. Such limit values are available in the toxic product literature and related data banks (/23/, /24/, /25/) but one should be aware of discrepancies existing among

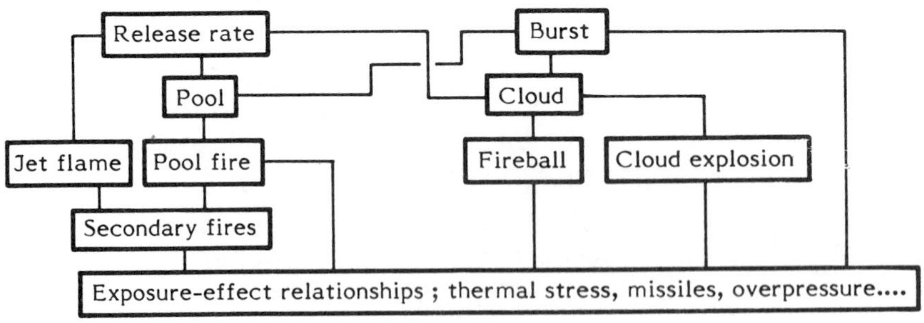

Figure 7

Event Tree for a LPG Tank Failure

the sources. To determine the dangerous zone, we use simple atmospheric transfer models, (Gaussian models /26/ and /27/). As the location of the accident is a priori unknown topography details need not be considered. It is worth noting that the determination of this zone requires knowledge of the released quantity lethal dose ratio only. Calculations carried out for a specific product, chlorine for example (see Figure 8), may thus be applied to all toxic products.

The outputs of the above described models consist in a set of areas associated with mortality rates. The preceding examples in urban environment (6,800 persons/km^2) correspond to the following number of deaths: about 5 in case of a tank burst, 55 in case of a fireball, 75 if the fireball is the result of a heated tank burst, and 400 if there is an air explosion. In the case of atmospheric dispersion in the same urban zone, a cloud of 10 tons of chlorine is likely to cause 6,00 deaths, while the release of the whole contents of uranium hexafluoride container would result in 20 immediate deaths and 2 radiation induced cancers. All these figures are to be viewed with caution. They may overestimate the effects, as shown by the release of a hundred tons of chlorine that resulted in only twenty deaths in Mexico in 1981. But they may also underestimate them as the Los Alfaques disaster shows. Generally, evacuation possibilities and protection offered by the environment are hard to assess quantitatively. Some authors however use attenuation factors, in the order of 10(/12/). Anyway the mere orders of magnitude allow to draw conclusions when comparing different types of events.

6. THE RESULTS AND THEIR RELIABILITY

The synthesis of the preceding steps allows us to assess the risk, knowing that an accident has occurred and according to the different scenarios (Table 1). In France there take place about 15 accidents caused by propane annually; we can then expect on the average 7.5 deaths due to the material and another 2.5 due to the collision itself. Assuming a quite conservative figure for the protection offered by the environment there still are 0.75 deaths per year due to the material.

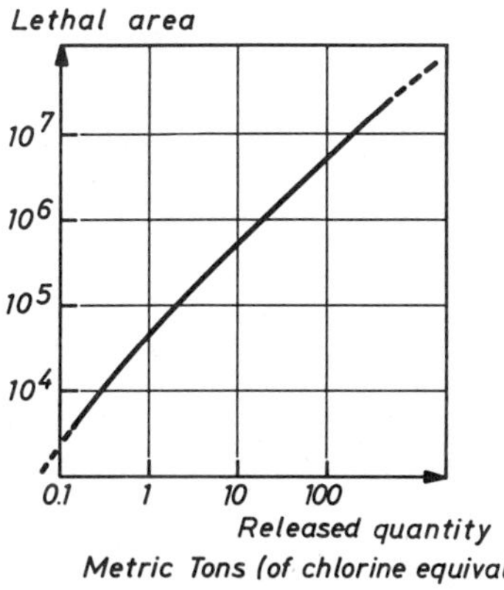

Figure 8

Lethal Areas and Taxic Gases Dispersion

Table 1

Risk Associated with Propane Transportation

Release category	Area category		Consequences of an event	Expected fatalities for one accident
	Urban	(0.04)	0.1 x 74.8 + 0	0.013
Moderate release	Semiurban	(0.32)	0.1 x 7.7 + 0	0.011
(0.044)	Rural	(0.64)	0.02 x 0.5 + 0	0
Total release	Urban	(0.04)	0.1 x 408 + 0.9 x 55	0.198
without fire	Semiurban	(0.32)	0.1 x 42 + 0.9 x 5.7	0.164
(0.055)	Rural	(0.64)	0.8 x 3 + 0	0.084
Total release	Urban	(0.04)	1 x 74.8	0.018
with fire	Semiurban	(0.32)	1 x 7.7	0.015
(0.006)	Rural	(0.64)	1 x 0.4	0.002
TOTAL				0.5

( ) : Probability knowing an accident has occurred.

However such results are not self explanatory neither for the public nor for the authorities. What is the reliability of these calculations? Is seven deaths per year a high or a low figure? How does one compare this figure with other risks and what action is to be taken? Indeed those three questions are interrelated. Obviously a high uncertainty is inherent in these calculations but it can only be meaningfully analyzed within a given decisional framework.

In dealing with probabilistic evaluations involving rare events, empirical verification of the final result is not possible. On the other hand observed data on even non severe accidents allow certain partial verifications. These may be qualitative: it could be verified that the failure modes used in the modeling are the right ones. They can also be quantitative: it could be verified that for the intermediate stages the theoretical evaluations coincide with the statistics. For example, although liquefied petroleum gas tank major ruptures are actually too rare for an empirical value to be deduced, statistical data could allow, nevertheless, to disprove theoretical values if their disagreement were too wide.

Sensitivity analyses on selected parameters can also be carried out. A priori, with each stage of the assessment (accident environment, container failure, physical behavior, exposure-effect relationship, human environment) is associated its own uncertainty. But given the interactions among these stages, there is no independence and the errors are not propagated. Modelling of the accident environment and of a tank failure matter whether the tank resists 5, 10 or 20 minutes as all such fires last at least half an hour. Therefore the uncertainty regarding thermal behavior can be neglected.

In the 10^6 to 10^7 joules range the failure energy threshold is more or less linearly related to the overall failure rate (cf Figure 5), and it is in this zone that propane tank and UF_6 container failures are located: beyond that zone, the failure probability decreases more rapidly than the

Table 2

Assessment of the Risk Associated with Road Transportation of Some Hazardous Materials -Fatalities Associated with French Yearly Traffic-

Cause of death	Gammagraphy sources	Radiopharmaceutics	Uranium Hexafluoride	Propane
Collision	$5\ 10^{-4}$	$8.6\ 10^{-2}$	$6.5\ 10^{-2}$	2.5
Material	$1.6\ 10^{-6}$	$2\ 10^{6}$	$6.3\ 10^{-4}$	7.5

threshold (between 10^7 and 2×10^7 J the probability is divided by 4), and furthermore the validity of the statistical extrapolation begin to be questionable. It seems then plausible to associate with these two stages an uncertainty of a factor of 2 to 3 in the case of propane or UF_6. The physical phenomena involve uncertainties of the same order, at least as long as they are averaged over weather conditions and topography. A more specific analysis might lead to higher uncertainty ranges. Depending on whether the "source term" is rather well known or not, the uncertainty concerning such models as the diffusion model can be negligible or dominant. Errors related to the use of the exposure-effect relationships and to the population density and vulnerability are the most difficult to quantify. The dose-effect relationship problem is common to all risk assessments, while vulnerability is more specific to the case of transports. The type of housing, of terrain, of vegetation and clothing can offer diverse forms of protection against thermal, chemical and mechanical aggression. The Los Alfaques disaster demonstrated the effect of this vulnerability. When comparing the transport of different products over the same route there is no need to much refine this type of modeling, but when looking for the best route for a given product, the problem becomes crucial.

The sensitivity analysis appears thus to involve an analysis of the goal of the assessment. Very often the results of those studies are used in a comparative way. Most of the sources of uncertainty are thus eliminated since the biases are the same, while others take on a greater importance especially those which are related to the features which distinguish two products, two routes or two safety options. Besides the assessment objective, the criterion chosen to quantify the risk (Mathematical expectation of death, occurrence of an "unacceptable" accident, annual frequency of events where a material is involved, expected annual cost of damage etc...) will also affect the relative importance of the uncertainty sources. In practice, the differences among the various results may reach several orders of magnitude and the model inaccuracies do not invalidate the results of the comparison.

7. APPLICATIONS OF THE RESULTS OF QUANTITATIVE ASSESSMENTS

Table II presents the results of some risk assessments carried out by the CEPN. When results are compared on the basis of annual traffic the differences are important. The low risks associated with nuclear material transportation considered here are in part due to the low traffic volume. However the overall annual risk might well be the right criterion upon which to base an analysis of the acceptability. In this respect, we may consider the "trivial" risk due to road traffic accidents themselves resulting from the transport studied, as a reference. It is worth noting that in the first three cases the risk due to the material is lower by two orders of magnitude. Lastly, the "risk spectrum" must be considered; here it does not change the hierarchy as propane is also the product associated with the most serious events.

From the authorities' point of view, there are other obvious criteria to consider when comparing the relative risk of these transports. The risk can be brought down to a traffic unit such as vehicle-km, package-km or even ton-km. In this last case the risk due to the transport of industrial radiography sources becomes the greatest, being from orders of magnitude above the propane risk, since about a hundred grams of the product are transported with each shipment. The health risk associated with the transport of propane is then only 2 to 3 times higher than the risk due to UF_6. These thoughts illustrate the difficulties of comparing risks related to different products, but they show the benefits of possessing results in figures when choosing the criteria for rationalizing the management of hazardous material transportation. In the simplier cases, when only options related to the same product are compared, assessments can give direct solutions. For example, bypassing the Rhône corridor for UF_6 trucks is ten times more cost-effective than adopting a protective overpack, or an escort /8/, but the cost per avoided death is 100 times greater than that used in the domain of conventional road safety.

Direct use of risk assessment studies in the field of hazardous material transportation is possible from now on whenever the decisional framework is clear. Quantitative assessment also provides useful criteria for a more rational approach to risk management, even though such an approach is not yet of widespread use. In both cases, the discrepancies are such that model imprecisions can be safety neglected. We hope furthermore that this study will show that the methodological tools, necessary for this type of assessment, exist and that their implementation does not necessarily imply the development of sophisticated and/or expensive studies.

REFERENCES

1. "Final environmental statement on the transport of radioactive materials by air and other modes", Rapport NUREG 0170, US NRC Washington, 1977.
2. Geffen, G.A., "An assessment of the risk of the transportation of propane by truck and train", PNL report 3308, Richland, 1980.
3. Andrews, W.B., "Assessment of transporting liquid chlorine by rail", PNL report 3376, Richland, 1980.
4. Rhoads, R.E. & al, "An assessment of transporting gasoline by truck", PNL report 2133, Richland, 1978.
5. R. Laautsaki, J. Fieandt, "Risk assessment of the transportation of hazardous gases in bulk", Loss Prevention Symposium, Bâle, Septembre 1980.
6. T. Meslin, "Assessment and management of risk in the transport of dangerous materials: the case of chlorine transport in France", Risk Analysis, Vol 1 n°2, pp. 137-141, 1981.
7. "Evaluation de lassûreté des transports d'hexafluorure d'uranium naturel", Rapport CEPN n±4, Paris, 1977.

8. P. Hubert, P. Pages, "Cost effectiveness of safety measures applying touranium hexafluoride transportation in France", in PATRAM 1983, New-Orleans, Mai 1983.
9. "Risk assessment and cost-benefit analysis of the transport of small radioactive material type B packages," Rapport CEPN n± 38, Paris, 1980.
10. "L' évaluation de la sûreté des transports de matieres radioactives: le cas des petits colis de type A", Rapport CEPN n± 42, Paris, 1982.
11. "Canvey: an investigation of the potential hazards from operations in the Canvey Island area", HMSO, Londres, 1978.
12. "Risk analysis of six potentially hazardous industrial objects the Rijnmond area", Reidel, Dordrecht, Pays-Bas, 1982.
13. Harter, G.A., "TRIP: a computer program using detailed route modelling to assess comparative risks from overland transport", in PATRAM'80, Berlin, Novembre 1980.
14. Baker, W.E., Strelhow, R.A., "Explosion hazards and evaluation", Elsevier, Amsterdam, 1983.
15. A. Lannoy, "Methodes probabilities et déterministes d'estimation du risque industriel appliqués a la prévision des effects des explosions", These docteur ingérieru a l'Université de Poitiers, 1983.
16. K. Gugan, "Unconfined vapor explosion", G. Godwin, Londres, 1980.
17. "Methods for the calculation of the physical effects of the escape of dangerous materials (Yellow Book)", TNO, Pays-Bas, 1979.
18. "Risk analysis of the transportation of Plutonium Dioxide", Rapport CEPN n± 49, Paris, 1981.
19. Geffen, C.A., "An assessment of the risk of transporting Uranium hexafluoride by truck and train", PNL report 2211, Richland, 1980.
20. B. Duret, J.C. Bonnard, "etude thermique d'un conteneur d'UF6 soumis à un feu accidental", Note CEA/TT/STML S1-16, Grenoble, 1981.
21. Fishburn, B., Shagg, N., Lu, P., "Blast effect of a pancake shaped fuel drop-air cloud;, in Journal of hazardous materials, 5, Elsevier, Amsterdam, 1981.
22. Roberts, F., "Target response to liquid fuel fire and to vapor cloud explosions", London Symposium on "Risks of fuel transport", London June 1982.
23. Sax, I., "Dangerous properties of hazardous materials", Van Nostrand Reinhold, New York, 1979.
24. "Registry of Toxic effects of chemical substances", NIOSH, 1976.
25. Banque de donnés OHM-TADS, "Oil and hazardous material, technical Assistance data systems" géré par le NIH/EPA.
26. A. Doury, "Une méthode de calcul pratique et généale pour la prévision numérique des pollutions véhiculés par l'atmosphere", Rapport CEA-R 4280, Fontenay-aux-Roses, 1972.
27. "Note d'information sur les modeles de dispersion atmosphérique", Cahiers de l'Union des Industries chimiques, Paris, 1982.

INDEX